SCHAUM'S OUTLINE OF

THEORY AND PROBLEMS

OF

INTRODUCTION
to
DIGITAL SYSTEMS

•

JAMES E. PALMER, Ph.D.
Professor of Electrical Engineering
Rochester Institute of Technology

DAVID E. PERLMAN
Associate Professor of Electrical Engineering
Rochester Institute of Technology

SCHAUM'S OUTLINE SERIES
McGRAW-HILL

New York San Francisco Washington, D.C. Auckland Bogotá
Caracas Lisbon London Madrid Mexico City Milan
Montreal New Delhi San Juan Singapore
Sydney Tokyo Toronto

JAMES E. PALMER is Professor of Electrical Engineering at the Rochester Institute of Technology, in Rochester, New York (R.I.T.). He received his B.Sc. from the University of Western Ontario, his M.S.E.E. from the University of Pennsylvania, and his Ph.D from Case Institute of Technology. His research interests are the design of digital systems with an accent on product design and its concurrent engineering aspects. From 1968 to 1974 he was Director of Engineering at Gannon University. From 1974 to 1978 he was Head of the Electrical Engineering Department at RIT. Presently he teaches courses in the areas of digital system design and test and also in the areas of control system design.

DAVID E. PERLMAN is an Associate Professor of Electrical Engineering at the Rochester Institute of Technology, in Rochester, New York. He received B.E.E. and M.E.E. degrees from Cornell University and, following ten years as a design engineer and researcher at the Eastman Kodak Company, he left to become one of the founders and Vice President of Advanced Development of Detection Systems, Inc., in Fairport, New York, a position he held for thirteen years. He holds twelve patents. In 1982, Mr. Perlman joined the faculty at R.I.T., where he has been teaching graduate and undergraduate courses in electronics as well as undergraduate courses in circuits and digital systems.

Schaum's Outline of Theory and Problems of
INTRODUCTION TO DIGITAL SYSTEMS

10 11 12 13 14 15 CUS/CUS 0 9 8 7 6 5

ISBN 0-07-048439-2

Sponsoring Editor: John Aliano
Production Supervisor: Leroy Young
Editing Supervisor: Meg Tobin
Front Matter Editor: Maureen Walker

Library of Congress Cataloging-in-Publication Data

Palmer, James (James E.)
Schaum's outline of theory and problems of introduction to digital systems / James Palmer and David Perlman.
p. c.m.—(Schaum's outline series)
Includes index.
ISBN 0-07-048439-2
1. Digital electronics. 2. Logic circuits. I. Perlman, David (David E.) II. Title. III. Series.
TK7868.D5P35 1993
621.381—dc20

91-46678
CIP

McGraw-Hill
A Division of The McGraw-Hill Companies

Preface

The goal of this book is to introduce a unified design methodology into the introductory course in digital systems. It is based on the course "Introduction to Digital Systems" which is offered to freshmen and incoming transfer students in the electrical engineering curriculum at the Rochester Institute of Technology.

As is usual in books on this subject, the first chapter describes number systems in general and the binary system in particular as a prelude to introducing the two-valued logical variable and signals which represent it in all computer and digital circuits.

The next three chapters describe a coherent design procedure for systems using combinatorial (or combinational) logic. Three different means of specifying a combinatorial problem—the truth table, Boolean equations, and logic diagrams—are discussed in Chapter 2, while Chapter 3 deals with the manipulations of Boolean algebra and contains additional material on the construction and interpretation of Karnaugh maps. Here, the design problem is analyzed at a purely logic level, independent of hardware considerations, and the relation between K maps, Boolean equations, and logic diagrams is explored. Chapter 4 presents a structured approach to the hardware implementation of logic using mixed-logic methodology. The result is a totally unambiguous design tool which yields functional logic circuitry while preserving the identity of the original underlying Boolean relations.

Chapter 5 offers a description of commonly used MSI and LSI combinatorial logic elements with emphasis placed on devices (such as multiplexers and ROMs) which can be programmed for specific applications.

The remainder of the book is primarily concerned with synchronous sequential logic. The construction and use of timing diagrams is developed in Chapter 6 where computer-aided design tools such as schematic capture and simulation software are introduced. The logical function of basic memory elements (flip-flops) is discussed in Chapter 7 and some important MSI and LSI combinations of flip-flops are covered in Chapter 8, which deals with registers, counters, and data storage devices. In Chapter 9, the basic operation of programmable devices containing both combinatorial logic and flip-flops is discussed. Chapter 10 illustrates both traditional design procedures and the use of algorithmic state machine charts as design tools for synchronous sequential logic and for simple state machines. Chapter 11 takes a nontraditional view of logic elements as control device components and provides an introduction to programmable gate arrays and their operation.

A word about symbols is appropriate here since, unfortunately, no single notation has achieved universal acceptance. While the bubble has been used for many years, in positive logic notation, to indicate logical inversion, it is also currently used as an alternative to the half-arrow to denote a low-TRUE signal in mixed logic systems. Since reserving the half-arrow to exclusively designate low-TRUE is less ambiguous, the authors have emphasized this notation for use in developing the unified design process presented in chapters 2–4. Bearing in mind, however, that most currently available schematic capture and simulation software packages produce bubbles and not half-arrows, we cannot arbitrarily banish the low-TRUE bubble. Its application is discussed in Section 4.3 and, it will be found scattered throughout the book (as in Figs. 4-76, 4-79 and 5-38) in order to present students with examples of the symbology that they are likely to come across in "the real world".

The situation is no less confused when it comes to denoting connections (or lack thereof) in programmable logic devices. In chapters 9 through 11, "x's", solid circles or solid rectangles are used to indicate connections while hollow circles or no symbol at all indicate the lack of a connection. Since, in all likelihood, the reader will come across all or several of the above conventions, it was decided to present a generous sprinkling of each in the examples and problems, taking care to avoid any ambiguity in the meaning of a symbol.

This book is designed to function as either a text for an introductory course in the design of digital systems or for use as a supplement to other textbooks. Typical of the Schaum's Outline Series, it contains numerous worked examples as well as supplementary problems with answers. It is important to note that in design there are often several valid solutions to a given problem. In these cases, the authors have used their best judgment in selecting the solution given and, when appropriate, have presented alternative approaches to a representative group of problems as new techniques are developed in successive chapters.

It will be noted that many of the logic and timing diagrams in this book have been computer generated and several generic observations concerning schematic capture and simulation appear in the text and in App. C. This is a natural consequence of the fact that it has become increasingly difficult to treat the design of digital systems without reference to Electronic Design Automation (EDA) software. The authors have chosen to use LogicWorks™, a somewhat scaled-down version of DesignWorks™ digital logic design software from Capilano Computing Systems Ltd., because it is an extremely user-friendly simulation package which is attractively priced for educators and students. Interested readers should write Capilano at 406-960 Quayside Drive, New Westminster, B.C., Canada V3M 6G2 or call (800) 444-9064 or (604) 522-6200 for further information and/or a demonstration disc.

The authors would like to express their appreciation for the helpful comments of Dr. Charles Schuler who reviewed the manuscript and kept our spirits up with constructive encouragement. We would also like to express our appreciation to several "generations" of undergraduate students at RIT who gave us invaluable feedback on the effectiveness of our pedagogy and the accuracy of problem solutions. In the not so distant past, the creation of hundreds of diagrams integrated with the text required the acknowledgment of gargantuan (and incredibly patient) effort by one or more harassed secretaries and/or artists. We need make no such mention here since the job was done "in house" with the aid of an Apple Macintosh II* computer running two screens—one for drawing and the other for text, which made the project reasonably manageable and usually enjoyable. We do, however, want to especially acknowledge the patience and support of our wives, Mary Palmer and Marjorie Lu Perlman, both of whom put up with a lot of lost cohabitation during the many late nights and weekends that went into this project.

JAMES E. PALMER
DAVID E. PERLMAN

Contents

Chapter 1 NUMBERS AND THE BINARY SYSTEM **1**

1.1 Introduction 1
1.2 Number Systems 1
1.3 Conversion between Bases 4
1.4 Basic Binary Arithmetic 5
1.5 Codes 7
1.6 Error Detection and Correction 11

Chapter 2 DESIGN OF COMBINATIONAL LOGIC I **24**

2.1 Combinational Logic 24
2.2 Truth Tables 24
2.3 Boolean Equations and Basic Logical Functions 26
2.4 The Relation between Boolean Equations and Truth Tables 29
2.5 Logic Diagrams 30

Chapter 3 DESIGN OF COMBINATIONAL LOGIC II: MANIPULATION **50**

3.1 Introduction 50
3.2 Boolean Algebra Basics 50
3.3 Hardware Implications 54
3.4 Basic K Maps 54
3.5 Further Applications of K Maps 57

Chapter 4 HARDWARE AND THE MIXED-LOGIC CONVENTION **86**

4.1 Introduction 86
4.2 Gate Hardware 86
4.3 Mixed Logic as a Design Tool 87
4.4 Mixed Logic as a Descriptive Convention 94
4.5 Uses of Mixed Logic in Troubleshooting 96

Chapter 5 MSI AND LSI ELEMENTS **128**

5.1 Introduction 128
5.2 Multiplexers 128
5.3 Decoders and Demultiplexers 133
5.4 The Read-Only Memory (ROM) 134

Chapter 6 TIMING DIAGRAMS **167**

6.1 Introduction 167
6.2 Microtiming Diagrams 167

6.3 Hazards 169
6.4 Macrotiming Diagrams 171
6.5 Timing Simulations 172
6.6 Feedback in Combinational Circuits 174

Chapter *7* THE FLIP-FLOP **194**

7.1 Introduction 194
7.2 The Basic Latch 194
7.3 The Chatterfree Switch 195
7.4 Clocked RS Flip-Flop 196
7.5 The JK Flip-Flop 198
7.6 JK Flip-Flop with Preset and Clear 200
7.7 Signal Propagation within the Flip-Flop 201
7.8 Other Flip-Flop Types 203
7.9 Flip-Flop Triggering and Timing 205
7.10 Metastability 207

Chapter *8* COMBINATIONS OF FLIP-FLOPS **233**

8.1 Registers 233
8.2 Parallel-Serial Conversion 236
8.3 Ripple Counters 237
8.4 Rate Multipliers 240
8.5 Random-Access Memory 241

Chapter *9* APPLICATION SPECIFIC DEVICES **269**

9.1 Introduction 269
9.2 Programming Technologies 269
9.3 Proms and EPRoms 270
9.4 Programmable Array Logic (PAL*) 272
9.5 The Programmed Logic Array (PLA) 277
9.6 Gate Arrays 279
9.7 Programmable Gate Arrays 281
9.8 Full Custom Design 281

Chapter *10* DESIGN OF SIMPLE STATE MACHINES **306**

10.1 Introduction 306
10.2 Traditional State Machine Design with D Flip-Flops 307
10.3 Design with JK Flip-Flops 309
10.4 Design for Programmable Logic Devices 312
10.5 The ASM Chart 313
10.6 Design from an ASM Chart: Boolean Implementation for Minimal Number of Flip-Flops 316
10.7 Design from an ASM Chart: One-Hot Controller Implementation 318
10.8 Design from an ASM Chart: State Table Entry to a Programmable Logic Device 319

10.9 Clock Skew in State Machines . . . 321
10.10 Initialization and Lockout in State Machines . . . 322

Chapter *11* ELECTRONICALLY PROGRAMMABLE FUNCTIONS **353**
11.1 Introduction . . . 353
11.2 Basic Components . . . 353
11.3 Programmable Gate Arrays . . . 354
11.4 Arithmetic Logic Units . . . 358
11.5 Programmable Registers . . . 362

Appendix *A* BASIC BOOLEAN THEOREMS AND IDENTITIES . . . 381

Appendix *B* STANDARD LOGIC SYMBOLS . . . 382

Appendix *C* SOME COMMENTS ON DIGITAL LOGIC SIMULATION . . . 388

INDEX . . . 391

Chapter 1

Numbers and the Binary System

1.1 INTRODUCTION

In modern digital systems it is necessary to electronically store and process large quantities of data in the presence of electrical noise and interfering signals. The data is usually in a binary (two valued) form since this allows the use of reliable and easily replicated storage and computational devices comprising large numbers of logically connected electronic switches fabricated within integrated circuits. Such devices, containing thousands (and in many cases, millions) of transistors, are inherently resistant to faults because the voltage or current levels representing the two binary states are far enough apart to prevent errors caused by spurious interference. Various data encoding and error checking schemes, such as Gray coding and parity checks, are often used to reduce the already low probability of undetected errors. Since the binary number system is universally employed in digital processing, it is useful to understand its relationship to other number systems, as well as the properties of number systems in general and the methods of conversion from one to another.

1.2 NUMBER SYSTEMS

In everyday use, numbers are represented in the decimal (base 10) system which has 10 symbols (0, 1, 2, 3, 4, 5, 6, 7, 8, 9). This system is *weighted* in that it makes use of a *positional notation* wherein the value assigned to a particular digit is determined by its position in the sequence of digits which represents a given number. Consider the base 10 number 853828. The digit 8 occurs three times in the sequence, but each occurrence has a different weight because the digit occupies a different position corresponding to a power of the base. This arrangement is shown below.

10^5	10^4	10^3	10^2	10^1	10^0	Column weights
8	5	3	8	2	8	Digits

$$853828 = 8 \times 100{,}000 + 5 \times 10{,}000 + 3 \times 1000 + 8 \times 100 + 2 \times 10 + 8 \times 1$$

The left-most 8 is weighted by 10^5, the next 8 by 10^2, and the last by 10^0. This positional notation is easily extended to decimal *fractions*, in which case, negative powers of the base 10 are used:

$$0.725 = 7 \times 10^{-1} + 2 \times 10^{-2} + 5 \times 10^{-3}$$

The Binary System

It is possible to express a number in any base. In the binary case, the base is 2 and only two symbols are needed (0 and 1). Each digit is called a "bit" and, again, positional notation is used. To find the decimal equivalent of any binary number, merely write the decimal equivalent of each of the powers of 2, multiply by the appropriate binary digit, and add the results.

EXAMPLE 1.1 Express the binary number 1100111.1101 as a decimal (base 10) number.

Since the integer part has seven digits (bits), the most significant has a weight of 2^6 or 64. Its decimal equivalent may be easily computed as

$$\begin{aligned} 1100111 &= 1 \times 2^6 + 1 \times 2^5 + 0 \times 2^4 + 0 \times 2^3 + 1 \times 2^2 + 1 \times 2^1 + 1 \times 2^0 \\ &= 1 \times 64 + 1 \times 32 + 0 \times 16 + 0 \times 8 + 1 \times 4 + 1 \times 2 + 1 \times 1 \\ &= 103_{10} \end{aligned}$$

For the decimal part,

$$
\begin{aligned}
.1101 &= 1 \times 2^{-1} + 1 \times 2^{-2} + 0 \times 2^{-3} + 1 \times 2^{-4} \\
&= 1 \times 0.5 + 1 \times 0.25 + 0 \times 0.125 + 1 \times 0.0625 \\
&= 0.8125_{10}
\end{aligned}
$$

Since binary numbers require only two symbols, they are ideally suited for representation by electronic devices since only two easily distinguishable states, such as ON and OFF (conducting and nonconducting), are required.

The advantages of binary are best illustrated by considering the effect of noise or interference on the performance of a data-processing system. In the binary case, when data is to be transmitted or retrieved from storage, it is necessary for the receiver to determine which of two levels a given signal is nearer. A threshold for decision can be set up midway between these two levels so that any additive noise which is less than the difference between the signal level and the threshold is ignored. With decimal storage on the other hand, a system with the same overall voltage range assigned to its signals has a much smaller noise immunity because the given range must be divided into 10 separate levels (see Fig. 1-1).

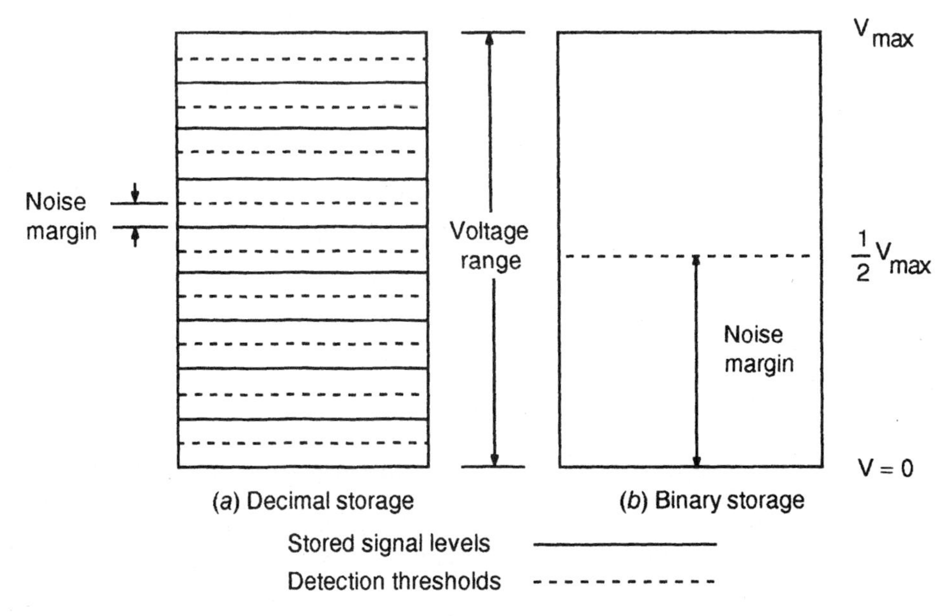

Fig. 1-1

In data systems, we speak of a figure of merit called *noise margin* which is defined as the maximum noise voltage (or current) which can be tolerated without causing an undesirable output change.

EXAMPLE 1.2 Compare the basic noise margins of binary and decimal data systems having ideal hardware components.

For the binary case, a 1 is stored as V_{max} [typically 5 volts (V)] and a 0 as approximately 0 V. The threshold would be set at $V_{max}/2$, and any noise less than this value is ignored. In the decimal system, there would be 10 equally spaced storage levels between 0 and V_{max} (0, $V_{max}/9$, 2 $V_{max}/9$, etc.) and there would be thresholds set up halfway between adjacent storage levels ($V_{max}/18$, 3 $V_{max}/18$, etc.). Any noise which is more than $V_{max}/18$ would result in an erroneous data reading. For the binary case with $V_{max} = 5$ V, the noise margin would be $5/2 = 2.5$ V. The decimal system with the same V_{max}, on the other hand, would have a noise margin of only $(5/9)/2 = 0.28$ V which is obviously less desirable than the binary case.

Octal and Hexadecimal Systems

While the binary system provides great practical advantages for the storage and processing of data in digital systems because it makes use of only two symbols, a given number expressed in binary consists of a much longer sequence of digits than the corresponding decimal number. If data is to be entered manually, only a two-key keyboard would be needed, but these keys have to be struck many times. This data-entry problem is often solved by treating binary numbers in groups.

Octal numbers make use of 3-bit groups in accordance with the following table:

Binary	Octal Digit
000	0
001	1
010	2
011	3
100	4
101	5
110	6
111	7

Each octal symbol represents the numerical equivalent of a binary 3-bit group, and the eight symbols constitute a base 8 number system. In this case, an eight-key keyboard is necessary for data entry, but it need be struck only one-third as often as a binary keyboard.

EXAMPLE 1.3 Express the octal number 247 as a decimal and a binary number.

The octal number is positional, with the lowest-order (right-most) digit being weighted by $8^0 = 1$ and the highest-order digit by $8^2 = 64$. Thus $247 = 2 \times 64 + 4 \times 8 + 7 \times 1 = 167_{10}$.

Reference to the preceding table indicates that conversion to binary is easily achieved by grouping:

```
     247
    / | \
010  100  111
```

This conversion is easily checked by determining the decimal equivalent of the resulting binary number, 10100111. Note that leading zeros may be dropped. The most significant bit is in the eighth place to the left and is therefore weighted by $2^7 = 128$. Thus, $10100111 = 128 + 32 + 4 + 2 + 1 = 167$.

A general method of converting between numbers of different bases is discussed in Sec. 1.3.

Hexadecimal notation extends the grouping idea to 4 bits and constitutes a base 16 number system. The table of corresponding bit groups and hexadecimal symbols is shown below.

Binary	Hex	Binary	Hex	Binary	Hex	Binary	Hex
0000	0	0100	4	1000	8	1100	C
0001	1	0101	5	1001	9	1101	D
0010	2	0110	6	1010	A	1110	E
0011	3	0111	7	1011	B	1111	F

The hexadecimal symbols 0 to 9 are the decimal equivalents of the first ten 4-bit binary groups. To represent the last six groups, we need new symbols since there are no single decimal digits which represent numbers larger than 9. The first six letters of the alphabet are used for this purpose as shown. In the hexadecimal system, 16 keys are needed for a keyboard, but the striking rate is only one-fourth of that required with a binary keyboard.

EXAMPLE 1.4 Hexadecimal-binary conversion. (*a*) Convert 1101011100110 into an equivalent hex number. (*b*) Convert 4B2F into binary.

(*a*) 1101011100110 = (0001)(1010)(1110)(0110) Group by 4s
= 1 A E 6 Convert individually
= $1AE6 The dollar sign is commonly used to indicate a hex number

(*b*) 4B2F = (0100)(1011)(0010)(1111) Convert individually
= 0100101100101111 Ungroup
= 100101100101111 Drop leading zero

1.3 CONVERSION BETWEEN BASES

The following is a general method that may be used to convert numbers between any pair of bases:

1. Integers and fractions are converted separately.
2. The integer portion is converted using *repeated division by the new base* and using the sequence of remainders generated to create the new number. *Arithmetic is done in terms of the old base.*
3. The fractional part is converted by repeated *multiplication by the new base*, using the generated integers to represent the converted fraction. Again, *the arithmetic is done in the old base.*

EXAMPLE 1.5 Convert the decimal number 278.632 into an equivalent binary number.

Step 1. The integer is 278. The fraction is 0.632.

Step 2. Integer conversion.

Division	Generated remainder	
2)278		
2)139	0	
2)69	1	
2)34	1	
2)17	0	Read *up* to form: 100010110
2)8	1	
2)4	0	
2)2	0	
2)1	0	
0	1	Most significant bit (MSB)

Note that once a remainder has been formed, it plays no further role in the arithmetic. The integer process will always terminate.

Step 3. Fractional conversion.

Multiplication	Generated integer	
0.632 × 2 = 1.264	1	MSB
0.264 × 2 = 0.528	0	
0.528 × 2 = 1.056	1	
0.056 × 2 = 0.112	0	Read *down* to form: .101000011
0.112 × 2 = 0.224	0	
0.224 × 2 = 0.448	0	
0.448 × 2 = 0.896	0	
0.896 × 2 = 1.792	1	
0.792 × 2 = 1.584	1	

Note that once an integer has been formed it plays no further role. This process *may not terminate*; it is usually carried on only until accuracy requirements have been satisfied.

EXAMPLE 1.6 Convert the decimal number 123.456 to an equivalent octal (base 8) number.

Integer conversion:

Division	Generated remainder	
8)123		
8)15	3	
8)1	7	Read *up* to form 173
0	1	

Fractional conversion:

Multiplication	Generated integer	
$0.456 \times 8 = 3.648$	3	
$0.648 \times 8 = 5.184$	5	Read *down* to form 0.3513
$0.184 \times 8 = 1.472$	1	
$0.472 \times 8 = 3.776$	3	

The process has been arbitrarily terminated.

$$123.456_{10} = 173.3513_8 \qquad \text{(approximately)}$$

Check:

$$173_8 = 1 \times 64 + 7 \times 8 + 3 \times 1 = 123_{10}$$

$$0.3513_8 = 3 \times 0.1250 + 5 \times 0.0156 + 1 \times 0.0020 + 3 \times 0.0002 = 0.4556_{10}$$

1.4 BASIC BINARY ARITHMETIC

All the number systems discussed previously are positionally weighted, making it possible to do arithmetic one digit at a time with the use of *carries*. Complete addition and multiplication tables can be developed by repetitive application of the rules for a single digit.

Binary Addition

The binary addition table is quite simple and is shown below, where the two digits involved are denoted by X and Y. C_i is the *carry-in* from a preceding lower-order addition.

This is the classic 1 + 1 = 2

Carry, C_i	0	0	0	0	1	1	1	1
X digit	0	0	1	1	0	0	1	1
Y digit	0	1	0	1	0	1	0	1
	0	1	1	10	1	10	10	11

Note the presence of a *carry-out* which is generated in all single-bit additions where the result exceeds 1.

EXAMPLE 1.7 Addition of two long digit strings.

	011110001
X number	1010111001
Y number	0011010101
Sum	1110001110
Carry out	0011110001

When added, each pair of digits produces a sum and a carry-out when the sum exceeds 1. This carry becomes the carry-in for the next higher order digit as shown. When, for example, X = 1, Y = 1, and the carry-in is also 1, the sum is 3 (binary 11). The left bit, having a decimal value of 2, is carried to the next higher order column, leaving a 1 in the sum position directly below.

Binary Subtraction

Subtraction could be discussed in a similar fashion, making use of a *borrow* and producing a *difference*. In practice, however, subtraction is accomplished by the same hardware which is used for addition through the use of *complementary arithmetic*. In the binary case, negative numbers are represented as the *2s complement* of the corresponding positive binary number (see Examples 1.8 and 1.9 below). Subtracting a given number X from another binary number Y is accomplished by taking the 2s complement of X to convert it to $-X$ and *adding* this to Y. In this method, the left-most digit is interpreted as a *sign bit* (0 for positive, 1 for negative) which is treated as any other bit except that a carry-out from the addition of sign bits is neglected.

The 2s complement of a binary number is obtained by exchanging the 1s and 0s of the original number and adding 1 to the result.

EXAMPLE 1.8 Subtract 185_{10} from 230_{10} by converting to binary and using 2s complement arithmetic.

How many binary digits will be required for the computation is determined by the largest number (including the answer). In this case, the number 230 is largest and requires 8 bits plus one additional for the sign bit. Thus, the binary equivalent of 185 is written 010111001. *Note that leading zeros have no effect on the value.*

We convert this to a negative number by taking its 2s complement:

Step 1: Invert the 1s and 0s.

$$010111001 \rightarrow 101000110$$

Step 2. Add 1.

$$\begin{array}{r} 101000110 \\ +\underline{\qquad\quad 1} \\ 101000111 \end{array} = -185_{10}$$

Next, again using nine places, convert 230 to binary and add this to the result of step 2 above:

$$\begin{array}{rr} +230 = & 011100110 \\ -185 = & \underline{101000111} \\ & 1000101101 \end{array}$$

Neglecting the sign bit carry (extra bit on the left) yields 000101101 whose left-most bit is 0, indicating that the result is positive.

Check: 000101101, converted to decimal, is +45.

EXAMPLE 1.9 Subtract 230_{10} from 185_{10} by converting to binary and using 2s complement arithmetic.

The binary equivalent of 230 is 011100110, and its 2s complement is obtained by inverting the 1s and 0s and adding 1:

$$-230 = 100011010$$

Next, we add this to the binary equivalent of 185:

$$\begin{array}{rr} -230 = & 100011010 \\ +185 = & \underline{010111001} \\ & 111010011 \end{array}$$

The left-most bit is a 1 indicating that the result is negative. To obtain the desired magnitude, we take the 2s complement of our result since $-(-X) = X$.

$$\begin{array}{r} 000101100 \\ +\underline{\qquad\quad 1} \\ 000101101 \end{array}$$

The decimal equivalent is 45 which we have already determined to be negative.

1.5 CODES

Binary-Coded Decimal

Binary-coded decimal (BCD) numbers are essentially decimal numbers encoded in a convenient two-valued (binary) form. Each decimal digit is represented, in order, by its 4-bit binary equivalent; 4 bits being the minimum number required to represent the decimal integers 0 to 9. Since there are 16 possible combinations of 4 bits, six of these are unused in the BCD system.

EXAMPLE 1.10 Compare the binary and BCD representations of the decimal number 278.

In Example 1.5, it was shown that the binary equivalent of 278 is 100010110. The conversion was achieved by treating the given decimal number as a whole. In BCD conversion, *each decimal digit is encoded separately:*

$$278_{10} = (\underbrace{0010}_{2})(\underbrace{0111}_{7})(\underbrace{1000}_{8}) = 001001111000 \quad \text{(BCD)}$$

Gray Code

Another two-valued code which has engineering significance is the Gray code, sometimes referred to as reflected binary code. It is not a positionally weighted code and, for this reason, is not suitable for arithmetic operations.

Single-bit Gray code is identical to a single-bit binary code:

0
1

Two-bit Gray code is obtained by "reflecting" the single-bit Gray code in an imaginary mirror as shown below. For the second digit, 0s are added above the reflection axis and 1s below it.

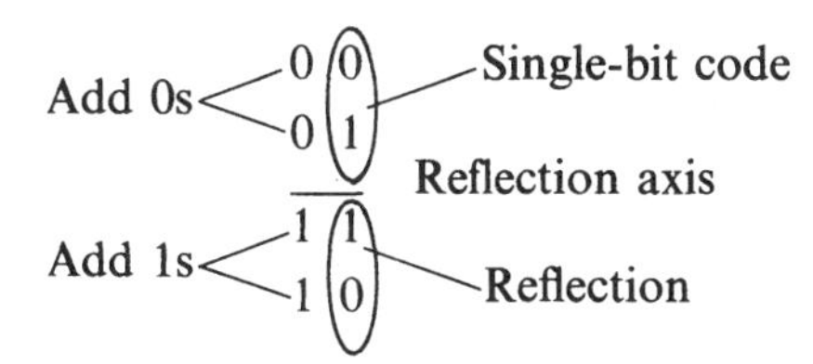

Three-digit Gray code is formed by using the two-digit code as a basis for reflection and again adding 0s above and 1s below.

000
001
011
010
——— Reflection axis
110
111
101
100

The process can be repeated for any number of digits.

The reflection process used in Gray code generation ensures that this code will have a *unit distance* property, meaning that successive code groups will differ in only 1 bit. For reference, a 4-bit Gray code is shown in Table 1.1 along with its decimal and binary equivalents. Observe the unit distance property of the Gray code as numbers progress through the sequence. Contrast this with the binary code where, for example, on passing from 7 to 8, all bits change.

Table 1.1

Decimal	Binary	Gray	Decimal	Binary	Gray
0	0000	0000	8	1000	1100
1	0001	0001	9	1001	1101
2	0010	0011	10	1010	1111
3	0011	0010	11	1011	1110
4	0100	0110	12	1100	1010
5	0101	0111	13	1101	1011
6	0110	0101	14	1110	1001
7	0111	0100	15	1111	1000

One of the main applications of the Gray code is in measurement, a typical example of which is described in Example 1.11.

EXAMPLE 1.11 Position Encoding Wheel. In a robotic system, the motion of the "arm" is often directed by a microprocessor which generates a signal commanding certain rotations of joints. This command is often applied to circuitry which controls the direction and speed of an electric motor. It is necessary for the computer to know the actual position of a joint and to compare it with the command in order to be sure that the desired motion has been carried out. Position measurement is often accomplished by connecting a small code wheel to the motor shaft. This wheel consists of concentric circular tracks which contain patterns of transparent and opaque sectors as shown in Fig. 1-2. Each track is individually associated with a light source and a light detector. When a transparent sector of the track is between the source and the detector, light is transmitted and an electric signal is produced by the detector. No output occurs when an opaque sector passes between the source and detector.

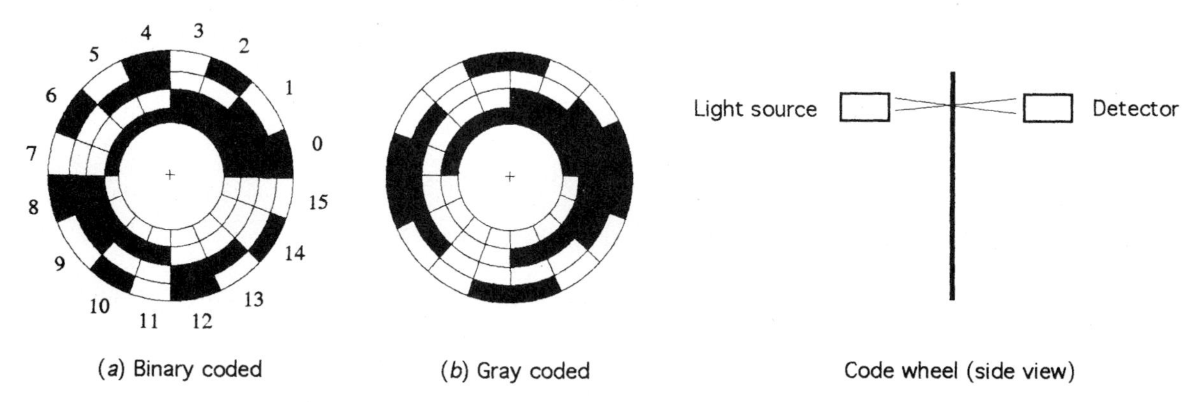

(*a*) Binary coded (*b*) Gray coded Code wheel (side view)

Fig. 1-2

The wheel in Fig. 1-2*a* is binary coded with the outermost track corresponding to the least significant digit. With a radial array of separate source-detector pairs aligned with each track, we see that as the wheel rotates, each sector passing the detector array will produce unique combinations of outputs which may be interpreted as binary numbers. For example, if the source-detector array is located along a vertical line at the top of the wheel and an illuminated detector is considered to produce a binary 1, then if the wheel rotates counterclockwise by somewhat more than one sector, detector outputs will indicate the binary number 0010. In the case of the wheel in Fig. 1-2*b*, sectors are identified by Gray-coded numbers.

Consider the binary-coded case where the wheel is positioned so that the detector array is located along the dividing line between sectors 7 and 8. Note that the sectors on each side of the boundary are different for all tracks. The light sources and detectors are not individually aligned with perfect tolerance, nor are the source emissions of zero width; the light spreads. If a light source is on the line, it may or may not cause a detector output. Thus, the binary number produced may be anywhere from all 0s to all 1s depending upon alignment and light spread. Gross errors can occur.

On the other hand, if we use a Gray-coded wheel, because of the unit distance property, there is only one track per sector where light transmission on each side of a sector boundary is different, so only one Gray-coded bit can be erroneous. The Gray-coding scheme can only produce numbers corresponding to adjacent sectors, and, consequently, no large errors are possible.

While Gray code is quite suitable for measurement, it is not useful for arithmetic because, as previously mentioned, it has no positional weighting.

Conversion Between Binary and Gray Codes

This problem is handled efficiently by noting that the Gray code can be considered to be a "differentiated" version of the equivalent binary. Conversion proceeds according to the following rules.

A. Conversion from binary to Gray

1. The left-most digits are the same in both systems.
2. Read the binary number from left to right. A change (0 to 1 or 1 to 0) generates a 1 in the Gray-coded number; otherwise a 0 is generated.

EXAMPLE 1.12 Binary-Gray conversion. Convert 01101001101 binary to Gray.

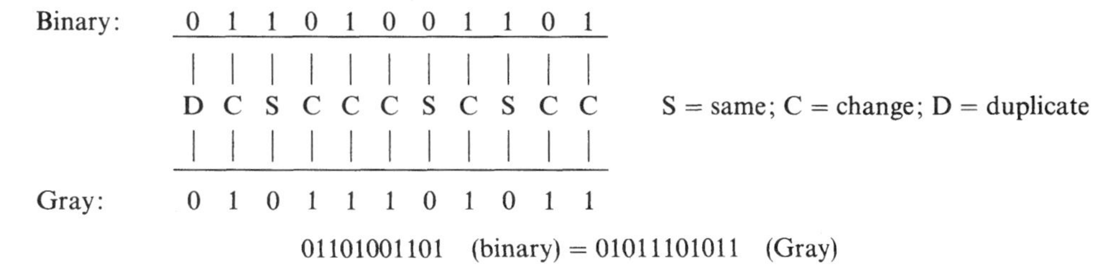

01101001101 (binary) = 01011101011 (Gray)

B. Conversion from Gray to binary

1. The left-most digits are the same in both systems.
2. Read the Gray number from left to right. A 1 means that the next binary digit must change; a 0 means the next binary digit is identical to the digit on its left.

EXAMPLE 1.13 Gray-binary conversion. Convert 1000110101010 Gray to an equivalent binary number.

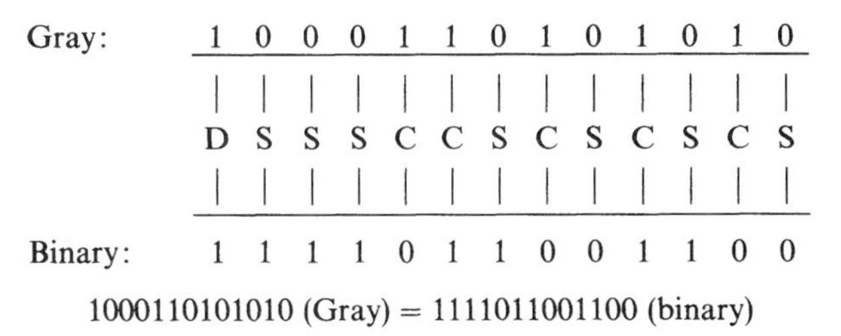

1000110101010 (Gray) = 1111011001100 (binary)

ASCII Code

Not all data stored and processed by computer is numerical. Because of the practical advantages of the binary system, other sorts of data are stored in two-valued (binary-like) form as well. The most commonly encountered *alphanumeric* code is the *American Standard Code for Information Interchange* (*ASCII*) of which some representative keyboard characters are presented in Table 1.2. This list is not intended to be exhaustive; merely illustrative. Note that decimal digits are listed as BCD-encoded digits preceded by 011 (9 = 011 1001). Other three-digit prefixes are used for nonnumeric data.

Table 1.2 Representative ASCII Code Set

Key	ASCII	Hexidecimal
A	100 0001	41
B	100 0010	42
C	100 0011	43
D	100 0100	44
E	100 0101	45
F	100 0110	46
G	100 0111	47
H	100 1000	48
I	100 1001	49
J	100 1010	4A
K	100 1011	4B
L	100 1100	4C
M	100 1101	4D
N	100 1110	4E
O	100 1111	4F
P	101 0000	50
Q	101 0001	51
R	101 0010	52
S	101 0011	53
T	101 0100	54
U	101 0101	55
V	101 0110	56
W	101 0111	57
X	101 1000	58
Y	101 1001	59
Z	101 1010	5A
Space	010 0000	20
(	010 1000	28
)	010 1001	29
+	010 1011	2B
0	011 0000	30
1	011 0001	31
2	011 0010	32
3	011 0011	33
4	011 0100	34
5	011 0101	35
6	011 0110	36
7	011 0111	37
8	011 1000	38
9	011 1001	39

1.6 ERROR DETECTION AND CORRECTION

Parity

In the electronic processing of data, it is impossible to avoid random noise with its consequent probability of the introduction of errors. One of the most common ways of dealing with this situation is to add redundant digits called *parity bits*. Each parity bit is associated with a group of data bits and is chosen to make the total number of 1s in the group either odd (*odd parity*) or even (*even parity*). At any point thereafter, the number of 1s in the entire group can be counted and compared with the parity originally created. This so-called simple parity check will always detect an odd number of errors among the group members, and it is highly effective in an environment where single errors are most prevalent. As an example, consider 4-bit BCD characters encoded with odd parity. The parity digit is the rightmost bit (refer to Table 1.3).

Table 1.3

Decimal	BCD	Parity Bit Added
0	0000	00001
1	0001	00010
2	0010	00100
3	0011	00111
4	0100	01000
5	0101	01011
6	0110	01101
7	0111	01110
8	1000	10000
9	1001	10011

If a 5 (0101 in BCD) is to be processed, the group 01011 in parity-checked BCD would be stored. If, at a later time, the group was retrieved as 11011 (first bit in error), the parity would now be even and the error would be detected.

Error Correction

Parity can also be used in special combinations which enable the correction as well as the detection of certain classes of errors. An example of the so-called *Hamming code* is shown in Fig. 1-3. In this case, 4 data bits (D_1, D_2, D_3, D_4) are used to create three additional parity bits (P_1, P_2, P_3) which are then added to the original four to form seven binary digits which are stored or transmitted as a group.

Fig. 1-3

Hamming code processing is done by a logic circuit which implements the procedure defined by Table 1.4. Three parity groups are formed from a different mix of three data digits and a single parity digit which is generated for each of these groups. Membership in a parity group is indicated by an X in the appropriate digit column. Note that each of the seven digits has a unique combination of parity group memberships. If it is assumed that only single digits will be in error from noise or interference, then a unique pattern of parity group failures is generated for each possible error. A knowledge of the parity failure pattern permits identification of the digit in error. Since this is a binary system, knowledge of which digit is erroneous makes correction possible.

Table 1.4 Hamming Code Generation

	D4	D3	D2	P3	D1	P2	P1
PARITY GROUP #1	X		X		X		X
PARITY GROUP #2	X	X			X	X	
PARITY GROUP #3	X	X	X	X			

Consider the case where data digits D_4, D_3, D_2, D_1 are 1011. Referring to Table 1.4, we see that for parity group 2, odd parity requires that P_2 be a 1 since D_1 and D_4 are group members, D_3 is 0, and D_2 (a nonmember) isn't counted. The rules of odd-parity generation combined with the algorithm defined by Table 1.4 will produce parity bits P_3, P_2, $P_1 = 110$. Hamming code is transmitted in the sequence D_4, D_3, D_2, P_3, D_1, P_2, P_1, so in the present example the logic circuit output will be transmitted as 1011110.

Suppose, now, that the second left-most digit is in error; i.e., the received sequence is 1111110. When retrieved, the parity groups are checked individually and reference to Table 1.4 shows that parity groups 2 and 3 will fail (show even parity). The only situation for which groups 2 and 3 fail and group 1 passes is when D_3 is in error. Its erroneous value is 1, and therefore, it should be corrected to a 0. If a parity bit is in error, only the single parity group with which it is affiliated will fail, thereby identifying it and allowing correction.

Nonparity Error-Detection Code

There are situations where techniques other than parity generation are used for error detection. For example, in some transatlantic radio communications, transmissions are encoded so that error-detection circuits in the receiver will generate an automatic request for retransmission (ARQ) if a character is garbled. The code used takes a data character (one of seventy possible) and configures it into an 8-bit format which has exactly four 1s and four 0s. At the receiver, the number of 1s in the character are counted, and if the sum is not four, an error is assumed. This code will detect all odd numbers of errors and will also detect those even numbers of errors where a 1 to 0 exchange is not accompanied by a 0 to 1 exchange.

Solved Problems

1.1 Convert the decimal number 234.567 to an equivalent binary number. Determine the fractional accuracy if nine conversion steps are undertaken.

Integer Conversion	Remainder	**Fractional Conversion**	Integer
2)234		0.567 × 2 = 1.134	1
2)117	0	0.134 × 2 = 0.268	0
2)58	1	0.268 × 2 = 0.536	0
2)29	0	0.536 × 2 = 1.072	1
2)14	1	0.072 × 2 = 0.144	0
2)7	0	0.144 × 2 = 0.288	0
2)3	1	0.288 × 2 = 0.576	0
2)1	1	0.576 × 2 = 1.152	1
0	1	0.152 × 2 = 0.304	0

The binary number is 11101010.100100010. Its right-most coefficient has a weight of 2^{-9} so that any error due to truncation will be less than 2^{-9} or 0.001953125.

1.2 Convert 345.678 (decimal) to an equivalent number in base 6 having a conversion error less than 0.001.

Since $6^{-3} = 0.0046$ (decimal) and $6^{-4} = 0.000771605$ (decimal), four fractional digits will be required.

Integer Conversion	Remainder	**Fractional Conversion**	Integer
6)345		0.678 × 6 = 4.068	4
6)57	3	0.068 × 6 = 0.408	0
6)9	3	0.408 × 6 = 2.448	2
6)1	3	0.448 × 6 = 2.688	2
0	1		

345.678 (decimal) = 1333.4022 (base 6)

1.3 Convert 217 (octal) to binary.

Apply the general conversion rule. Divide the number to be converted (217) by the new base (2). Do the arithmetic in the old base (8).

Examples of base 8 division:

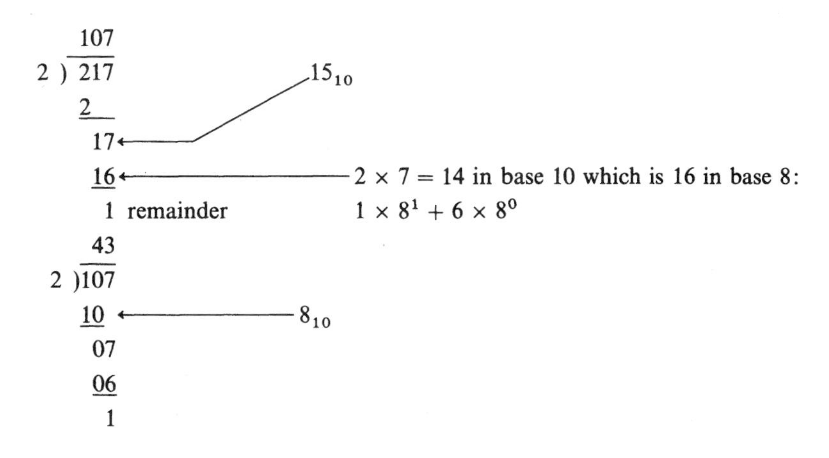

Perform the conversion, noting that there is no fractional part.

	Remainder
2)217	
2)107	1
2)43	1
2)21	1
2)10	1
2)4	0
2)2	0
2)1	0
0	1

217 (octal) = 10001111 (binary)

1.4 Express the hexadecimal number B73D as an equivalent octal number.

Change the hex to binary (groups of 4 bits).

$$\begin{aligned}\$B73D &= (1011)(0111)(0011)(1101) \\ &= 1011011100111101 \quad \text{(binary)}\end{aligned}$$

Change binary to octal (groups of 3 bits). Start from the right and add leading zeros as necessary.

$$\begin{aligned}1011011100111101 \quad \text{(binary)} &= (001)(011)(011)(100)(111)(101) \\ &= \;\;1 \quad\; 3 \quad\; 3 \quad\; 4 \quad\; 7 \quad\; 5 \\ &= 133475 \quad \text{(octal)}\end{aligned}$$

1.5 Express the octal number 23456 as a decimal number and as a BCD number.

$$\begin{aligned}23456 \quad \text{(octal)} &= 2 \times 8^4 + 3 \times 8^3 + 4 \times 8^2 + 5 \times 8^1 + 6 \times 8^0 \\ &= 2 \times 4096 + 3 \times 512 + 4 \times 64 + 5 \times 8 + 6 \times 1 \\ &= 8192 + 1536 + 256 + 40 + 6 \\ &= 10030\end{aligned}$$

Convert the individual decimal digits to binary

$$\begin{aligned}10030 \quad \text{(decimal)} &= (0001)(0000)(0000)(0011)(0000) \\ &= 0001\ 0000\ 0000\ 0011\ 0000 \quad \text{(BCD)}\end{aligned}$$

1.6 Convert the BCD number 0010 0111 1001 0101 into hexadecimal.

First, convert the BCD to decimal:

$$(0010)(0111)(1001)(0101) = 2795$$

Second, convert the decimal to binary.

2)2795	
2)1397	Rem = 1
2)698	Rem = 1
2)349	Rem = 0
2)174	Rem = 1
2)87	Rem = 0

2)43	Rem = 1
2)21	Rem = 1
2)10	Rem = 1
2)5	Rem = 0
2)2	Rem = 1
2)1	Rem = 0
0	Rem = 1

Reading up yields 101011101011 (binary).

Third, group by fours to form hex code.

$$101011101011 = (1010)(1110)(1011) = \text{AEB}$$

1.7 Convert F5.3B (hex) into equivalent octal.

Use an intermediate conversion into binary for convenience.

$$\$\text{F5.3B} = (1111)(0101).(0011)(1011)$$
$$= 11110101.00111011$$

Group by threes, relative to the binary point, to convert to octal.

$$11110101.00111011 = (011)(110)(101).(001)(110)(110)$$
$$= 365.166 \quad \text{(octal)}$$

1.8 Add $A47 to $854. Do the arithmetic in hex, noting that hex is a weighted code.

Adding digit by digit:

```
  A47
 +854
 129B
```

$7 + 4 = 11$ (decimal) = B (hex)
$4 + 5 = 9$ (decimal) = 9 (hex)
$A + 8 = 18$ (decimal) = 2 plus 1 to carry (hex)

1.9 Add 234 (decimal) to 189 (decimal) and do the arithmetic in binary.

Both numbers are positive; the process is simple addition. First convert to binary.

234 (decimal) = 0011101010 (binary)
189 (decimal) = 0010111101 (binary)

Add as shown in Sec. 1.4.

Carry-in	011111000 ←
234	0011101010
189	0010111101
Sum	0110100111
Carry-out	0011111000

The sum is positive (sign bit = 0) and can be converted to +423 (decimal), which checks.

1.10 Complementary addition can be used to perform subtraction in any base, including base 10. Use 9s complement addition to subtract 1462_{10} from 1937_{10}.

Add a leading zero to each number to serve as a sign digit. A number is complemented by subtracting it from $B^N - 1$ where B is the base and N is the number of places. In the present case, B = 10, N = 4 and the selected number (1462) will be subtracted from 9999.

```
 9999
-1462
─────
98537  ← There will never be a borrow since all numbers are less than 9
   +1
─────
98538  ← 9s complement of 1462
```

The left-most digit (sign digit) is a 9 indicating that the number is negative. That the 9s complement of 1462 is equivalent to −1462 may be shown by taking the 9s complement again to get back the original number. It should be evident that this process is equivalent to X = −(−X).

```
 99999
-98538
──────
 01461  ← The sign digit is 0, indicating a positive result.
    +1
──────
  1462
```

The desired subtraction is now performed.

```
  01937
 +98538
 ──────
 100475
```

Excluding the extraneous sign digit carry, we have 475 which is interpreted as a positive number since its left-most digit is 0.

1.11 Subtract 176 from 204. Do the work in binary using 2s complement notation.

The largest of these numbers requires 8 bits, so, with the sign bit, N = 9.

The complement of an N-digit base 2 number is obtained by subtracting the number from a number made up of N 1s. In the binary case, this is easily achieved by simply inverting the 1s and 0s.

```
Complement 176:          010110000 → 101001111    Invert the bits
Add 1:                               000000001
                                     ─────────
The 2s complement is:                101010000    This is equivalent to −176

            Add −176 to +204:     204 →  011001100
                                 −176 →  101010000
                                        ──────────
                                        1000011100
                                        |
                                  Carry out from
                                  sign-bit addition
                                  is neglected

            Sum: 000011100
```

The sign digit is 0, so the number is positive. Direct conversion leads to +28 (decimal), which checks.

1.12 Add −176 (decimal) to −204 (decimal); do the arithmetic in binary using 2s complement notation.

The answer will have a magnitude of 380_{10} which requires 9 binary bits. Including a sign bit, *we will need at least 10 digits* for the computation. Express the negative numbers in 2s complement form:

$$-176_{10} = 1101010000$$
$$-204_{10} = 1100110100$$

Add:

$$\begin{array}{r} 1101010000 \\ 1100110100 \\ \hline 11010000100 \end{array}$$

Ignore the left-most carry out and note that the next digit (sum sign digit) is 1, indicating a negative result. To find the magnitude of the negative number, take the 2s complement:

$$\begin{array}{rr} 1010000100 \rightarrow & 0101111011 \\ + & 1 \\ \hline & 0101111100 = 380_{10} \end{array}$$

The decimal answer is -380.

1.13 Subtract 204 from 176 in binary using 2s complement arithmetic.

Since the result will have a magnitude less than either given number, N = 9 bits. Take the 2s complement of the number to be subtracted (this is equivalent to making it negative).

$$-204_{10} \rightarrow 100110100$$

Add:

$$\begin{array}{rr} +176 \rightarrow & 010110000 \\ -204 \rightarrow & 100110100 \\ \hline & 111100100 \end{array}$$

There is no carry to ignore; the sign bit is negative. Take the 2s complement to determine the magnitude.

$$\begin{array}{rr} 111100100 \rightarrow & 000011011 \\ + & 1 \\ \hline & 000011100 = 28_{10} \end{array}$$

The decimal answer is -28.

1.14 Subtract 365 (octal) from 173 (octal). Use 8s complement addition to perform the subtraction.

Add a left-most (highest-order) digit as a sign digit; 7 for negative and 0 for positive. Take the complement by subtracting the selected octal number from 7777.

$$\begin{array}{r} 7777 \\ -0365 \\ \hline 7412 \\ +\quad 1 \\ \hline 7413 \end{array} \quad \text{(8s complement of 365)}$$

Add:

$$\begin{array}{r} 0173 \\ +7413 \\ \hline 7606 \end{array} \quad \text{(addition is done in octal)}$$

The octal addition was performed as follows:

$$3 + 3 = 6$$
$$7 + 1 = 8 \quad \text{(decimal)}$$
$$= 10 \quad \text{(octal)} \qquad \text{(the 1 is carried one column to the left)}$$
$$4 + 1 + 1\text{(carry)} = 6$$
$$7 + 0 = 7$$

There is no sign-bit carry-out to ignore.

Interpretation of the result: The left-most digit (sign digit) is 7; therefore the number is negative. The 8s complement of 7606 = 0171 + 1 = 0172; thus the magnitude is 172. The answer in octal is *either* 7606 *or* −172.

1.15 Produce the binary equivalent of the Gray-coded number 11011010011010.

1. The most significant digit is the same.
2. Read the Gray-coded number from left to right and recall that a 1 signifies a change in the equivalent binary as it proceeds from left to right.

```
Gray     1  1  0  1  1  0  1  0  0  1  1  0  1  0
            |  |  |  |  |  |  |  |  |  |  |  |  |     S = same
            C  S  C  C  S  C  S  S  C  C  S  C  S
            |  |  |  |  |  |  |  |  |  |  |  |  |     C = change
Binary   1  0  0  1  0  0  1  1  1  0  1  1  0  0
```

1.16 Give the Gray-coded equivalent of the hex number 3A7.

First, convert the hex to binary.

$$\$3A7 = (0011)(1010)(0111) = 1110100111 \quad \text{(binary)}$$

Convert binary to Gray.

```
Binary   1  1  1  0  1  0  0  1  1  1
            |  |  |  |  |  |  |  |  |     S = same
            S  S  C  C  C  S  C  S  S
            |  |  |  |  |  |  |  |  |     C = change
Gray     1  0  0  1  1  1  0  1  0  0
```

1.17 Find the Gray code equivalent of the octal number 527.

$$527 \quad \text{(octal)} = (101)(010)(111) = 101010111 \quad \text{(binary)} = 111111100 \quad \text{(Gray)}$$

1.18 When a block of data is stored on magnetic tape, sometimes parity is computed on both the rows and the columns. Create the row and column parity bits for the data group shown below, using odd parity.

Data
10110
10001
10101
00010
11000
00000
11010

In the parity-bit row, the highest-order zero indicates that there are an odd number of ones reading vertically down the left-most data column. The next bit, a 1, tells us that its associated data column has an even number of ones, etc.

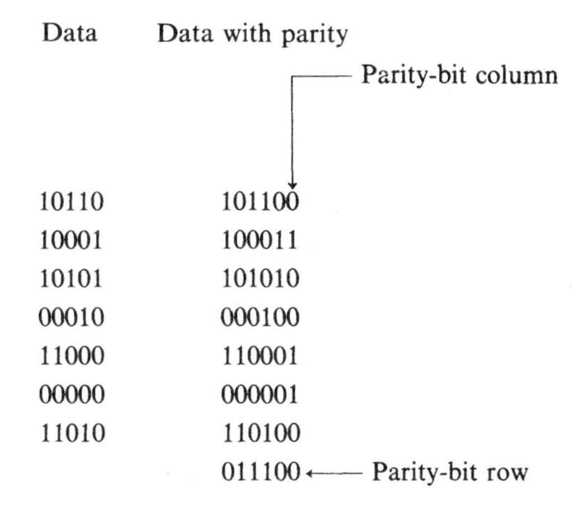

Data	Data with parity
10110	101100
10001	100011
10101	101010
00010	000100
11000	110001
00000	000001
11010	110100
	011100 ← Parity-bit row

1.19 Encode the data $D_4, D_3, D_2, D_1 = 1001$ into a 7-bit Hamming code. Use odd parity.

Refer to Table 1.4. Parity group 1 uses D_4, D_2, D_1, and P_1. Membership data bits are 101 and P_1 is 1. Parity group 2 uses D_4, D_3, D_1, and P_2. Membership data bits are 101 and P_2 is 1. Parity group 3 uses D_4, D_3, D_2, and P_3. Membership data bits are 100 and P_3 is 0. The final sequence is then $D_4, D_3, D_2, P_3, D_1, P_2, P_1 = 1000111$.

1.20 Suppose that the sequence encoded in Prob. 1.19 is received as 1010111. Which bit is in error?

Parity group 1 uses D_4, D_2, D_1, and $P_1 \rightarrow 1111$	(even)	Check fails
Parity group 2 uses D_4, D_3, D_1, and $P_2 \rightarrow 1011$	(odd)	Check passes
Parity group 3 uses D_4, D_3, D_2, and $P_3 \rightarrow 1010$	(even)	Check fails

Referring to the parity group membership chart, the only single digit which is a member of groups 1 and 3 but not of group 2 is digit D_2. Therefore, D_2 is the erroneous digit.

1.21 It is known that 4 bits of data were encoded in single-error-detecting code (odd parity) as defined in Table 1.4. If it is assumed that only single errors may occur, what is the correct data if the received sequence is $D_4, D_3, D_2, P_3, D_1, P_2, P_1 = 1011111$? What is the correct data?

Parity group 1 = 1111	Fails
Parity group 2 = 1011	Passes
Parity group 3 = 1011	Passes

The only digit which occurs solely in parity group 1 is parity bit P_1 and, therefore, P_1 is in error. The data digits are all correct. Note that the failure of only a single parity group means that the error is in a parity digit.

1.22 Interpret the following ASCII-coded sequence:

1000001 1000100 1000100 0100000 0110011 0110100 0111001

Using Table 1.2:

1000001 → A
1000100 → D
1000100 → D
0100000 → Blank
0110011 → 3
0110100 → 4
0111001 → 9

The answer is, "ADD 349."

1.23 Compare the noise margins of a decimal storage system with that of a hypothetical three-state system. The two systems make use of the current in a transistor as the stored variable.

Assume that the physical states are equally spaced between zero current and full current (I_{max}) for both systems. As described in Sec. 1.2, the spacing between signal levels and thresholds for the decimal system is $I_{max}/18$. For the three-state system, the signal levels would be 0, $I_{max}/2$, and I_{max}. With decision thresholds set halfway between these levels (at $I_{max}/4$ and $3\ I_{max}/4$), the separation of the threshold from a signal level is $I_{max}/4$. Thus, the three-state system has a noise margin which is 4.5 times larger than that of the decimal system $(I_{max}/4)/(I_{max}/18)$.

1.24 Devise a single error-correcting code for an 11-bit data group.

There must be a unique combination of parity group members for each bit of the coded group so that at least four parity groups are required. The total number of bits will be 11 plus 4, or 15 bits. Since four parity group members can be arranged in 16 distinct combinations, four parity groups are adequate and each bit can be "tagged." It is convenient to order the groupings in binary form as shown in Table 1.5 (refer to Sec. 2.2 on truth table construction).

Table 1.5

Bit	PG4	PG3	PG2	PG1
P1				X
P2			X	
D1			X	X
P3		X		
D2		X		X
D3		X	X	
D4		X	X	X
P4	X			
D5	X			X
D6	X		X	
D7	X		X	X
D8	X	X		
D9	X	X		X
D10	X	X	X	
D11	X	X	X	X

1.25 Suppose the single-error-detecting code in Prob. 1.24 is used in a system with odd parity. The data, ordered with D_{11} as the most significant bit, is

D_{11}	D_{10}	D_9	D_8	D_7	D_6	D_5	D_4	D_3	D_2	D_1
1	0	1	0	0	1	1	1	1	0	0

What is the encoded sequence?

D_{11}	D_{10}	D_9	D_8	D_7	D_6	D_5	P_4	D_4	D_3	D_2	P_3	D_1	P_2	P_1
1	0	1	0	0	1	1	1	1	1	0	1	0	1	1

The P_4 entry is 1 because members of this group (D_5 to D_{11}) have an even number of 1s. This process is repeated for the remaining parity entries.

1.26 What action will the system of Prob. 1.25 take if the retrieved sequence is 100001111111011?

Note that there are two digits received in error (D_9 and D_2). Since, however, the system is designed to detect only a single error, it will respond as if this were the case. We see that parity-check PG4 alone fails and the detection circuitry will "assume" that the digit which belongs only to PG4 (P_4) is the single bit in error. It will "correct" P_4 to 0 leaving the retreived message still erroneous.

1.27 The ARQ coding technique is described in Sec. 1.6. Show that this technique will detect all odd numbers of errors in a code word and *some* of the even numbers of errors. Use the code word 11001010 as an example.

The key to the code is that exactly four 1s are sent in every word. A single error must cause one of the 1s to become 0 or one of the 0s to become 1 and, in either case, the 1s count is changed from four. For any odd number of errors, the number of 1 to 0 transitions will always be different than the number of 0 to 1 transitions and an erroneous 1s count will always result. In the case of an even number of errors, it is possible that the 1 to 0 transitions could be balanced by an equal number of 0 to 1 transitions thereby keeping the number of 1s constant and resulting in an error-detection failure. For example, see the following table in which the transmitted code is 11001010.

Received	Errors	1s Count
11001110	One	Five
10001010	One	Three
11110010	Three	Five
11111110	Three	Seven
10100010	Three	Three
10011010	Two	*Four* (undetected)
11111010	Two	Six

Supplementary Problems

1.28 Convert the binary number 100110111 to decimal, octal, and hexadecimal.

1.29 Convert the decimal number 416 into binary, octal, and hexadecimal.

1.30 Write the decimal digits 0 through 10 in base 4.

1.31 Express the decimal number 250.5 in base 7 and base 16.

1.32 Convert the following numbers to their decimal equivalents:
(*a*) 1032.2_4 (*b*) 0.342_6 (*c*) 60_7 (*d*) 188_{12}

1.33 Convert the hexadecimal number AB6 into binary, octal, and decimal.

1.34 Express 205 (decimal) as a binary and as a hex number.

1.35 Convert the BCD number 10010110 to hexadecimal.

1.36 Given the hexadecimal number $A38, convert it to octal.

1.37 Given the hexadecimal number $D3C, convert it to BCD.

1.38 Given the octal number 1216 convert it to BCD.

1.39 Add the octal numbers 27 and 42 and express the results in binary.

1.40 Find the 2s complement equivalent of −119 (decimal) assuming a 10-bit binary word length.

1.41 Subtract 100010 from 10100 using the 2s complement method.

1.42 Add −105 (base 10) and −56 (base 8). *Note that both numbers are negative.* Do the arithmetic in 2s complement binary using 9 bits, including the sign bit. Express the magnitude of your answer in hexadecimal.

1.43 Subtract 583_{10} from 736_8. Do the arithmetic in 2s complement binary and express the result in complemented base 6.

1.44 Find the 16s complement of $0B7 (hex).

1.45 Convert 1001101100011 (Gray code) to binary.

1.46 Convert 1100101101101 (binary) to Gray code.

1.47 A shaft encoder similar to that shown in Fig. 1-2*b* is Gray coded into 32 equal sectors. Determine the output Gray code if the shaft is rotated 36° clockwise.

1.48 Express the Gray-coded number 101101101011 as equivalent hexadecimal and binary numbers.

1.49 If, on an ASCII keyboard, the keys ADD 21 are pressed, what binary signal enters the computer?

1.50 The data 1011 is properly encoded into a single error-correcting code according to the rules of Table 1.4. It is received, after contamination by noise, with data bits D_3 and D_2 in error. The receiver is expecting a single error and processes the received code group accordingly. Show the complete transmitted and received words and indicate the decision made by the receiver.

Answers to Supplementary Problems

1.28 311 (decimal), 467 (octal), $137 (hex)

1.29 110100000 (binary), 640 (octal), $1A0 (hex)

1.30 0, 1, 2, 3, 10, 11, 12, 13, 20, 21, 22

1.31 $505.333\ldots_7$, $FA.8

1.32 (*a*) 78.5 (*b*) 0.62037 (*c*) 42 (*d*) 248

1.33 101010110110 (binary), 5266 (octal), 2742 (decimal)

1.34 11001101, $CD

1.35 $60

1.36 5070_8

1.37 0011 0011 1000 1000

1.38 0110 0101 0100

1.39 $71_8 = 111001$

1.40 1110001001

1.41 The result is the negative with a magnitude of 01110.

1.42 The result is negative. Its magnitude in binary is 10010111 which is equal to $97.

1.43 −105 (base 10); the magnitude is 5303_6.

1.44 $F49

1.45 1110110111101

1.46 1010111011011

1.47 The encoder turns 3.2 sectors; if the first is numbered 0, the Gray code will be 00010 (5 bits are required).

1.48 $DB2, 110110110010 (binary)

1.49

A	D	D	space	2	1
(1000001)	(1000010)	(1000010)	(0100000)	(0110010)	(0110001)

1.50 1011110 is the transmitted word, and 1101110 is the received word. The receiver will assume that bit D_1 is in error, and the "corrected" data will be 1100, which is incorrect.

Chapter 2

Design of Combinational Logic I

2.1 COMBINATIONAL LOGIC

Combinational (or combinatorial) logic is defined as that class of digital circuits where, at any given time, the state of all outputs depends only upon the values of the inputs at that time and not upon any previous input states. A combinational logic circuit may be regarded as a *black box* having N input lines and P output lines, each of which carries a digital (or logic) function which can have only two possible values, commonly denoted as 1 and 0 or TRUE and FALSE.

Starting from a verbal or symbolic statement of the relation between output and input, it is the job of the logic designer to "fill the box" with a circuit comprised of suitable interconnections of fundamental logical components called *gates.*

2.2 TRUTH TABLES

A truth table is a convenient way to symbolically represent a logic function. All possible combinations of input variable values (usually in ascending bindary order) are presented in tabular form, and, for each unique combination of inputs, the output variable values are listed with a separate column assigned to each variable. The truth table thus constitutes a complete specification of the combinational logic to be designed.

To illustrate the use of a truth table for problem specification, consider the case of a logic block which is to *convert 4-bit Gray-coded numbers to binary numbers.* The box to be designed is shown in Fig. 2-1.

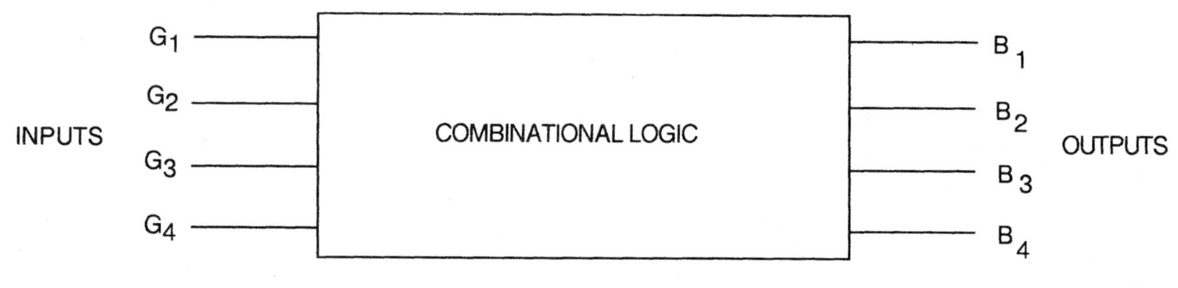

Fig. 2-1

Subscripts indicate the position of each digit in the sequence, 4 being the most significant and 1 the least. The relation between binary and Gray codes was presented in Sec. 1.5, and, from this listing, a truth table may be constructed.

Truth Table Construction

With reference to Table 2.1.

1. Set up a table having a separate column for each input and output variable.
2. Fill out the left side by listing every possible combination of input values in ascending binary order (see Example 2.1 below). If there are N input variables, there will be 2^N rows, each specifying a unique combination.

Table 2.1

All possible combinations of input variable values

INPUTS				OUTPUTS			
G_4	G_3	G_2	G_1	B_4	B_3	B_2	B_1
0	0	0	0	0	0	0	0
0	0	0	1	0	0	0	1
0	0	1	0	0	0	1	1
0	0	1	1	0	0	1	0
0	1	0	0	0	1	1	1
0	1	0	1	0	1	1	0
0	1	1	0	0	1	0	0
0	1	1	1	0	1	0	1
1	0	0	0	1	1	1	1
1	0	0	1	1	1	1	0
1	0	1	0	1	1	0	0
1	0	1	1	1	1	0	1
1	1	0	0	1	0	0	0
1	1	0	1	1	0	0	1
1	1	1	0	1	0	1	1
1	1	1	1	1	0	1	0

3. For each row, fill out the corresponding desired value of each of the P output variables.

The truth table, having N + P columns and 2^N rows, is a complete symbolic specification of a combinational logic function, which in the present case is the relationship between binary and Gray codes.

EXAMPLE 2.1 The following describes a simple *method of listing all possible combinations of the values of N binary variables* for use in the input side of a truth table (refer to Table 2.1):

1. There will be 2^N rows (always an even number).
2. The right-most column (G_1 in the table) contains alternating 1s and 0s beginning with a 0 in the first row: 0, 1, 0, 1, 0, 1, ...
3. The second column to the left is comprised of two 0s followed by two 1s repeating alternatively: 0, 0, 1, 1, 0, 0, 1, 1, ...
4. Column 3 starts with four 0s followed by four 1s, repeating alternatively.
5. In general, the kth column has $2^{(k-1)}$ alternating 0s and 1s.

In many cases, constructing the truth table from a problem specification can be quite difficult, and consequently, the process to be implemented in logic must often be studied in detail until patterns are observed. The implementation of a *binary adder* illustrates the technique. In this case, the designer might, with a scratch pad or calculator, repeatedly add pairs of binary numbers. Since the binary number system is positional, it soon becomes evident that only the addition of single digits need be investigated because the general procedure is the same for each digit. Thus, a process which works for one digit will also work for any other digit, and any hardware designed need only be replicated to handle multidigit numbers.

The sum digit produced depends on the two digits being added and also on the presence of a carry from the preceding lower-order digit. The designer will conclude that there are three input variables involved in the generation of a binary sum digit: the X digit, the Y digit, and a carry-in C_i. There will be *two* output variables since, in addition to the sum digit S, a carry-out C_o must be generated to serve

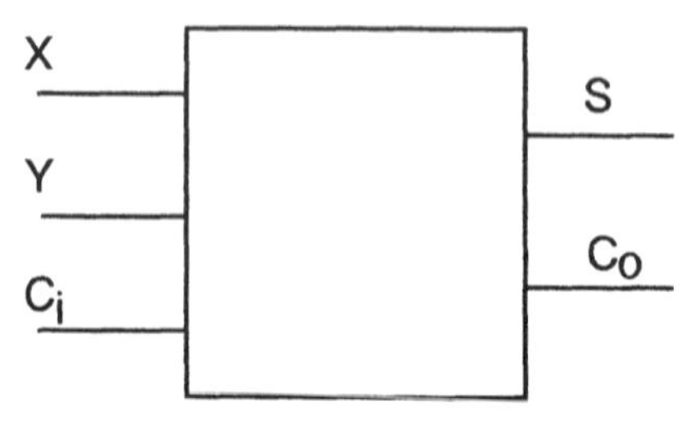

Fig. 2-2

as the carry-in for the next highest order digit. When all required variables have been identified, the logic implementation may be symbolically represented by a black box as shown in Fig. 2-2.

Once the single-digit addition process has been understood, it may be described by a truth table (Table 2.2) which lists all possible input combinations and the corresponding outputs. We now have a compact symbolic specification of the combinational logic which must be designed to go into our black box.

Table 2.2

INPUT			OUTPUT	
X	Y	C_i	S	C_o
0	0	0	0	0
0	0	1	1	0
0	1	0	1	0
0	1	1	0	1
1	0	0	1	0
1	0	1	0	1
1	1	0	0	1
1	1	1	1	1

2.3 BOOLEAN EQUATIONS AND BASIC LOGICAL FUNCTIONS

As demonstrated above, the complete design specification for combinational logic can be expressed in the form of a truth table. All the information in this truth table can also be formulated as a set of logical equations; one for each output. Significantly, we may utilize the efficient functional structures of a large body of mathematics called *Boolean algebra*, which deals with variables having only two values (usually termed TRUE or FALSE), if we consider the 1s and 0s of the truth table equivalent to the TRUEs and FALSEs, respectively, of Boolean algebra.

Structure of Boolean Equations

Boolean equations have the following basic characteristics:

1. Each equation has the form F = f(A, B, C, ...) where F, A, B, C, etc., are variables.
2. All the variables are *logical variables* which are characterized by having only *two possible values:* TRUE or FALSE, HIGH or LOW, 1 or 0, etc.
3. Logical variables are related in the equations by means of *logical operators* or connectives: LOGICAL EQUIVALENCE, AND, OR, and LOGICAL INVERSION.

Logical Equivalence

Two logical variables are *equivalent* if they have *identical truth values*. The symbol for logical equivalence is an equals sign (=).

EXAMPLE 2.2 The expression A = B means that if A is TRUE, then B must also be TRUE, and if A is FALSE, then B must also be FALSE. No other possibilities can exist. Note that logical variables A and B can, in turn, represent complicated logical expressions.

The Boolean AND Function

The AND function is defined such that the output is TRUE if and only if *all* inputs are TRUE. The symbol for logical AND is a dot (·).

EXAMPLE 2.3 The expression F = A · B states that F is TRUE if both A and B are TRUE. In common usage, *the dot is omitted*, so we simply write, F = AB. The AND function may be represented by a truth table:

A	B	F
0	0	0
0	1	0
1	0	0
1	1	1

The Boolean OR Function

The OR function is defined such that the output is TRUE if any one *or* more of the inputs is TRUE. The symbol for logical OR is a plus sign (+).

EXAMPLE 2.4 The expression F = A + B states that F is TRUE if either A or B or both are TRUE. The truth table representation for the OR function is

A	B	F
0	0	0
0	1	1
1	0	1
1	1	1

Logical Inversion

The expression A′ means that if A is TRUE, then A′ will be FALSE and vice versa.

A	A′
TRUE	FALSE
FALSE	TRUE

A	A′
1	0
0	1

Note that in practice, either $\bar{A}$ or A′ may be used to indicate logical inversion, the latter form being preferred by those using word processors.

EXAMPLE 2.5 In the expression F = A′B, F is TRUE if A is FALSE and B is TRUE.

Exclusive OR

The connective called exclusive OR, though very useful for the logic design engineer, is not, strictly speaking, a basic Boolean function. It has only two input variables and produces a TRUE output if, and only if, one of the inputs is TRUE and the other is FALSE.

The exclusive OR function is *often referred to as XOR* and its symbol is $\oplus$.

EXAMPLE 2.6 The truth table representation of the XOR function, $F = A \oplus B$ is

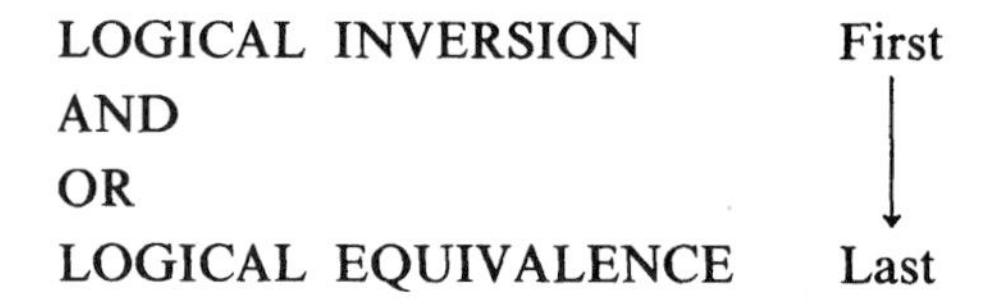

A	B	F
0	0	0
0	1	1
1	0	1
1	1	0

Hierarchy of Operations

As in any algebra, the sequence of operation is very important. For Boolean algebra, this hierarchy of operations is ordered as follows:

LOGICAL INVERSION	First
AND	↓
OR	↓
LOGICAL EQUIVALENCE	Last

The order of execution can be modified by the use of parentheses in the same manner as in conventional algebra. Operations within the innermost parentheses are done first and proceed outward until all parentheses are cleared.

Reading a Boolean Equation

Given $F = AB + B'C' + A'(C + D')$, we read the equation as follows:

(*a*) F will be 1 (true) if the right-hand side of the equation (RHS) is also true.
(Uses equality)

(*b*) RHS will be 1 if any combination of AB or B′C′ or A′(C + D′) is 1.
(Uses definition of OR)

(*c*) AB will be 1 if both A and B are 1.
(Uses definition of AND)

(*d*) B′C′ will be 1 if both B′ and C′ are 1.
B′ will be a 1 if B is a 0.
C′ will be a 1 if C is a 0.
Thus, B′C′ will be a 1 if B and C are both 0.
(Uses AND and LOGICAL INVERSION)

(*e*) A′(C + D′) will be a 1 if A′ is a 1 and (C + D′) is also 1.
A′ will be a 1 if A is a 0.
(C + D′) will be a 1 if C is a 1 or if D is a 0 or if both of these conditions hold. Thus, A′(C + D′) will be a 1 if A is 0 and if either D is 0 or C is 1 (or both).
(AND, OR, LOGICAL INVERSION and use of parentheses)

Note how parentheses can dramatically affect the value of an expression. For example, the student should confirm that $A'C + D'$ is quite different than $A'(C + D')$ or $A'(C + D)'$.

2.4 THE RELATION BETWEEN BOOLEAN EQUATIONS AND TRUTH TABLES

Consider, again, a module that is designed to perform 1-bit binary addition. The circuit within the block in Fig. 2-2 forms the sum of binary inputs X, Y, and the carry-in C_i from a previous module. There are two output variables: the sum S and a carry-out C_o. The truth table which specifies the addition process occurring within the box is reproduced in Table 2.3 for convenience.

Table 2.3

INPUT			OUTPUT	
X	Y	Ci	S	Co
0	0	0	0	0
0	0	1	1	0
0	1	0	1	0
0	1	1	0	1
1	0	0	1	0
1	0	1	0	1
1	1	0	0	1
1	1	1	1	1

In order to recast the information in the truth table in Boolean equation form, we develop one equation for each of the two output columns. Note that the truth table rows correspond to unique combinations of input values. For example, the bottom row corresponds to the situation where X is 1 *and* Y is 1 *and* C_i is 1. If we make 1 equivalent to TRUE, then we see from the truth table's bottom row that one condition for S (and C_o) being TRUE is that X, Y, and C_i are all true. Thus, from Example 2.3, we have

$$S = XYC_i \tag{2.1}$$

The Boolean AND function requires that all the variables involved be TRUE. In order to apply it to a truth table, which has numerous 0 (FALSE) entries, we make use of LOGICAL INVERSION as shown in Example 2.5. Thus, we see that it is possible to write an AND expression for each truth table row. For example, the equation

$$S = X'YC_i' \tag{2.2}$$

corresponds to the third row of Table 2.3 and can be read as "S will be TRUE when (if) X′ and Y and C_i' are all TRUE." This is equivalent to stating that "S will be true when X and C_i are FALSE and Y is TRUE."

We see from the truth table that there is more than one input situation (row) where the corresponding output is 1 (TRUE). Here, the Boolean OR function proves useful since we see that the sum (S) is a 1 (TRUE) if any of the physical situations corresponding to row 2, 3, 5, or 8 occur. Expressed algebraically,

$$S = X'Y'C_i + X'YC_i' + XY'C_i' + XYC_i \tag{2.3}$$

Similarly, the output C_o can be expressed as:

$$C_o = X'YC_i + XY'C_i + XYC_i' + XYC_i \tag{2.4}$$

These two equations, taken together, contain all the information that is in the truth table (Table 2.3) since they specify which input combinations (rows) produce 1s in the two output columns. The column entries which are not 1s must be 0s; they can have no other value since Boolean algebra is two-valued.

EXAMPLE 2.7 It is also possible to write a Boolean equation from a truth table using all rows where the output column contains a 0. For the adder under consideration, the resulting Boolean equations would be for $S' = 1$ and $C_o' = 1$ (equivalent to 0 values for S and C_o). This yields

$$S' = X'Y'C_i' + X'YC_i + XY'C_i + XYC_i'$$

$$C_o' = X'Y'C_i' + X'Y'C_i + X'YC_i' + XY'C_i'$$

2.5 LOGIC DIAGRAMS

It is often helpful to draw a pictorial schematic diagram of the logic to be designed. This diagram, which makes use of standard logic symbols, is useful in making the translation from logic equations to functioning circuits.

Basic Functions and Their Symbols

Components which accomplish the fundamental logical operations OR and AND are called *gates* and they can be used as building blocks to construct extremely complex logical functions. Their symbols, pictured in Fig. 2-3, have three inputs, but, in practice, any number may be used, as required.

A third basic symbol very useful to logic designers, though not universally adapted, is the slash mark* used to denote logical inversion (see Fig. 2-4).

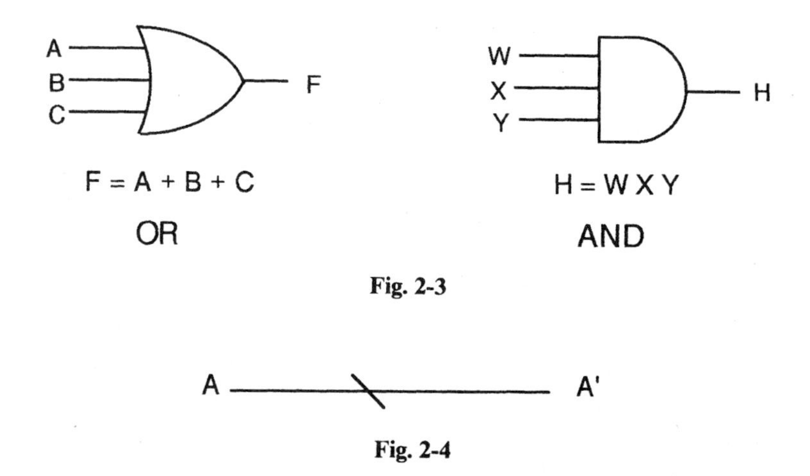

Fig. 2-3

Fig. 2-4

XOR

As mentioned previously in Sec. 2.3, the exclusive OR function is not fundamental; its logic can be expressed in Boolean form as

$$F = AB' + BA' \tag{2.5}$$

and synthesized with two AND gates and one OR gate as we can observe from the truth table in Example 2.6. The XOR function appears so often, however, that a standard logic symbol has been assigned to it (see Fig. 2-5).

EXAMPLE 2.8 The XOR gate may be used to produce a LOGICAL INVERSION as shown in Fig. 2-6.

A given variable A may be either passed through the gate unchanged or inverted, depending upon the logical value of a second variable which serves as a control switch. This behavior is easily demonstrated by observing, in the truth table of Example 2.6, that when $B = 0$, F is identical to A and when $B = 1$, F is the logical inverse of A.

---

* The symbol shown in Fig. 2-4 is a *logic* symbol only; its hardware implementation is discussed in Sec. 4.3.

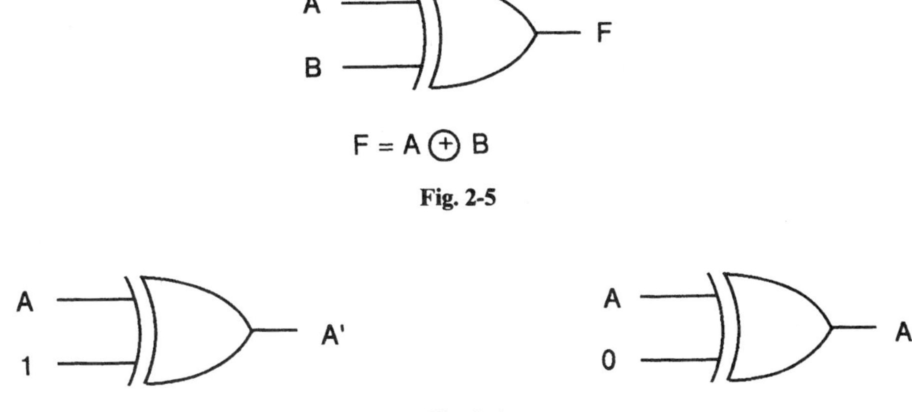

Fig. 2-5

Fig. 2-6

Combining Symbols

It is possible to combine the basic symbols graphically to obtain a picture of a logic function. For example, the equation $F = A'BC + AB'D + A'B'CD'$ may be depicted as in Fig. 2-7.

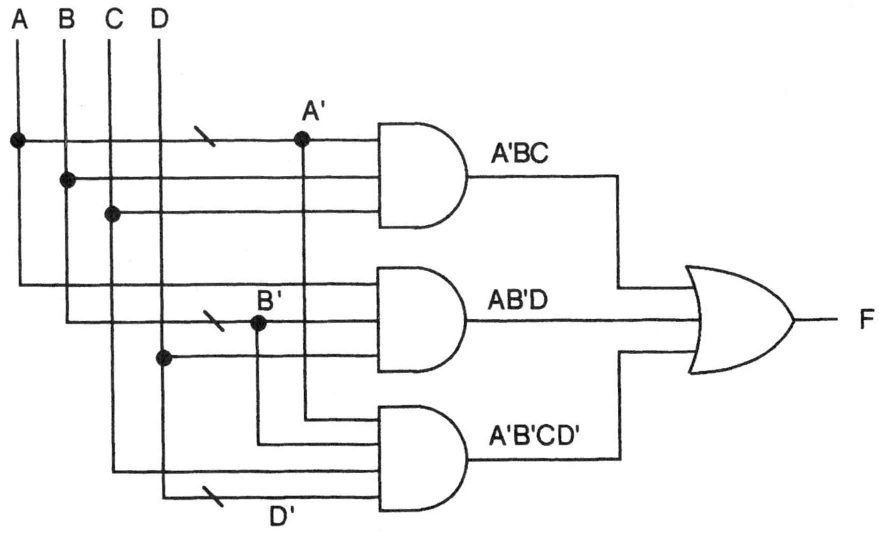

Fig. 2-7

On the other hand, it is sometimes required to obtain the underlying Boolean equation from a given logic diagram. Consider the circuit shown in Fig. 2-8. The output F is produced by OR gate G_1 which has three inputs: variable E and the outputs of AND gates G_2 and G_3. For convenience, we will use the same designation for a gate and its output. Thus, depending upon the context, G_2 may be taken to be the identification of a particular gate or the output variable produced by this gate. With this convention in mind, it is a simple matter to write logic equations directly from a diagram.

Proceeding from right to left, we have,

$$F = E + G_2 + G_3$$
$$G_2 = AB'D$$
$$G_3 = ABC'$$

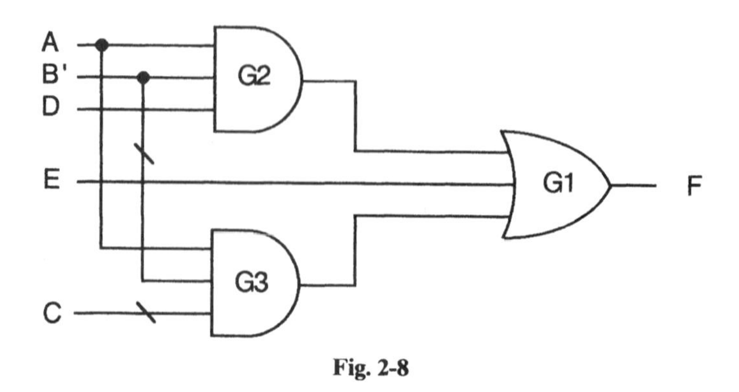

Fig. 2-8

Simple algebraic substitution yields

$$F = E + AB'D + ABC'$$

The importance of the logic diagram arises from the fact that digital hardware manufacturers designate electronic circuits in accordance with the logic accomplished by them (e.g., OR gates); the circuit symbol is generally a close approximation to, or variation of, the logic symbol. Thus, a logic diagram can be used to estimate hardware needs by equating logic symbols with their corresponding electronic gates. Additionally, the connections between symbols on the logic diagram can lead directly to the specification of wiring between hardware elements. Since interconnection of physical components requires adherence to compatibility rules at various interfaces, the logic diagram can also be an important aid to the designer in identifying and solving potential connection problems.

Solved Problems

2.1 Create the truth table for a box that accepts four data digits and creates three additional parity digits for a single-error-correcting Hamming code as defined in Table 1.4. Call the data digits D_4, D_3, D_2, D_1 and denote the parity digits by P_3, P_2, P_1.

Using the method shown in Example 2.1, fill out the left side of the truth table with all possible input combinations. Next, assign values to the three output variables following the rules defined by Table 2.4. For example, consider the fourth row, 0011. Parity group 1 is seen to have an even number of 1s since $D_1 = 1$, $D_2 = 1$, $D_4 = 0$, and D_3 is not a group member. To achieve odd parity, a 1 must be added, and this is achieved by making parity-bit P_1 a 1. Continuing the analysis, D_1 is the only member of parity group 2 which is a 1 so the group's parity is odd and $P_2 = 0$. Bit $P_3 = 0$ because D_2 is the only 1 in parity group 3. The process is continued for each row until the table is completed as shown in Table 2.4.

2.2 Most calculators and digital watches make use of seven-segment displays for output. Each digit of these displays is comprised of seven bars of light-emitting semiconductor (or light-absorbing liquid crystal) material arranged as shown in Fig. 2-9(*b*). These bars are energized selectively to provide visual display of a desired digit. For example, if the decimal number 2 is to be displayed, then segments a, b, g, e, and d are energized. Create the truth table for logic that receives a BCD digit as input and provides seven outputs to drive a corresponding display digit.

The truth table (Table 2.5) is created by arranging the input side in increasing binary order corresponding to decimal digits 0 through 9. The output side lists the segments which are to be "lit" to form the selected digit; a 1 indicates an energized (visible) segment.

Note that because the inputs are BCD digits, there are six unused binary combinations on the input side.

Table 2.4

D4	D3	D2	D1	P3	P2	P1
0	0	0	0	1	1	1
0	0	0	1	1	0	0
0	0	1	0	0	1	0
0	0	1	1	0	0	1
0	1	0	0	0	0	1
0	1	0	1	0	1	0
0	1	1	0	1	0	0
0	1	1	1	1	1	1
1	0	0	0	0	0	0
1	0	0	1	0	1	1
1	0	1	0	1	0	1
1	0	1	1	1	1	0
1	1	0	0	1	1	0
1	1	0	1	1	0	1
1	1	1	0	0	1	1
1	1	1	1	0	0	0

2.3 Show how a truth table may be re-created from Boolean equations by constructing the truth table that corresponds to the following:

$$F = A'B'C' + A'BC' + AB'C' + AB'C$$
$$G = A'BC + ABC'$$

Note that this is the inverse process to that used to obtain Eqs. (2.3) and (2.4).

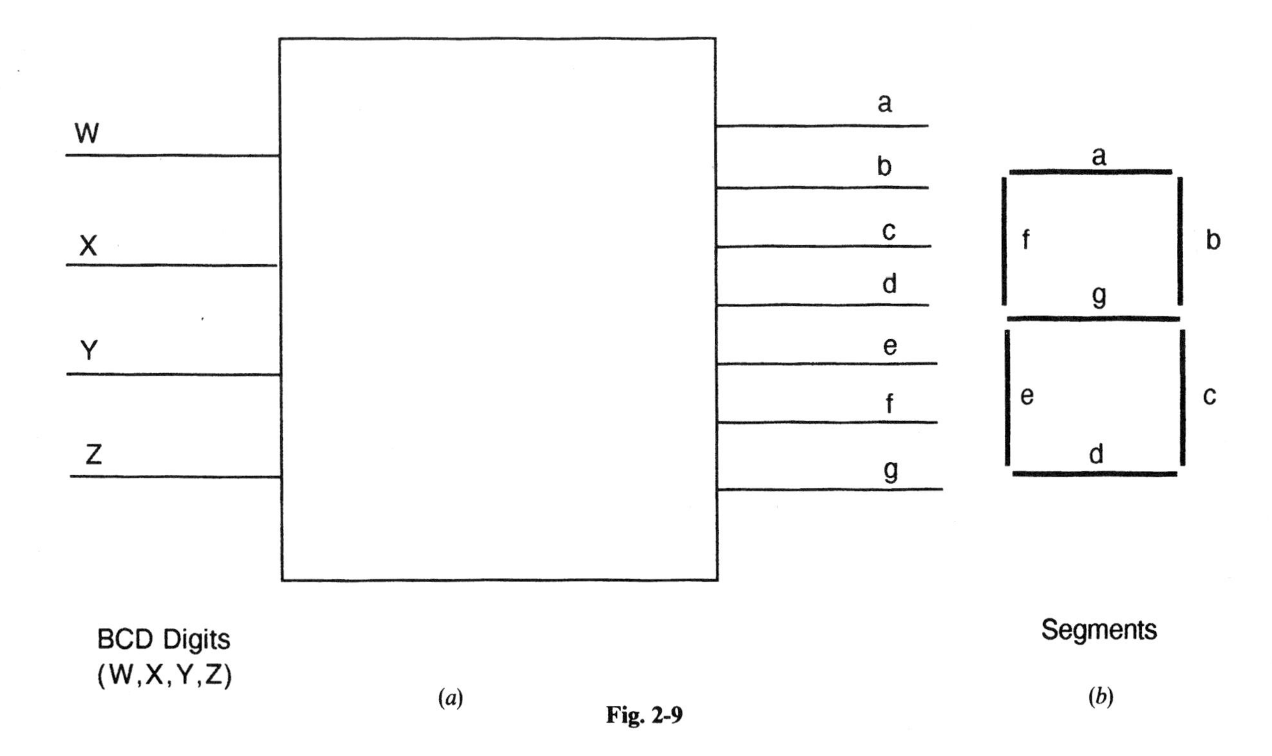

(*a*) (*b*)

Fig. 2-9

Table 2.5

Decimal Digit	INPUT W	X	Y	Z	OUTPUT a	b	c	d	e	f	g
0	0	0	0	0	1	1	1	1	1	1	0
1	0	0	0	1	0	1	1	0	0	0	0
2	0	0	1	0	1	1	0	1	1	0	1
3	0	0	1	1	1	1	1	1	0	0	1
4	0	1	0	0	0	1	1	0	0	1	1
5	0	1	0	1	1	0	1	1	0	1	1
6	0	1	1	0	1	0	1	1	1	1	1
7	0	1	1	1	1	1	1	0	0	0	0
8	1	0	0	0	1	1	1	1	1	1	1
9	1	0	0	1	1	1	1	1	0	1	1

The general method is as follows:

1. Tabulate all possible combinations of the independent variables in ascending binary order. This establishes the rows.
2. Create an output column for each dependent variable.
3. From the Boolean equation for each output column, determine in which rows the 1s are located.
4. Fill in the remaining output column slots with 0s.

If the equation for F is examined, we see that there are three independent variables A, B, C, and four AND terms, each of which corresponds to a truth table row where the output column F contains a 1. In each of these AND terms, subsitute 1 for a direct variable (unprimed) and a 0 for a logically inverted variable to obtain the desired row codes. The four rows where F = 1 are thus seen to correspond to the input variable combinations ABC = 000, 010, 100, and 101. After using the same process to identify the rows corresponding to G = 1, the truth table may be easily completed (see Table 2.6).

Table 2.6

INPUT A	B	C	OUTPUT F	G
0	0	0	1	0
0	0	1	0	0
0	1	0	1	0
0	1	1	0	1
1	0	0	1	0
1	0	1	1	0
1	1	0	0	1
1	1	1	0	0

2.4 Write the Boolean equivalent of the exclusive OR function $A \oplus B$ using the basic Boolean functions AND, OR, LOGICAL INVERSION.

The desired function may be written directly from the truth table given in Example 2.6:

A	B	$A \oplus B$
0	0	0
0	1	1
1	0	1
1	1	0

The function $F = A \oplus B$ will be true if A and NOT B are true OR if B and NOT A are true. The key property of XOR which distinguishes it from ordinary OR is that *F is not true if A and B are both true.*

In symbolic form, $A \oplus B = AB' + BA'$.

2.5 Another common function is the inverse exclusive OR, $\overline{A \oplus B}$. Show that its Boolen representation is $AB + A'B'$ and that the function defines *logical equivalence.*

Create the truth table (see Table 2.7).

Table 2.7

A	B	$A \oplus B$	$\overline{A \oplus B}$
0	0	0	1
0	1	1	0
1	0	1	0
1	1	0	1

We see that the function $\overline{A \oplus B}$ is TRUE if A and B have the same value (are logically equivalent). From the truth table, we have:

$$\overline{A \oplus B} = AB + A'B'$$

Note that the same result can be obtained by applying the theorems of Boolean algebra (refer to Prob. 3.4).

2.6 Create the logical equations corresponding to the truth table shown in Table 2.8.

F is TRUE for the combinations specified by rows 4, 6, and 8, and G is TRUE for rows 1, 2, 6, 7, and 8.

$$F = A'BC + AB'C + ABC$$
$$G = A'B'C' + A'B'C + AB'C + ABC' + ABC$$

Table 2.8

A	B	C	F	G
0	0	0	0	1
0	0	1	0	1
0	1	0	0	0
0	1	1	1	0
1	0	0	0	0
1	0	1	1	1
1	1	0	0	1
1	1	1	1	1

Table 2.9

A	B	C	W	X	Y
0	0	0	0	0	1
0	0	1	1	1	1
0	1	0	0	0	0
0	1	1	0	0	0
1	0	0	0	0	1
1	0	1	1	1	1
1	1	0	1	1	0
1	1	1	0	1	1

2.7 Given the following output equations, determine the system truth table.

$$W = AB'C + A'B'C + ABC'$$
$$X = A'B'C + AB'C + ABC' + ABC$$
$$Y = A'B'C' + A'B'C + AB'C + AB'C' + ABC$$

W will be TRUE for the following cases of ABC: 001, 101, and 110. X will be TRUE for ABC = 001, 101, 110, and 111. Y will be true for ABC = 000, 001, 100, 101, and 111. See Table 2.9.

2.8 A system of logic is to be designed which has two outputs (F, G) and three inputs (W, X, Y). One output F will be TRUE if an odd number of inputs are TRUE. The other output G will be TRUE if only one input alone is TRUE. Draw the truth table and write the corresponding Boolean equations.

Create the input side of the truth table from the known number of inputs (3) and apply the design criteria to each row to obtain the output entries as shown in Table 2.10.

The equations are determined by setting down the input row combinations (AND terms) which correspond to output 1s:

$$F = W'X'Y + W'XY' + WX'Y' + WXY$$
$$G = W'X'Y + W'XY' + WX'Y'$$

Table 2.10

W	X	Y	F	G
0	0	0	0	0
0	0	1	1	1
0	1	0	1	1
0	1	1	0	0
1	0	0	1	1
1	0	1	0	0
1	1	0	0	0
1	1	1	1	0

Table 2.11

A	B	C	F	G
0	0	0	0	0
0	0	1	1	0
0	1	0	0	1
0	1	1	0	1
1	0	0	1	0
1	0	1	0	0
1	1	0	1	0
1	1	1	1	1

2.9 Given the truth table shown in Table 2.11, create the governing Boolean equations using those terms where the output columns contain 0s.

Equations corresponding to 0s in the output columns:

$$F' = A'B'C' + A'BC' + A'BC + AB'C$$
$$G' = A'B'C' + A'B'C + AB'C' + AB'C + ABC'$$

2.10 The following function represents a typical XOR structure:

$$F(A, B, C, D) = (A \oplus B) \oplus (C \oplus D)$$

Construct the truth table.

The desired truth table (Table 2.12) can be obtained by systematically filling out columns for each term.

Careful inspection of the truth table reveals that F will be 1 only when there are an *odd number of 1s among the inputs.* This is a *general property of the exclusive-OR* function, and it is very useful for the implementation of parity-generation and error-checking systems.

2.11 Create the Boolean equations for the seven-segment decoder of Prob. 2.2.

Refer to the solution for Prob. 2.2 and recall that the rows in the truth table correspond to AND terms and that the columns in each output correspond to an ORing of these ANDs for each output.

$$a = W'X'Y'Z' + W'X'YZ' + W'X'YZ + W'XY'Z + W'XYZ' + W'XYZ + WX'Y'Z' + WX'Y'Z$$
$$b = W'X'Y'Z' + W'X'Y'Z + W'X'YZ' + W'X'YZ + W'XY'Z' + W'XYZ + WX'Y'Z' + WX'Y'Z$$
$$c = W'X'Y'Z' + W'X'Y'Z + W'X'YZ + W'XY'Z' + W'XY'Z + W'XYZ' + W'XYZ + WX'Y'Z' + WX'Y'Z$$
$$d = W'X'Y'Z' + W'X'YZ' + W'X'YZ + W'XY'Z + W'XYZ' + WX'Y'Z' + WX'Y'Z$$
$$e = W'X'Y'Z' + W'X'YZ' + W'XYZ' + WX'Y'Z'$$
$$f = W'X'Y'Z' + W'XY'Z' + W'XY'Z + W'XYZ' + WX'Y'Z' + WX'Y'Z$$
$$g = W'X'YZ' + W'X'YZ + W'XY'Z' + W'XY'Z + W'XYZ' + WX'Y'Z' + WX'Y'Z$$

Table 2.12

A	B	C	D	$A \oplus B$	$C \oplus D$	F
0	0	0	0	0	0	0
0	0	0	1	0	1	1
0	0	1	0	0	1	1
0	0	1	1	0	0	0
0	1	0	0	1	0	1
0	1	0	1	1	1	0
0	1	1	0	1	1	0
0	1	1	1	1	0	1
1	0	0	0	1	0	1
1	0	0	1	1	1	0
1	0	1	0	1	1	0
1	0	1	1	1	0	1
1	1	0	0	0	0	0
1	1	0	1	0	1	1
1	1	1	0	0	1	1
1	1	1	1	0	0	0

2.12 Draw the logic diagram for output P_1 of the parity generator of Prob. 2.1.

Relevant inputs are obtained directly from the truth table rows. Refer to Fig. 2-10.

2.13 Draw the logic diagram of the binary-bit adder of Fig. 2-2.

The two relevant Boolean expressions have already been derived from the truth table [Eqs. (2.3) and (2.4)].

$$S = X'Y'C_i + X'YC_i' + XY'C_i' + XYC_i$$
$$C_o = X'YC_i + XY'C_i + XYC_i' + XYC_i$$

The logic diagram can be drawn by inspection as shown in Fig. 2-11.

2.14 Write the Boolean equation which corresponds to the logic diagram shown in Fig. 2-12.

Operation outputs	Function
G_1	AB'
G_2	$G_1 + C = AB' + C$
G_5	CD'
G_4	$G_5 + A' + B' = CD' + A' + B'$
$G_3 = F$	$D \cdot G_2 \cdot G_4 = D(AB' + C)(CD' + A' + B')$

2.15 The logic shown in Fig. 2-13*a* is termed a *half adder*. Determine its truth table. Also shown is the combination of two half adders to form a full-bit adder (Fig. 2-13*b*). Determine the truth table for the logic of Fig. 2-13*b*, and show that it is indeed that of a complete adder capable of accepting a carry-in (C_i) from a lower-order digit and generating a carry-out (C_o) as required for the next highest order digit.

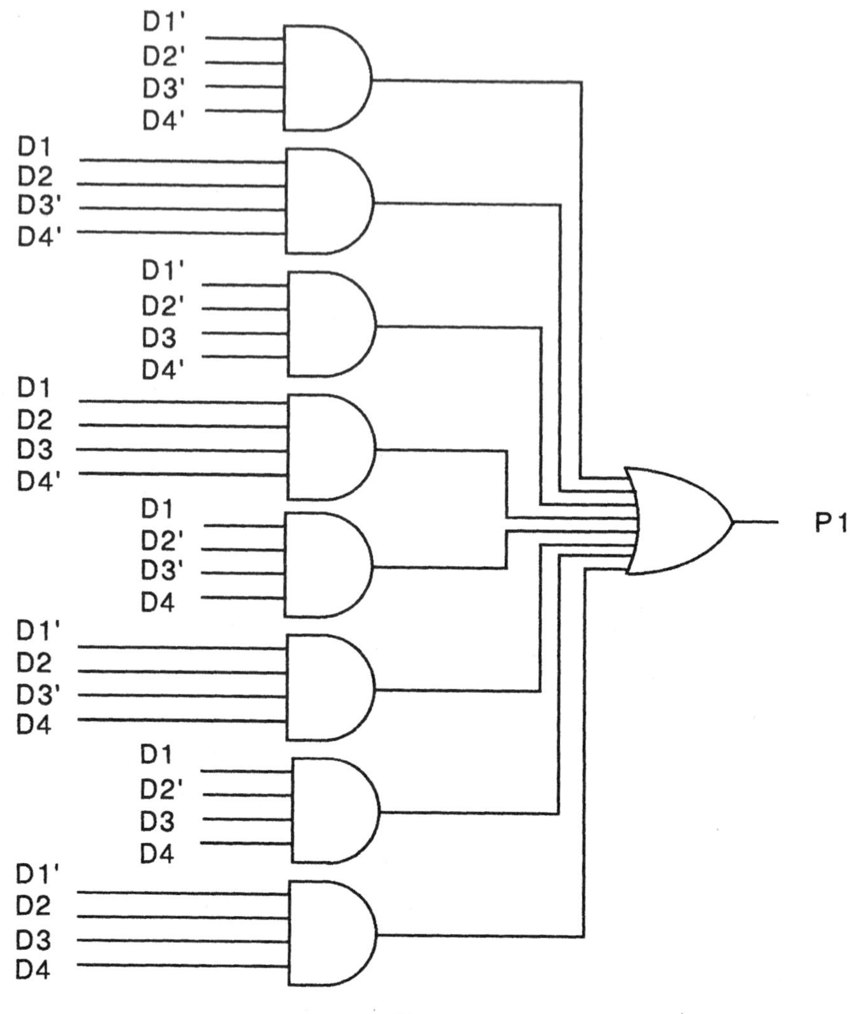

Fig. 2-10

(*a*) The half adder truth table is

A	B	C	S
0	0	0	0
0	1	0	1
1	0	0	1
1	1	1	0

C (carry) = 1 only when both A and B are 1. S (sum) = 1 only if A or B, but not both, are 1s.

We see that the logic follows the rules of single-bit addition, with a carry being generated only when two 1s appear simultaneously at the inputs A and B.

(*b*) The full adder truth table (Table 2.13) is drawn using the following intermediate variables to facilitate the process:

$$X = AB \qquad Z = C_i Y \qquad Y = A \oplus B \qquad S = C_i \oplus Y \qquad C_o = Z + X$$

Table 2.13 shows that a carry-out is generated any time two or more input 1s are present and that the sum output is 1 only when there is an odd number of input 1s. This satisfies the rules for binary addition as described in Sec. 1.4.

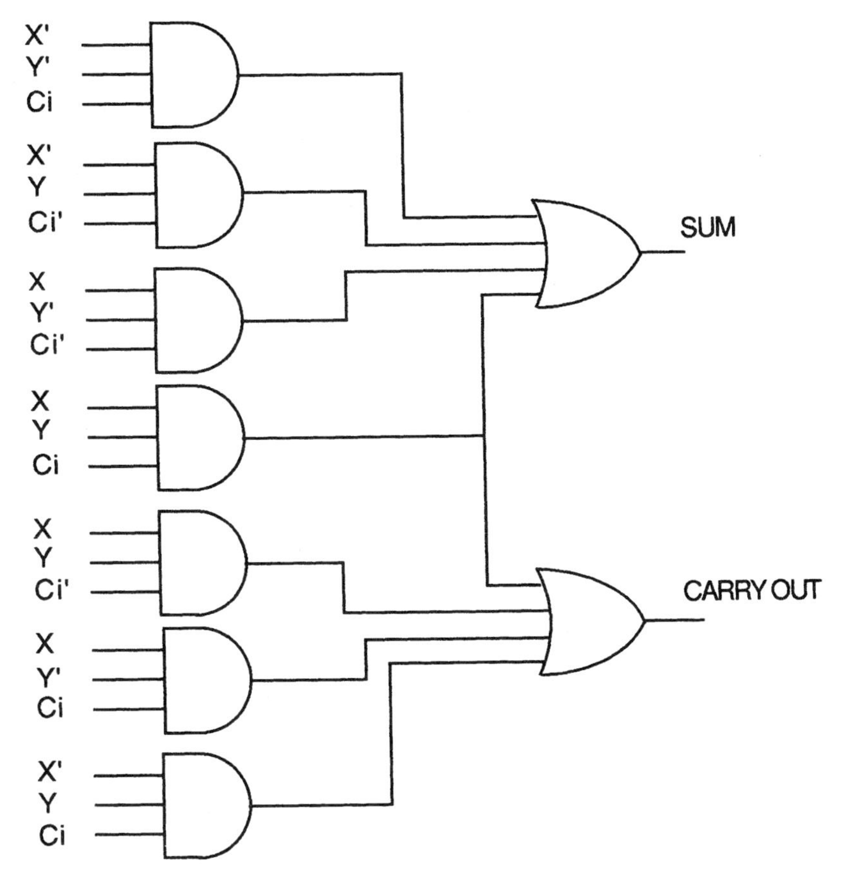

Fig. 2-11

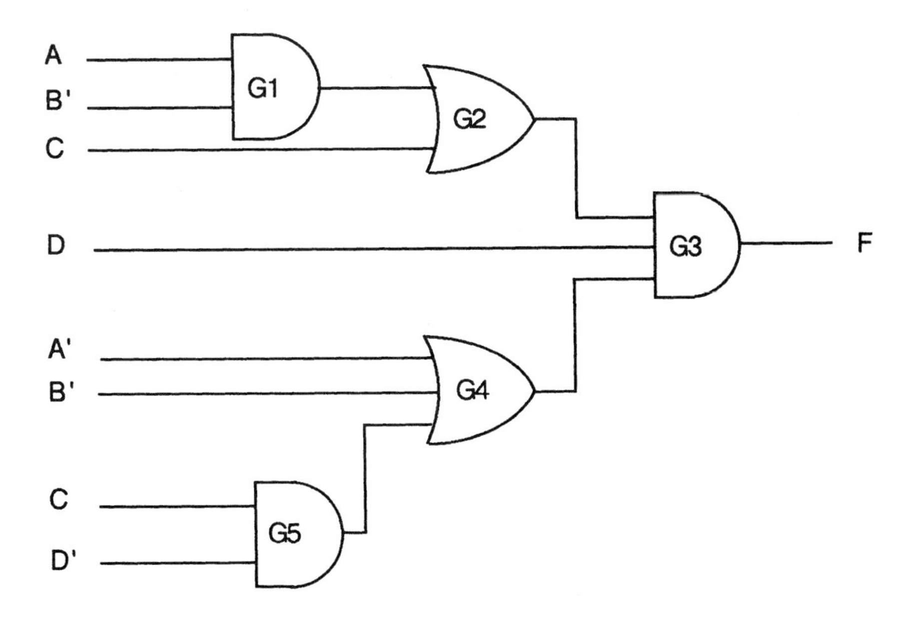

Fig. 2-12

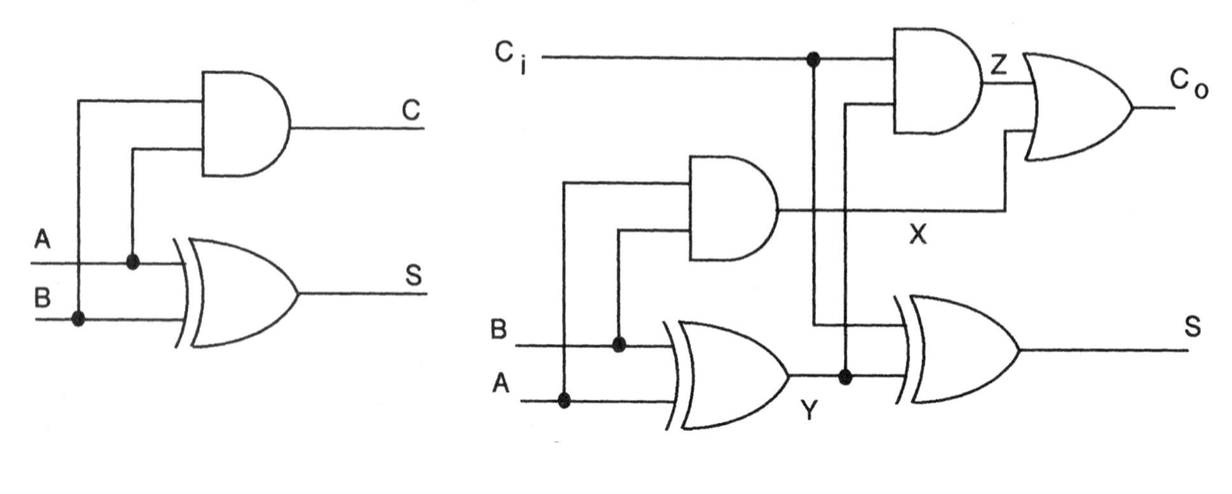

(a) Half adder

(b) Full adder

Fig. 2-13

2.16 In the diagram shown in Fig. 2-14, the box represents a logical element which produces a TRUE output if and only if a majority of the three inputs are TRUE. Draw the equivalent logic diagram for the box.

Draw the truth table.

A	B	C	F
0	0	0	0
0	0	1	0
0	1	0	0
0	1	1	1
1	0	0	0
1	0	1	1
1	1	0	1
1	1	1	1

The Boolean equation for F follows by inspection,

$$F = A'BC + AB'C + ABC' + ABC$$

from which we obtain the desired logic, as shown in Fig. 2-15.

Table 2.13

A	B	C_i	X	Y	Z	C_o	S
0	0	0	0	0	0	0	0
0	0	1	0	0	0	0	1
0	1	0	0	1	0	0	1
0	1	1	0	1	1	1	0
1	0	0	0	1	0	0	1
1	0	1	0	1	1	1	0
1	1	0	1	0	0	1	0
1	1	1	1	0	0	1	1

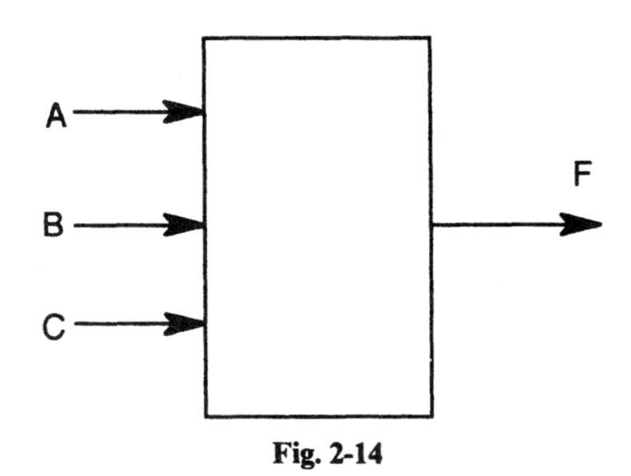

Fig. 2-14

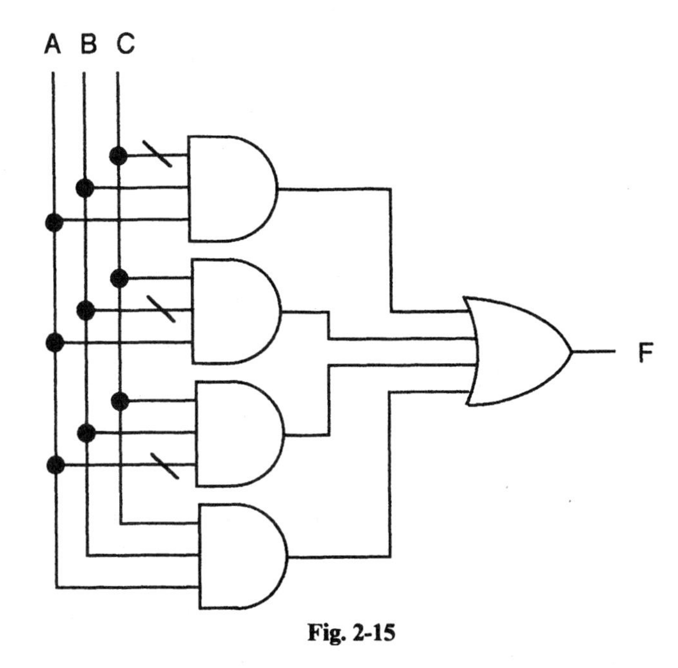

Fig. 2-15

2.17 In the circuit diagram shown in Fig. 2-16, the boxes represent logical elements identical to those described in Prob. 2.16. Write the Boolean expression for G.

Circuit 1 behaves as described in Prob. 2.16:

$$F_1 = A'BC + AB'C + ABC' + ABC$$

Circuit 2 behaves as an OR gate for C′ and D because of the hard-wired 1 at an input:

$$\begin{aligned} F_2 &= CD \cdot 1 + C'D' \cdot 1 + C'D \cdot 0 + C'D \cdot 1 \\ &= CD + C'D' + C'D \\ &= C' + D \end{aligned}$$

This implementation may be confirmed with a truth table or by reference to basic Boolean relationships discussed in Chapter 3.

Circuit 3 functions as an AND gate for F_1 and F_2 because of the hard-wired 0:

$$\begin{aligned} G &= F_1' F_2 \cdot 0 + F_1 F_2' \cdot 0 + F_1 F_2 \cdot 1 + F_1 F_2 \cdot 0 \\ &= F_1 F_2 = (C' + D)(A'BC + AB'C + ABC' + ABC) \end{aligned}$$

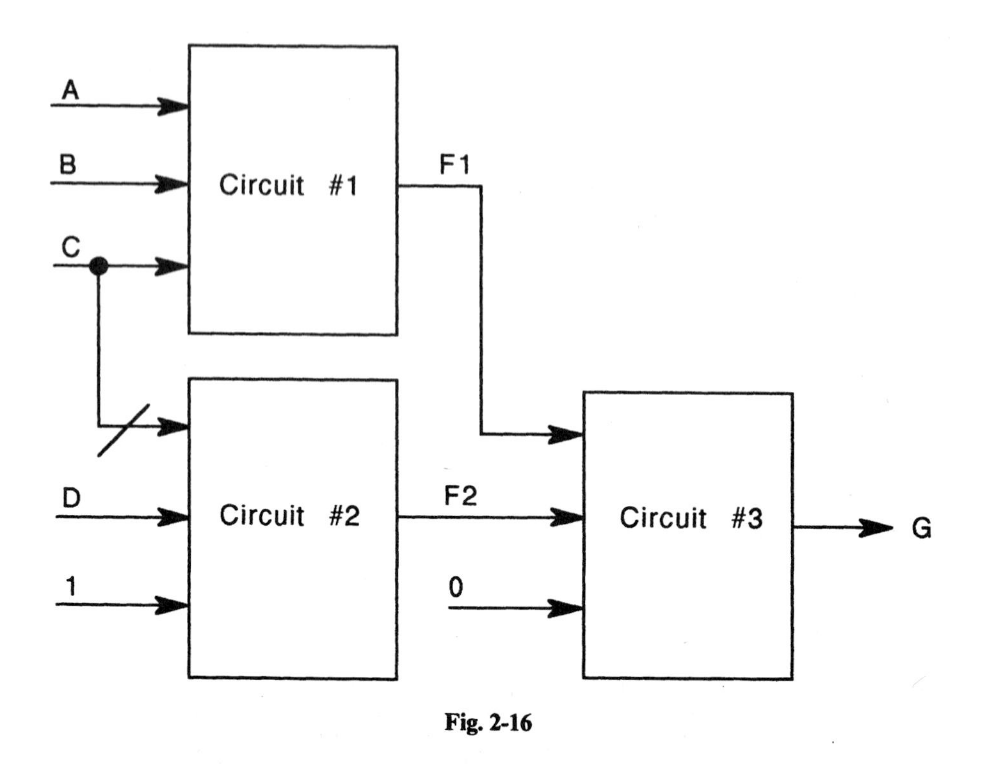

Fig. 2-16

Supplementary Problems

2.18 Draw a logic diagram which represents the function F = (A'B + AD')'C. The available inputs are A, B, C, and D.

2.19 Write the Boolean expression that the logic diagram in Fig. 2-17 represents.

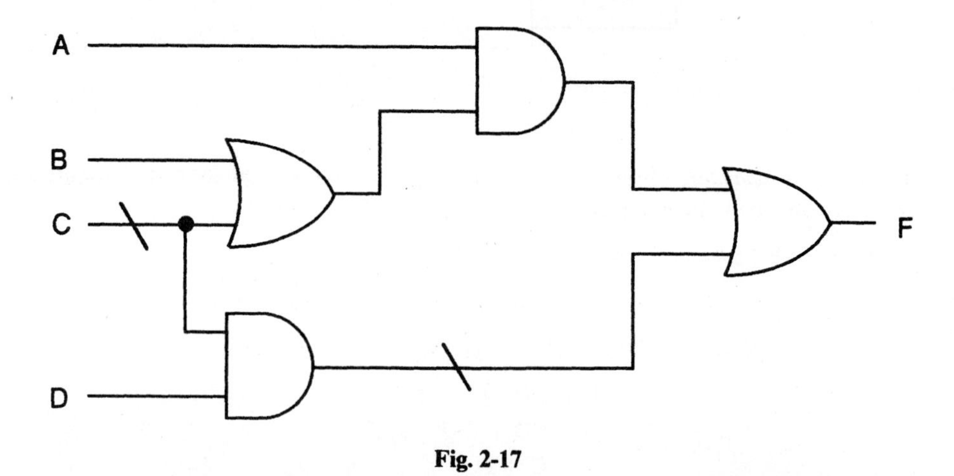

Fig. 2-17

2.20 Create the truth table for a circuit which multiplies two 2-bit numbers (A_1, A_0, and B_1, B_0).

2.21 Write a set of Boolean equations for the multiplier of Prob. 2.20.

2.22 Draw a logic diagram for the equations of Prob. 2.21.

2.23 Create the truth table and draw the logic diagram for the system specified by the following Boolean equations:

$$F = A'B'C + A'BC + AB'C$$
$$G = A'BC' + A'BC + ABC'$$

2.24 Create the truth table and logic equations for a circuit which generates the parity digits for a Hamming code as described in Table 1.4.

2.25 Create the truth table and the logic equations for a circuit which controls a light (L) through two switches (A and B). The light is to be on (TRUE) when both A and B are TRUE, and either switch can independently turn the light off and on.

2.26 Determine the Boolean equations and the truth table for the logic diagram shown in Fig. 2-18.

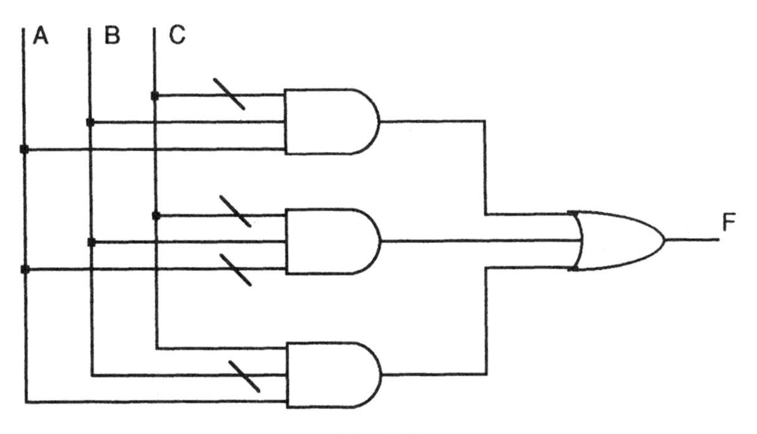

Fig. 2-18

2.27 Create the truth table for a single-bit binary subtractor which performs the operation $X - Y = D$. Proceed in a manner analogous to that used to develop the single-bit binary adder described in Sec. 2.4. Use the concept of "borrow" to replace "carry" and assume that there will always be something available to borrow from.

2.28 Using the truth table for a single-bit binary subtractor created in Prob. 2.27, obtain the Boolean equations for the outputs.

2.29 Using the truth table for a single-bit binary subtractor created in Prob. 2.27 or the equations of Prob. 2.28, draw the corresponding logic diagram.

2.30 Create the truth table for a system with three inputs (A, B, C) and two outputs (D, E) where D is to be TRUE if and only if an odd number of inputs are TRUE and E is to be true whenever at least two inputs are TRUE.

2.31 Create the truth table for the system described by the logic diagram of Fig. 2-19.

2.32 Write the Boolean equations for the logic of Prob. 2.30.

2.33 A digital system has four outputs and five inputs, the latter comprising a 4-bit binary number and an up/down (U/D) bit. The input number ranges between the binary equivalent of 3 and 12. If U/D is 0, the output is the next highest binary number and, if U/D = 1, the output is the next lowest. The number sequence closes on itself so that an input of 12 will produce a 3 at the output and an input of 3 will produce 12. Call the inputs A, B, C, D, U/D and the outputs A_n, B_n, C_n, D_n and create the truth table for all possible input combinations.

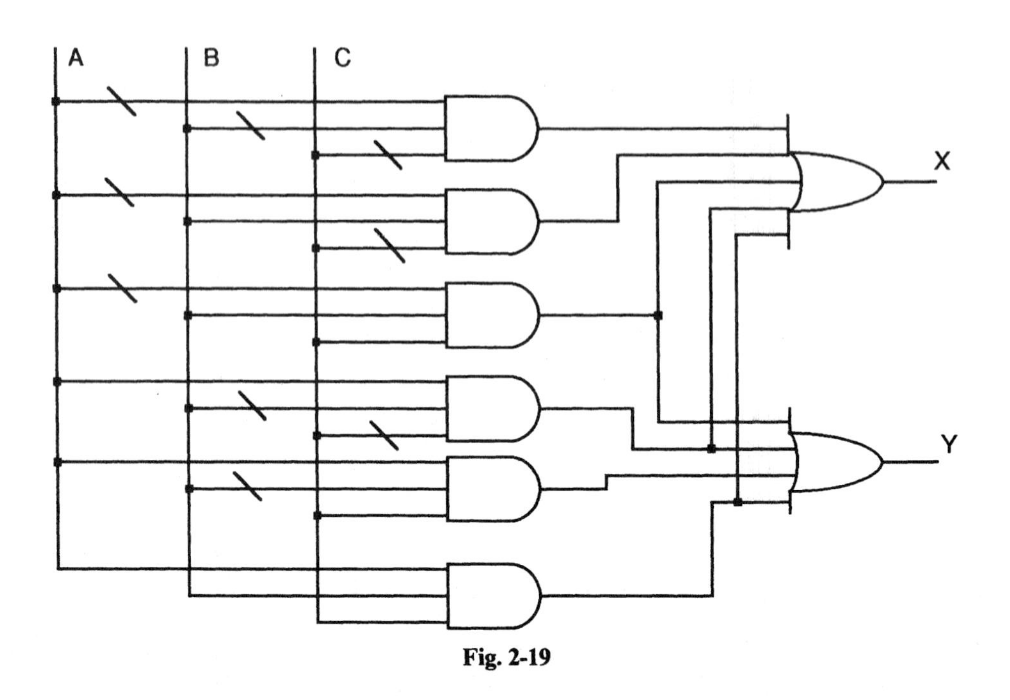

Fig. 2-19

Answers to Supplementary Problems

2.18 See Fig. 2-20.

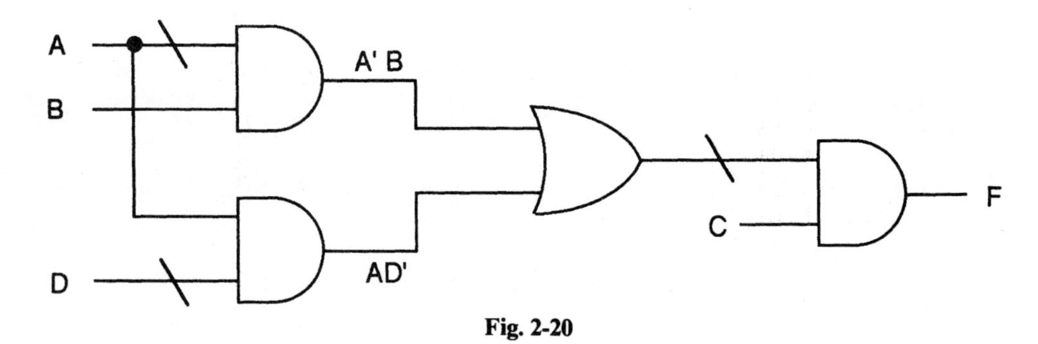

Fig. 2-20

2.19 $A(B + C') + (C'D)'$

2.20 The largest 2-bit number is 11 (decimal 3); thus the largest product will be decimal 9 (binary 1001). Four bits will be required for the product. Denoting them by C_3, C_2, C_1, and C_0, we obtain the truth table shown in Table 2.14.

2.21 $C_3 = A_1 A_0 B_1 B_0$

$C_2 = A_1 A_0 B_1 B_0' + A_1 A_0' B_1 B_0 + A_1 A_0' B_1 B_0'$

$C_1 = A_1' A_0 B_1 B_0' + A_1' A_0 B_1 B_0 + A_1 A_0' B_1' B_0 + A_1 A_0' B_1 B_0 + A_1 A_0 B_1' B_0 + A_1 A_0 B_1 B_0'$

$C_0 = A_1' A_0 B_1' B_0 + A_1' A_0 B_1 B_0 + A_1 A_0 B_1' B_0 + A_1 A_0 B_1 B_0$

2.22 See Fig. 2-21. Note that some AND terms occur in more than one expression but need only be realized once.

2.23 See Fig. 2-22.

Table 2.14

INPUTS				OUTPUTS			
A1	A0	B1	B0	C3	C2	C1	C0
0	0	0	0	0	0	0	0
0	0	0	1	0	0	0	0
0	0	1	0	0	0	0	0
0	0	1	1	0	0	0	0
0	1	0	0	0	0	0	0
0	1	0	1	0	0	0	1
0	1	1	0	0	0	1	0
0	1	1	1	0	0	1	1
1	0	0	0	0	0	0	0
1	0	0	1	0	0	1	0
1	0	1	0	0	1	0	0
1	0	1	1	0	1	1	0
1	1	0	0	0	0	0	0
1	1	0	1	0	0	1	1
1	1	1	0	0	1	1	0
1	1	1	1	1	0	0	1

2.24 Since the parity bits depend on the data bits, use the latter as inputs. The truth table is shown in Table 2.15.

$$\begin{aligned}
P_3 &= D'_4 D'_3 D'_2 D'_1 + D'_4 D'_3 D'_2 D_1 + D'_4 D_3 D_2 D'_1 + D'_4 D_3 D_2 D_1 \\
&\quad + D_4 D'_3 D_2 D'_1 + D_4 D'_3 D_2 D_1 + D_4 D_3 D'_2 D'_1 + D_4 D_3 D'_2 D_1 \\
P_2 &= D'_4 D'_3 D'_2 D'_1 + D'_4 D'_3 D_2 D'_1 + D'_4 D_3 D'_2 D_1 + D'_4 D_3 D_2 D_1 \\
&\quad + D_4 D'_3 D'_2 D_1 + D_4 D'_3 D_2 D_1 + D_4 D_3 D'_2 D'_1 + D_4 D_3 D_2 D'_1 \\
P_1 &= D'_4 D'_3 D'_2 D'_1 + D'_4 D'_3 D_2 D_1 + D'_4 D_3 D'_2 D'_1 + D'_4 D_3 D_2 D_1 \\
&\quad + D_4 D'_3 D'_2 D_1 + D_4 D'_3 D_2 D'_1 + D_4 D_3 D'_2 D_1 + D_4 D_3 D_2 D'_1
\end{aligned}$$

Table 2.15

D4	D3	D2	D1	P3	P2	P1
0	0	0	0	1	1	1
0	0	0	1	1	0	0
0	0	1	0	0	1	0
0	0	1	1	0	0	1
0	1	0	0	0	0	1
0	1	0	1	0	1	0
0	1	1	0	1	0	0
0	1	1	1	1	1	1
1	0	0	0	0	0	0
1	0	0	1	0	1	1
1	0	1	0	1	0	1
1	0	1	1	1	1	0
1	1	0	0	1	1	0
1	1	0	1	1	0	1
1	1	1	0	0	1	1
1	1	1	1	0	0	0

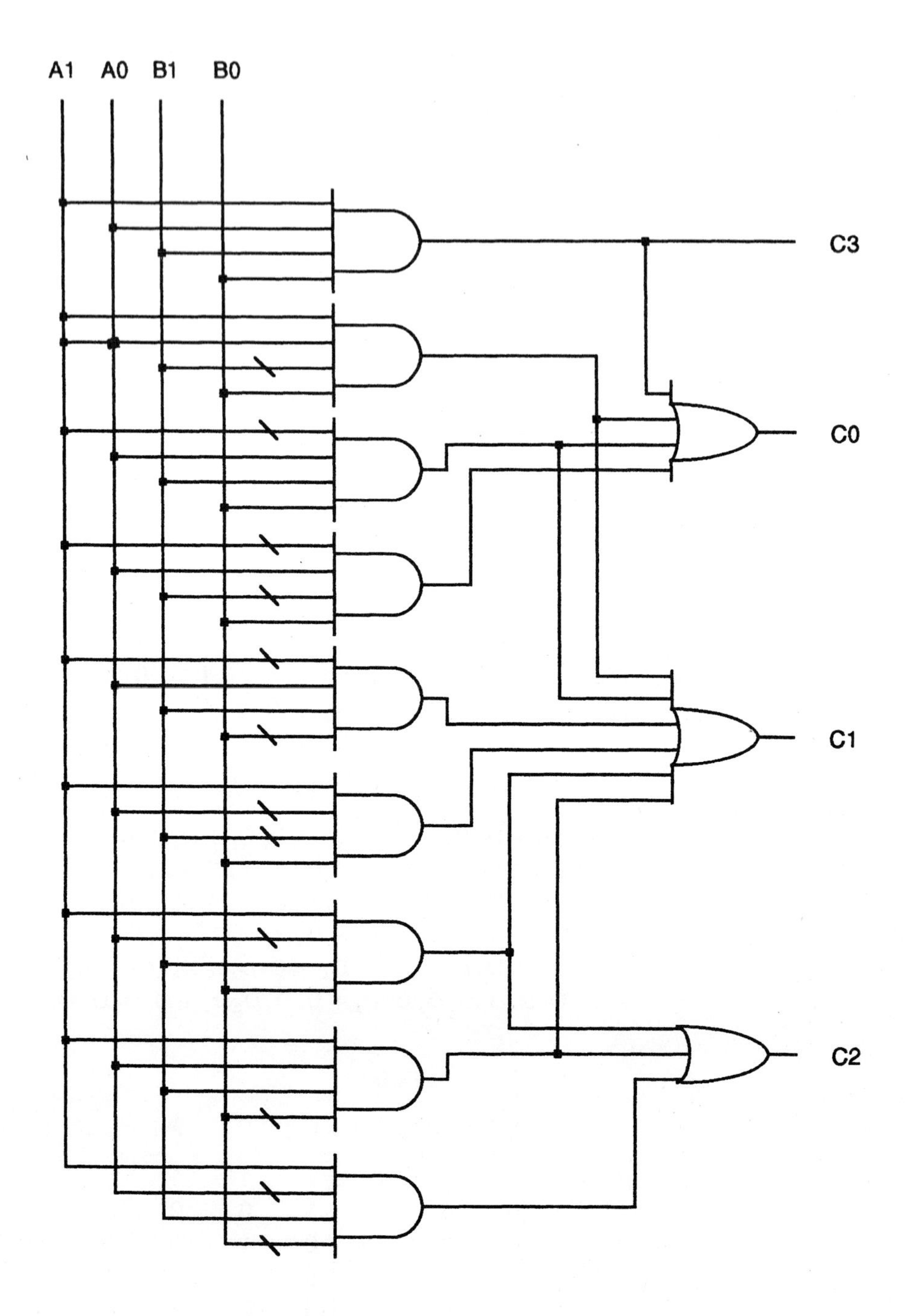

2.25 L = AB; either AB′ or A′B must turn the light off (L′ = AB′ + A′B). Also, a single-switch operation produces AB or A′B′ to turn the light back on. Either of two Boolean equations can be used to describe the operation and produce the corresponding truth table shown in Table 2.16.

$$L = AB + A'B' \qquad L' = A'B + AB'$$

Table 2.16

A	B	L
0	0	1
0	1	0
1	0	0
1	1	1

A	B	C	F	G
0	0	0	0	0
0	0	1	1	0
0	1	0	0	1
0	1	1	1	1
1	0	0	0	0
1	0	1	1	0
1	1	0	0	1
1	1	1	0	0

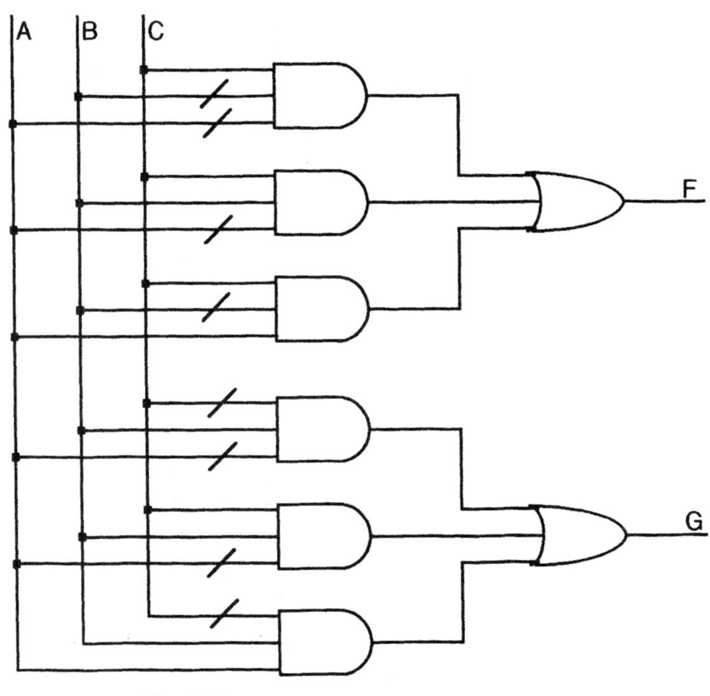

Fig. 2-22

2.26 F = ABC′ + A′BC′ + AB′C. The truth table is shown in Table 2.17.

Table 2.17

A	B	C	F
0	0	0	0
0	0	1	0
0	1	0	1
0	1	1	0
1	0	0	0
1	0	1	1
1	1	0	1
1	1	1	0

2.27 Let B_0 be a "borrow-out," i.e., a borrow required from a higher-order digit. Let B_i represent a "borrow-in" meaning that the bit has been borrowed from. The truth table is shown in Table 2.18.

Table 2.18

X	Y	Bi	D	Bo
0	0	0	0	0
0	0	1	1	1
0	1	0	1	1
0	1	1	0	1
1	0	0	1	0
1	0	1	0	0
1	1	0	0	0
1	1	1	1	1

2.28 $D = X'Y'B_i + X'YB_i' + XY'B_i' + XYB_i$

$B_o = X'Y'B_i + X'YB_i' + X'YB_i + XYB_i$

2.29

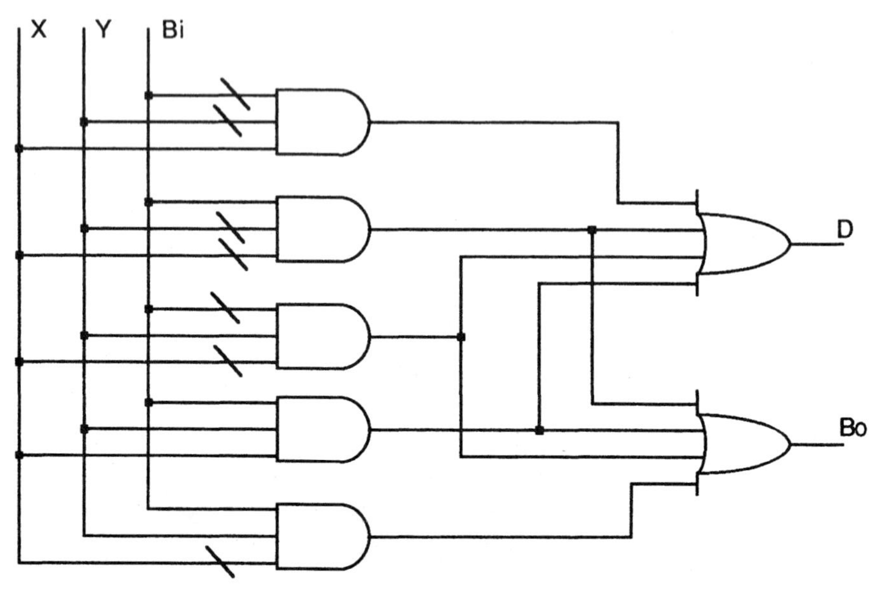

Fig. 2-23

2.30

Table 2.19

A	B	C	D	E
0	0	0	0	0
0	0	1	1	0
0	1	0	1	0
0	1	1	0	1
1	0	0	1	0
1	0	1	0	1
1	1	0	0	1
1	1	1	1	1

2.31

Table 2.20

A	B	C	X	Y
0	0	0	1	0
0	0	1	0	0
0	1	0	1	0
0	1	1	1	1
1	0	0	1	1
1	0	1	0	1
1	1	0	0	0
1	1	1	1	1

2.32 D = A′B′C + A′BC′ + AB′C′ + ABC
E = A′BC + AB′C + ABC′ + ABC

2.33

Table 2.21

U/D	A	B	C	D	An	Bn	Cn	Dn
0	0	0	1	1	0	1	0	0
0	0	1	0	0	0	1	0	1
0	0	1	0	1	0	1	1	0
0	0	1	1	0	0	1	1	1
0	0	1	1	1	1	0	0	0
0	1	0	0	0	1	0	0	1
0	1	0	0	1	1	0	1	0
0	1	0	1	0	1	0	1	1
0	1	0	1	1	1	1	0	0
0	1	1	0	0	0	0	1	1
1	0	0	1	1	1	1	0	0
1	0	1	0	0	0	0	1	1
1	0	1	0	1	0	1	0	0
1	0	1	1	0	0	1	0	1
1	0	1	1	1	0	1	1	0
1	1	0	0	0	0	1	1	1
1	1	0	0	1	1	0	0	0
1	1	0	1	0	1	0	0	1
1	1	0	1	1	1	0	1	0
1	1	1	0	0	1	0	1	1

Chapter 3

Design of Combinational Logic II: Manipulation

3.1 INTRODUCTION

As demonstrated in Chap. 2, the complete design specification for combinational logic can be expressed in the form of a truth table and all the information in that truth table can be expressed as a set of logical equations. Since truth table and Boolean variables share singular two-valued properties, we may appropriate useful theorems and reduction rules from established Boolean algebra. These techniques allow us to manipulate a set of logical equations to obtain equivalent expressions which often result in significantly simpler circuit realizations.

A truth table or its related Boolean equation(s) can also be displayed graphically in the form of a Karnaugh map (K map). This method permits a designer to achieve rapid simplification of complicated Boolean equations by exploiting the excellent pattern recognition capabilities of the human brain.

3.2 BOOLEAN ALGEBRA BASICS

Fundamental Theorems

There are several useful theorems that arise from the definitions of the basic Boolean operations presented in Chap. 2. In the following listing, A is any Boolean (logical) variable which can have values of either 1 (TRUE) or 0 (FALSE):

1. Any logical variable OR'd with 1 produces 1.

$$A + 1 = 1 \tag{3.1}$$

2. Any logical variable OR'd with 0 remains unchanged.

$$A + 0 = A \tag{3.2}$$

3. Any logical variable AND'd with 1 remains unchanged.

$$A \cdot 1 = A \tag{3.3}$$

4. Any logical variable AND'd with 0 produces 0.

$$A \cdot 0 = 0 \tag{3.4}$$

5. Any logical variable OR'd or AND'd with itself remains unchanged.

$$A + A = A \tag{3.5a}$$

$$A \cdot A = A \tag{3.5b}$$

6. A logical variable OR'd with its inverse produces 1.

$$A + A' = 1 \tag{3.6}$$

7. A logical variable AND'd with its inverse produces 0.

$$A \cdot A' = 0 \tag{3.7}$$

EXAMPLE 3.1 The various Boolean theorems can be more easily understood and remembered by making use of a simple electric circuit analogy. If a logical 1 is thought of as a continuous conducting path like a closed switch contact, then it seems reasonable that a 0 would be a nonconducting or open switch contact. Each AND function may then be regarded as a series connection of two *normally open* switch contacts which close (become conducting) when the switches are operated (see Fig. 3-1). Switches A and B must *both* be operated for the light (L) to be on (TRUE). If $B = 0$ (unactuated and thus nonconducting), then the light must be off (FALSE) regardless of what A does, and we see that $A \cdot 0 = 0$ (Theorem 4).

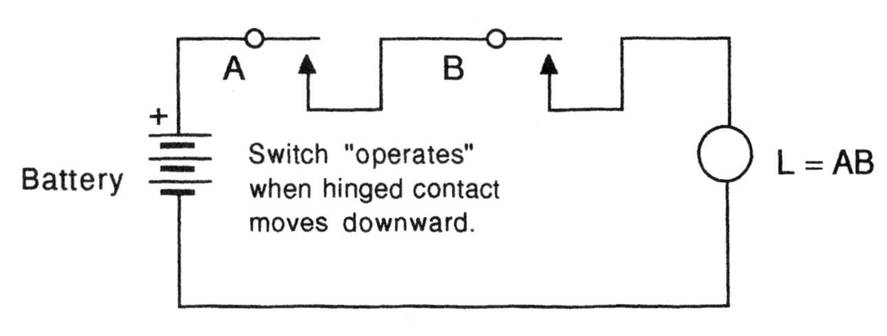

Fig. 3-1

Similarly, the OR function may be envisioned as two normally open switches in parallel (see Fig. 3-2). Switches A OR B alone can light the lamp. If $B = 1$ (actuated and conducting), then the light must be on (TRUE) regardless of what A does, and we see that $A + 1 = 1$ (Theorem 1).

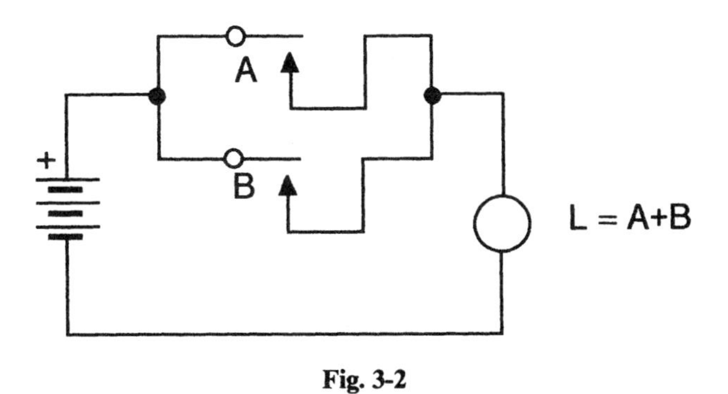

Fig. 3-2

The inverse function is represented by a *normally closed* switch which becomes nonconductive (FALSE) when operated (see Fig. 3-3). In the circuit of Fig. 3-3, it should be clear that B must operate AND A must NOT operate (A′) in order for the lamp to be on (TRUE).

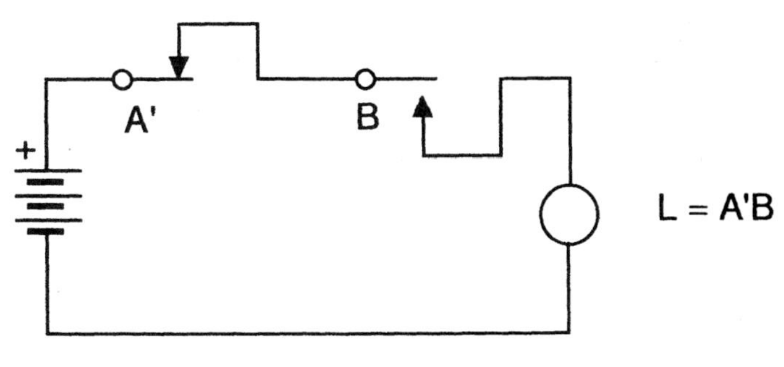

Fig. 3-3

If we wish, *we can represent any combinational logic function by an appropriate interconnection of normally open and normally closed switches.* For example, the Boolean expression $F = AB + B'C' + A'(C + D')$ may be simulated by the circuit shown in Fig. 3-4 which, when conductive, represents the condition $F = \text{TRUE}$.

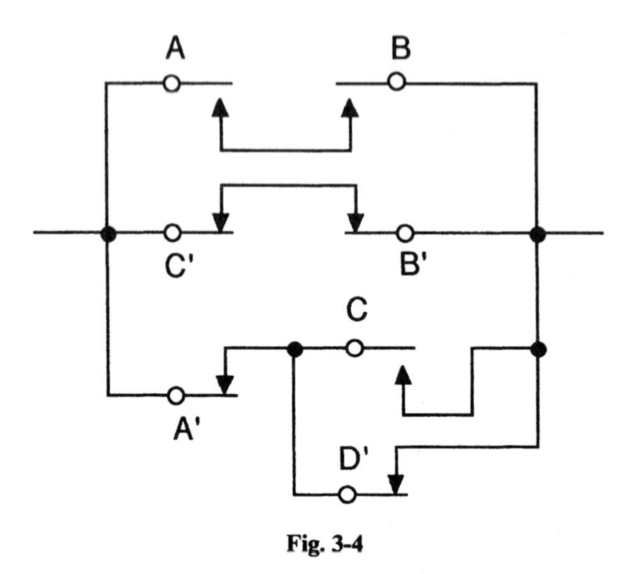

Fig. 3-4

The switch analogy to logic, besides its use as a visualization aid, has important historical significance. Before the invention of solid-state devices and integrated circuits (ICs), logic was actually constructed with electromechanical switches (relays) and early computers contained literally thousands of these devices interconnected by mazes of wires. The term "bug," as applied to a hardware or software failure, traces its origin back to the 1940s when a computer malfunction was found to be caused by a dead insect lodged between a pair of relay contacts!

Distributive Law

$$A(B + C) = AB + AC \quad \text{(as in ordinary algebra)} \tag{3.8}$$

Commutative Law

$$AB = BA \tag{3.9a}$$

$$A(B + C) = (B + C)A \tag{3.9b}$$

Absorption Theorem

$$ABC + ABC' = AB(C + C') = AB \tag{3.10}$$

This theorem is useful in reducing the number of terms in a Boolean expression. Its proof makes use of fundamental theorems 3 and 6. A related useful theorem is,

$$A + AB = A \tag{3.11}$$

which may be easily proven by using the distributive law and fundamental theorems 1 and 3.

De Morgan's Theorem

One of the most important theorems of Boolean algebra may be expressed in two related forms:

$$AB = (A' + B')' \tag{3.12a}$$

$$A + B = (A'B')' \tag{3.12b}$$

Alternatively, the student may be familiar with the equivalent notation:

$$AB = \overline{(\bar{A} + \bar{B})} \tag{3.13a}$$

$$A + B = \overline{\bar{A}\bar{B}} \tag{3.13b}$$

Refer to the Solved Problems section for the proof of this theroem. An often overlooked consequence of De Morgan's theorem is that if a designer is willing to intelligently add a few logic inversions, then ANDs may be exchanged for ORs and vice versa. This tells us that inside every AND function there is also an OR, and it follows that *any hardware that can implement an AND can also do an OR provided that sufficient logic inversions can be used.*

Proving Theorems by Exhaustion

Theorems of Boolean algebra, expressed as logical equations, can be easily proven by the *method of exhaustion.*

1. Construct a 2^N-row truth table, where N is the number of variables.
2. Assign a separate column to the various elements on both sides of the equation.
3. Evaluate these elements for each input variable combination (row).
4. If for every input combination the two sides of the equation yield identical truth values, then the theorem is proven.

EXAMPLE 3.2 Proof by exhaustion of the distribution law, A(B + C) = AB + AC, is shown in Table 3.1. Note that the two boxed columns in Table 3.1 are identical for all possible combinations of the input variables (i.e., for every truth table row); thus the corresponding Boolean functions A(B + C) and AB + AC must be logically equivalent.

Table 3.1

A	B	C	B+C	A(B+C)	AB	AC	AB+AC
0	0	0	0	0	0	0	0
0	0	1	1	0	0	0	0
0	1	0	1	0	0	0	0
0	1	1	1	0	0	0	0
1	0	0	0	0	0	0	0
1	0	1	1	1	0	1	1
1	1	0	1	1	1	0	1
1	1	1	1	1	1	1	1

A useful theorem which often comes in handy in the manipulations of Boolean algebra is

$$A + A'B = A + B \tag{3.14a}$$

which is easily proven by the exhaustion method.

Simple variable exchange yields a complementary form of Eq. (3.14*a*) which is also very useful:

$$A' + AB = A' + B \tag{3.14b}$$

Simplification of Boolean Expressions by Manipulation

Complicated Boolean expressions often contain redundant terms which can be identified and removed by the systematic application of fundamental theorems and identities.

EXAMPLE 3.3 Simplify (ab + bc′ + cd + bd′ + bc).

f = ab + bc′ + cd + bd′ + bc	
= ab + b(c′ + c) + cd + bd′	(Distributive law)
= ab + b + cd + bd′	(Fundamental theorems 6 and 3)
= b + cd + bd′	[Eq. 3.11)]
= b + cd	[Eq. (3.11) again]

3.3 HARDWARE IMPLICATIONS

The AND and OR operations of a combinational logic design are implemented in hardware by electronic circuit elements called gates whose inputs and outputs are voltage levels which can be associated with the logic states, TRUE and FALSE. With appropriate connections, these gates can also be used to produce LOGICAL INVERSIONs. Sections 4.2 and 4.3 of Chap. 4 discuss the relationship between hardware and logic in some detail. At this point, it is sufficient to point out that logic simplifications obtainable through the use of Boolean algebra can affect the amount of hardware required to implement a given logic design. Since gates use up space on a silicon chip, a manipulation of the Boolean equations into logically equivalent, but simpler forms, can often have a significant impact on the size and therefore speed and cost of an integrated circuit (IC). Effective algebraic simplification requires skill and experience and, for complicated logic designs, optimal reductions are as much art as science and often require the use of powerful computer-aided design software.

3.4 BASIC K MAPS

The K map, by taking advantage of the human brain's exceptional pattern recognition capabilities, can be used as a powerful *tool in the simplfication of Boolean expressions.* K maps are directly related to truth tables and are derived from them as follows:

1. One map is created for each output variable of the logic to be designed.
2. Map coordinates, appearing along two orthogonal sides, embody a complete set of all possible input variable combinations ordered in Gray code form.
3. Entries in the body of the map are the output values corresponding to each truth table row.

Consider the maps associated with the sum and carry outputs of the single-bit adder defined by Table 2.2.

In the body of the map of Fig. 3-5, each entry corresponds to a truth table row and each 1 corresponds to an AND term in the Boolean equation for the output variable with which the map is associ-

	INPUT			OUTPUT	
Row	X	Y	Ci	S	Co
0	0	0	0	0	0
1	0	0	1	1	0
2	0	1	0	1	0
3	0	1	1	0	1
4	1	0	0	1	0
5	1	0	1	0	1
6	1	1	0	0	1
7	1	1	1	1	1

(*a*) Truth table

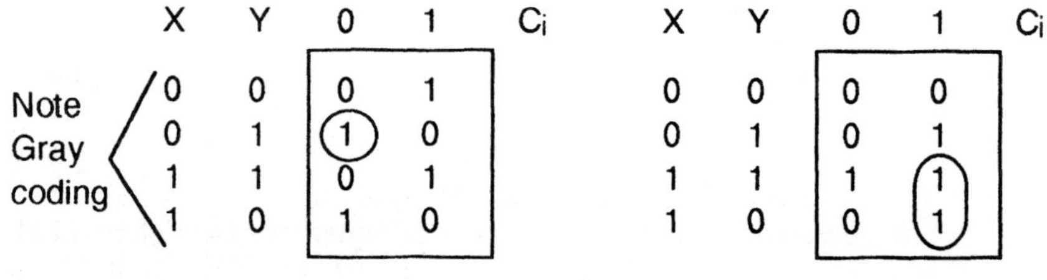

X	Y	Ci = 0	Ci = 1
0	0	0	1
0	1	1	0
1	1	0	1
1	0	1	0

Note Gray coding

(*b*) K-map for S

X	Y	Ci = 0	Ci = 1
0	0	0	0
0	1	0	1
1	1	1	1
1	0	0	1

(*c*) K-map for C_o

Fig. 3-5

ated. For example, the circled term in the S map has the coordinates $X = 0$, $Y = 1$, and $C_i = 0$. This corresponds to truth table row 2 and the AND term $X'YC_i'$ in the Boolean expression for S (refer to Section 2.4).

The purpose of Gray coding the axes is to ensure that adjacent map entries will correspond to inputs that differ in the value of only one variable. The utility of this arrangement will be discussed shortly.

EXAMPLE 3.4 The complete Boolean equation for C_o may be read directly from the corresponding K map by simply writing down the coordinates of each of the 1s in the body of the map and connecting them by ORs. Thus, we obtain by inspection of the C_o map in Fig. 3-5:

$$C_o = X'YC_i + XYC_i' + XYC_i + XY'C_i$$

Using the K Map to Simplify Boolean Expressions

The circled pair of adjacent 1s in the C_o map corresponds to the two-term Boolean expression $XYC_i + XY'C_i$. The absorption theorem [Eq. (3.10)] shows that this can be reduced to a single term (XC_i) because the two original terms differ only in one variable (Y).

In general, the absorption theorem may be applied directly by inspection of the map by noting the adjacent 1s, considering them as a block of two, and determining the *minimal set of coordinates that can define this block.* Clearly, in Fig. 3-5c, both 1s are in the second column ($C_i = 1$) and, for both these 1s, X is 1 also. Thus, the corresponding Boolean term is XC_i, and we have eliminated Y which is the only independent variable which takes on both possible values in the circled block of two.

There are two other blocks of two in the K map for C_o as shown in Fig. 3-6. Block b yields the simplified term XY, eliminating C_i, and block c produces YC_i, eliminating X. Note that *it is permissible for blocks to overlap*; in the current example, one of the 1s is used three times.

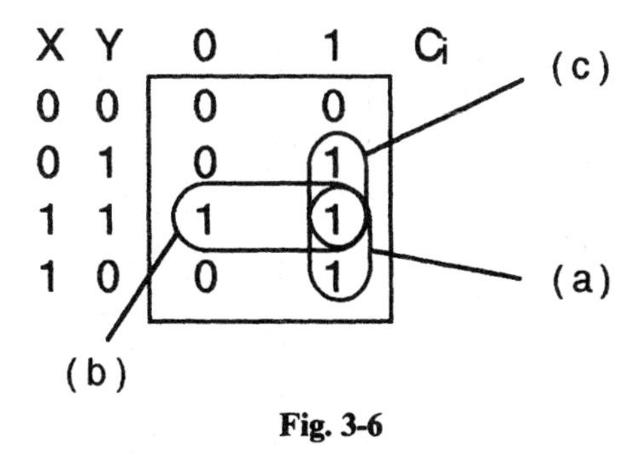

Fig. 3-6

We may now write the reduced Boolean expression for C_o:

$$Co = XCi + XY + YCi$$

Block(a) Block (b) Block (c)

Compare this to the unsimplified expression in Eq. (2.4) to which it is completely equivalent.

EXAMPLE 3.5 **Algebraic confirmation of the K-map simplification for C_o.** Fundamental theorem 5 (Sec. 3.2) permits us to add the term XYC_i twice to the right side of the Boolean equation in Example 3.4 without altering its truth:

$$C_o = XYC_i' + XYC_i + XY'C_i + XYC_i + X'YC_i + XYC_i$$

Using the distributive law,

$$C_o = XY(C_i + C_i') + XC_i(Y + Y') + YC_i(X + X')$$

Since A + A′ = 1 (Theorem 6),

$$C_o = XY \cdot 1 + XC_i \cdot 1 + YC_i \cdot 1$$

Finally, since A · 1 = A (Theorem 3), we have

$$C_o = XY + XC_i + YC_i$$

Note the relative simplicity of the K-map method with its use of pattern recognition.

The Four-Variable Map

Consider the following function:

$$F(W, X, Y, Z) = XY'Z + X'Y'Z' + XZ' + W'XZ'$$

This function requires a four variable map whose coordinates are arranged in Gray-coded pairs as shown in Fig. 3-7. To understand how mapping progresses, let us consider individual terms. The term XY′Z states that X = 1, Y = 0, and Z = 1, while W is irrelevant. The value X = 1 is identified with the middle two rows, while Y = 0 and Z = 1 are uniquely associated with the second column. Thus, the term XY′Z is located at the intersection of the middle rows and the second column as defined by the encircled 1s in Fig. 3-7*a*. Similarly, the term XZ′ is identified by the intersection of the middle two rows (X = 1) and the first and fourth columns (Z = 0). This maps into a rectangular block of four 1s as shown in Fig. 3-7*b*. Note that the K map may be regarded as being wrapped into a tube with its first and last columns or first and last rows adjacent. Figure 3.7*c* shows the entire function displayed on a single map.

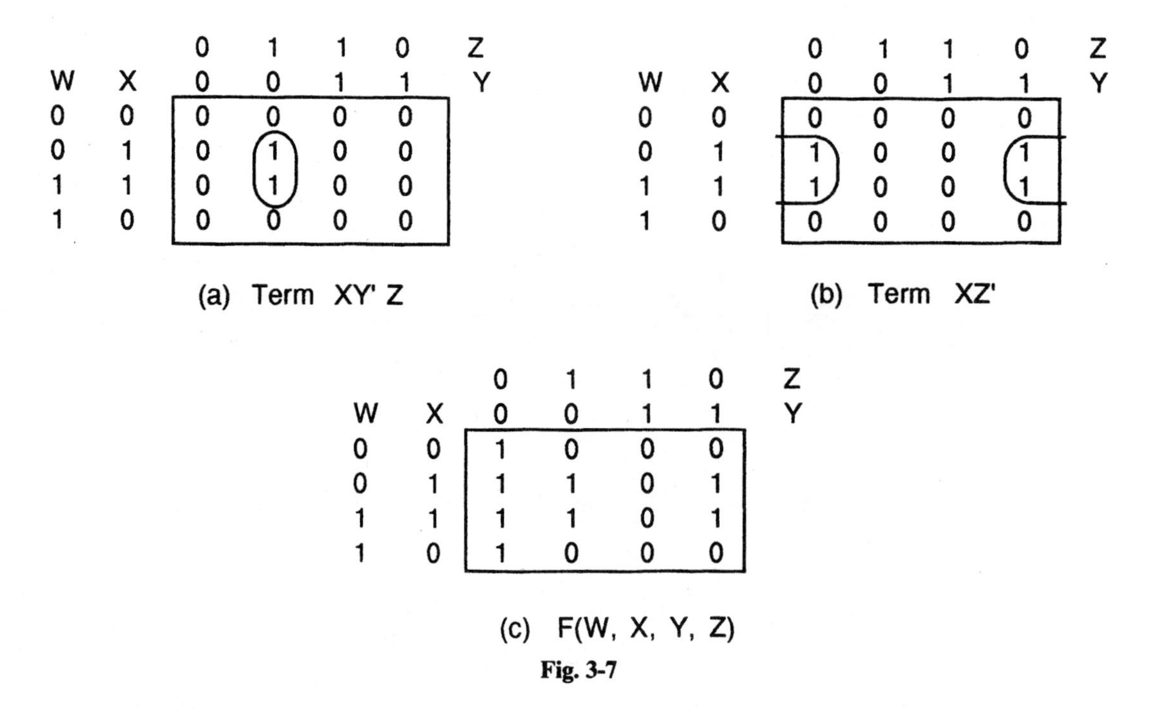

(a) Term XY' Z

(b) Term XZ'

(c) F(W, X, Y, Z)

Fig. 3-7

Simplification Rules

Study the map and identify as many blocks of *adjacent* 1s as possible while adhering to the following rules:

1. Any encirclement or block must contain only 2^N 1s, where N is an integer.
2. No 0s can be included in any block.

3. Blocks can overlap; i.e., 1s can be included in more than one block.
4. Blocks can be formed by considering opposite map boundaries to be adjacent; i.e., the top and bottom rows and the first and last columns.
5. The best simplification is *usually* obtained by seeking the largest blocks first.

Applying these rules to the map for F yields the encirclements shown in Fig. 3-8.

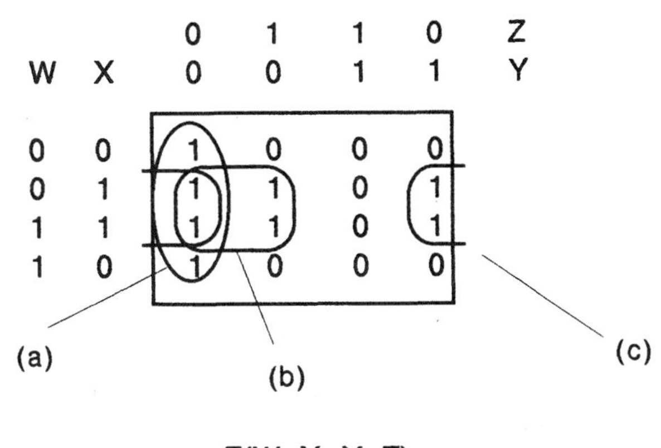

F(W, X, Y, Z)

Fig. 3-8

The simplified Boolean expression is now obtained by inspection. Each block (encirclement) of 1s represents a term; thus there will be three terms having variables that are easily identified by their corresponding block's coordinates. Encirclement (*a*), for example, being confined to the first column, corresponds to $Y = 0$ and $Z = 0$. Since W and X have both possible values included, they are ignored by virtue of the absorption theorem. The remaining two terms are obtained in a similar manner, yielding the following equation in which the blocks corresponding to each term are indicated by letters keyed to Fig. 3-8:

$$F = \underbrace{Y'Z'}_{(a)} + \underbrace{XY'}_{(b)} + \underbrace{XZ'}_{(c)}$$

The method described above is called a *1s covering.* Recall that the Boolean algebra may be used to create an equation for the logical inverse of a function by simply writing an expression containing terms from those truth table rows where the output is 0. In this case, a simplification can be obtained from the K map by encircling blocks of 0s instead of 1s and proceeding as before. This method is called a 0s *covering* and, in the present example, yields a simplified expression for F′ instead of for F.

EXAMPLE 3.6 Obtain the simplified function F′ from a 0s covering of the map in Fig. 3-7*c*.
See Fig. 3-9.

3.5 FURTHER APPLICATIONS OF K MAPS

Don't Cares

In many applications of digital logic there are often combinations of the independent variables which never occur. A system having BCD inputs is a case in point: the BCD digits will always be 4-bit binary equivalents of the decimal numbers 0 through 9. Since there are 16 unique combinations of 4

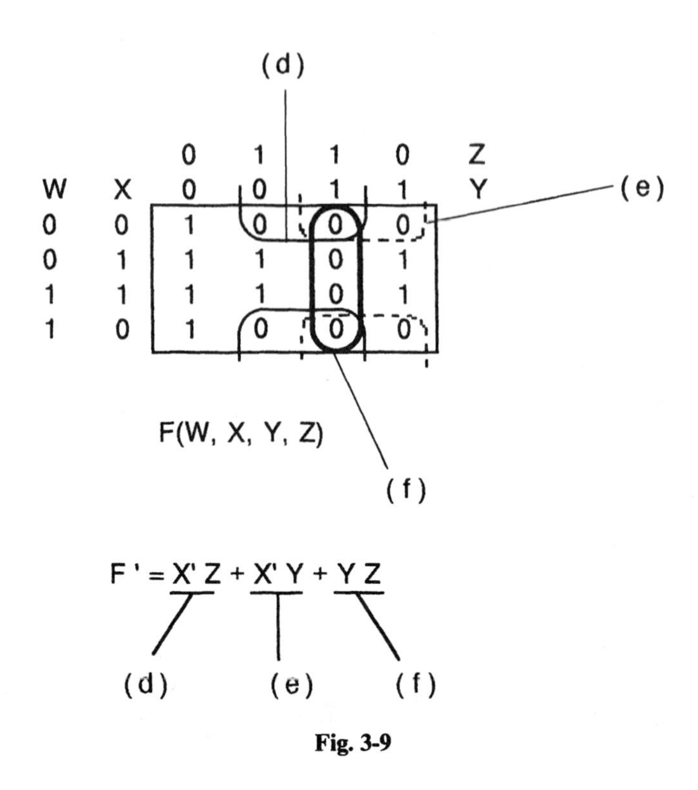

Fig. 3-9

bits, there are six combinations (representing the decimal numbers 10 through 15) which are not used. *Boolean terms which represent input combinations that cannot occur are called "don't cares,"* probably because the designers "don't care" what the outputs are for these terms.

Consider implementation of the logic for an interface circuit that accepts BCD inputs and produces outputs to drive a seven-segment visual display (a truth table for this block was developed in Prob. 2.2). The specification for segment a is mapped in Fig. 3-10 where the BCD bits have been labeled W, X, Y, Z.

	Z	0	1	1	0
W	X \ Y	0	0	1	1
0	0	1	0	1	1
0	1	0	1	1	1
1	1	x	x	x	x
1	0	1	1	x	x

"a" Map

[x indicates "don't care" terms]

Fig. 3-10

A conventional 1s covering yields the "a" map in Fig. 3-11, and the resulting function is

$$a = W'Y + X'Y'Z' + W'XZ + WX'Y'$$

Since input combinations containing "X"s can never occur, we may assign any value to X that we like. It is obviously to our advantage to let X = 1 since this creates additional opportunities to extend map encirclements. A 1s covering that takes advantage of the "don't cares" and the resulting simplified Boolean equation is shown in Fig. 3-12.

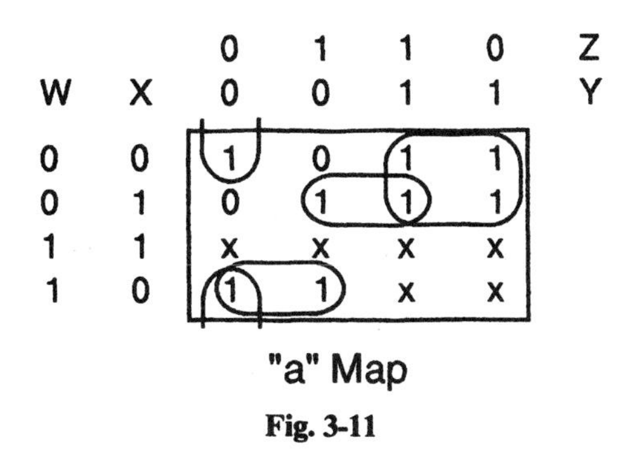

Fig. 3-11

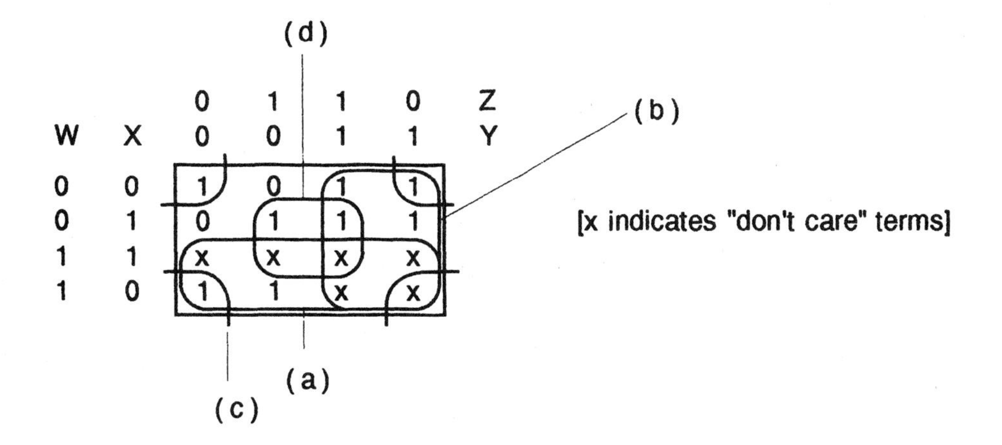

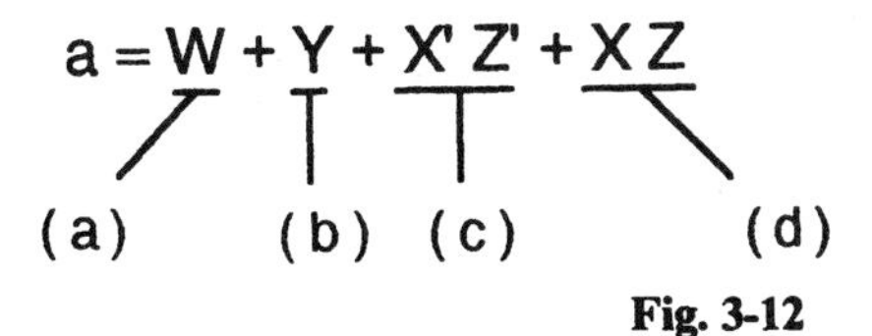

Fig. 3-12

Since this equation is obviously simpler than the one obtained without the use of "don't cares," it seems reasonable to add an additional covering rule to our list of simplification rules in Sec. 3.4:

6. Don't cares *may* be used to extend block coverings but need not be covered themselves.

It is important to note that there is no need to cover all the Xs, or any at all, for that matter. They may be treated as either 1s or 0s and used as the designer sees fit.

Five- and Six-Variable Maps

As the number of variables in the logic expression increases beyond four, the usefulness of K maps for pattern recognition begins to decrease. Designers can extend the map's utility, however, by making use of mirror symmetry. Consider the five-variable map shown in Fig. 3-13. Note that the coordinate axes are still Gray coded which ensures that adjacent squares (map locations) will differ in only one variable. This does *not*, however, imply that *all* squares which differ in only one variable will be physically adjacent. Non-physically adjacent terms relevant to the simplification process may be located by creating a reflection axis halfway down the three-variable coordinate listing as shown; those squares which have mirror symmetry about this axis are also adjacent in the sense that they differ in only one variable.

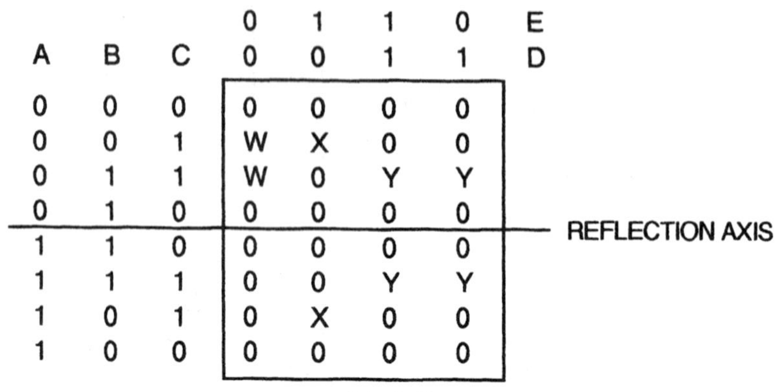

			0	1	1	0	E
A	B	C	0	0	1	1	D
0	0	0	0	0	0	0	
0	0	1	W	X	0	0	
0	1	1	W	0	Y	Y	
0	1	0	0	0	0	0	
1	1	0	0	0	0	0	
1	1	1	0	0	Y	Y	
1	0	1	0	X	0	0	
1	0	0	0	0	0	0	

Fig. 3-13

The entries WW are adjacent in the usual sense and yield a reduced term A′CD′E′. The entries XX are symmetric about the reflection axis. Inspection shows that they differ in only the variable A, so the reduced term is B′CD′E. Entries YYYY form a block of four when the reflection axis is taken into account and yield the reduced term BCD.

It is possible to extend the reflection concept to six variables and beyond, but the identification of patterns becomes increasingly difficult.

EXAMPLE 3.7 A six-variable map with two axes of reflection is shown in Fig. 3-14. The XXXX entries are symmetric about both axes and correspond to B′CE′F. The eight Y entries are also symmetric about both axes and correspond to BCE. The two Ws are only symmetric about the vertical reflection axis and contribute the slightly simplified term AB′C′EF.

			0	1	1	0	0	1	1	0	F
			0	0	1	1	1	1	0	0	E
A	B	C	0	0	0	0	1	1	1	1	D
0	0	0	0	0	0	0	0	0	0	0	
0	0	1	0	X	0	0	0	0	X	0	
0	1	1	0	0	Y	Y	Y	Y	0	0	
0	1	0	0	0	0	0	0	0	0	0	
1	1	0	0	0	0	0	0	0	0	0	
1	1	1	0	0	Y	Y	Y	Y	0	0	
1	0	1	0	X	0	0	0	0	X	0	
1	0	0	0	0	W	0	0	W	0	0	

Fig. 3-14

Six-Variable Map Slices

An alternate approach to the six-variable map is to create 4 four-variable maps and to mentally consider them as being stacked on top of each other (see Example 3.8). If these mental gymnastics are accomplished, the basic rules of adjacency can be applied in the usual manner.

EXAMPLE 3.8 An alternate approach to six-variable mapping is shown in Fig. 3-15. After stacking the four map slices in the order EF = 00, EF = 01, EF = 11, and EF = 10, we find the following adjacencies:

WW	Adjacent on the bottom slice only (yields the term B′C′DE′F′)
YYYYYYYY	Adjacent within and between two slices (yields the term BDF)
XXXXXXXX	Adjacent within and between the top and bottom slices (yields B′D′F′). Note that the two X entries in the EF = 11 map are not included in this grouping
ZZ	Adjacent between the middle two slices (yields A′BCD′F)

		0	1	1	0	D
A	B	0	0	1	1	C
0	0	X	W	0	X	
0	1	0	0	0	0	
1	1	0	0	0	0	
1	0	X	W	0	X	

EF = 00 (Bottom)

		0	1	1	0	D
A	B	0	0	1	1	C
0	0	0	0	0	0	
0	1	0	Y	Y	Z	
1	1	0	Y	Y	0	
1	0	0	0	0	0	

EF = 01

		0	1	1	0	D
A	B	0	0	1	1	C
0	0	0	0	0	0	
0	1	0	Y	Y	Z	
1	1	0	Y	Y	0	
1	0	X	0	0	X	

EF = 11

		0	1	1	0	D
A	B	0	0	1	1	C
0	0	X	0	0	X	
0	1	0	0	0	0	
1	1	0	0	0	0	
1	1	X	0	0	X	

EF = 10 (Top)

Fig. 3-15

Computer-Aided Reduction

As we can see from the preceding examples, identifying map adjacencies can become very tedious as the number of variables increases to five or more. Fortunately, there is a considerable amount of software in existence for the reduction of Boolean functions which permits us to use a computer to obtain simplification of multivariable expressions. This software is usually of the brute force variety which employs algorithms that apply intelligently guided trial and error.

Mapping the Exclusive OR

Exclusive-OR functions tend to map into a checkerboard pattern. Consider the map of the Boolean expression, $F = (A \oplus B) \oplus (C \oplus D)$ in Prob. 2.10 as shown in Fig. 3-16. Since there are no adjacencies, we conclude that the given function is in its simplest expressible form.

Quite often, if a map can be manipulated into a checkerboard or checkerboard-like pattern, the associated function can be very effectively simplified using XORs.

A	B	D=0, C=0	D=1, C=0	D=1, C=1	D=0, C=1
0	0	0	1	0	1
0	1	1	0	1	0
1	1	0	1	0	1
1	0	1	0	1	0

F

Fig. 3-16

EXAMPLE 3.9 Given the map in Fig. 3-17, we wish to obtain an optimum simplification. Proper assignment of 1s and 0s to appropriate "don't cares" in the last two rows can create the checkerboard pattern corresponding to the function mapped in Fig. 3-16. It is interesting to compare this result with the expression obtained by adjacency reductions shown in Fig. 3-18.

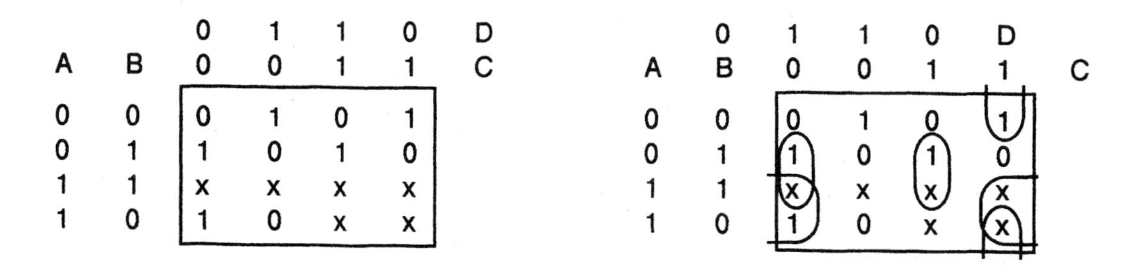

F = A D' + A' B' C' D + B' C D' + B C D + B C' D'

Fig. 3-17 **Fig. 3-18**

Solved Problems

3.1 Demonstrate that $A + A' = 1$.

Since the logical variable A can be only 1 or 0, if $A = 0$, then $A' = 1$ and vice versa. Thus, the left side of the given expression can never have two 0s, and the result is obvious from fundamental theorem 1.

3.2 Prove both forms of De Morgan's theorem.

Use proof by exhaustion.

1. $AB = (A' + B')'$. As seen in Table 3.2, the two boxed columns are identical for every combination of the variables A and B.

Table 3.2

A	B	AB	A'	B'	A' + B'	(A' + B')'
0	0	0	1	1	1	0
0	1	0	1	0	1	0
1	0	0	0	1	1	0
1	1	1	0	0	0	1

2. A + B = (A'B')'. See Table 3.3. Again, the two boxed columns prove the theorem.

Table 3.3

A	B	A + B	A'	B'	A'B'	(A'B')'
0	0	0	1	1	1	0
0	1	1	1	0	0	1
1	0	1	0	1	0	1
1	1	1	0	0	0	1

3.3 Prove the following Boolean identity by exhaustion: AB + BC + A'C = AB + A'C.

Set up the truth table with columns for each variable and term (Table 3.4). If, for every combination of the independent variables, the left- and right-hand sides of the given equation have the same logical value, then the two expressions are identical.

Table 3.4

A	B	C	A'	AB	BC	A'C	AB + A'C + BC	AB + A'C
0	0	0	1	0	0	0	0	0
0	0	1	1	0	0	1	1	1
0	1	0	1	0	0	0	0	0
0	1	1	1	0	1	1	1	1
1	0	0	0	0	0	0	0	0
1	0	1	0	0	0	0	0	0
1	1	0	0	1	0	0	1	1
1	1	1	0	1	1	0	1	1

Identical

Another (and simpler) method of proving the given identity is to employ a K-map simplification. (Refer to Prob. 3.7.)

3.4 Show that the two given Boolean expressions are equivalent, respectively, to the exclusive-OR function and its inverse as stated in Eqs. (3.15) and (3.16).

$$AB' + BA' = A \oplus B \tag{3.15}$$

$$AB + A'B' = \overline{A \oplus B} \tag{3.16}$$

Equation (3.15) is easily verified by inspecting the truth table for A ⊕ B. Equation (3.16) may be proven by making use of De Morgan's theorem [Eqs. (3.12*a*) and (3.12*b*):

$$\overline{A \oplus B} = \overline{AB' + BA'} = (A' + B)(B' + A)$$
$$= A'B' + A'A + BB' + BA$$

Since A'A and BB' are both 0 [Eq. (3.7)], we are left with the identity expressed by Eq. (3.16).

3.5 Prove or disprove the following identity. If it is proven, does it follow that ACD = 0?

$$A'C + CD + AB'C' = A'C + CD + AB'C' + ACD$$

Set up a truth table with columns for each of the terms as shown in Table 3.5. LS stands for left side (of the identity) and RS for right side. The identity is proven by exhaustion since LS = RS for all possible combinations of the independent variables. Note that this does *not* imply that ACD is 0. Rather, the term is *redundant* since it is only equal to 1 when one of the remaining terms is also 1.

Table 3.5

A	B	C	D	A'C	CD	AB'C'	ACD	LS	RS
0	0	0	0	0	0	0	0	0	0
0	0	0	1	0	0	0	0	0	0
0	0	1	0	1	0	0	0	1	1
0	0	1	1	1	1	0	0	1	1
0	1	0	0	0	0	0	0	0	0
0	1	0	1	0	0	0	0	0	0
0	1	1	0	1	0	0	0	1	1
0	1	1	1	1	1	0	0	1	1
1	0	0	0	0	0	1	0	1	1
1	0	0	1	0	0	1	0	1	1
1	0	1	0	0	0	0	0	0	0
1	0	1	1	0	1	0	1	1	1
1	1	0	0	0	0	0	0	0	0
1	1	0	1	0	0	0	0	0	0
1	1	1	0	0	0	0	0	0	0
1	1	1	1	0	1	0	1	1	1

3.6 Prove the following Boolean identities.

(*a*) A(B′ + C) = AB′ + AC

(*b*) A(A + B′C) = A

(*c*) A′B′ + AB′ + AB = A + B′

(*a*) A(B′ + C) = AB′ + AC. Using exhaustion we get Table 3.6

Table 3.6

A B C	B'	(B' + C)	A (B' + C)	A B'	A C	A B' + A C
0 0 0	1	1	**0**	0	0	**0**
0 0 1	1	1	**0**	0	0	**0**
0 1 0	0	0	**0**	0	0	**0**
0 1 1	0	1	**0**	0	0	**0**
1 0 0	1	1	**1**	1	0	**1**
1 0 1	1	1	**1**	1	1	**1**
1 1 0	0	0	**0**	0	0	**0**
1 1 1	0	1	**1**	0	1	**1**

The identity is proven since the indicated boldface entries in Table 3.6 match for all possible combinations of the independent variables A, B, and C.

(*b*) A(A + B′C) = A. Using exhaustion we get Table 3.7.

Table 3.7

A	B	C	B' C	A + B' C	A (A + B' C)
0	0	0	0	0	0
0	0	1	1	1	0
0	1	0	0	0	0
0	1	1	0	0	0
1	0	0	0	1	1
1	0	1	1	1	1
1	1	0	0	1	1
1	1	1	0	1	1

By algebraic manipulation:

$$\begin{aligned} A(A + B'C) &= A + AB'C && \text{(Distributive law and Theorem 5)} \\ &= A(1 + B'C) && \text{(Distributive law)} \\ &= A && \text{(Theorems 1 and 3)} \end{aligned}$$

(*c*) A′B′ + AB′ + AB = A + B′. Using exhaustion we get Table 3.8.

Table 3.8

A	B	C	A'	B'	A' B'	AB'	AB	LS	RS
0	0	0	1	1	1	0	0	1	1
0	0	1	1	1	1	0	0	1	1
0	1	0	1	0	0	0	0	0	0
0	1	1	1	0	0	0	0	0	0
1	0	0	0	1	0	1	0	1	1
1	0	1	0	1	0	1	0	1	1
1	1	0	0	0	0	0	1	1	1
1	1	1	0	0	0	0	1	1	1

By algebraic manipulation:

$$\begin{aligned} A'B' + AB' + AB &= (A' + A)B' + AB && \text{(Distributive law)} \\ &= B' + AB && \text{(Theorem 6)} \\ &= B' + A && \text{[Eq. (3.14}b\text{)]} \end{aligned}$$

3.7 Use a K map to prove that (AB + BC + A′C) = (AB + A′C).

The left-hand expression is mapped as shown in Fig. 3-19.

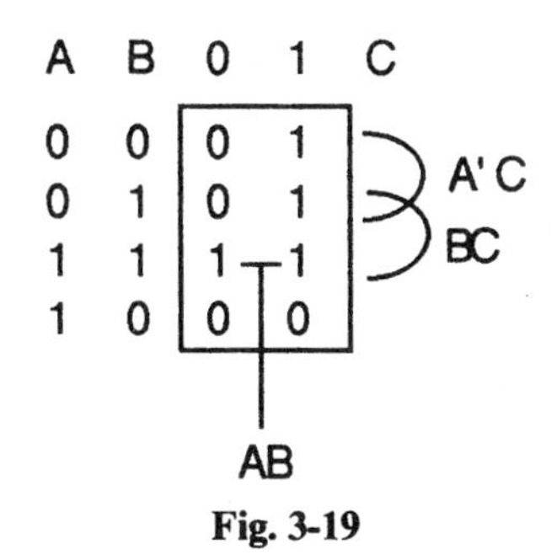

Fig. 3-19

The first 1 in row 3 is defined by the coordinates ABC′ and the adjacent 1 by ABC. Since ABC′ + ABC = AB(C + C′) = AB, the two 1s in row 3 define term AB. Variable C is eliminated because it takes on both possible values. (Refer to the discussion in Sec. 3.4.)

Similarly, BC is mapped as two 1s in the second column and located so that A is 0 in one case and 1 in the other. A′C is also located in column 2 and located so as to eliminate B. We may now encircle adjacencies to obtain a simplification as shown in Fig. 3-20.

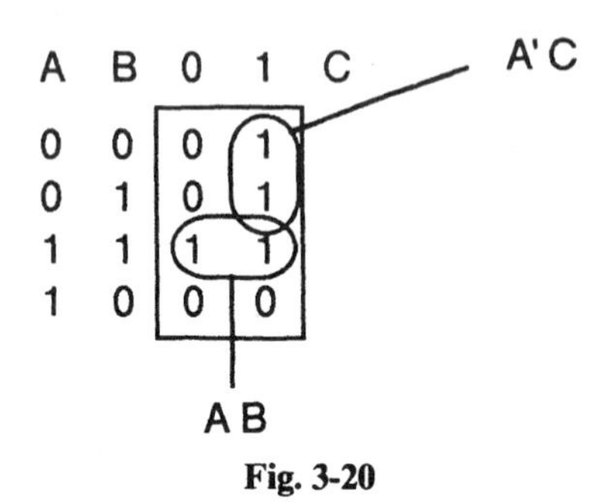

Fig. 3-20

All 1s are accounted for by the two coverings shown so that the function may be expressed by AB + A′C. We see that the term BC is redundant as demonstrated in Prob. 3.3.

3.8 Map the function F(A, B, C, D) = AB′C + B′D′ + BCD′ + AB′C′D.

Refer to Fig. 3-21. AB′C is the intersection of the fourth row (AB′) and the third and fourth columns (C). B′D′ is the intersection of the first and fourth columns (D′) and the first and fourth rows (B′). BCD′ is the intersection of the second and third rows (B) and the fourth column (CD′). AB′C′D is the intersection of the fourth row (AB′) and the second column (C′D).

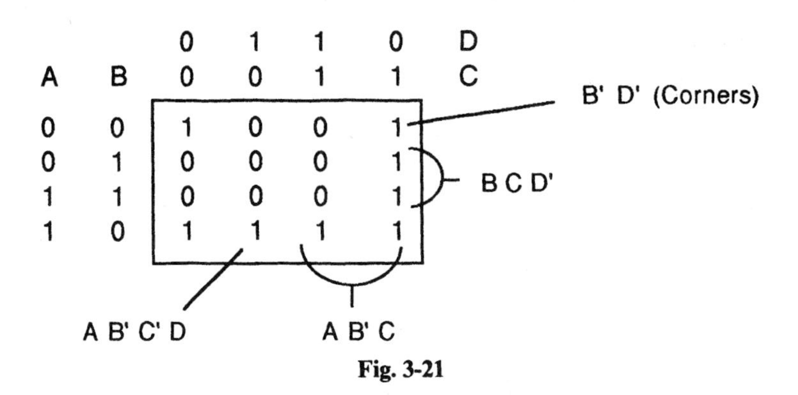

Fig. 3-21

3.9 Use the map developed for Prob. 3.8 to obtain a simplified alternative form for F.

Refer to Fig. 3.22. CD′ is the rectangular 4 × 1 block consisting of the fourth column. AB′ is the rectangular 1 × 4 block consisting of the fourth row. B′D′ is defined by the four corners, i.e., the intersection of the first and fourth rows with the first and fourth columns. The function is

$$F = CD' + AB' + B'D'$$

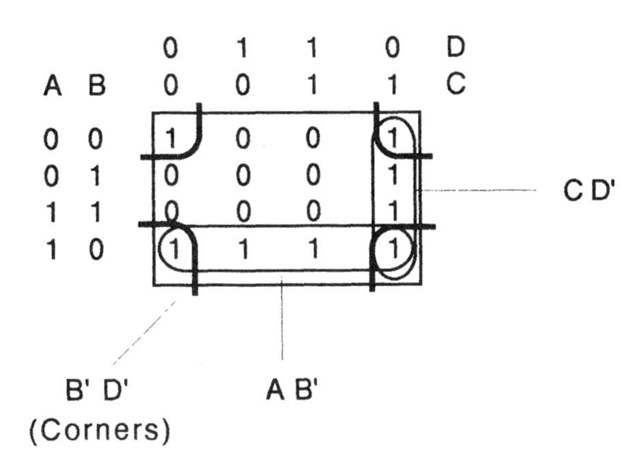

Fig. 3-22

3.10 Use the map developed for Prob. 3.8 to obtain a simplified expression for F′.

Refer to Fig. 3-23. The 0s covering shown above yields the desired expression. Individual terms are keyed to the diagram to aid in identification.

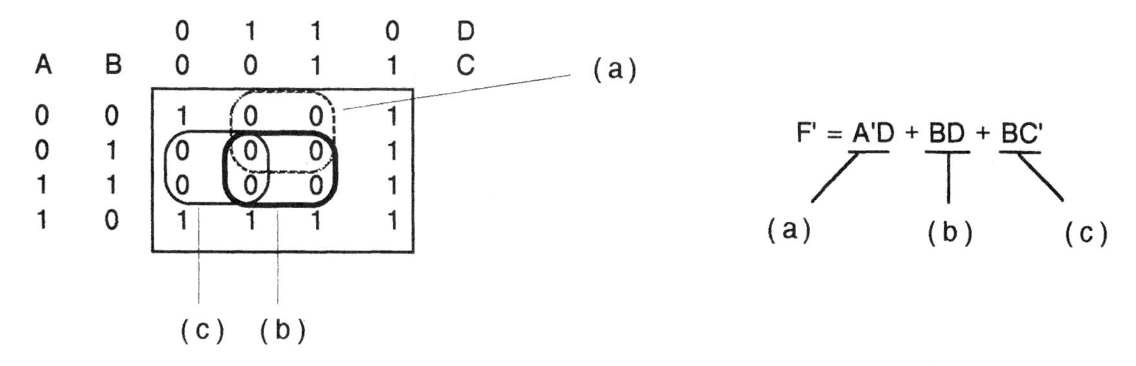

Fig. 3-23

3.11 Map the function of four variables which is TRUE whenever three or more of the variables are TRUE, and write the simplest Boolean expression you can.

Draw the map axes using A, B, C, D as input variables and count the number of 1s in the coordinates of each square. See Fig. 3-24. The best simplification we can do is four coverings of two 1s each as shown in Fig. 3-25. This yields the function:

$$\begin{aligned} F &= BCD + ABC + ACD + ABD \\ &= AB(C + D) + CD(A + B) \end{aligned}$$

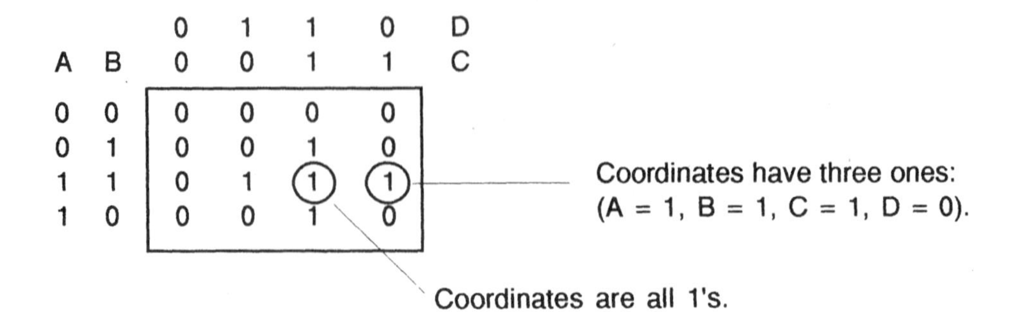

Fig. 3-24

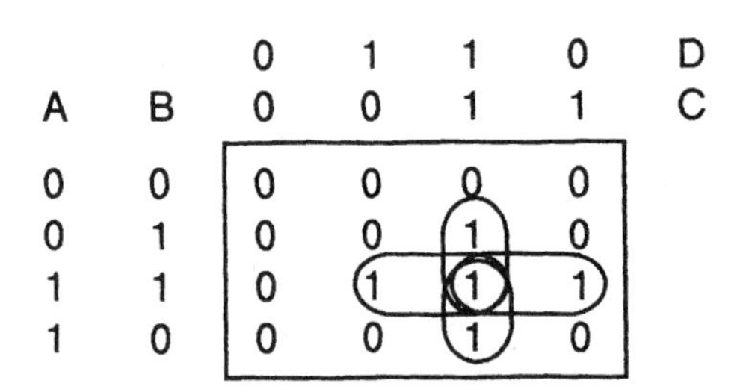

Fig. 3-25

3.12 Using the don't cares, obtain a reduced Boolean expression for the b segment output of a seven-segment display driver having BCD inputs. Refer to Prob. 2.2.

The map is derived from the truth table of Prob. 2.2 where W, X, Y, Z represent the 4 bits of a given BCD digit. See Fig. 3-26.

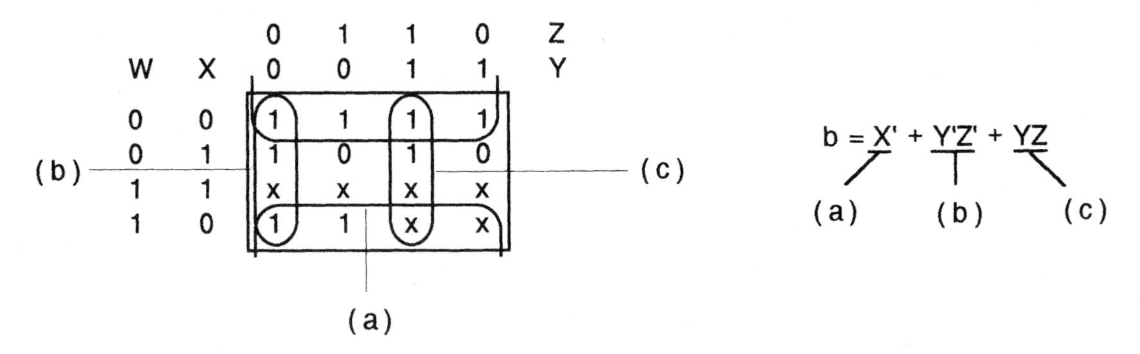

Fig. 3-26

3.13 The Boolean expression for the b output of a BCD to seven-segment display circuit contains the term Y'Z' + YZ. Referring to Sec. 3.5, we see that the a segment expression contains this structure also. Show that instead of using three gates, the term can be realized with the logical inversion of an exclusive OR.

Draw the truth table for F = Y'Z' + YZ as shown in Table 3.9.

Table 3.9

Y	Z	F	F'
0	0	1	0
0	1	0	1
1	0	0	1
1	1	1	0

Note that F′ is the XOR function (see Sec. 2.3), so its inverse is the desired function F. The b segment logic diagram may thus be drawn in the simplified form of Fig. 3-27.

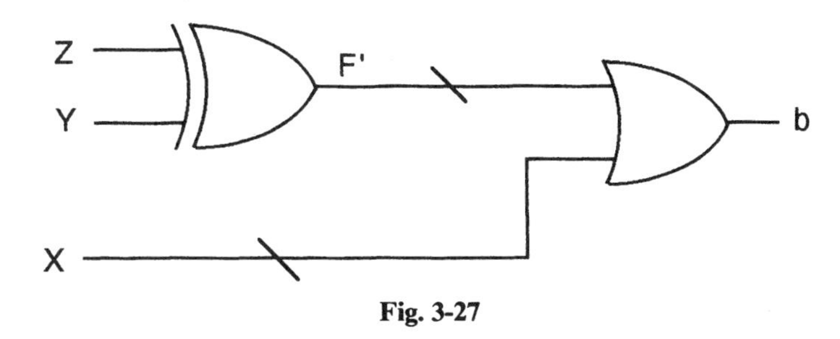

Fig. 3-27

3.14 The map of a five-variable function is shown in Fig. 3-28. Obtain a simplified expression for F(A, B, C, D, E).

A B C \ E D	0 0	1 0	1 1	0 1
0 0 0	0	0	0	1
0 0 1	1	1	0	0
0 1 1	0	0	1	1
0 1 0	1	0	0	0
1 1 0	0	0	0	0
1 1 1	0	0	1	1
1 0 1	0	0	0	0
1 0 0	0	0	0	1

Fig. 3-28

Use a reflection axis and 1s covering to obtain the function below whose terms are keyed to the map shown in Fig. 3-29.

$$F = \underset{(c)}{\underline{B'C'DE'}} + \underset{(a)}{\underline{A'B'CD'}} + \underset{(d)}{\underline{A'BC'D'E'}} + \underset{(b)}{\underline{BCD}}$$

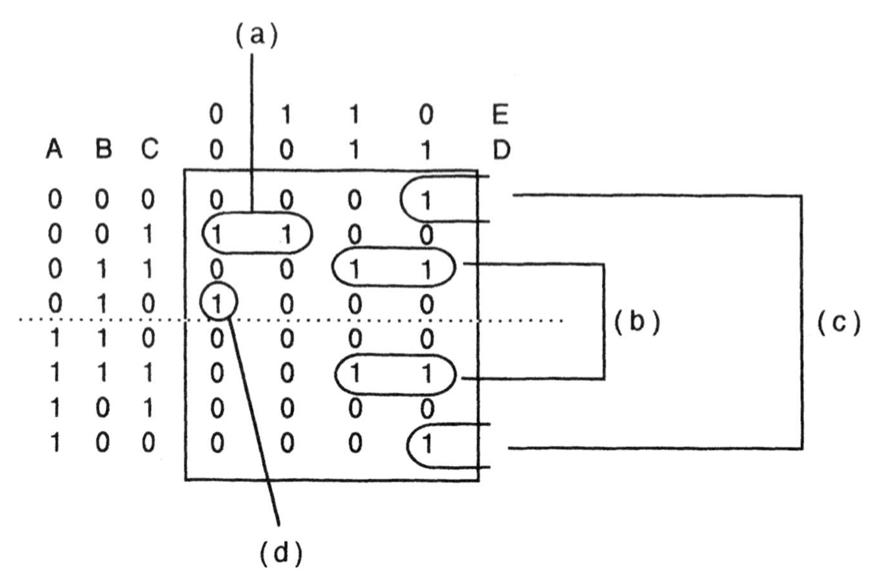

Fig. 3-29

3.15 Given the five-variable map shown in Fig. 3-30 which contains some "don't cares," obtain a simplification for both 1s and for 0s.

A	B	C	E=0, D=0	E=1, D=0	E=1, D=1	E=0, D=1
0	0	0	x	0	0	1
0	0	1	1	1	0	0
0	1	1	x	x	1	1
0	1	0	1	0	0	x
1	1	0	0	0	0	0
1	1	1	x	x	1	1
1	0	1	0	0	0	0
1	0	0	x	x	0	1

F

Fig. 3-30

With reference to Fig. 3-31, the 1s covering yields

$$F = \underset{(c)}{\underline{BC}} + \underset{(d)}{\underline{A'CD'}} + \underset{(b)}{\underline{B'C'E'}} + \underset{(a)}{\underline{A'C'E'}}$$

Note that each half of the map may be treated independently; i.e., coverings need not be symmetrical about the reflection axis [(a) and (d) for example].

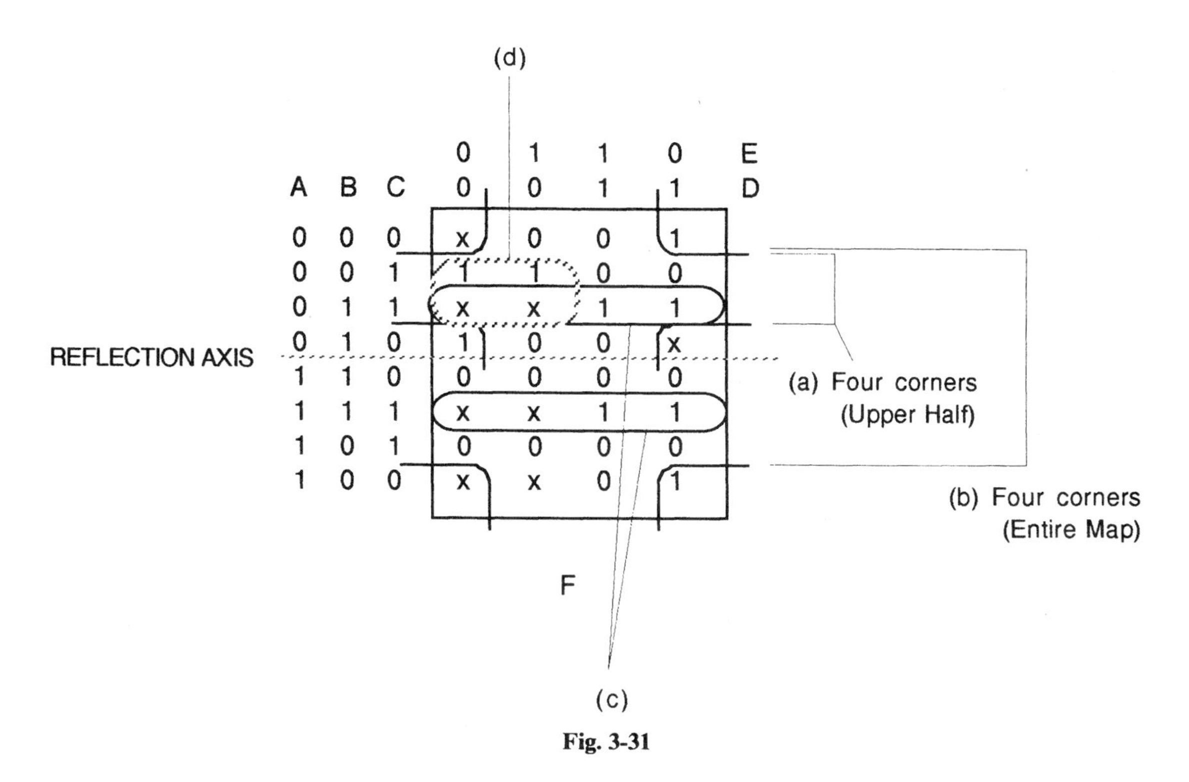

Fig. 3-31

The 0s covering is shown in Fig. 3-32. Because of the symmetry, coverings (a) and (c) may be treated as a single block of eight 0s, which yields the term C′E. To confirm this, consider them separately:

$$\text{(a) } BC'E \qquad \text{(c) } B'C'E$$

$$BC'E + B'C'E = C'E(B + B') = C'E$$

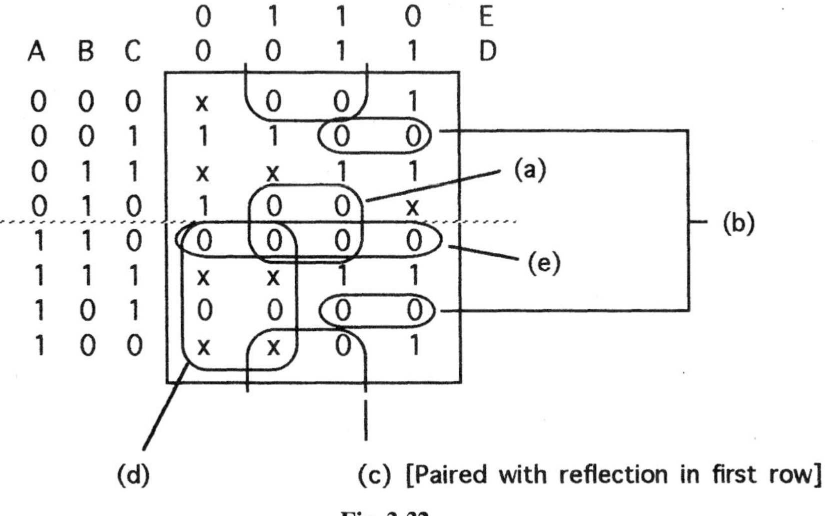

Fig. 3-32

Thus, the desired Boolean expression is seen to be

$$F' = \underbrace{AD'}_{(d)} + \underbrace{B'CD}_{(b)} + \underbrace{C'E}_{(a),(c)} + \underbrace{ABC'}_{(e)}$$

3.16 It is desired to design a digital combinational logic system having the truth table shown in Table 3.10. Inputs are A, B, C, and D and the output is F. The Xs correspond to input combinations which never occur and are therefore "don't care" states.

Table 3.10

	A	B	C	D	F
0	0	0	0	0	1
1	0	0	0	1	0
2	0	0	1	0	X
3	0	0	1	1	1
4	0	1	0	0	0
5	0	1	0	1	0
6	0	1	1	0	X
7	0	1	1	1	1
8	1	0	0	0	1
9	1	0	0	1	0
10	1	0	1	0	X
11	1	0	1	1	1
12	1	1	0	0	X
13	1	1	0	1	X
14	1	1	1	0	X
15	1	1	1	1	X

(*a*) Draw the K map, simplify as much as you can, and write the resulting Boolean expression for F.

(*b*) Draw the logic diagram corresponding to the simplified function F.

(*a*) The K-map coordinates of each 1 and each "don't care" are easily read directly from the truth table. Row 6, for example, tells us that an X should be placed at A = 0, B = 1, C = 1, and D = 0, which is located at the intersection of the second row and fourth column of the map shown in Fig. 3-33. Simplifying, we get Fig. 3-34.

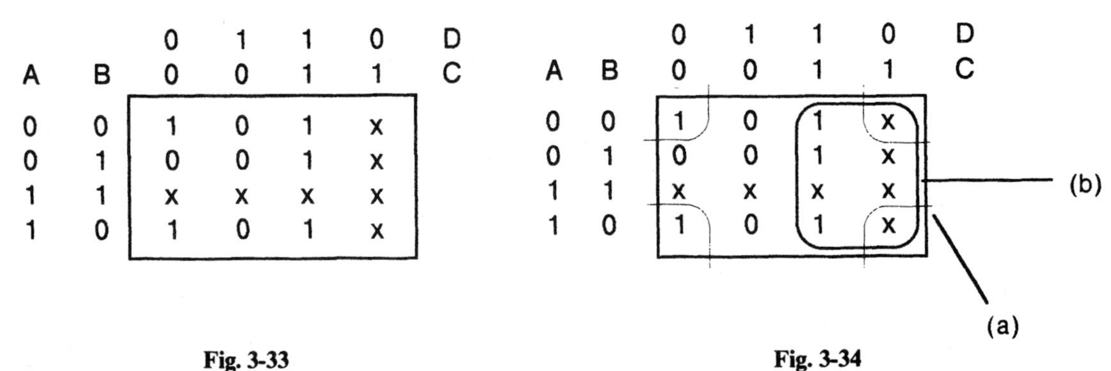

Fig. 3-33

Fig. 3-34

The resulting equation is,

$$F = C + B'D'$$

(b) (a)

(*b*) The logic diagram is drawn by inspection (see Fig. 3-35).

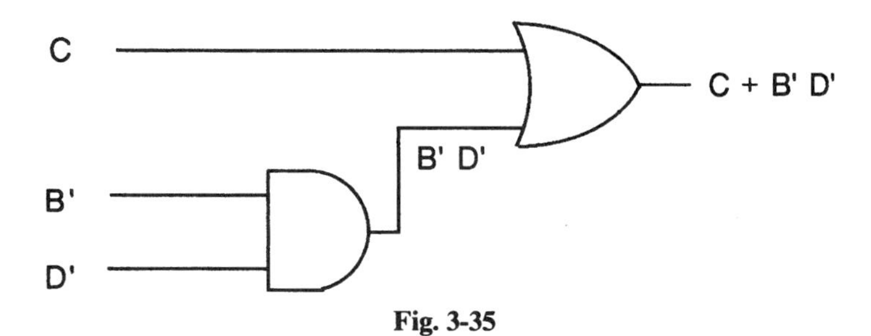

Fig. 3-35

3.17 Given the logic diagram shown in Fig. 3-36,

(*a*) Write the logic equation for F.

(*b*) Simplify the logic as much as possible.

(*c*) Draw the simplified logic diagram.

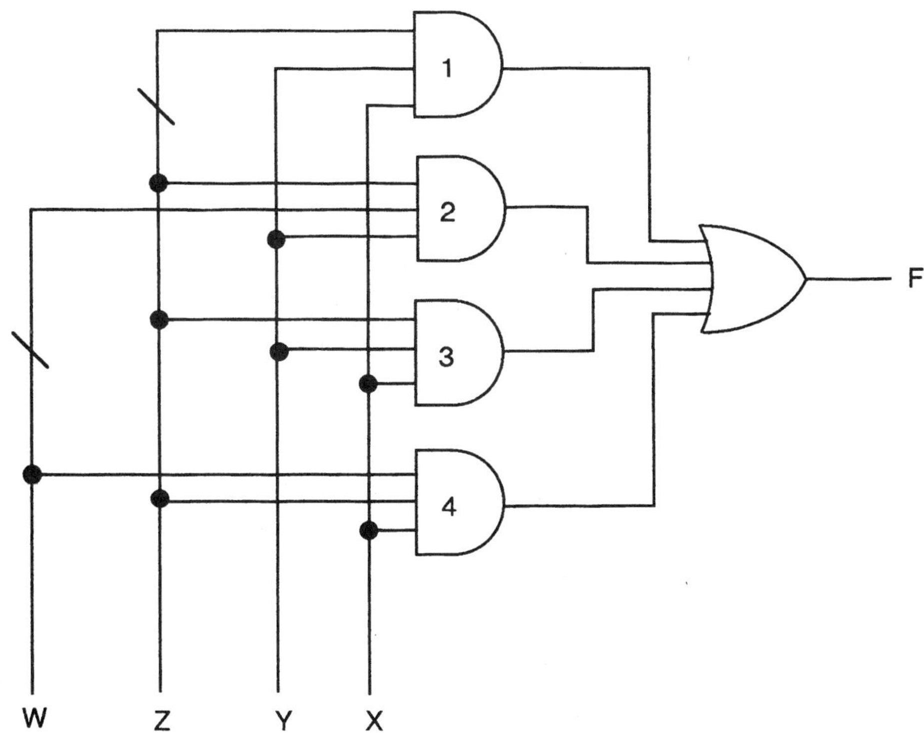

Fig. 3-36

(*a*) $F = XYZ' + W'YZ + XYZ + WXZ$. The AND gates associated with each term are indicated by numbers 1 to 4.

1 2 3 4

(*b*) Each term in the above equation maps as a pair of 1s as shown in Fig. 3-37.

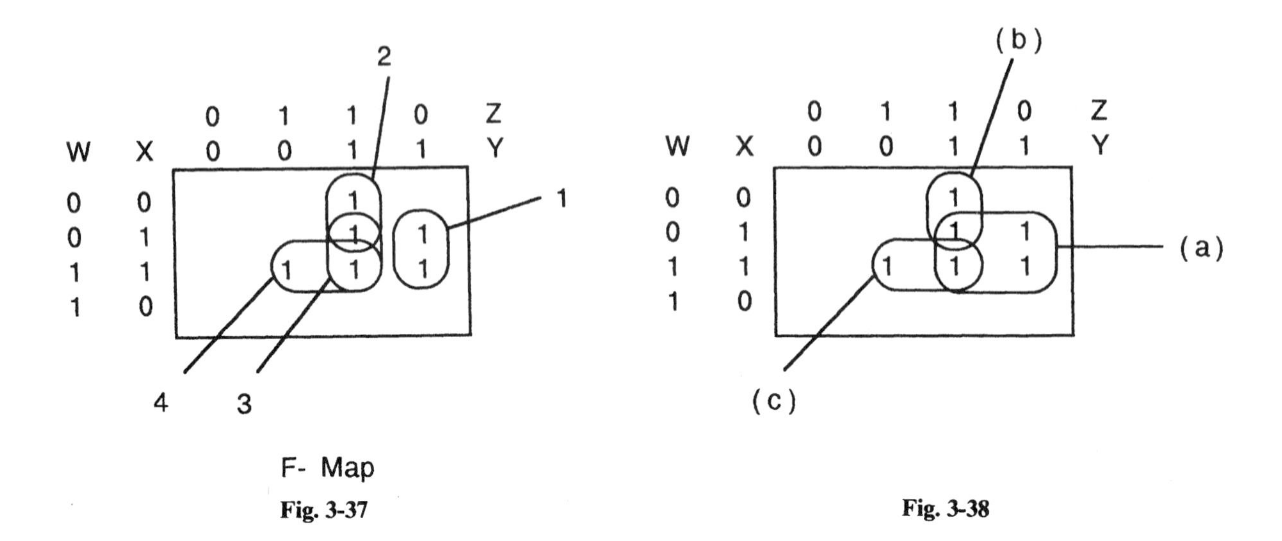

Fig. 3-37

Fig. 3-38

It is apparent that simplification can be achieved by combining terms 1 and 3 in a covering that encompasses four 1s as shown in Fig. 3-38. The resulting equation is

$$F = \underset{(a)}{XY} + \underset{(b)}{W'YZ} + \underset{(c)}{WXZ}$$

(c) The three terms indicated above correspond to the outputs of AND gates in the logic diagram of Fig. 3-39.

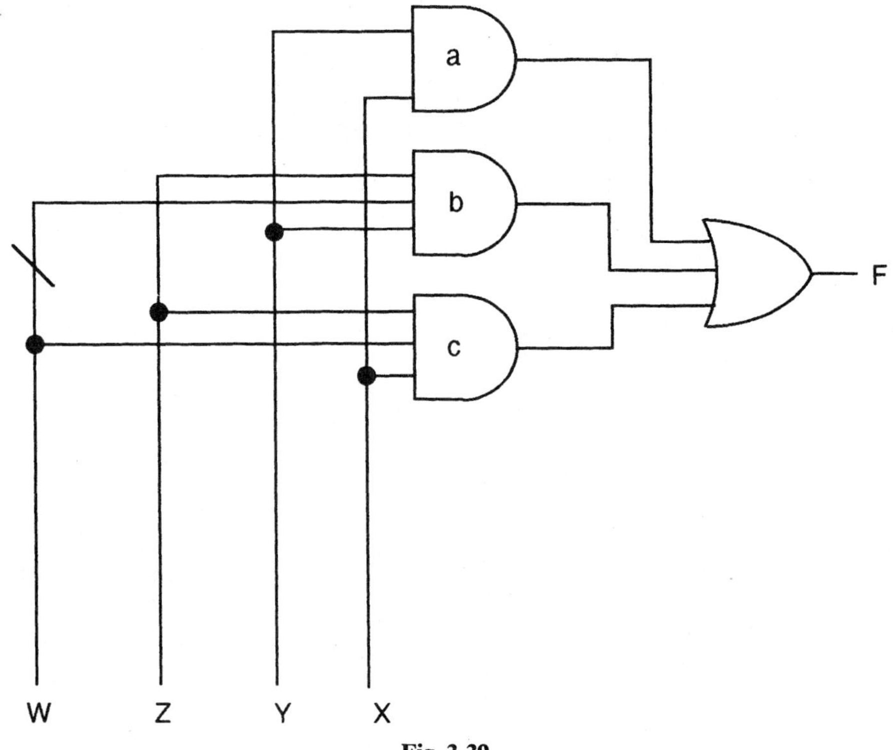

Fig. 3-39

3.18 It is required to design a logic circuit which will check for parity. Demonstrate the use of the exclusive-OR gate for this purpose.

Checking parity involves determining whether there are an even or odd number of 1s in a given binary word. The absence of odd parity must imply either a binary 0 or even parity; no other possibilities exist.

Consider the XOR circuit shown in Fig. 3-40. A + B yields a 1 when the two inputs are different, or equivalently, when there are an odd number of 1s at the input to gate G_1. Next, consider G_2. Output F will be a 1 if there are an odd number of 1s at its input, meaning that G_1 must produce a 1 when C is 0, or, if G_1 is 0, C must be 1. There are two possibilities:

1. If $G_1 = 1$, then A, B is odd (has an odd number of 1s) and F can only be 1 if C = 0.
2. If $G_1 = 0$, then A, B is even or 0 and F can only be 1 if C = 1.

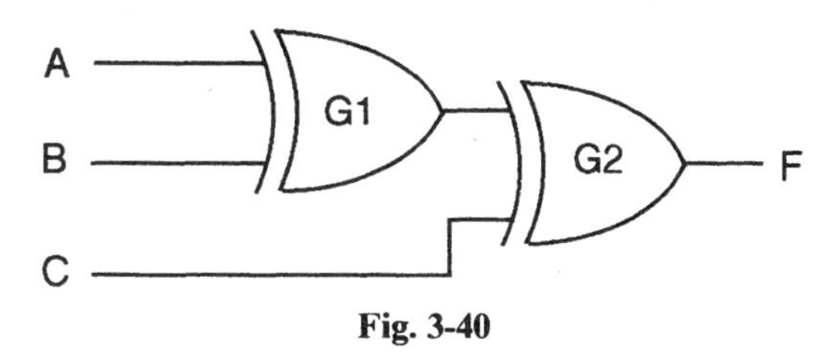

Fig. 3-40

In both cases, F can be 1 only if the input A, B, C has an odd number of 1s (odd parity). This is clearly demonstrated by constructing the K map for F as shown in Fig. 3-41. Consider the effect of adding a third XOR gate.

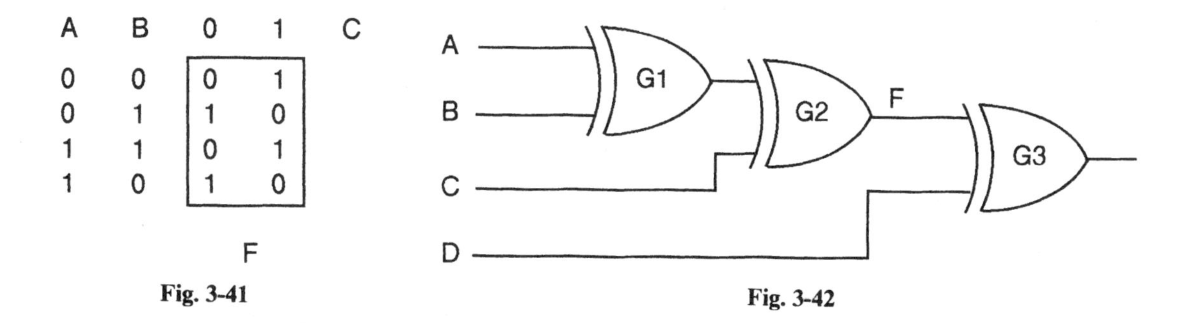

A	B	C = 0	C = 1
0	0	0	1
0	1	1	0
1	1	0	1
1	0	1	0

Fig. 3-41 **Fig. 3-42**

As shown in Fig. 3-42, F will be 1 if A, B, C is odd. G_3 will produce a 1 if F = 1 and D = 0 [(A, B, C, D) odd] or if F = 0 and D = 1. In the latter case A, B, C must be even or zero, and the necessity that D = 1 requires that A, B, C, D has odd parity. It should be clear that the circuit can be extended indefinitely by cascading XOR gates so that the parity of a binary number of any length can be checked.

3.19 The XOR circuit of Prob. 3.18 produces a 1 when its input has odd parity. Show that a logical inversion of this output will produce a 1 for even-parity input.

Logical inversion of the output means that for each combination of input variables, output 1s become 0s and 0s become 1s. The resulting K map is shown in Fig. 3-43. It can be seen at a glance that this arrangement produces a 1 when A, B, C is even or 0 and the corresponding Boolean expression represents the *inverse XOR* function $F' = A \oplus B \oplus C$ (refer to Prob. 3.4).

If we have a hardware means of inverting the output, it need only be added to the circuit of Prob. 3.18 to convert 1 = odd parity to 1 = even parity. Hardware realization of the logical inversion will be discussed in Sec. 4.3.

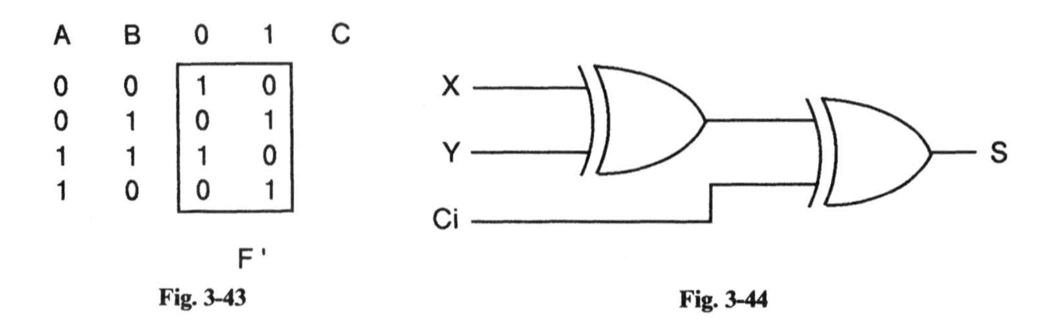

Fig. 3-43

Fig. 3-44

3.20 Show that the sum output of the single-bit adder in Sec. 3.4 can be realized with exclusive ORs.

The K map for S is identical to that of the three-variable XOR function in Prob. 3.18 which means that the two functions have the same truth table and must be equivalent. The realization for the sum output is therefore as shown in Fig. 3-44.

3.21 Using the boxes defined in Prob. 2.16, find a simplified expression for the logic accomplished by the system shown in Fig. 3-45.

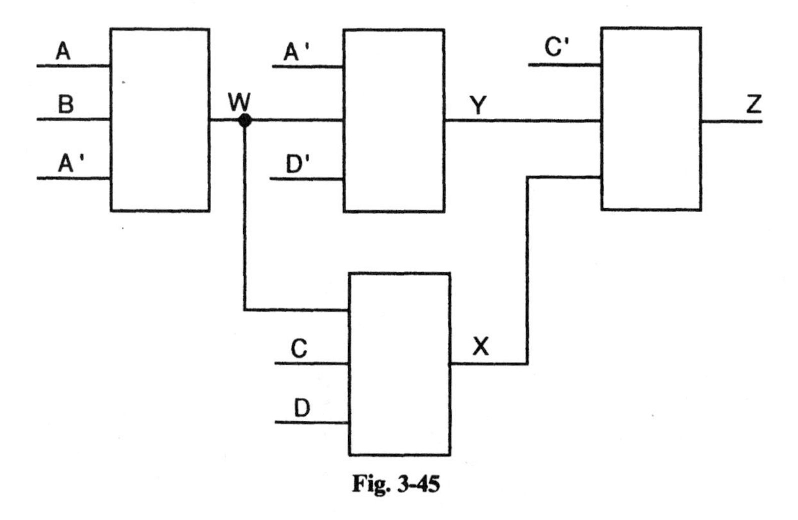

Fig. 3-45

Using the truth table obtained in Prob. 2.16 and creating the appropriate K map yields Fig. 3-46. After making the indicated variable substitutions, the resulting logic expressions for the box outputs are simplified using Boolean algebra:

$$
\begin{aligned}
W &= AB + A'B + AA' = AB + A'B + 0 = B(A + A') \\
&= B \cdot 1 = B \\
X &= CD + CW + DW \\
&= CD + BC + BD \\
Y &= A'W + A'D' + D'W \\
&= A'B + A'D' + BD' \\
Z &= XY + XC' + YC' \\
&= (CD + BC + BD)(A'B + A'D' + BD') + (CD + BC + BD)C' + (A'B + A'D' + BD')C' \\
Z &= T_1 + T_2 + T_3
\end{aligned}
$$

where,

$$
\begin{aligned}
T_1 &= (CD + BC + BD)(A'B + A'D' + BD') \\
&= A'BCD + A'BC + A'BD + A'BCD' + BCD'
\end{aligned}
$$

$$
\begin{aligned}
T_2 &= (CD + BC + BD)C' = BC'D \\
T_3 &= (A'B + A'D' + BD')C' = A'BC' + A'C'D' + BC'D'
\end{aligned}
$$

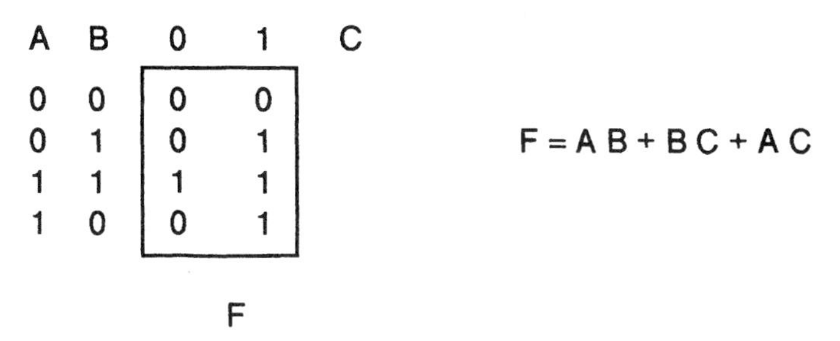

Fig. 3-46

The expression for Z can be simplified by mapping it. Start with T_1, as shown in Fig. 3-47. Inserting the terms T_2 and T_3 into the T_1 map yields the composite map shown in Fig. 3-48 from which we obtain the desired function:

$$Z = A'C'D' + A'B + BD' + BC'$$

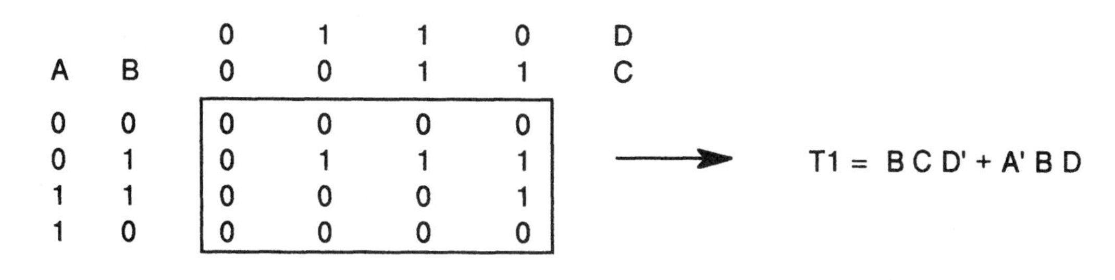

Fig. 3-47

		0	1	1	0	D
A	B	0	0	1	1	C
0	0	1	0	0	0	
0	1	1	1	1	1	
1	1	1	1	0	1	
1	0	0	0	0	0	

Z

Fig. 3-48

3.22 Using the truth table for the seven-segment display driver developed in Prob. 2.2, simplify the logic using K maps.

The K maps are shown in Fig. 3-49. One set of solutions is

$$a = W + Y + XZ + X'Z'$$
$$b = X' + YZ + Y'Z'$$
$$c = X + Y' + Z$$
$$d = W + XY'Z + X'Y + X'Z' + YZ'$$
$$e = X'Z' + YZ'$$
$$f = W + Y'Z' + XY' + XZ'$$
$$g = W + YZ' + X'Y + XY'$$

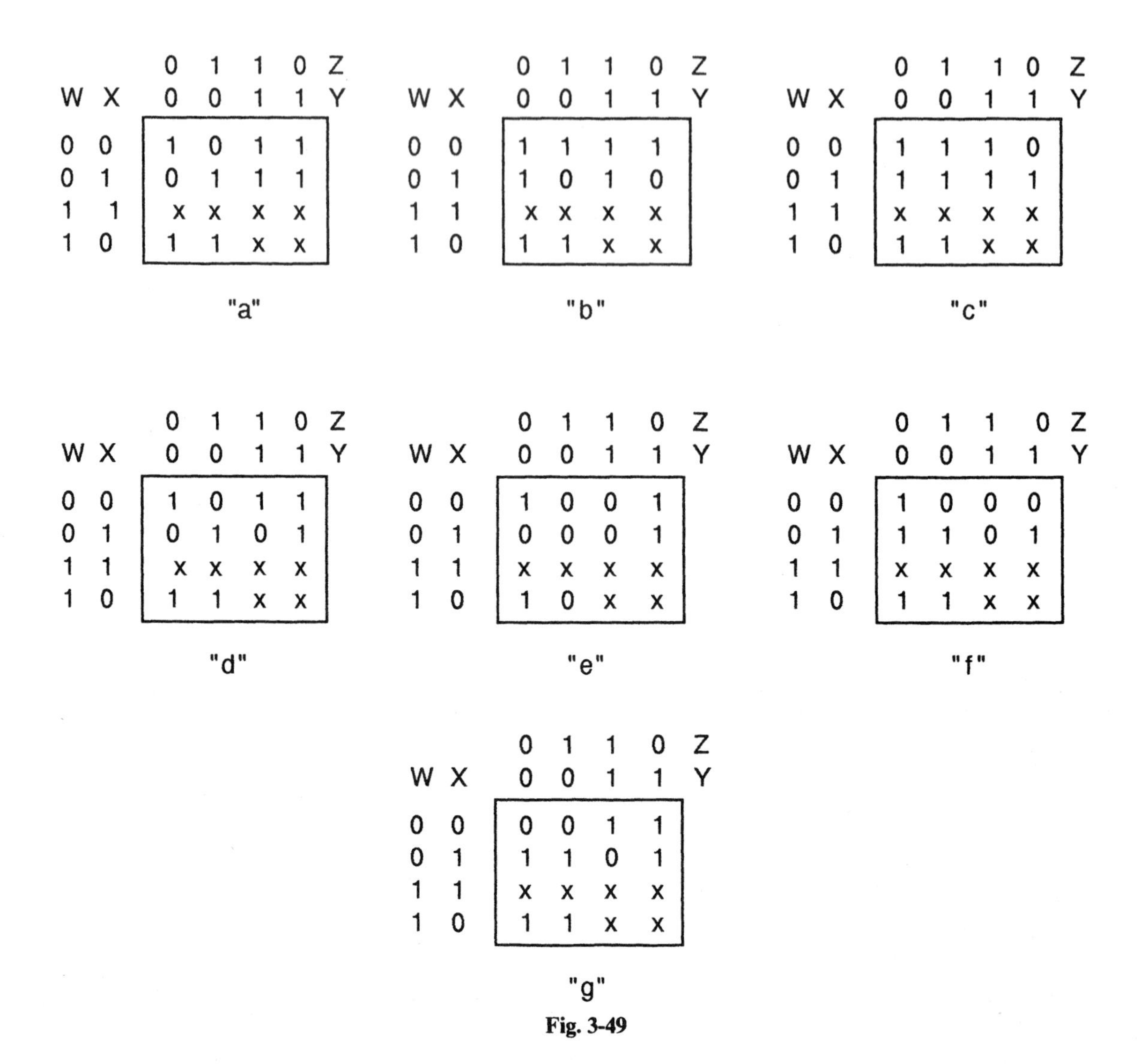

Fig. 3-49

Because of the relatively large number of don't cares, there are many possible arrangements. The particular set of solutions chosen takes advantage of multiple appearances of a single term which may be helpful in reducing gate count. Experience is often an important factor in obtaining an optimum simplfication.

Supplementary Problems

3.23 Simplify the following Boolean functions to minimize the number of variables and terms.

(*a*) BCD + B′C + BCD′

(*b*) (A + B)′(A′ + B′)′

(*c*) (A + AB′ + CD)′

(*d*) B′(B + C) + D′ + CD

(*e*) {A[B + C(A′ + B)]}′

3.24 For each of the following Boolean functions, find the complement and simplify it as much as you can.

(*a*) $[(AA' + B)']B$

(*b*) $(AB + AC')$

(*c*) $A'BC + ACD + A'BC' + B'D$

(*d*) $A'B' + DC'$

3.25 For the two-digit multiplier described in Prob. 2.20, simplify the design equations using K maps and determine the logic gates required.

3.26 In Prob. 2.24, Boolean equations for the parity digits of a simple Hamming code are obtained. Find an exclusive-OR implementation for P_3.

$$P_3 = D'_4 D'_3 D'_2 D'_1 + D'_4 D'_3 D'_2 D_1 + D'_4 D_3 D_2 D'_1 + D'_4 D_3 D_2 D_1$$
$$+ D_4 D'_3 D_2 D'_1 + D_4 D'_3 D_2 D_1 + D_4 D_3 D'_2 D'_1 + D_4 D_3 D'_2 D_1$$

3.27 In Prob. 2.25, the truth table for a lighting problem was developed. Show the K map and sketch a logic diagram.

3.28 Simplify the Boolean equations for the subtractor of Prob. 2.28.

3.29 Using the K map in Fig. 3.50, make the best use you can of the "don't cares" to simplify the expression for F. Assume that a minimal number of maximal coverings is desired.

		0	1	1	0	Z
W	**X**	0	0	1	1	Y
0	0	1	0	1	1	
0	1	0	1	1	1	
1	1	x	x	x	0	
1	0	1	0	x	x	

"F" Map

Fig. 3-50

3.30 Using the truth table developed in Prob. 2.33, produce K maps for the outputs A_n and C_n and make the best use you can of the "don't cares" to simplify the Boolean expressions. Use the minimal number of maximal coverings.

3.31 It is desired to design a combinational logic circuit that will cause a light to go on each time the decimal equivalent of a 4-bit binary input is divisible by 3. It is known that the numbers 0, 1, 7, 11, and 14 will never occur as inputs. Map the function, simplify it as much as you can with a Karnaugh 1s covering, and draw the logic diagram.

3.32 A system has four inputs and is to produce an output which is TRUE if and only if an odd number of the inputs are TRUE. Design a logic circuit that will meet the specification.

3.33 A six-variable map for a function Z is shown in Fig. 3-51. Obtain a simplified Boolean expression for Z using a minimal number of maximal coverings.

3.34 Create the K maps for a 4-bit binary to Gray code converter. Denote the binary digits by D, C, B, A and the Gray bits by W, X, Y, Z.

			0	1	1	0	0	1	1	0	F
			0	0	1	1	1	1	0	0	E
A	B	C	0	0	0	0	1	1	1	1	D
0	0	0	0	1	0	0	0	0	0	0	
0	0	1	0	1	0	0	0	0	1	0	
0	1	1	0	0	1	1	X	1	0	0	
0	1	0	0	0	0	X	X	0	0	0	
1	1	0	0	0	0	0	0	0	0	0	
1	1	1	0	0	1	1	1	X	X	0	
1	0	1	0	1	0	0	0	0	1	0	
1	0	0	0	X	1	0	0	1	0	0	

Fig. 3-51

3.35 Create the simplified Boolean equations for the 4-bit binary to Gray code converter of Prob. 3.34.

3.36 Using K maps, demonstrate that A'B + C'D' + A'B'C' = C'A' + A'BC + AC'D'.

3.37 Show that A'B + C'D' + A'B'C' = (B'C + AC + AD)'.

3.38 Obtain an alternate form for F = (B'C + AC + AD)' by using DeMorgan's theorem and other Boolean algebra.

3.39 Obtain the truth table for a device which produces a TRUE output if and only if a majority of its three inputs are TRUE. Write the Boolean equation, and simplify it.

Answers to Supplementary Problems

3.23 (*a*) C (*b*) 0 (*c*) A'(C' + D') (*d*) C + D' (*e*) A' + B'

3.24 (*a*) 1 (*b*) A' + B'C (*c*) B'D' + A(BC' + D') (*d*) (D' + C)(A + B)

3.25 See Fig. 3-52.

3.26

		0	1	1	0	D1
D4	D3	0	0	1	1	D2
0	0	1	1	0	0	
0	1	0	0	1	1	
1	1	1	1	0	0	
1	0	0	0	1	1	

P3

A1 A0	B1 B0 = 00	01	11	10
0 0	0	0	0	0
0 1	0	0	0	0
1 1	0	0	1	0
1 0	0	0	0	0

C3 = A1 A0 B1 B0 (One four input AND gate)

A1 A0	B1 B0 = 00	01	11	10
0 0	0	0	0	0
0 1	0	0	0	0
1 1	0	0	0	1
1 0	0	0	1	1

C2 = A1 A0' B1 + A1 B1 B0'
(Two three input AND gates and one two input OR gate)

A1 A0	B1 B0 = 00	01	11	10
0 0	0	0	0	0
0 1	0	0	1	1
1 1	0	1	0	1
1 0	0	1	1	0

C1 = A1 B1' B0 + A1' A0 B1 + A0 B1 B0' + A1 A0' B0
(Four three input AND gates and one four input OR gate)

A1 A0	B1 B0 = 00	01	11	10
0 0	0	0	0	0
0 1	0	1	1	0
1 1	0	1	1	0
1 0	0	0	0	0

C0 = A0 B0 (One two input AND gate)

Fig. 3-52

Since none of the 1s coverings involve D_1, we may remap the function with three variables:

D4	D3	0	1 D2
0	0	1	0
0	1	0	1
1	1	1	0
1	0	0	1

P3

When the inputs D_4, D_3, D_2 have an even number of 1s, then 1s occur as entries. Thus, we must use the XOR configuration with an added inverter as discussed in Prob. 3.19 and shown in Fig. 3-53. An additional XOR gate with one input tied to logic 1 acts as the required inverter.

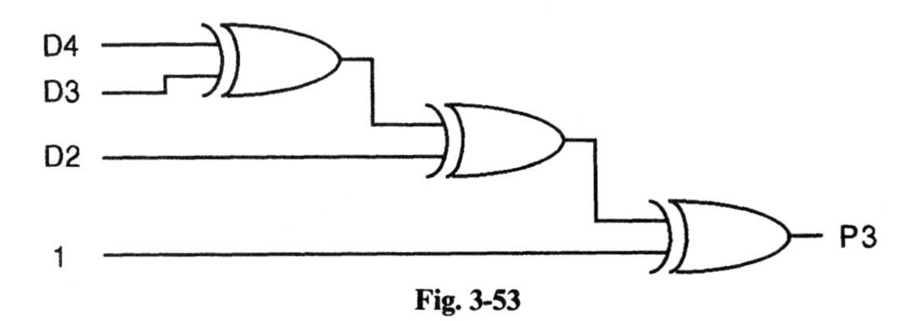

Fig. 3-53

3.27 See Fig. 3-54. Noting the checkerboard pattern.

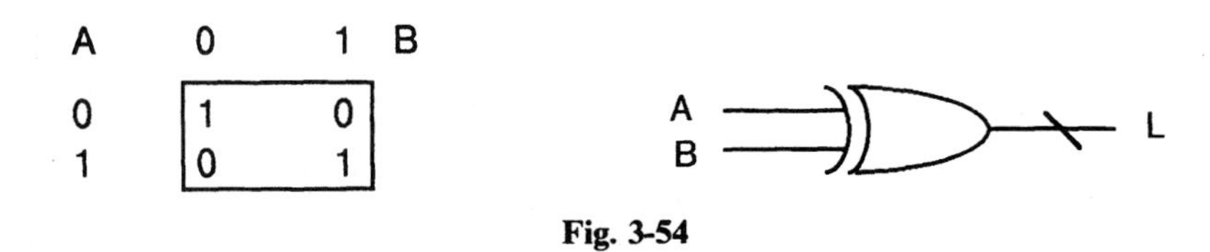

Fig. 3-54

3.28 $D = X \oplus Y \oplus B_i \qquad B_o = X'Y + X'B_i + YB_i$

3.29 Since Boolean algebra is not a calculus for minimization problems, several solutions are possible. What follows is the author's best effort (see Fig. 3-55).

	Z	0	1	1	0
W	X \ Y	0	0	1	1
0	0	1	0	1	1
0	1	0	1	1	1
1	1	0	1	1	0
1	0	1	0	0	1

"F" Map

$F = XZ + X'Z' + W'Y$

Fig. 3-55

3.30 See Fig. 3-56.

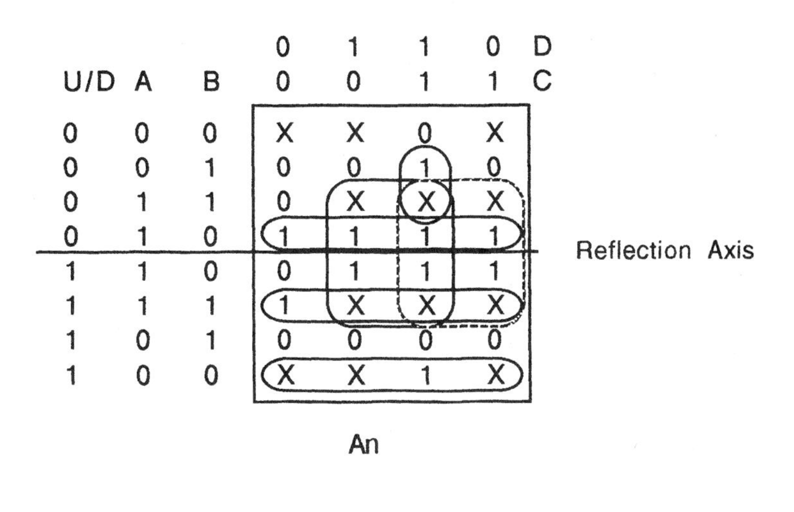

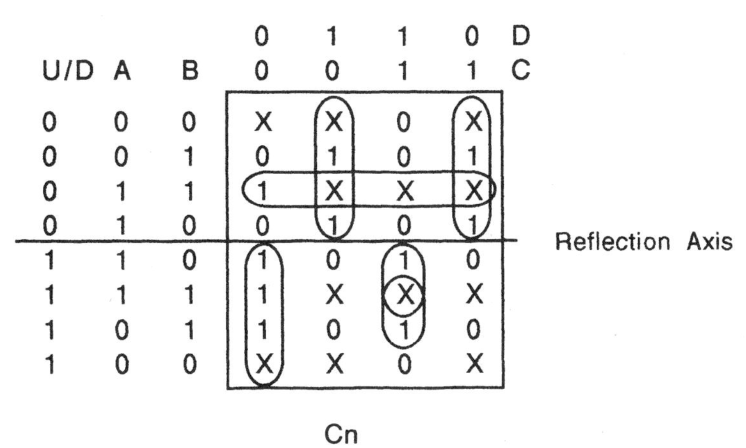

Fig. 3-56

3.31 See Fig. 3-57.

3.32 An XOR solution is particularly compact (see Fig. 3-58).

3.33 $Z = BCEF + BCE + AB'C'EF + B'CE'F + B'D'E'F$

3.34 See Fig. 3-59.

3.35 $W = D$

$X = CD' + C'D = C \oplus D$

$Y = BC' + B'C = B \oplus C$

$Z = AB' + A'B = A \oplus B$

3.36 See Fig. 3-60.

3.37 Map each function. The left-hand side is a 1s covering for F (as in Fig. 3-60). The right-hand function (inside the brackets) is an expression for F′ and, hence, a 0s covering. Again, the two maps are identical indicating the logical equivalency.

	D	0	1	1	0
A	B \ C	0	0	1	1
0	0	X	X	1	0
0	1	0	0	X	1
1	1	1	0	1	X
1	0	0	1	X	0

L

$L = B'D + BC + ABD'$

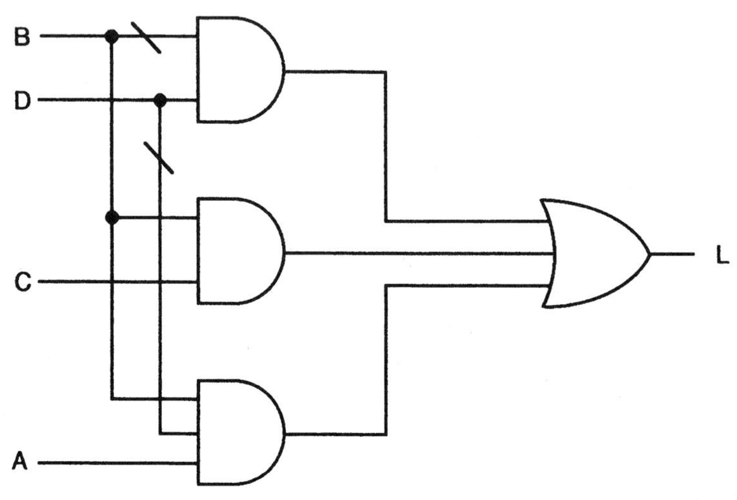

Fig. 3-57

	D	0	1	1	0
A	B \ C	0	0	1	1
0	0	0	1	0	1
0	1	1	0	1	0
1	1	0	1	0	1
1	0	1	0	1	0

F

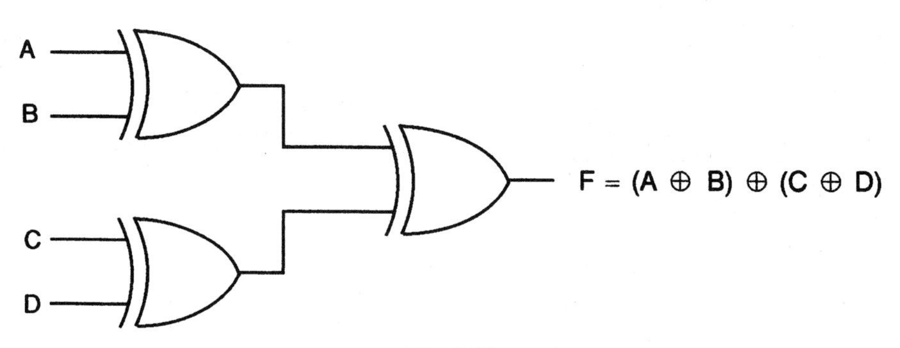

Fig. 3-58

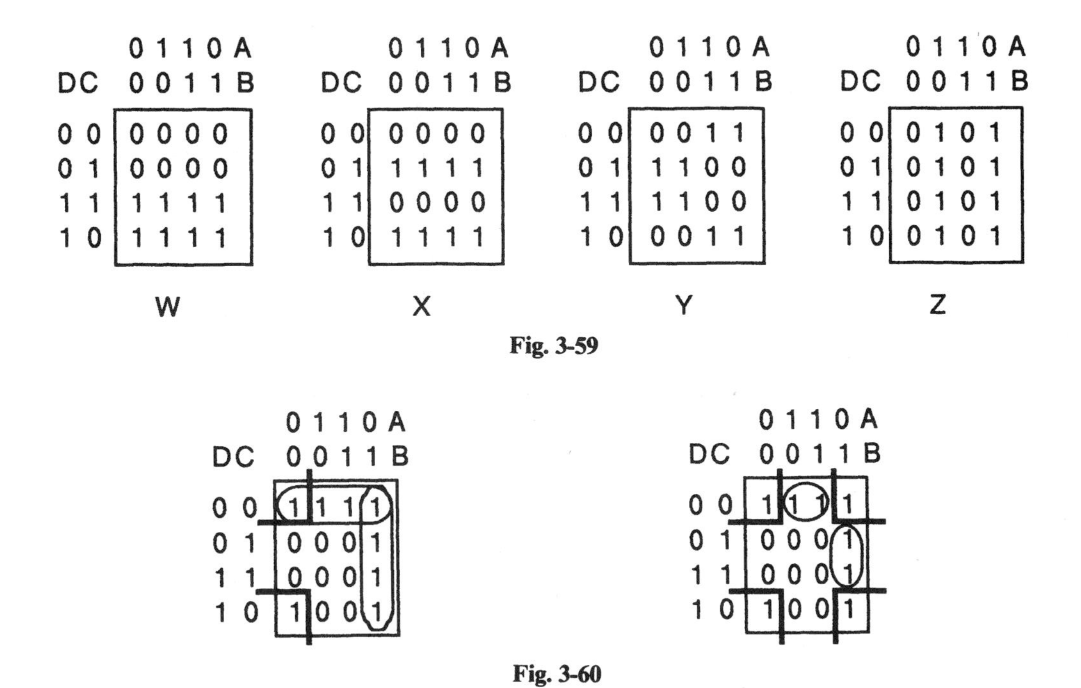

Fig. 3-59

Fig. 3-60

3.38 Several forms are given:

$$
\begin{aligned}
(B'C + AC + AD)' &= (B'C)'(AC)'(AD)' \\
&= (B + C')(A' + C')(A' + D') \\
&= (A'B + C')(A' + D') \\
&= A'C' + C'D' + A'B
\end{aligned}
$$

3.39 See Table 3.11.

Table 3.11

Inputs			Output
A	B	C	F
0	0	0	0
0	0	1	0
0	1	0	0
0	1	1	1
1	0	0	0
1	0	1	1
1	1	0	1
1	1	1	1

$F = A'BC + ABC' + AB'C + ABC$ (from the truth table); $F = AB + BC + AC$ (simplified version).

Chapter 4

Hardware and the Mixed-Logic Convention

4.1 INTRODUCTION

The linkage between designed combinational logic and its hardware implementation is usually described using some form of mixed-logic convention in which the relationship between logical and voltage values is chosen by the designer and symbols representing logical and electrical behavior are combined. Mixed-logic conventions are used both as design aids in the transition from logic to hardware (synthesis) and as a means of describing the logic implemented by hardware (analysis). Both applications are discussed.

The basic hardware unit for the physical implementation of combinational logic is the gate which, as mentioned in Sec. 2.5, serves as the elemental building block of digital logic systems. Gate circuits (hardware) operate with two distinct electrical signal levels, usually termed HIGH (H) and LOW (L), which are used to represent logical 1s and 0s (or TRUEs and FALSEs). It is the responsibility of the logic designer to arrange a suitable matchup between gate signals and the logic truth values of the design.

4.2 GATE HARDWARE

Gate hardware is available in the form of ICs which contain anywhere from a few to hundreds of thousands of gates in a single silicon chip. The functional characteristics of a particular gate type are usually specified by the manufacturer in the form of a voltage or current input/output (I/O) table which describes the *electrical behavior* of the physical device.

EXAMPLE 4.1 One of the most common types of hardware is the so-called NAND gate whose I/O table is shown in Fig. 4-1. Though only three inputs are shown, these gates are manufactured with various numbers of inputs (see App. B). The I/O table consists of an exhaustive listing of all possible combinations of input voltage levels and their corresponding output voltage levels.

INPUT			OUTPUT
A	B	C	F
L	L	L	H
L	L	H	H
L	H	L	H
L	H	H	H
H	L	L	H
H	L	H	H
H	H	L	H
H	H	H	L

Voltage I/O Table

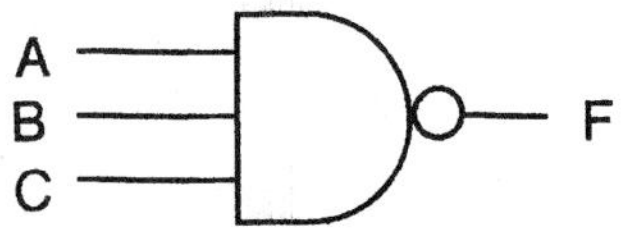

A
B
C
F

NAND Hardware Symbol

Fig. 4-1

If we equate the voltage levels of Fig. 4-1 with logic values TRUE or FALSE, then the voltage I/O table may be interpreted as a logical truth table (refer to Sec. 2.2). Since the early days of logic design, logical values have most commonly been assigned so as to make the highest voltage equal TRUE (H = 1 or HT). This assignment is termed *positive logic.*

EXAMPLE 4.2 Positive logic mapping of three-input NAND hardware. Assigning H = 1 (and, by implication, L = 0) yields the truth table shown in Fig. 4-2*a*. Note that this truth table does not meet the definition of either AND or OR. It is interpreted as an AND followed by a logical inversion (a negated AND or NAND for short). The NAND symbol in Fig. 4-1 indicates this inversion by a small circle (bubble).

INPUT			OUT
A	B	C	F
0	0	0	1
0	0	1	1
0	1	0	1
0	1	1	1
1	0	0	1
1	0	1	1
1	1	0	1
1	1	1	0

(a) Positive Logic Hardware Truth Table

INPUT			OUT
A	B	C	G
0	0	0	0
0	0	1	0
0	1	0	0
0	1	1	0
1	0	0	0
1	0	1	0
1	1	0	0
1	1	1	1

(b) AND Truth Table

OUTPUT
G'
1
1
1
1
1
1
1
0

(c) Negated AND

Fig. 4-2

4.3 MIXED LOGIC AS A DESIGN TOOL

Discretionary Assignment of Logical Values to Hardware

Many designers have abandoned positive logic and begun to work with a concept called mixed logic. In this approach, the assignment of logical values to voltage values is not fixed but is, instead, left to the designer's discretion.

Consider mixed-logic mapping of the NAND hardware gate of Example 4.1 where we assign H = 1 [HIGH-TRUE (HT)] *for the inputs* and L = 1 [LOW-TRUE (LT)] *for the output.* See Table 4.1.

Table 4.1

INPUT (HT)			OUT (LT)
A	B	C	F
0	0	0	0
0	0	1	0
0	1	0	0
0	1	1	0
1	0	0	0
1	0	1	0
1	1	0	0
1	1	1	1

This truth table clearly satisfies the definition of a logical AND function; i.e., the output is TRUE if and only if all the inputs are TRUE. Alternatively, if an assignment is made where L = 1 (LT) at the inputs and H = 1 (HT) at the output, we obtain a truth table where the output is TRUE if any one or more of the inputs is TRUE. See Table 4.2.

Table 4.2

INPUTS (LT)			OUT (HT)
A	B	C	F
1	1	1	1
1	1	0	1
1	0	1	1
1	0	0	1
0	1	1	1
0	1	0	1
0	0	1	1
0	0	0	0

This is recognized as a description of the logical OR function.

Note that the same piece of hardware can be used to implement two different logical operations, thereby providing considerable flexibility to the designer.

Mixed-Logic Symbols

To prevent confusion, two separate logic symbols for the NAND *hardware* gate have been developed, one to indicate its use for logical AND and the other for logical OR. These are shown in Fig. 4-3.

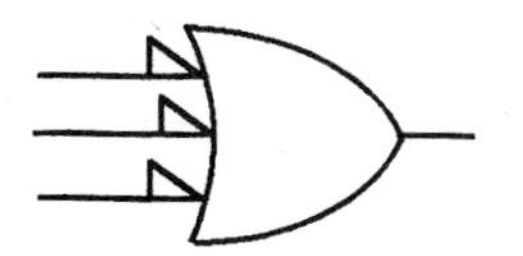

NAND hardware used as OR

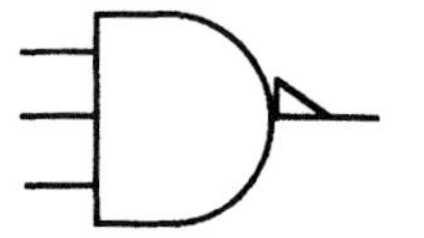

NAND hardware used as AND

Fig. 4-3

The *half arrow* is used to indicate LT at the point in the circuit where it appears. Conversely, the absence of a half arrow indicates HT. Some designers prefer to use a small circle or bubble as a LT indicator, and, quite often, a redundant double labeling is used whereby signal variable designators are suffixed either .L or .H to indicate their associated truth values. These alternate forms are shown in Fig. 4-4.*

A term often used synonymously with LOW-TRUE is active-LOW. In this case, the symbols in Fig. 4-4 would be interpreted as depicting an OR gate with active-LOW inputs or an AND gate with an active-LOW output.

* A truth table analysis and symbology for NOR hardware is presented in Solved Problem 4.1.

Fig. 4-4 NAND logic.

In order to facilitate the use of mixed logic as a design tool, many designers use a *slash mark* to indicate logic inversion. The slash (first introduced in Sec. 2.5) is a special logic symbol which we interpret as indicating a shift of interest from a variable to its logical inverse. *Note that this symbol does not represent hardware*; it is a *logical component* only. It does, however, have a relationship with the half-arrow convention. If the horizontal line in Fig. 4-5 is regarded as a piece of high-conductivity electrical wire (hardware), then, because of the properties of conductors, if any given voltage is applied to the left end of the wire, the right end must be at this same voltage. Suppose, however, that we assign 0 V to represent FALSE and some higher voltage, say 5 V, to represent TRUE. Suppose, also, that we apply 5 V to the left end of the wire. There will then be an apparent conflict since the slash mark forces us to interpret the right side as FALSE when the left side is TRUE, while, at the same time, we are being urged by the laws of nature to expect a high voltage on both sides. The only way that these two requirements can be reconciled is to designate the left side as HT and the right side as LT, or vice versa. This implies that *every slash carries with it an associated half arrow* which serves as a bridge between logic and hardware (see Fig. 4-6).

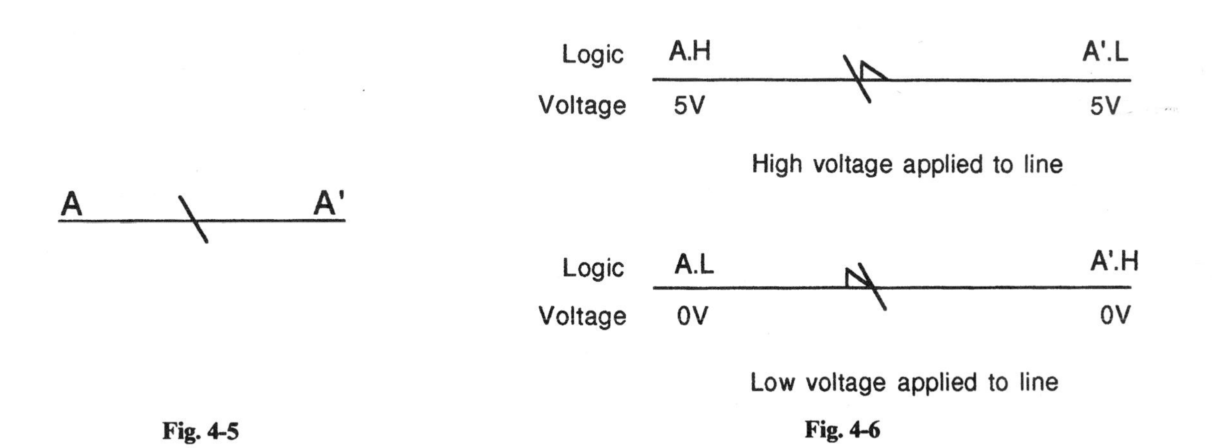

Fig. 4-5 **Fig. 4-6**

EXAMPLE 4.3 Given logical variables A, B, and C which are all HT, we wish to draw a mixed-logic diagram for the functon F = A'BC using NAND hardware. The slash in Fig. 4-7 indicates the necessity for creating A' in the final hardware realization.

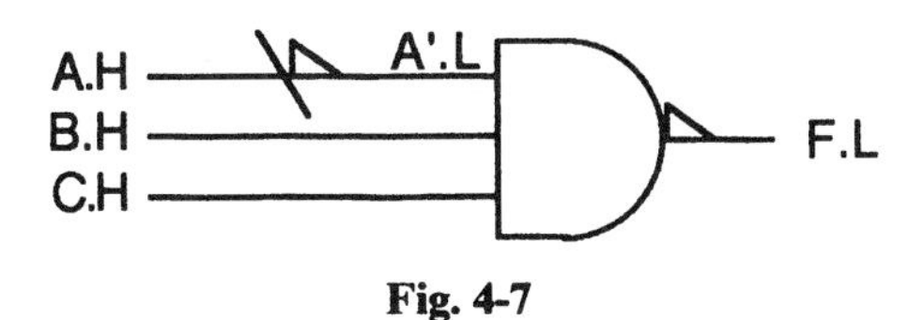

Fig. 4-7

There is a problem with Fig. 4-7 if it is to be used to represent hardware. The gate is supposed to accomplish AND logic. When implemented in NAND hardware, it will produce a LOW output (interpreted as TRUE) when provided with HIGH inputs. The B and C inputs, designated HT, satisfy input requirements, but the third gate input cannot be HIGH and interpreted as A′ at the same time since the slash mark requires A′ to be LT. This problem can be readily identified when inspecting the diagram by noting that the line segment between the gate input and the slash mark contains an unbalanced half arrow; i.e., there is a half arrow on one end of the segment and no half arrow on the other. This indicates that even though the logic is correct, the truth value of the input is wrong.

The Voltage Inverter

The truth value problem may be solved by postulating a *hardware component* called a *voltage inverter* which *does no logic.* Its sole function is to produce a LOW output when its input is HIGH and a HIGH output when its input is LOW thereby *converting a variable from LT to HT and vice versa.* Mixed-logic symbols used to represent this component are shown in Fig. 4-8 where, as in the case of the slash mark, there is an associated half arrow.

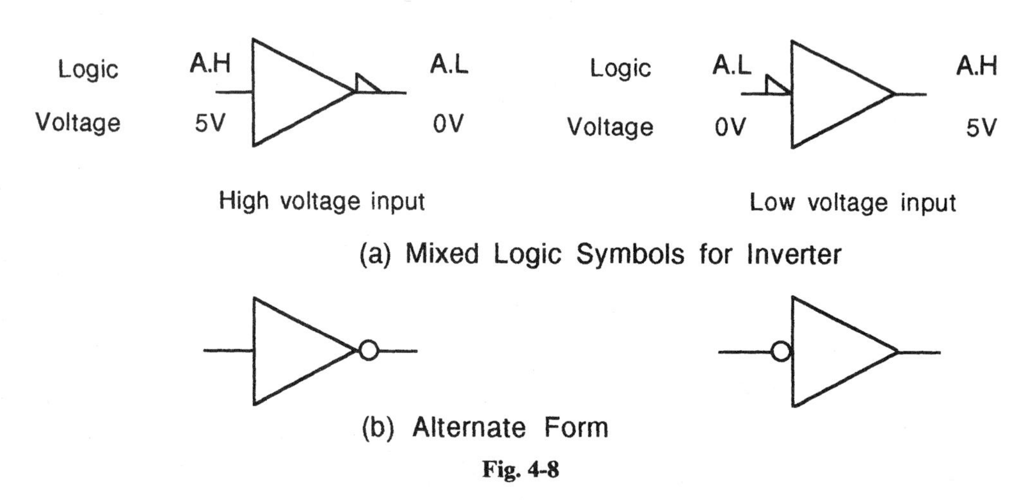

Fig. 4-8

As pointed out previously, some designers prefer the bubble to represent LT as shown in Figs 4-4 and 4-8*b*. Since there is no general agreement on this convention, we will use bubbles to indicate LT (rather than logic inversion) in some problems and half arrows in others. We have taken great care to avoid using the bubble to indicate logical inversion unless it is clearly stated that a positive logic convention is being used.

In the mixed-logic convention, we separate hardware and logic functions so that the operation which changes A.H to A′.H or A.L to A′.L is schematically represented by two entities: the slash mark (logic) indicating logical inversion and the triangle (hardware) indicating the required LT–HT conversion. The applicable mixed-logic symbols are shown in Fig. 4-9.

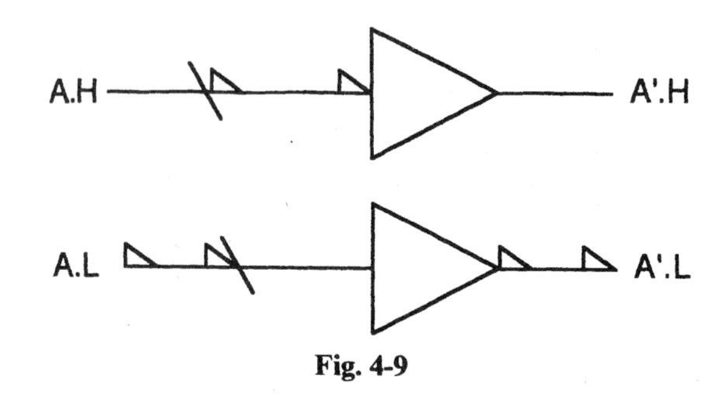

Fig. 4-9

Note that in addition to the half arrows associated with the slash and inverter, additional half arrows have been added to lines identified with LT variables. This results in a pairing of half arrows on each line segment where they appear. It is an important rule that in a properly constructed mixed-logic diagram, *half arrows will always be balanced on any given line segment.*

Either NAND or NOR hardware can be used to implement voltage inverters as shown in Fig. 4-10. Each of the circuits shown meets the requirements of a voltage inverter since it has HT inputs, an LT output, and performs no logic. The latter characteristic may be demonstrated by applying the Boolean identities $A \cdot A = A$, $A \cdot 1 = A$, $(A + A) = A$, and $(A + 0) = A$.

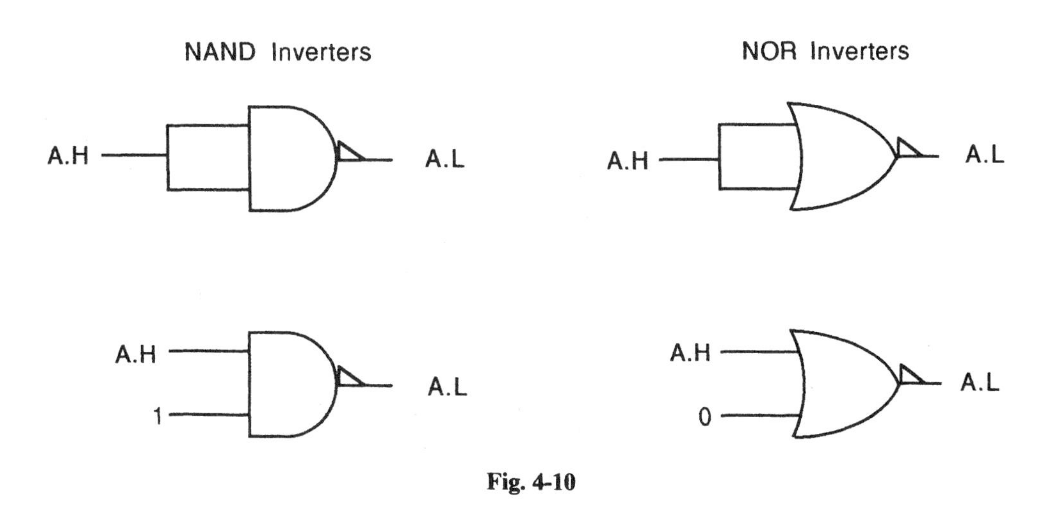

Fig. 4-10

EXAMPLE 4.4 Use of inverters. Suppose that it is desired to implement the logic function $F = A'BC + (BCD)'$ where the variables A, B, C, and D are all available in HT form only and F is to be HT. NAND hardware as described earlier in Sec. 4.3 is to be used.

Step 1. Construct the logic diagram using slashes where applicable (see Fig. 4-11).

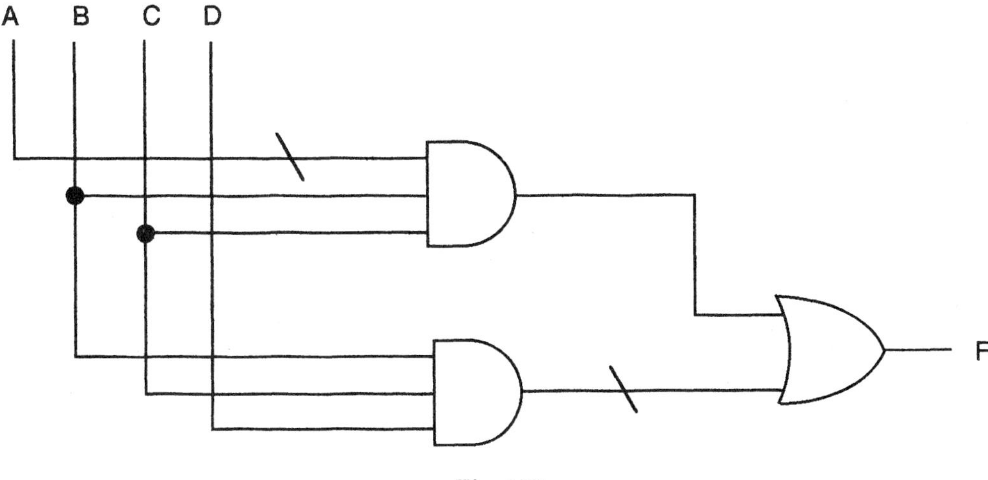

Fig. 4-11

Step 2. Insert half arrows at inputs and/or outputs appropriate to the selected hardware. Recall that there is an implicit half arrow associated with each slash (see Fig. 4-12).

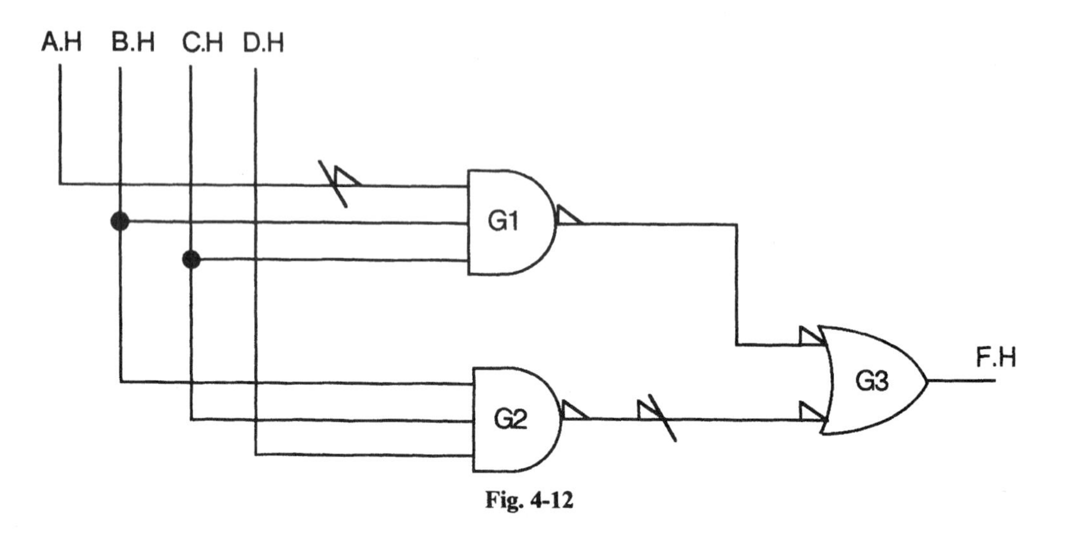

Fig. 4-12

Step 3. Identify the need for voltage inverters by noting half-arrow imbalances on any connecting line segments. There are two such cases in this example: the A′ inversion on the input line to G_1 and the segment between the G_3 input and the slash. These imbalances are resolved by adding voltage inverters G_4 and G_5 as shown in Fig. 4-13. Note that as far as the final result is concerned, the half arrows associated with slashes could have been placed on either side; there would be the need for a hardware inverter in either case. In general, it is good practice to maintain the association of half arrows with LT variables.

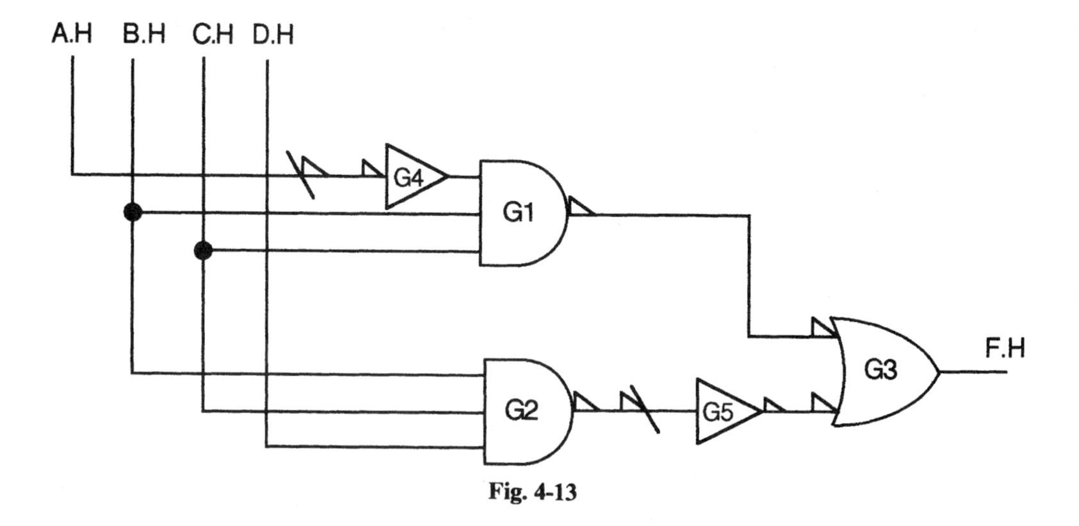

Fig. 4-13

Step 4. Convert the mixed-logic diagram into a wiring diagram. *Remove the slash marks* since they represent logic operations only and are not hardware elements. As shown in App. B, the pins on the dual-in-line package (DIP) are connected to the inputs and outputs of hardware logic elements. Gates G_1, G_2, and G_3 are selected as elements in the 74HC10 triple three-input NAND chip, while G_4 and G_5 are two of the six inverters on the 74HC04. When pin numbers are added to the drawing, the circuit is ready to be wired (see Fig. 4-14). Note that G_3 is a three-input NAND gate which has been converted to two inputs by connecting pins 9 and 10 together.

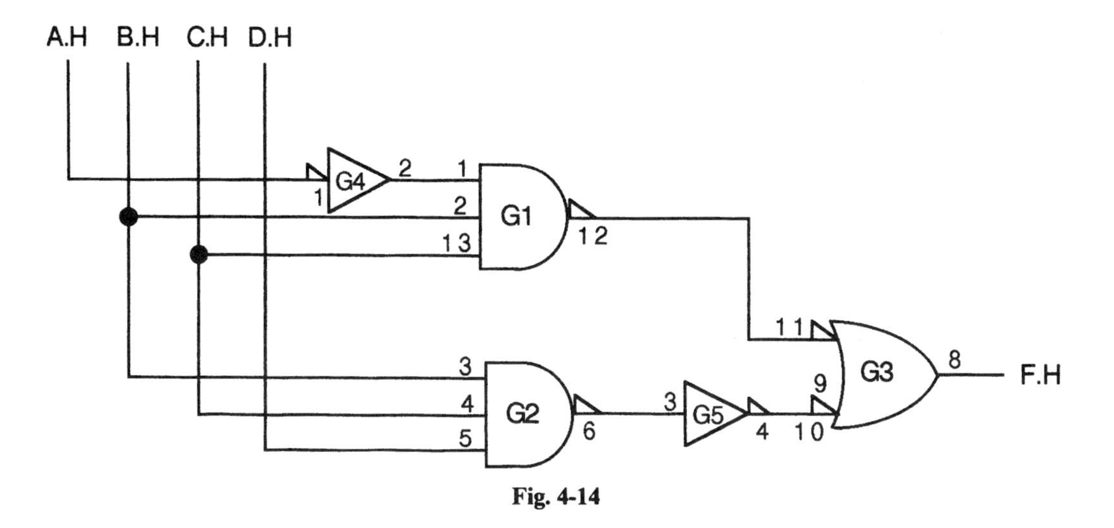

Fig. 4-14

EXAMPLE 4.5 Implement the function F = E + A'B(C + D)' in NAND hardware. The variables A, B, D, and E are to be HT, while C and F are to be LT.

Step 1. Draw a logic diagram using the slash convention (see Fig. 4-15).

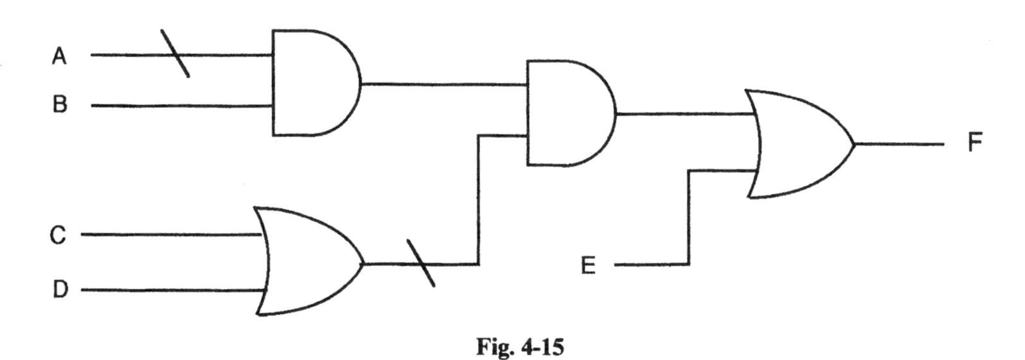

Fig. 4-15

Step 2. Add appropriate half arrows associated with logical inversions, the selected hardware, and any LT inputs and outputs, which will be redundantly labeled. (See Fig. 4-16.)

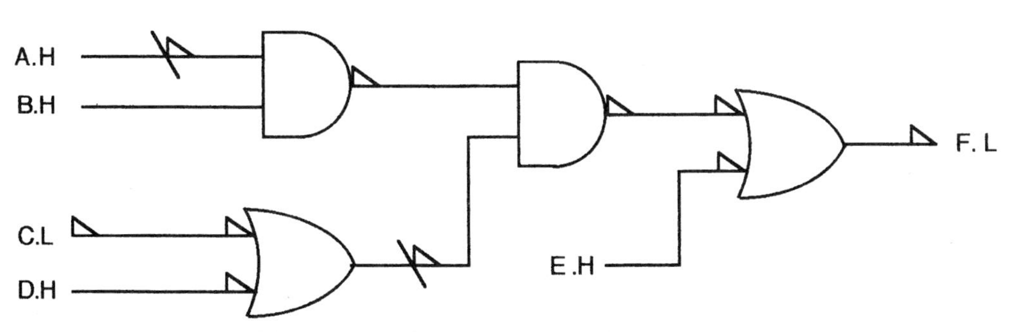

Fig. 4-16 Logic diagram with half arrows added.

Step 3. Balance the half arrows with voltage inverters (see Fig. 4-17).

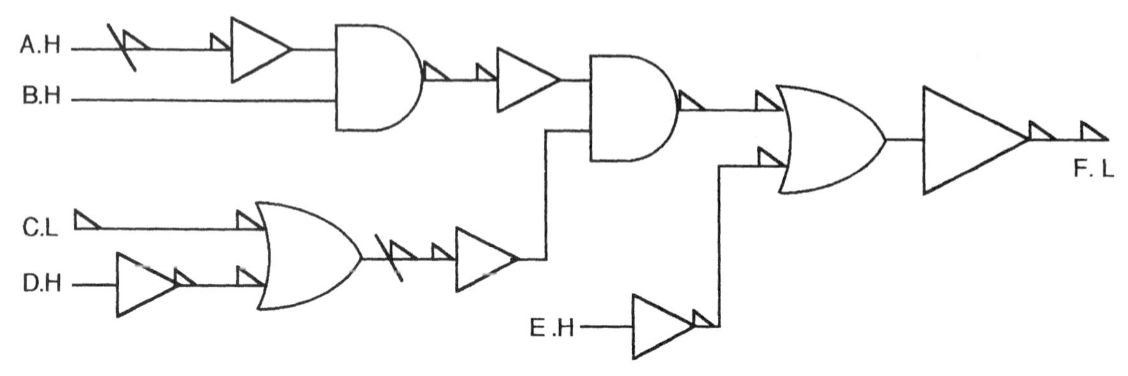

Fig. 4-17 Logic diagram with voltage inverters added.

Step 4. Convert to a wiring diagram (see Fig. 4-18).

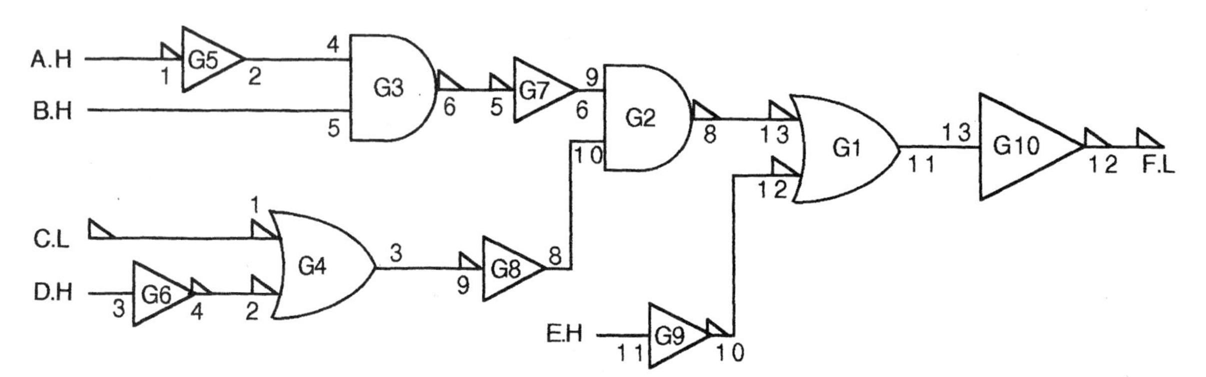

Fig. 4-18 Wiring diagram with slashes removed.

G_1 through G_4 are taken from the same 7400 quad two-input NAND chip; G_5 through G_{10} are taken from a single 7404 hex inverter.

It is important to note that G_6, G_7, G_9, and G_{10} illustrate the *use of voltage inverters where there is no corresponding logic inversion.*

Implementation Using Exclusive ORs

When hardware components other than standard gates are used, the logic designer must ascertain from hardware vendors what the truth values associated with the inputs and outputs are. The designer will then draw the half arrows as required and proceed as above. Common exclusive-OR hardware generally has HT inputs and output whereas exclusive-NOR devices will most likely have HT inputs and a LT output.

4.4 MIXED LOGIC AS A DESCRIPTIVE CONVENTION

One of the advantages of the mixed-logic convention is that it is easy to determine what logic is being accomplished by a circuit directly from the wiring diagram. This is particularly useful if one is trying to determine what some other designer had in mind before he or she left the company! *To recover the underlying Boolean expression from a mixed-logic diagram, simply remove all half arrows and inverters and then read the logic directly from the diagram.* The process may be easily traced by observing Figs. 4-15 through 4-17 in reverse order.

Not all designers use the technique described in Sec. 4.3 though they often make use of mixed-logic concepts in the description of logic circuits. In such cases, only circuitry which exists in hardware is shown in the diagram, thus eliminating logical inversion as a separate symbol.

EXAMPLE 4.6 Consider the wiring diagram of Example 4.4 (Fig. 4-14) in which the slashes have been eliminated. The underlying Boolean logic may be determined by assuming a logical inversion takes place on any line segment with unbalanced half arrows. Thus, referring to Fig. 4-19, we see that logical inversion takes place on the line segment between A and G_4 and between G_2 and G_5, whereas no logical inversion takes place between G_1 and G_3 or G_5 and G_3. The point to bear in mind is that in the mixed-logic convention, *an unbalanced set of half arrows may be interpreted as a logical inversion.*

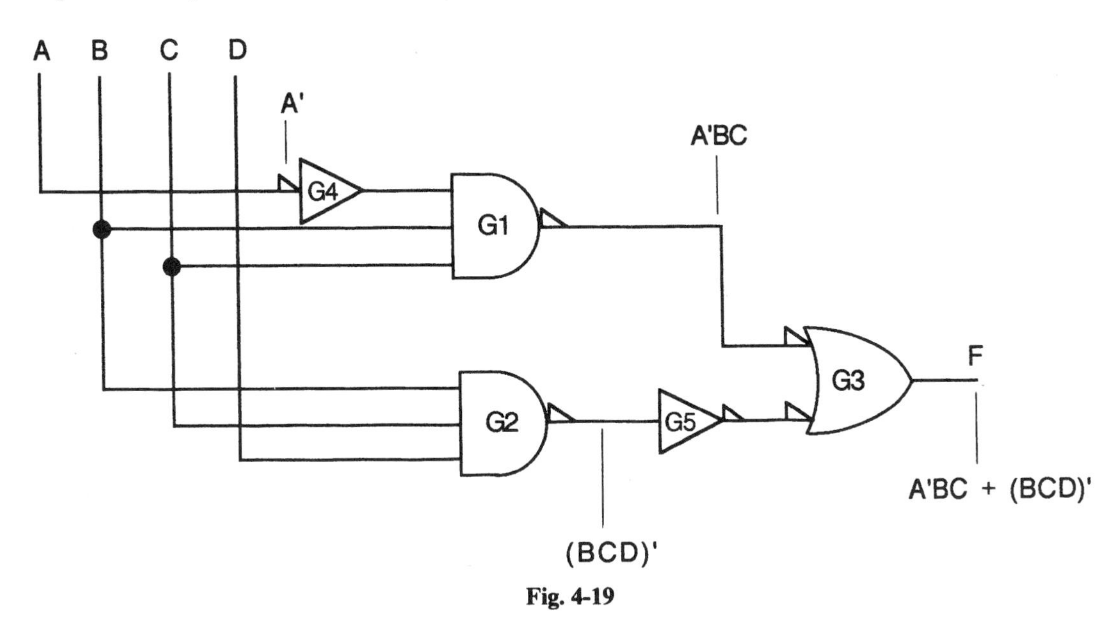

Fig. 4-19

EXAMPLE 4.7 The *positive logic* convention depicts hardware only, assumes that all signals are HT, and uses bubbles to indicate voltage inversion. In this scheme, which is quite common, NAND hardware is depicted by a single symbol, as shown in Fig. 4-1, irrespective of its logical purpose. Applying this alternate convention to the circuit of Fig. 4-19, we obtain the result shown in Fig. 4-20. The expected expression for function F can be obtained by applying De Morgan's theorem.

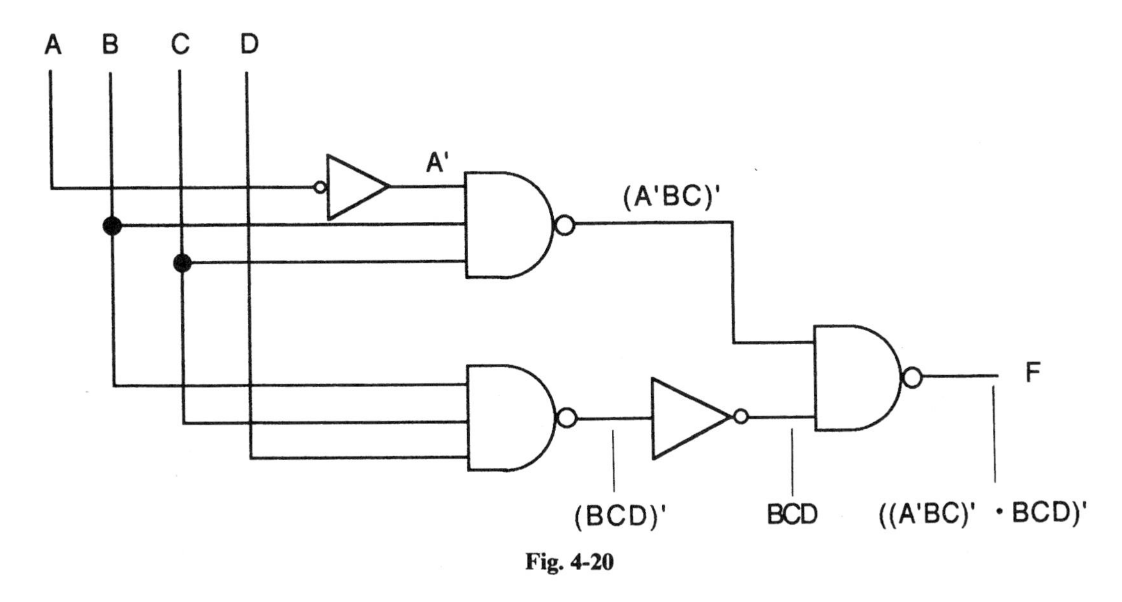

Fig. 4-20

Note that *in the positive logic convention, all voltage inverters perform logical inversion.* Because it required pen-and-paper Boolean manipulations to obtain OR functions, it should be clear that the positive logic hardware-oriented convention shown in Example 4.7 does not lend itself particularly well to identifying underlying logic.

It is important to note that there are several logic conventions currently in use, and the student should take care to understand the particular one being used in any document. The time involved is well spent in terms of error prevention.

4.5 USES OF MIXED LOGIC IN TROUBLESHOOTING

The mixed-logic convention, in addition to helping determine the placement of voltage inverters in a physical logic implementation, also permits preservation of the logic diagram throughout the design process. The half arrow is particularly valuable because it helps to determine the voltage associated with TRUE at a particular point in the circuit. This is important information since troubleshooting is generally done in terms of voltages rather than logic.

EXAMPLE 4.8

(*a*) In Fig. 4-21, input *logic* signals W, X, and Y vary with time as shown. Sketch the *logic* outputs at the points A, B, and F as functions of time.

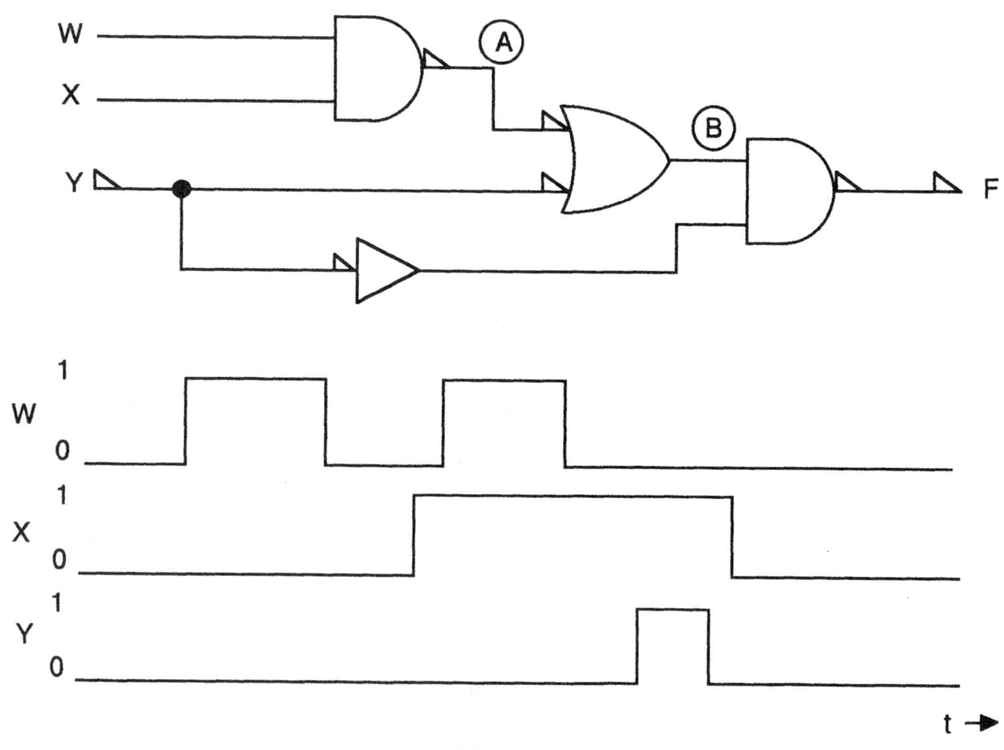

Fig. 4-21

Stripping away half arrows and the inverter (which doesn't perform any logic), we see that A = WX, B = WX + Y, and F = Y(WX + Y). A bit of Boolean algebra tells us that the function F may be reduced to F = Y which, the astute student will realize, can be accomplished in hardware by a single piece of wire! The reader should appreciate such circuits for their illustrative value alone and not grumble too loudly about academics and their isolation from "the real world." The required waveforms, aligned with the inputs for reference, are shown in Fig. 4-22.

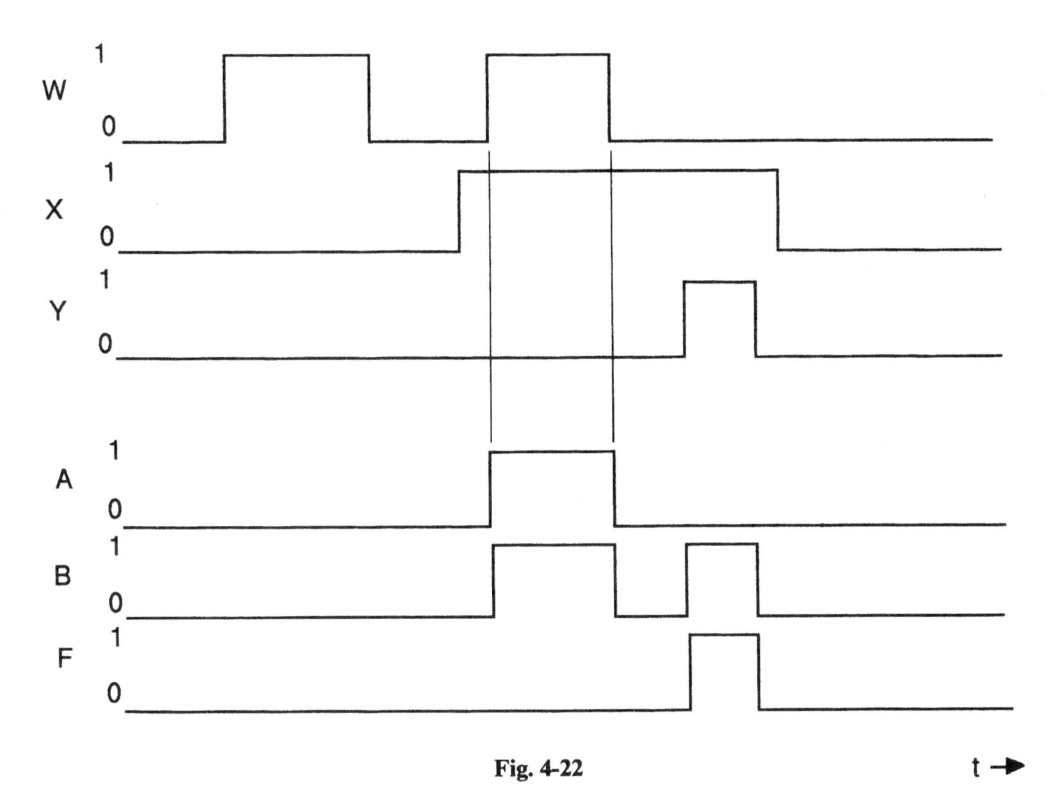

Fig. 4-22

(*b*) Sketch the *voltage* waveforms that would be observed at the inputs and at points A, B, and F if a logic analyzer, which treats all signs as HIGH-TRUE (HT), were used.

Since Y, A, and F are LT, the voltage waveshapes at these points will be inverted relative to their logic equivalents (see Fig. 4-23).

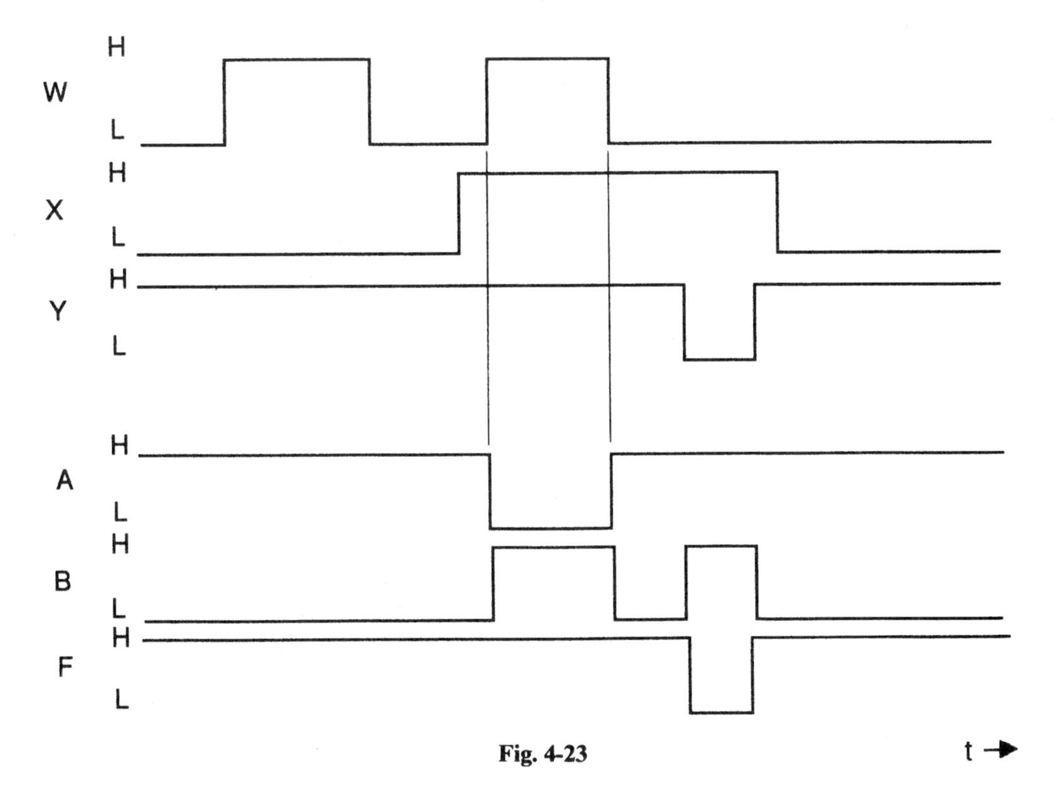

Fig. 4-23

Solved Problems

4.1 NOR hardware is defined as producing a HIGH output if and only if all of the inputs are LOW. Draw the voltage I/O table for the three-input case. Also, create appropriate truth tables and symbols (using the half-arrow convention) for the device used as an AND gate and as an OR gate.

The NOR I/O table is shown in Fig. 4-24. Mapping LT inputs and an HT output yields the logic truth table shown in Fig. 4-25 which is seen to meet the definition of AND. Mapping with HT inputs and the LT output yields the truth table corresponding to an OR (see Fig. 4-26).

INPUTS A B C	OUT F
L L L	H
L L H	L
L H L	L
L H H	L
H L L	L
H L H	L
H H L	L
H H H	L

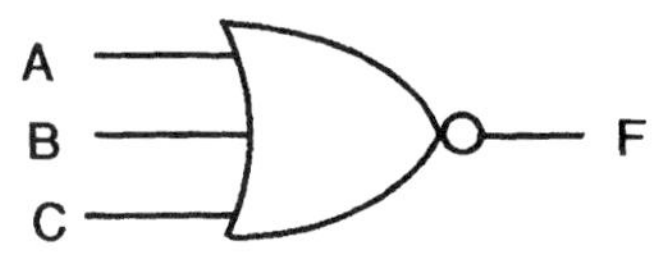

NOR HARDWARE SYMBOL

Fig. 4-24

INPUTS A B C	OUT F
1 1 1	1
1 1 0	0
1 0 1	0
1 1 0	0
0 1 1	0
0 1 0	0
0 0 1	0
0 0 0	0

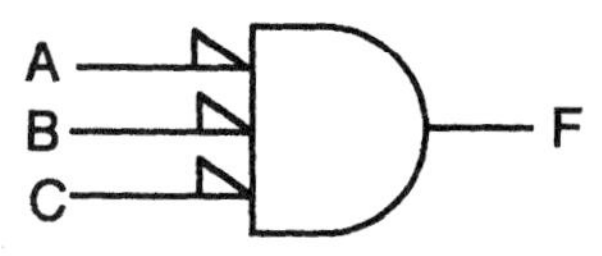

NOR HARDWARE USED AS AND

Fig. 4-25

INPUTS A B C	OUT F
0 0 0	0
0 0 1	1
0 1 0	1
0 1 1	1
1 0 0	1
1 0 1	1
1 1 0	1
1 1 1	1

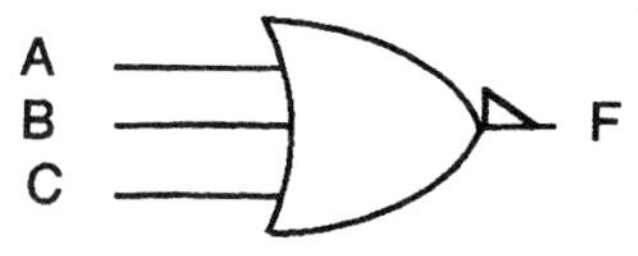

NOR HARDWARE USED AS OR

Fig. 4-26

4.2 The behavior of a digital hardware element is defined by the voltage chart shown in Table 4.3. Show how the truth values may be assigned for the circuit to behave as a logical AND, a logical OR, and as a voltage inverter.

Table 4.3

INPUTS A	INPUTS B	OUTPUT G
L	L	L
L	H	L
H	L	H
H	H	L

For the AND function, there is only one case where the output is TRUE, and thus row 3 must be selected. Function G will obviously be HT. It follows that we must assign LT to input B and HT to input A, creating the AND truth table shown in Table 4.4. Referring to Table 4.3, we see that for the OR function, G and A must be LT, while B is HT. Lastly, if A is held HIGH, then G will be the opposite voltage value of B, meeting the requirements for an inverter. (See Table 4.5.)

Table 4.4

INPUTS A	INPUTS B	OUTPUT G
F	T	F
F	F	F
T	T	T
T	F	F

Table 4.5

INPUTS A	INPUTS B	OUTPUT G
H	L	H
H	H	L

4.3 Three-input NOR gates are packaged three to a chip. Combine these gates as efficiently as possible to create an equivalent five-input OR function. Use the most convenient truth values for inputs and outputs.

Two three-input ORs can be combined into an equivalent five-input OR (Fig. 4-27). NOR implementation is obtained as shown in Fig. 4-28. The inverter required to balance the half arrows between the two gates can be obtained by tying the inputs of a NOR gate together as shown in Fig. 4-29. One chip is required; inputs are HT and outputs are LT.

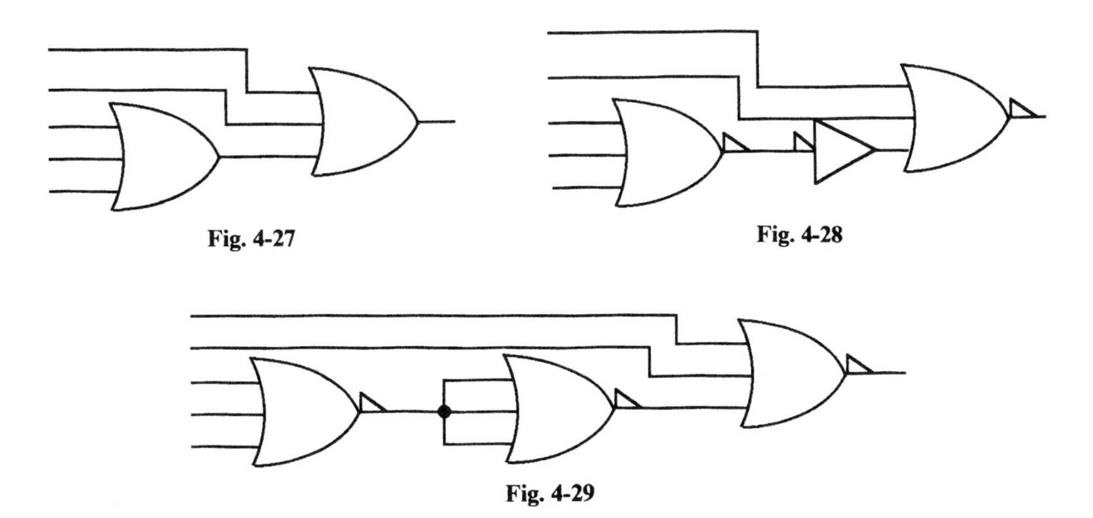

Fig. 4-27

Fig. 4-28

Fig. 4-29

4.4 Implement the logic function of Example 4.4 in NOR hardware.

A pure logic diagram is hardware independent. A hardware-specific mixed-logic diagram, however, incorporates half arrows to indicate appropriate truth values as shown in Fig. 4-30. In order to complete the implementation, we balance half arrows to determine the positions of any required voltage inverters (see Fig. 4-31). Note that the NOR implementation requires three more inverters than does the NAND implementation of Example 4.4.

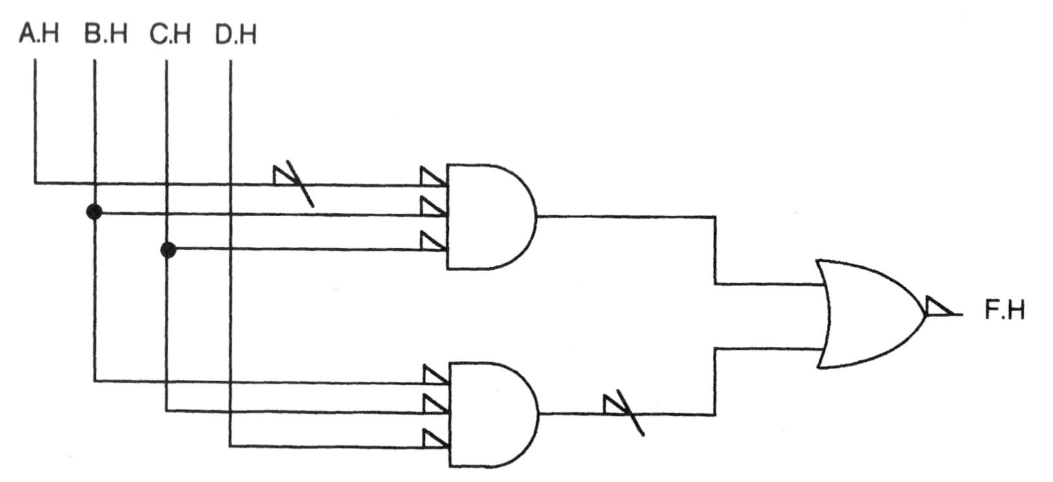

Fig. 4-30 Initial step in the creation of a NOR mixed-logic diagram.

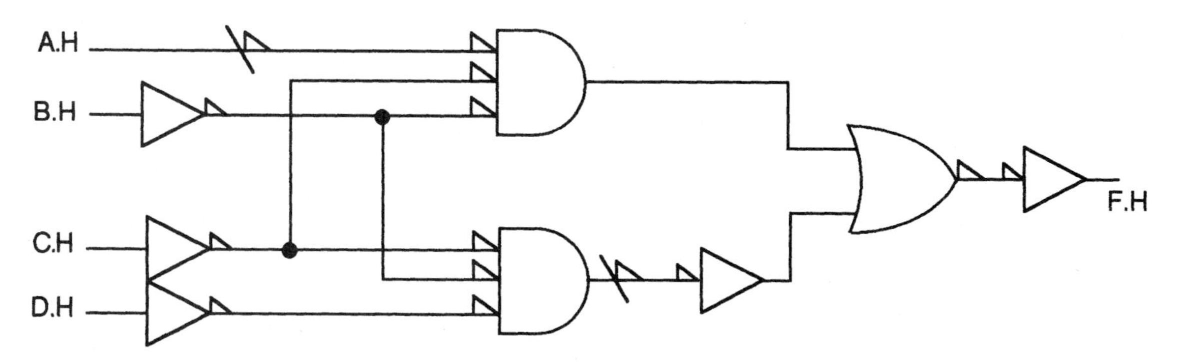

Fig. 4-31 Final NOR mixed-logic diagram.

4.5 Implement the full adder of Prob. 2.13 using NAND hardware.

See Fig. 4-32.

4.6 Implement the full adder of Prob. 2.15 using NAND hardware and making use of exclusive-OR hardware where applicable. Assume that A, B, C_i, S, and C_o are all HT.

Refer to Fig. 2-13 and add half arrows, as required (see Fig. 4-33). Since all the half arrows are balanced, no voltage inverters are needed.

4.7 Repeat Prob. 4.6 with exclusive NORs replacing the exclusive ORs. Recall that exclusive NORs have an LT output.

See Fig. 4-34. Note the requirement of two additional voltage inverters.

4.8 Repeat Prob. 4.6 with NOR hardware for the ANDs and ORs.

See Fig. 4-35. Note that five voltage inverters are needed in this case.

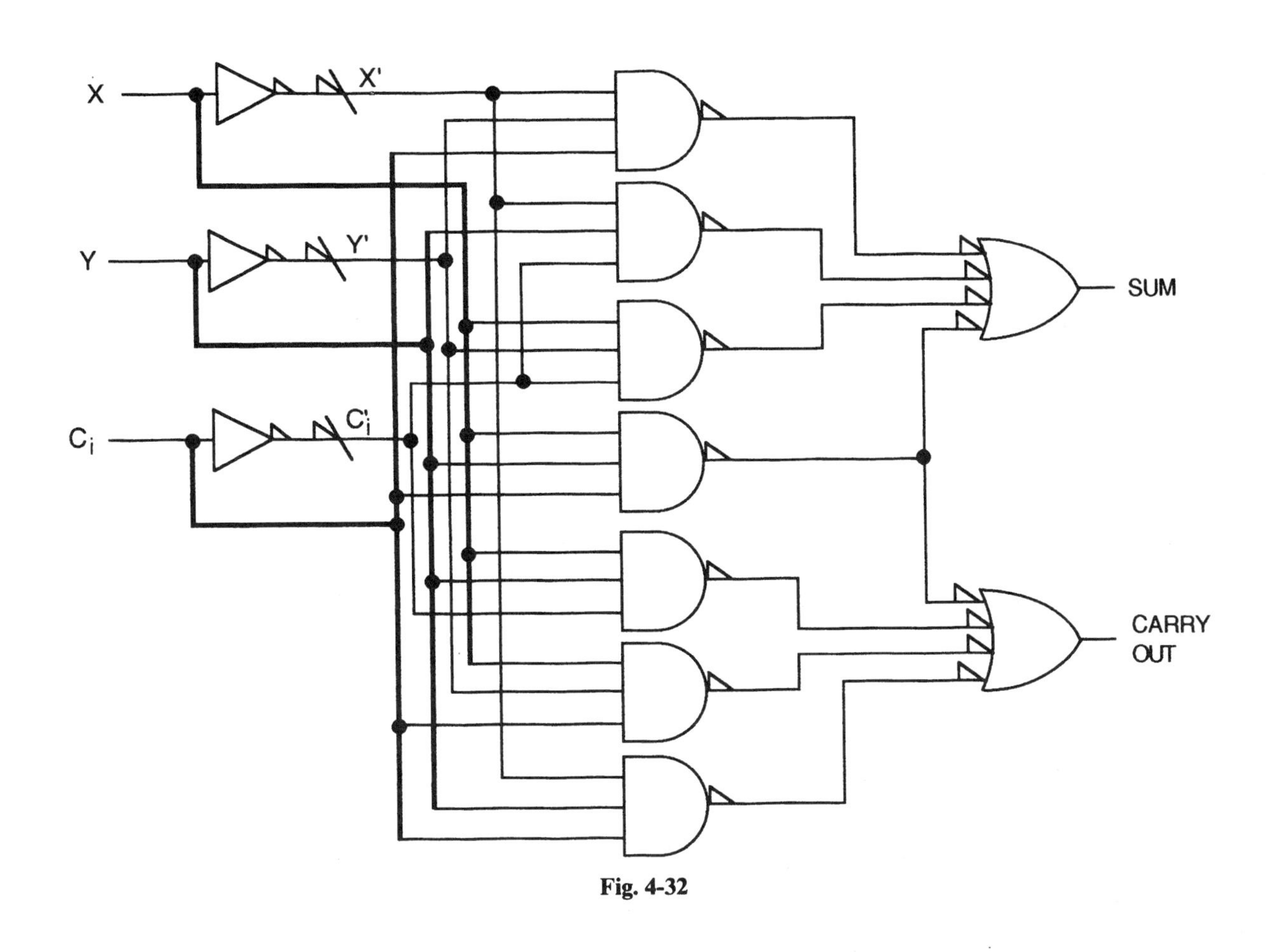

Fig. 4-32

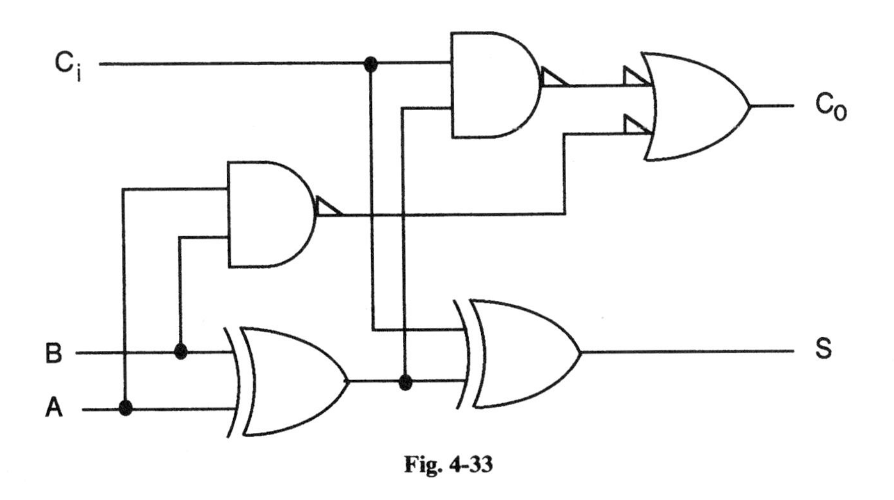

Fig. 4-33

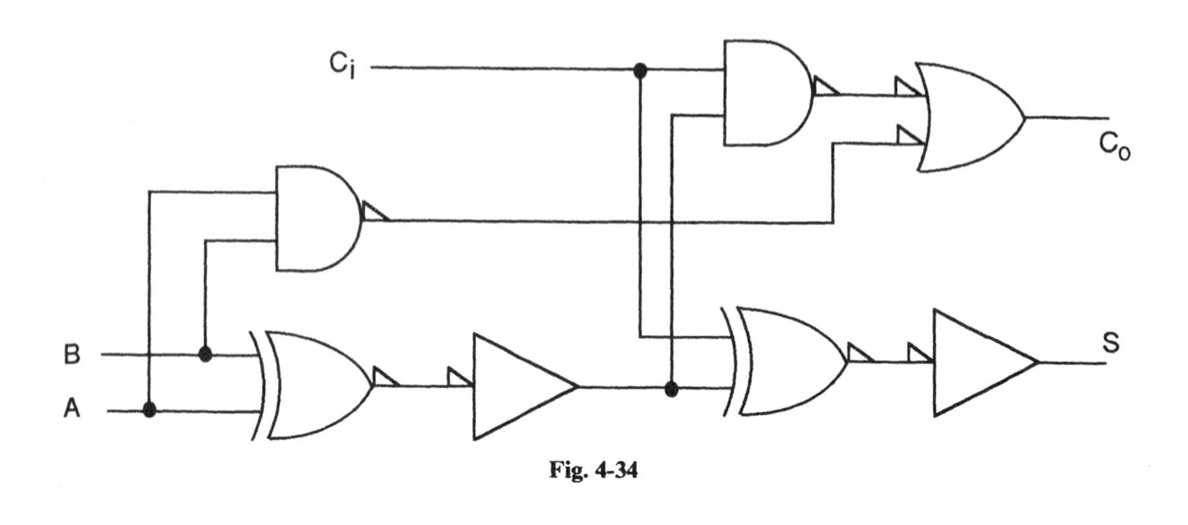

Fig. 4-34

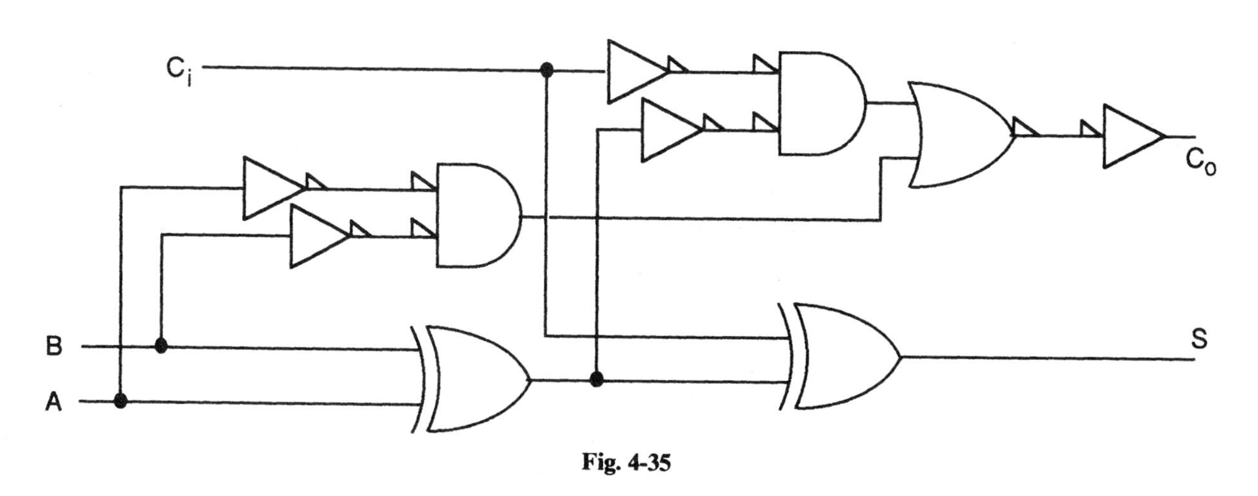

Fig. 4-35

4.9 Repeat Prob. 4.6 with A, B, and C_o all LT, while C_i and S are HT.

See Fig. 4-36.

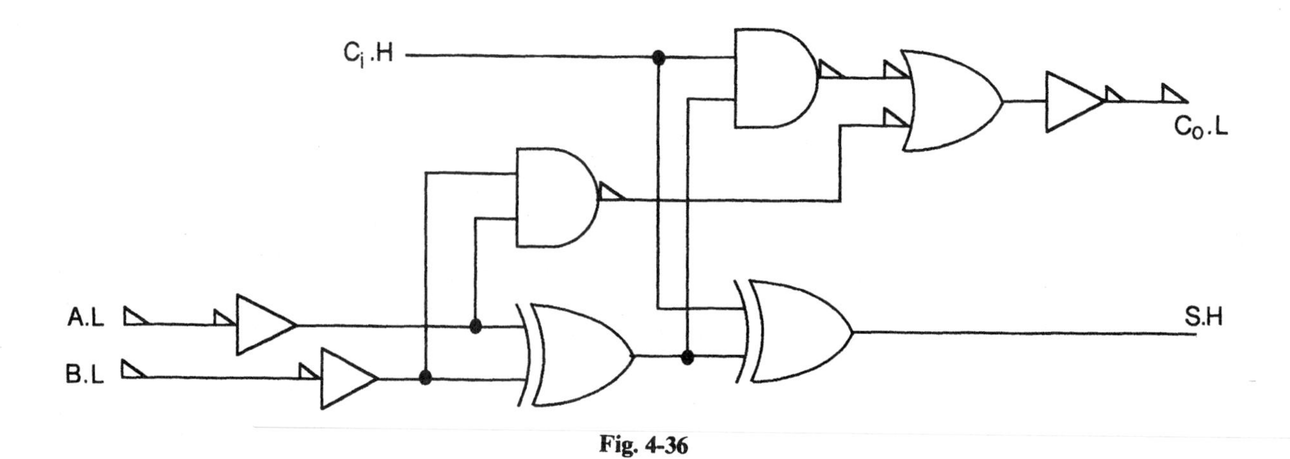

Fig. 4-36

4.10 Given the positive logic hardware diagram shown in Fig. 4-37, draw three different mixed-logic equivalents.

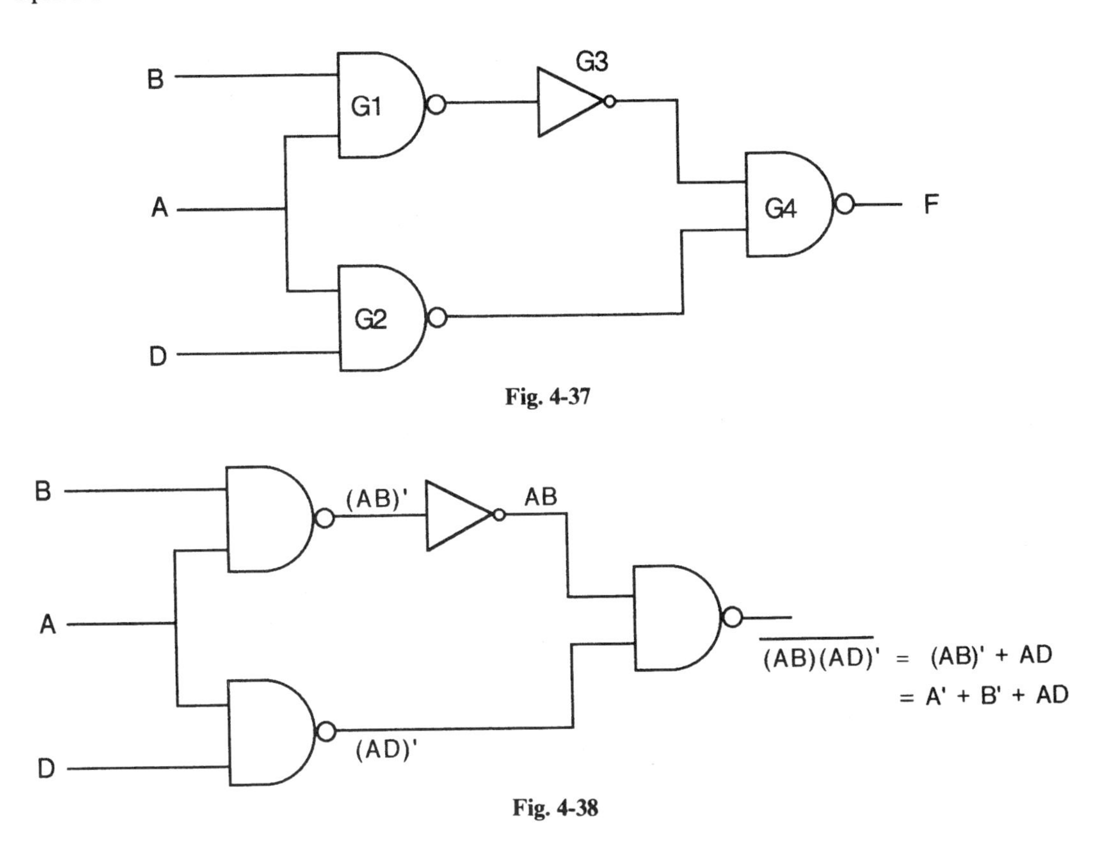

Fig. 4-37

Fig. 4-38

We can, if we wish, determine the underlying logic with the aid of De Morgan's theorem as shown in Fig. 4-38. Though this is not a necessary step, it is handy for comparative purposes and to emphasize how the output may be expressed in more than one way. Recall that all hardware inverters may be replaced, in mixed-logic representations, by the combined slash mark and LT to HT converter:

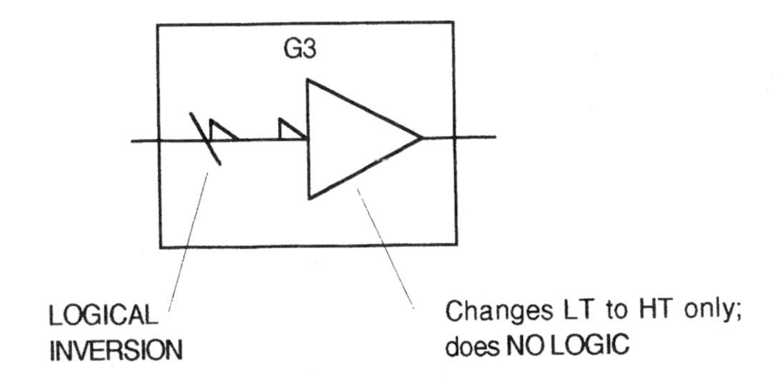

We now have several choices, three of which will be discussed.

1. Assume that all the NAND gates are performing the OR function. This leads to the arrangement shown in Fig. 4-39. Note that the slash mark associated with G_3 adds an extra half arrow between G_1 and G_4, while on the other hand, there is a missing half arrow in the circuit branch directly below. To correct the imbalances, we move this slash, add others at the inputs, and adjust half-arrow positioning as shown in Fig. 4-40. This case clearly illustrates how the dual interpretations of the inversion process are utilized. Inverter G_3 serves only to convert from HT to LT without a logic change, while the solitary

slash mark below it indicates the implicit logical inversion process which occurs between G_2 and G_4. Here, we are converting, in our minds, the variable $G_2.H$ to its equivalent $G_2'.L$, both of which are, in actuality, different interpretations of the same voltage level.

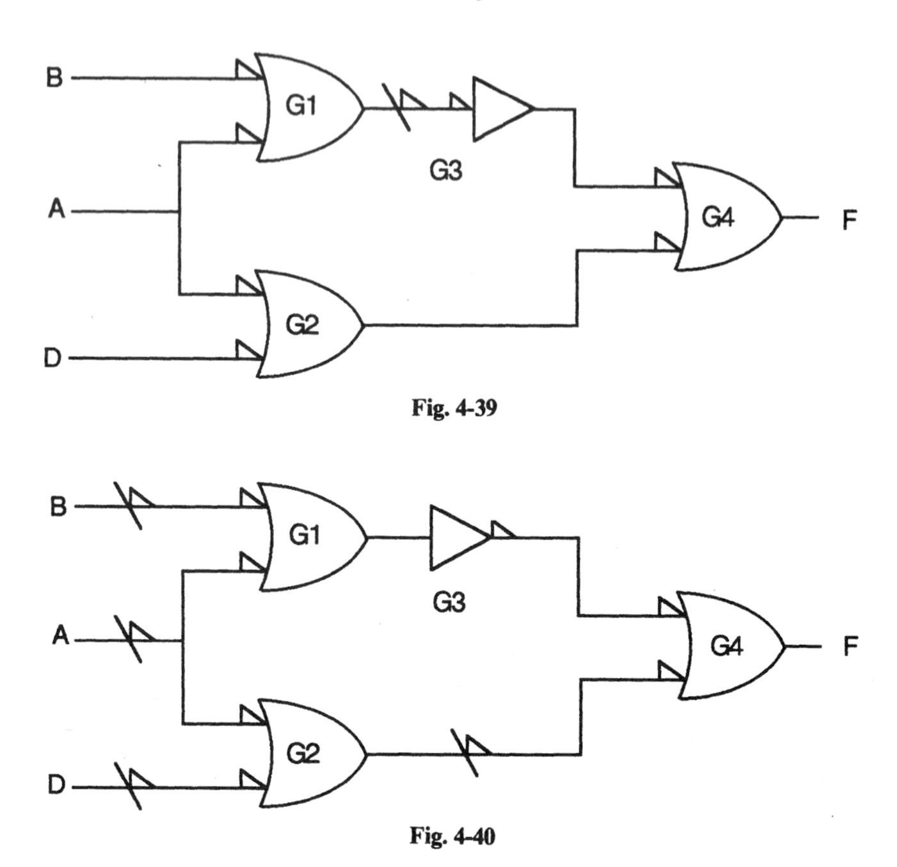

Fig. 4-39

Fig. 4-40

Stripping away half arrows and the inverter uncovers the underlying logic diagram (Fig. 4-41) from which we may read

$$F = (A' + B') + \overline{A' + D'} = A' + B' + AD$$

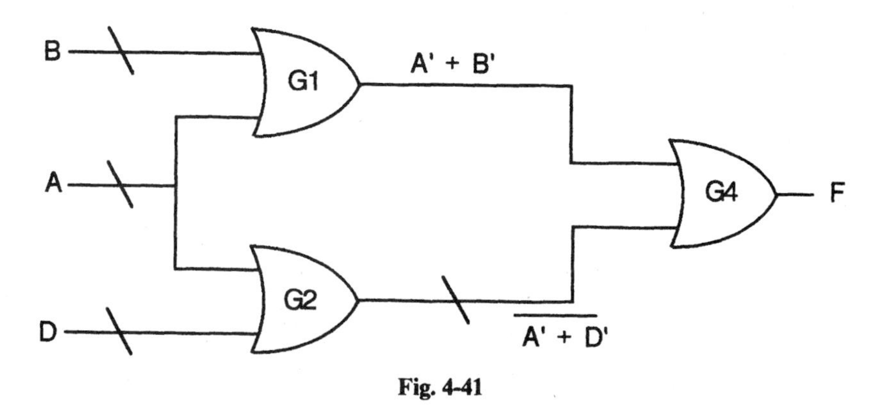

Fig. 4-41

2. Assume that all NAND gates are performing the AND function. This leads to the arrangement shown in Fig. 4-42. We now *add or remove slash marks* as required to balance the half arrows. Note that these elements are logical only and do not alter the basic hardware configuration. Refer to Fig. 4-43.

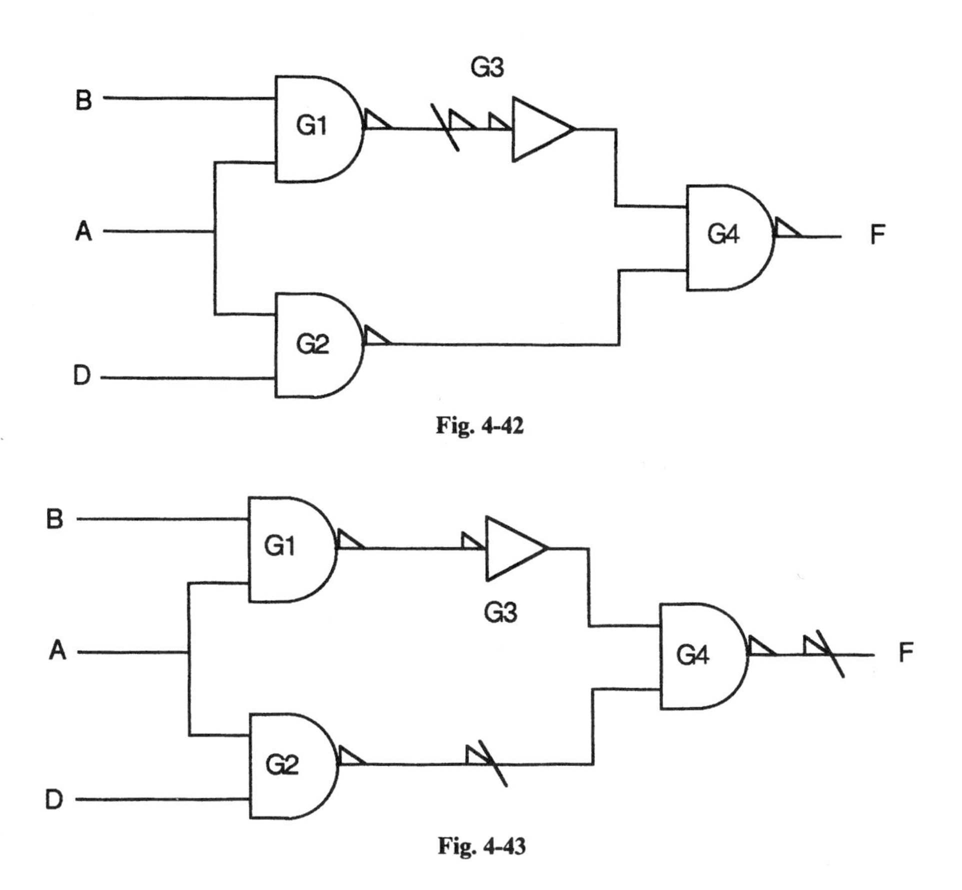

Fig. 4-42

Fig. 4-43

After removal of the inverter and all half arrows, the Boolean expression for the output is seen to be $F = \overline{(AB) \cdot (AD)'} = (AB)' + AD$, which is identical to the previous results.

3. Assume that G_1 and G_2 perform AND functions and G_4 is an OR, as we might have deduced from the result of Fig. 4-38. As can be seen in Fig. 4-44, all half arrows are balanced and no further manipulations are required. This is the simplest transformation of all and yields the expected function by inspection. The ability to perform transformations between positive logic hardware and mixed logic is particularly useful when it is desired to predict waveforms and study the effects of circuit delays (see Chap. 6).

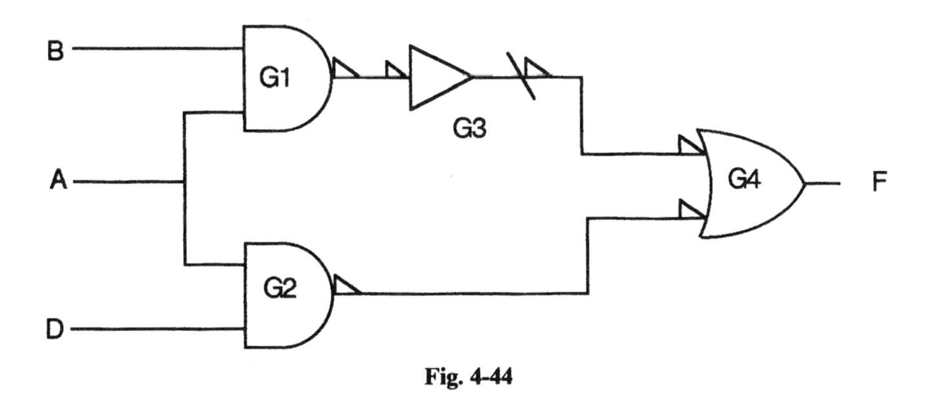

Fig. 4-44

4.11 The relative difficulty of deciphering the underlying logic from a positive logic hardware diagram becomes more evident as the circuit complexity increases. Figure 4-45 shows the positive logic version of the circuit in Example 4.5, Fig. 4-17. Determine the logic implemented.

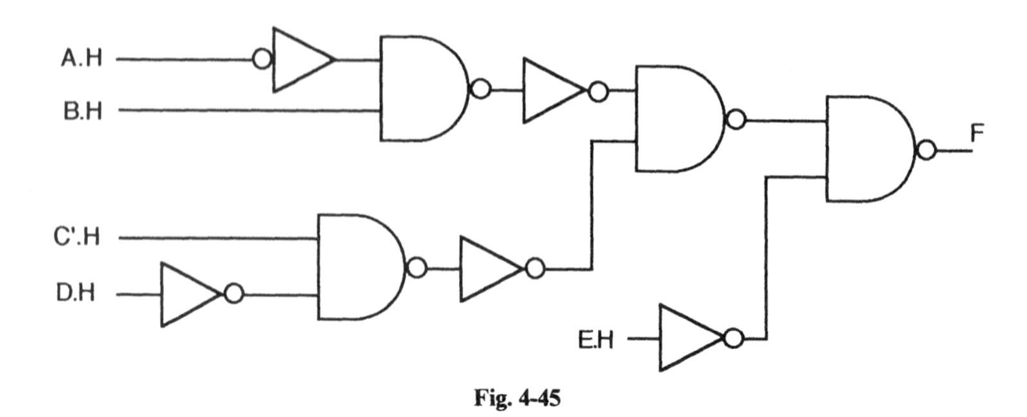

Fig. 4-45

Note that the input variable C.L has been replaced by its equivalent C'.H, and an HT output has been specified because only HT variables are permitted in the positive logic convention. Writing intermediate expressions on the diagram, we obtain the results shown in Fig. 4-46 which may be used to obtain the desired function as follows:

$$F = E'(\overline{\overline{A'B} + \overline{C'D'}}) = E + \overline{A'B} + \overline{C'D'} = E + A'B \cdot C'D' = E + A'B(C + D)'$$

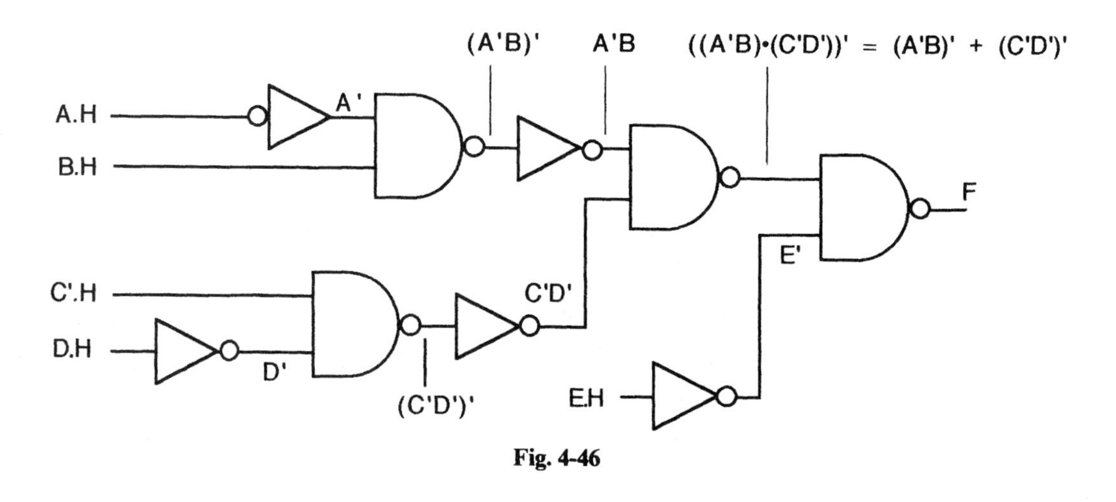

Fig. 4-46

It is instructive to compare the relative ease of deciphering the logic of Fig. 4-15 to the current exercise.

4.12 The logic of Fig. 4-47 has inputs which vary with time as shown in Fig. 4-48. Sketch the output *voltage* waveshapes assuming that the circuits respond instantaneously to input changes. Variables A, B, C, and D are all HT.

After removing inverters and half arrows from the logic diagram, we can easily determine that $F = AB + (C + D)'$ which, after application of De Morgan's theorem to the second term, yields $F = AB + C'D'$. That is, F will be TRUE if either A and B are both TRUE or C and D are both FALSE. Also, from the logic diagram, $G = B'(C + D)'$ or, equivalently, $G = B'C'D'$. Thus, G will be TRUE only if B, C, and D are all FALSE.

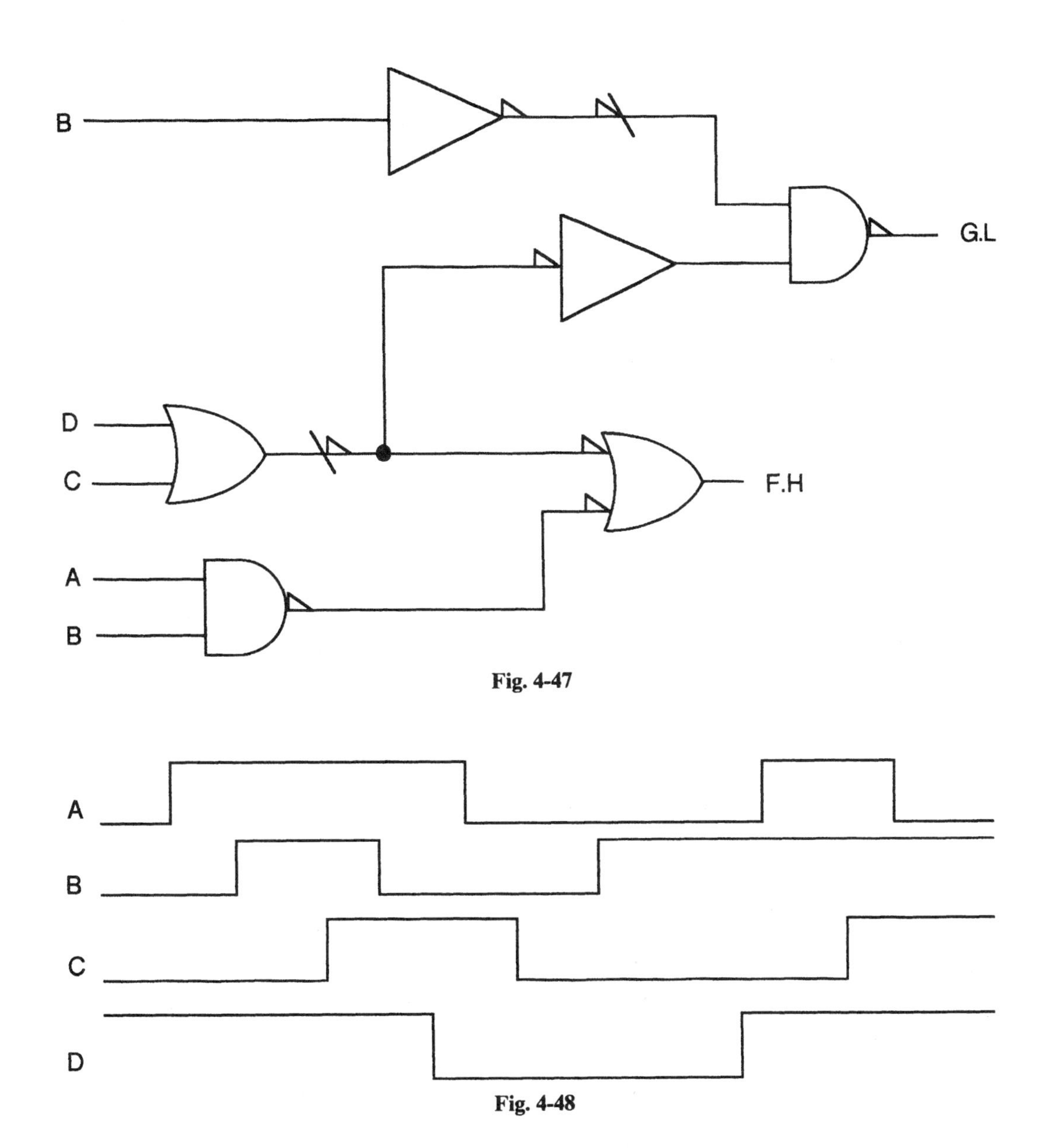

Fig. 4-47

Fig. 4-48

The required output *logic* waveforms are obtained by setting 1 = TRUE, 0 = FALSE, and applying the Boolean equations on a point-by-point basis as shown in Fig. 4-49. Since G is specified as LT, the G waveform must be inverted to obtain the desired *voltage* output.

4.13 It is desired to design a circuit which will produce a TRUE output (F) whenever any two or more of its three inputs A, B, C are TRUE. Assume that A and C are HT, while B and F are LT. See Fig. 4-50.

(*a*) Draw a K map and select the best 1s covering.

(*b*) Implement the design of part (*a*) using NOR hardware.

(*a*) Sum the 1s in the coordinates of each map square and place a 1 in any square whose coordinates add to 2 or more. See Fig. 4-51.

(*b*) Adding half arrows and the necessary voltage inverters we get Fig. 4-52.

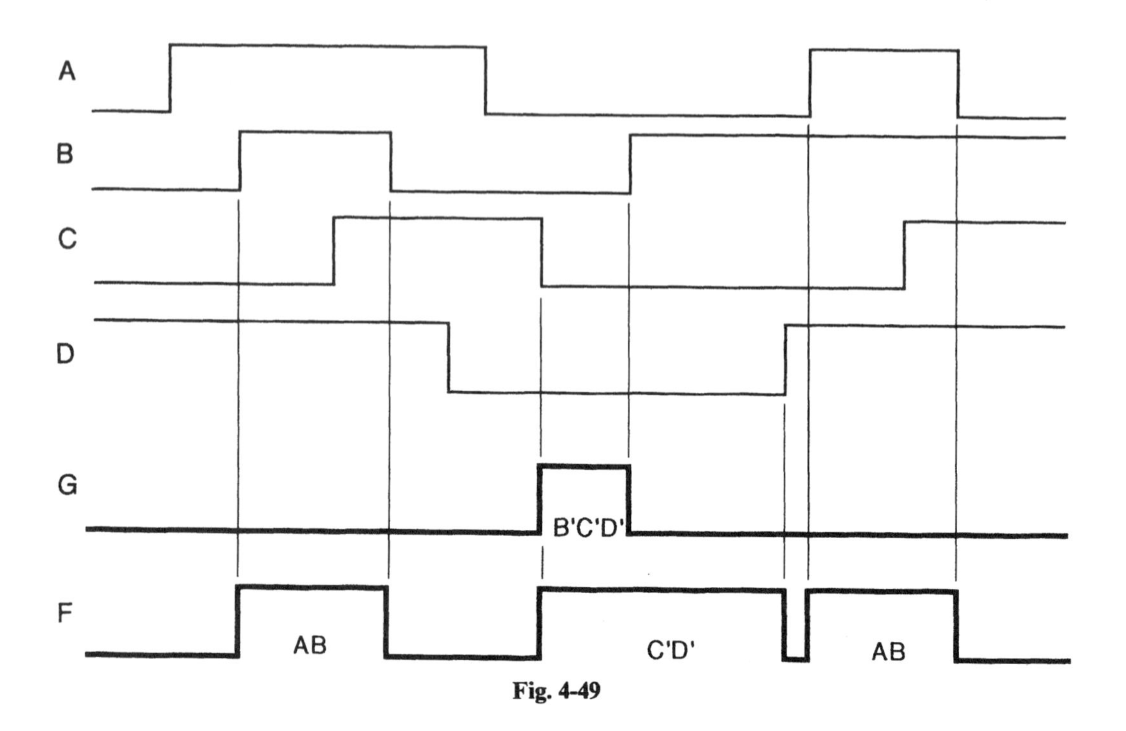

Fig. 4-49

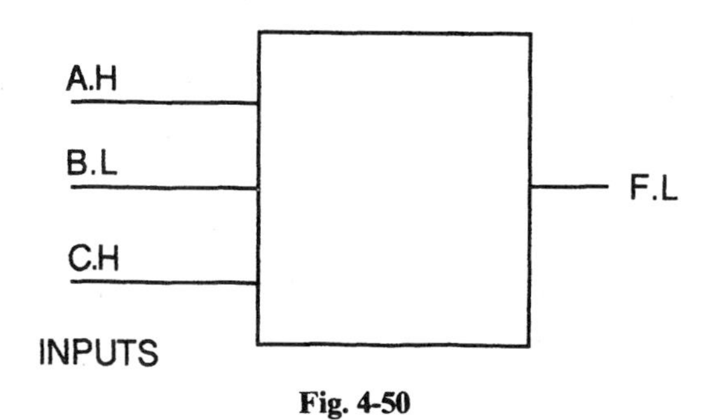

Fig. 4-50

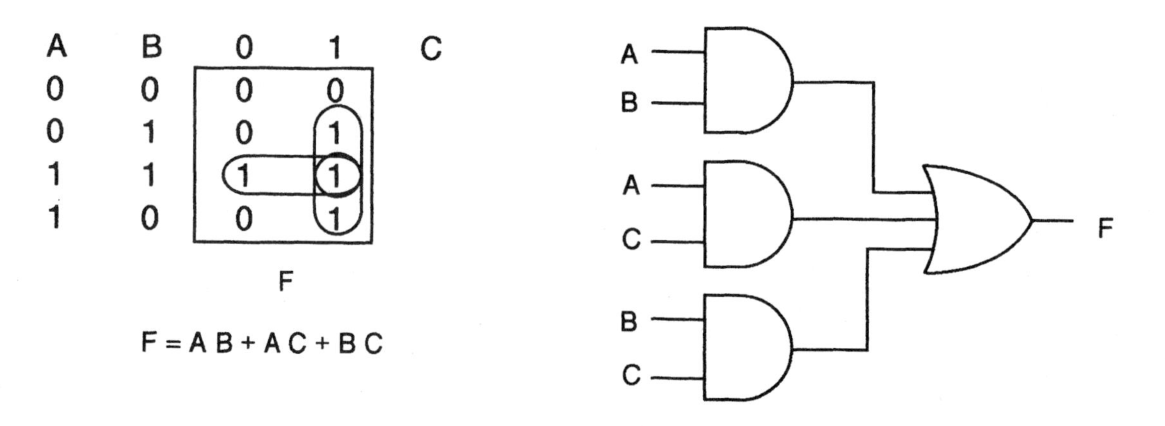

Fig. 4-51

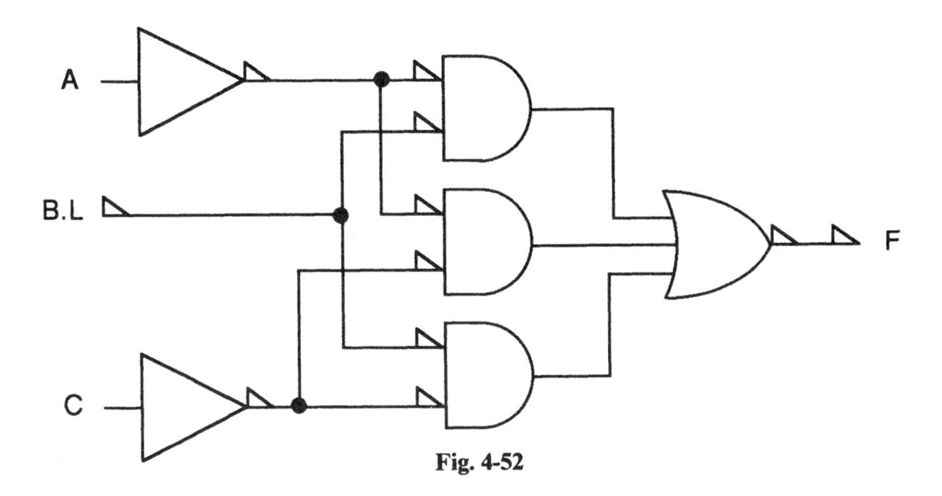

Fig. 4-52

4.14 Repeat Prob. 4.13 with

(*a*) A 1s covering NAND implementation.

(*b*) A 0s covering NAND implementation.

(*a*) The logic diagram is the same as the one shown in Fig. 4.51; its NAND implementation is shown in Fig. 4-53.

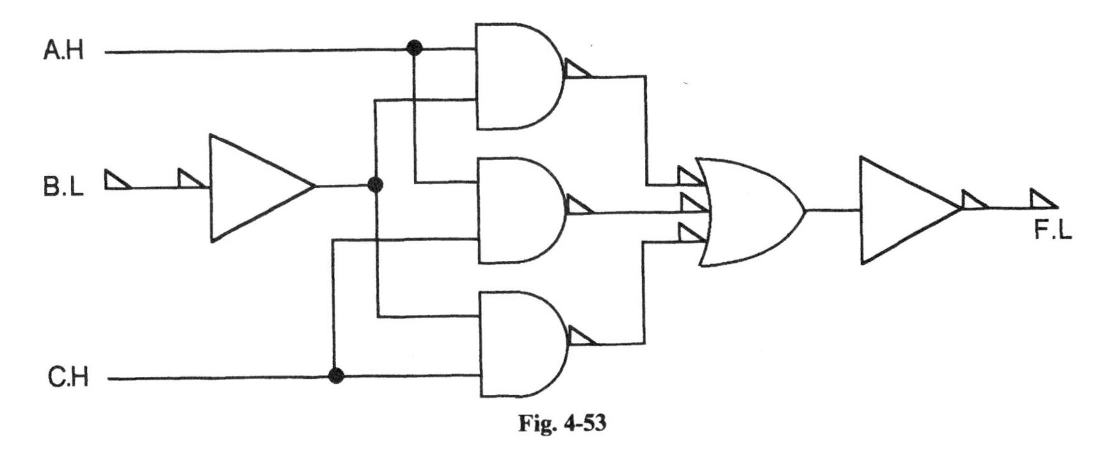

Fig. 4-53

(*b*) The simplest expression obtainable from a 0s covering is $F' = A'B' + A'C' + B'C'$. The corresponding logic diagram is shown in Fig. 4-54 and its NAND implementation in Fig. 4-55. Notice,

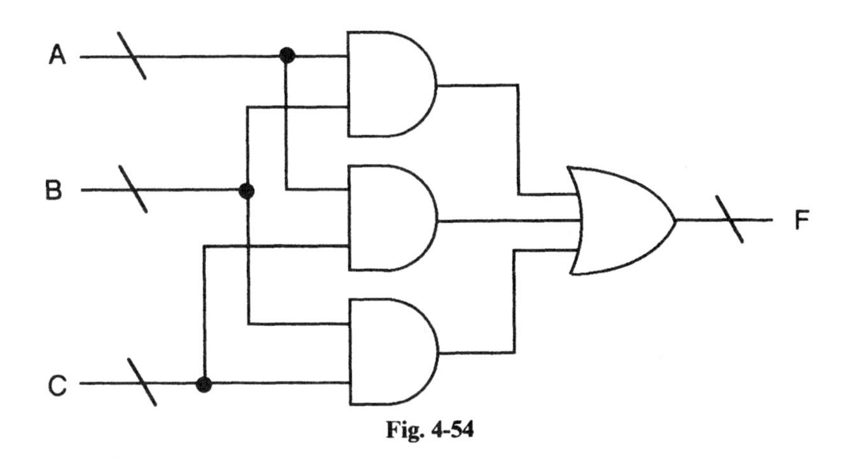

Fig. 4-54

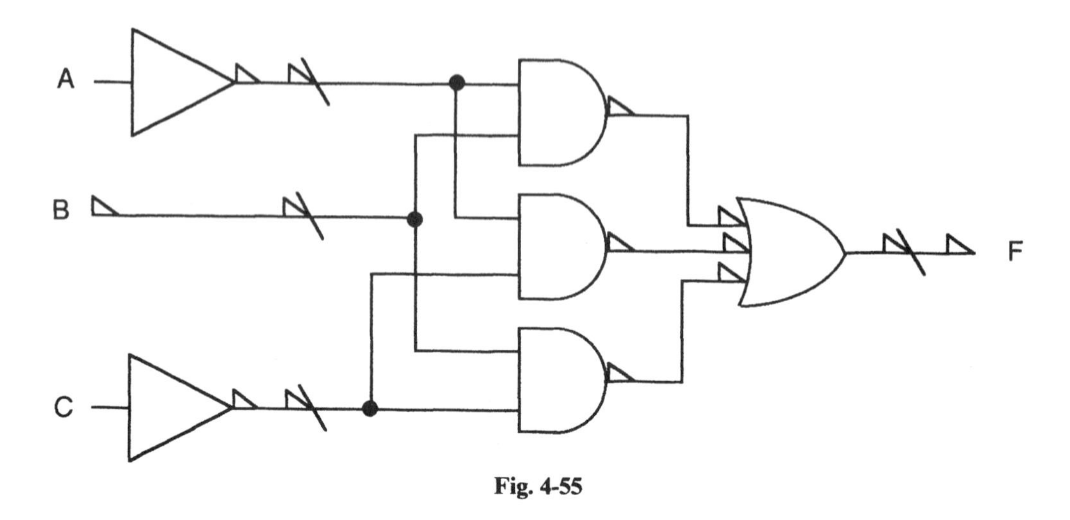

Fig. 4-55

by comparing Figs. 4-52, 4-53, and 4-54, how hardware selection and the choice between coverings influences the number and placement of inverters.

4.15 Design the logic to examine two 2-bit binary numbers (A_2A_1 and B_2B_1) and produce a TRUE output if the number B_2B_1 is greater than or equal to the number A_2A_1. The variables A_1 and A_2 are both LT, while B_2, B_1 and the output are HT. Implement in NAND hardware.

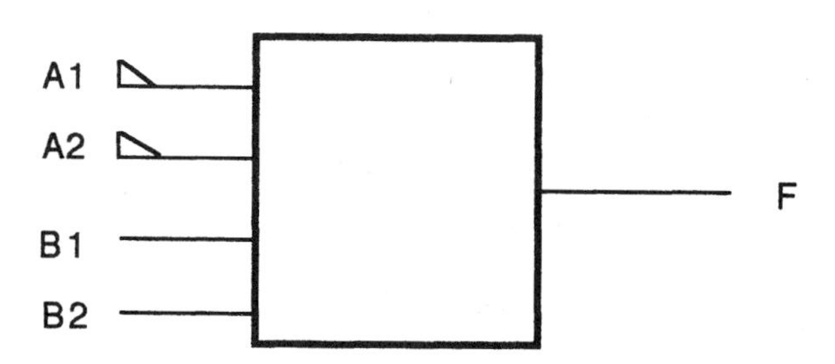

The specified requirements can be entered directly into a K map, and the required function obtained by an appropriate 1s covering as shown in Figs. 4-56 and 4-57. The corresponding mixed-logic diagram is shown in Fig. 4-58.

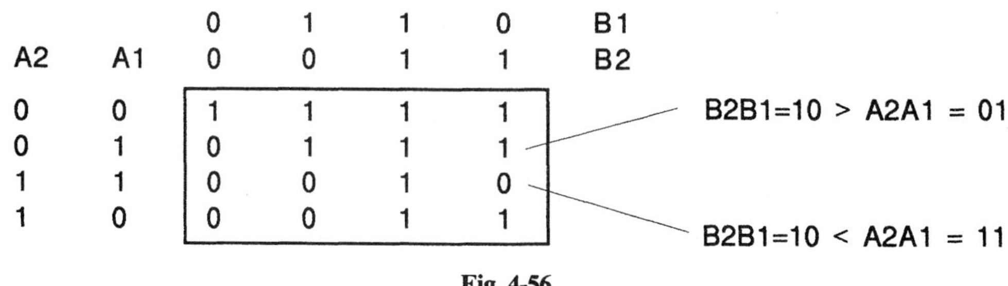

A2	A1	B1=0, B2=0	B1=1, B2=0	B1=1, B2=1	B1=0, B2=1
0	0	1	1	1	1
0	1	0	1	1	1
1	1	0	0	1	0
1	0	0	0	1	1

Fig. 4-56

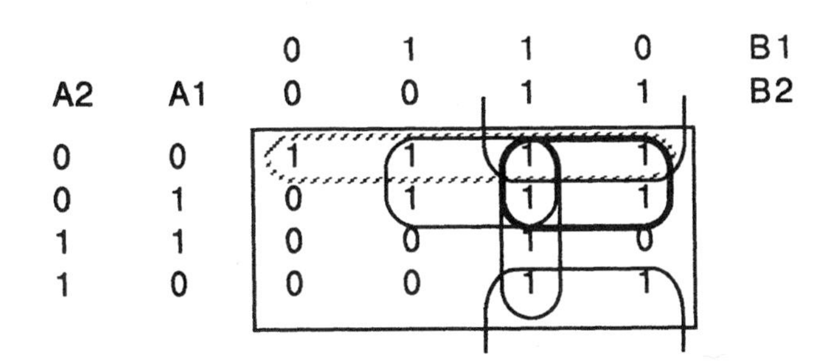

F = A2' A1' + B2B1 + A2' B1 + A1' B2 + A2' B2

Fig. 4-57

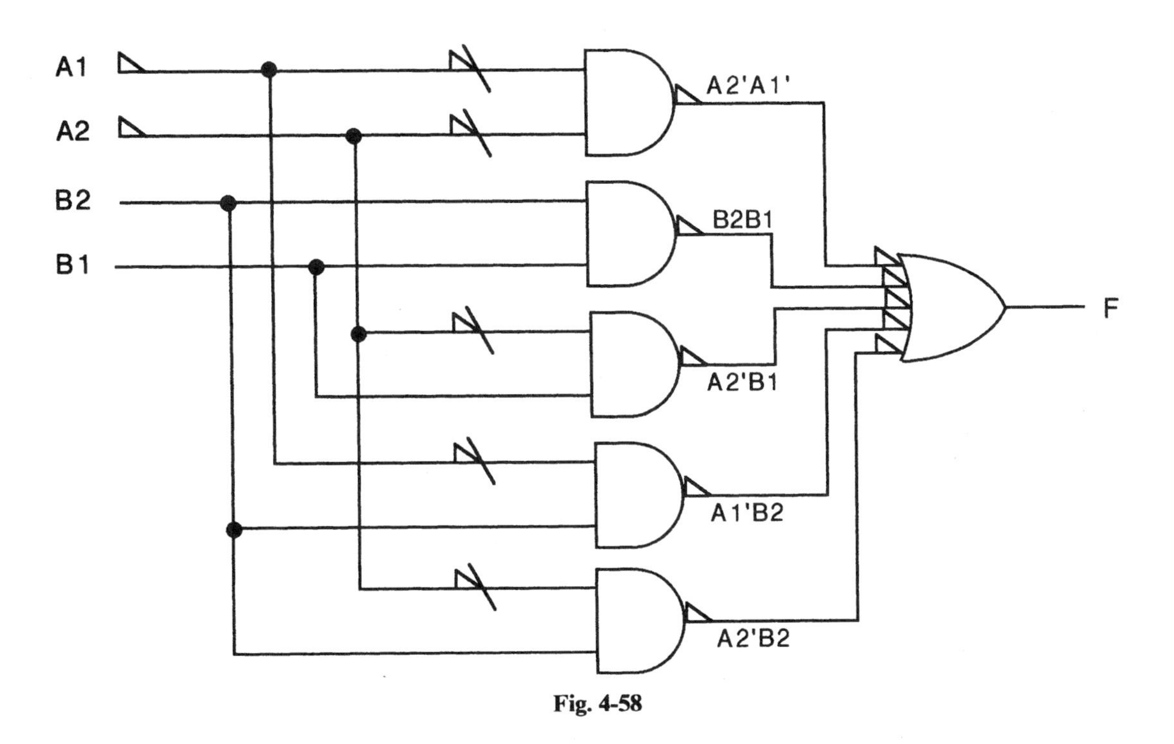

Fig. 4-58

4.16 Given the Boolean function F(A, B, C, D) = [D(A + B′C)(AB + C′)]′ where A, B, and D are HT, while C and F are LT, implement the logic with physical elements which have a voltage I/O table described as follows: the output is HIGH if and only if all the inputs are HIGH.

The specified physical device characteristics may be represented in tabular form as in Table 4.6. Mapping H = T for both inputs and output, we obtain an AND function as shown in Fig. 4-59. The OR

Table 4.6

INPUTS		OUTPUT
X	Y	Z
L	L	L
L	H	L
H	L	L
H	H	H

function is obtained by mapping inputs and output as L = T (Fig. 4-60). The logic diagram is shown in Fig. 4-61, and its hardware implementation is presented in Fig. 4-62 using the mixed-logic convention. Note the proliferation of inverters. With this choice of hardware, a voltage inverter is needed every time an AND is connected to an OR and vice versa.

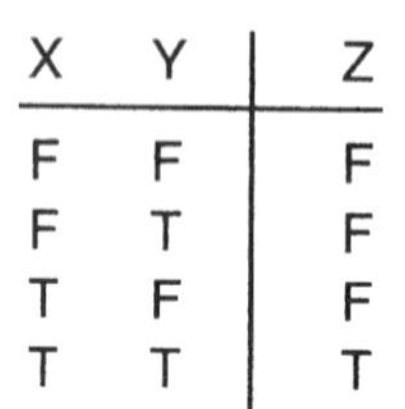

X	Y	Z
F	F	F
F	T	F
T	F	F
T	T	T

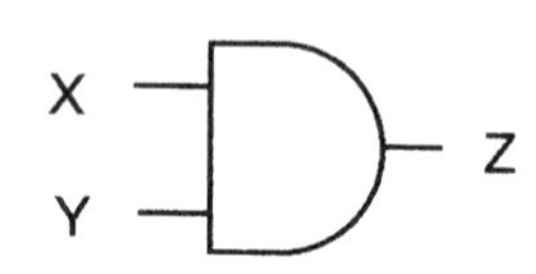

Fig. 4-59

X	Y	Z
T	T	T
T	F	T
F	T	T
F	F	F

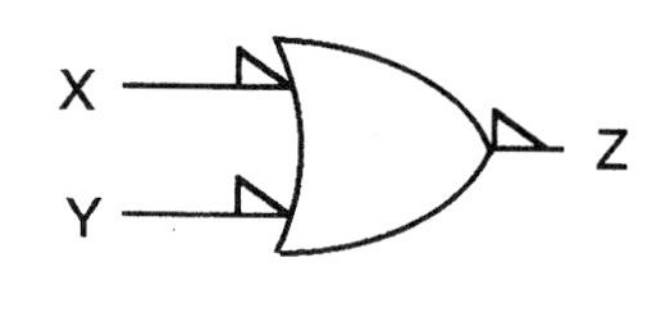

Fig. 4-60

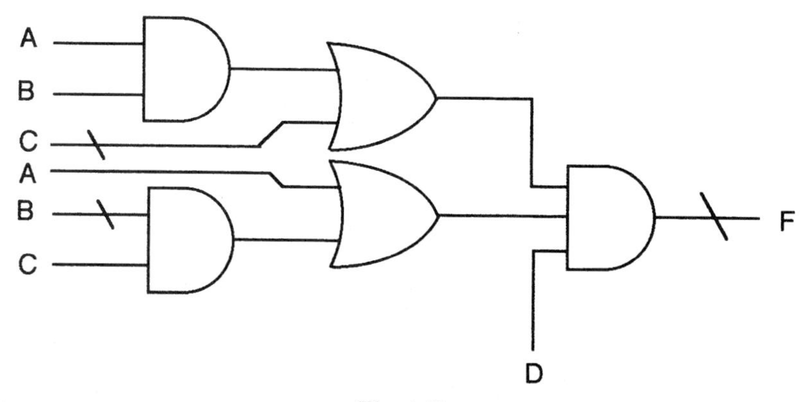

Fig. 4-61

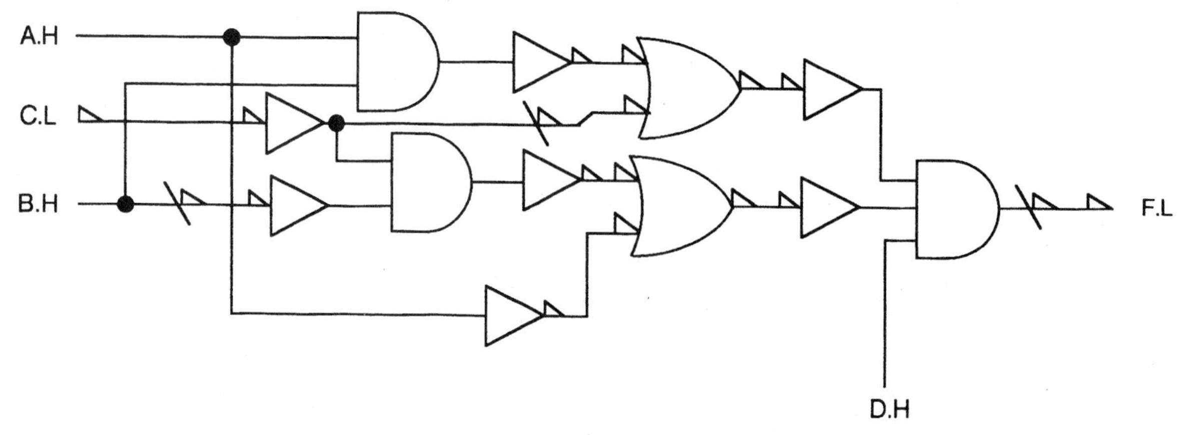

Fig. 4-62

4.17 Given the map shown in Fig. 4-63, implement the function first using NAND hardware and inverters and then XOR gates. Assume that the output and all inputs are HT.

		0	1	1	0	D
A	B	0	0	1	1	C
0	0	1	0	1	0	
0	1	1	0	1	0	
1	1	X	X	0	1	
1	0	X	1	0	1	

F

Fig. 4-63

(*a*) NAND realization: A 1s covering making use of all the "don't cares" yields F = AC′ + C′D′ + AD′ + A′CD. The NAND implementation is shown in Fig. 4-64.

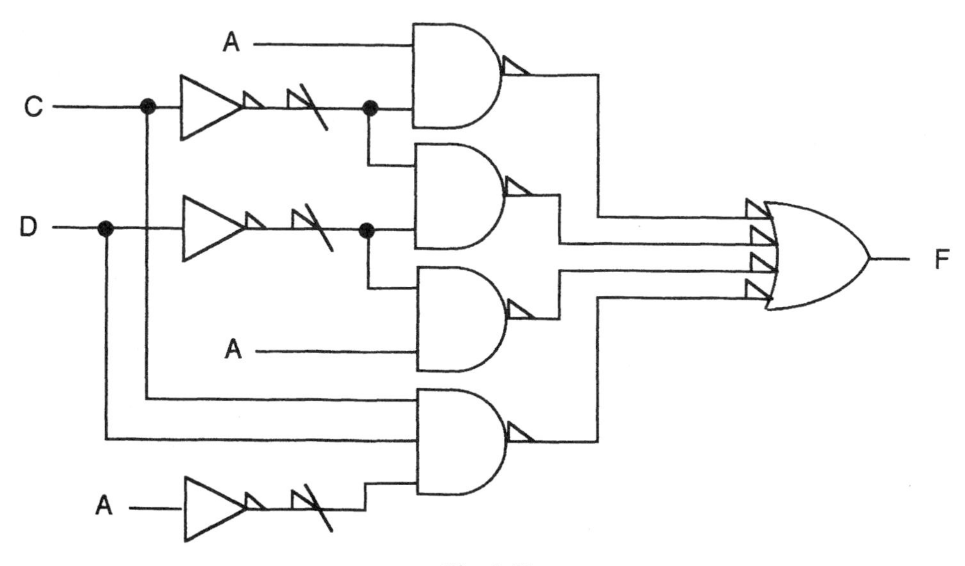

Fig. 4-64

(*b*) XOR realization: The map can be put into a checkerboard form, indicative of XOR functions, by proper choice of "don't cares"; i.e., 0s in the first column and a 1 in the second. The resulting function is

$$F = A'C'D' + AC'D + A'CD + ACD'$$
$$= A'(C'D' + CD) + A(C'D + CD')$$

The terms in parentheses are recognized as being equal to $(C \oplus D)'$ and $(C \oplus D)$, respectively, as shown in Prob. 3.4. Let $X = C \oplus D$. Then,

$$F = A'X' + AX$$
$$= (A \oplus X)'$$

Substituting for X, we obtain $F = (A \oplus C \oplus D)'$, which may be realized by the circuit of Fig. 4-65. Note how the third XOR gate is being used to obtain logical inversion as described in Example 2.8.

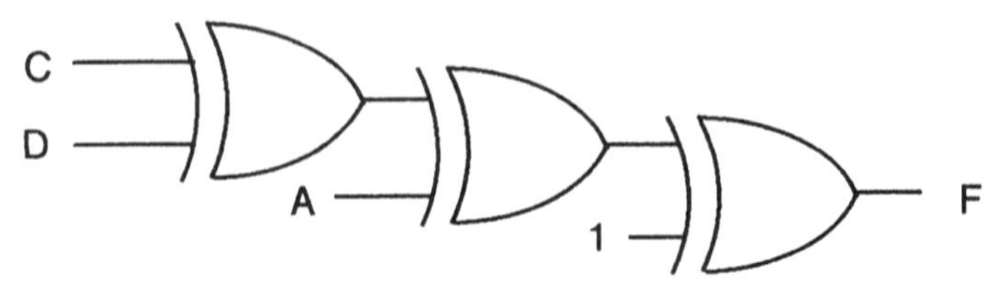

Fig. 4-65

Supplementary Problems

4.18 Implement the function F = A′C′D′ + BD′ + BC′ in NAND hardware.

4.19 Given the K map of the function F shown in Fig. 4-66, assign values to the don't cares so as to make the best 1s covering and draw the logic diagram. Specify LT or HT input and output variables as required to minimize the number of voltage inverters needed for a NAND implementation.

		0	1	1	0	D
A	B	0	0	1	1	C
0	0	1	0	1	X	
0	1	0	0	1	X	
1	1	X	X	X	X	
1	0	1	0	1	X	

F

Fig. 4-66

4.20 Implement the function F = A(B + C′) with NAND hardware using inverters where required. It is specified that all variables (both input and output) are to be HT.

4.21 Suppose that all the inverters in the NAND hardware circuit of Prob. 4.20 are removed and replaced by wires. Write a Boolean expression for the positive logic that this circuit would produce, again assuming HT inputs and output.

4.22 Given the function F = ABC + A′BD′ + BC′D + ABC′, simplify the logic assuming that the first four truth table rows are "don't cares" and that the variables are arranged in alphabetical order with A most significant. Draw a NOR hardware logic diagram.

4.23 Implement the logic for the seven-segment decoder driver which was developed in Prob. 3.22.

4.24 Design the logic to examine two 2-bit binary numbers (A_2A_1 and B_2B_1) and produce a TRUE output if the number B_2B_1 is greater than the number A_2A_1. Implement in NAND hardware, and define the variables as LT or HT so that no inverters are required.

4.25 Given the K map shown in Fig. 4-67, simplify the logic using a 1s covering, and implement in NAND hardware. Use mixed logic with the bubble convention to indicate LT.

A B	0 0	0 1	1 1	1 0
(D)	0	1	1	0
(C)	0	0	1	1
0 0	1	0	1	1
0 1	0	1	0	0
1 1	1	1	X	0
1 0	1	X	1	1

Fig. 4-67

4.26 Implement the logic of Prob. 4.25 using a 0s covering and NOR hardware.

4.27 Implement the simplified multiplier of Prob. 3.25 in NAND hardware and determine the small-scale integration (SSI) chip count for the design.

4.28 Using the equation for L, implement the lamp control circuit of Prob. 2.25 in NAND hardware and with XORs. Assume that the switches A and B are HT and that L is LT. Compare the SSI chip counts.

4.29 Repeat Prob. 4.28, this time using the equation for L′.

4.30 Implement, using NAND hardware, the binary-bit subtractor designed in Probs. 2.27–2.29, assuming that the inputs and outputs are all HT. Estimate the chip count.

4.31 Using the simplified Boolean equations for the binary-bit subtractor developed in Prob. 3.28, implement the circuit in NOR and XOR hardware, using HT inputs and outputs. Compare the chip count to the results of Prob. 4.30.

4.32 Repeat Prob. 4.31 if the inputs and outputs are LT.

4.33 Determine the logic equation represented by the circuit shown in Fig. 4-68.

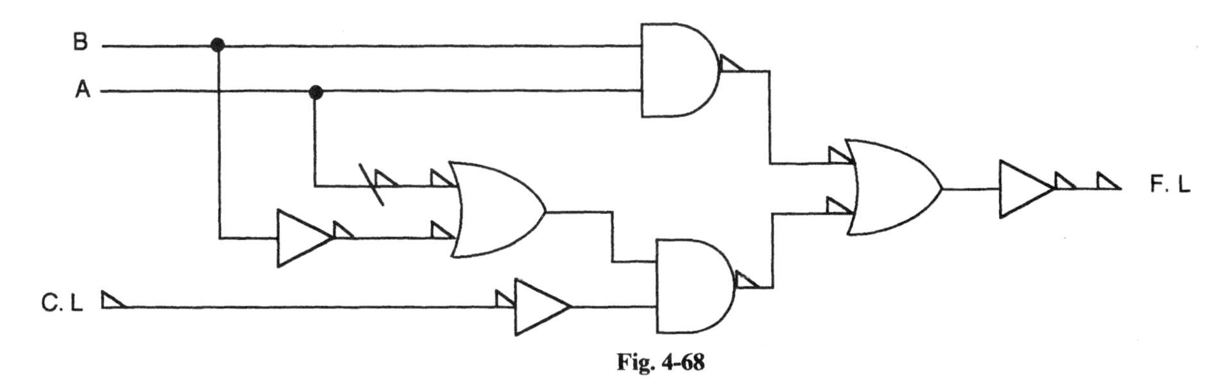

Fig. 4-68

4.34 For the 4-bit binary to Gray code converter discussed in Probs. 3.34 and 3.35, obtain an XOR representation for the outputs and comment on the estimated chip counts.

The following four problems demonstrate how hardware and truth value assignments affect circuit topology and complexity.

4.35 Implement the 0s covering of Prob. 3.38 using NOR hardware. Assume that input and output variables are all HT.

4.36 Implement the circuit of Prob. 4.35 using NAND hardware and, again, assume that input and output variables are all HT. Compare the results.

4.37 Two logically equivalent Boolean forms of Probs. 4.35 and 4.36 are:

$$F = (B'C + AC + AD)' = A'C' + C'D' + A'B$$

Implement both of them using NAND hardware, assuming that B and F are LT, while all other variables are HT. Compare the results with one another and with the circuits obtained in Probs. 4.35 and 4.36.

4.38 Repeat Prob. 4.37 using NOR hardware.

4.39 Write the Boolean expression represented by the positive logic circuit shown in Fig. 4-69.

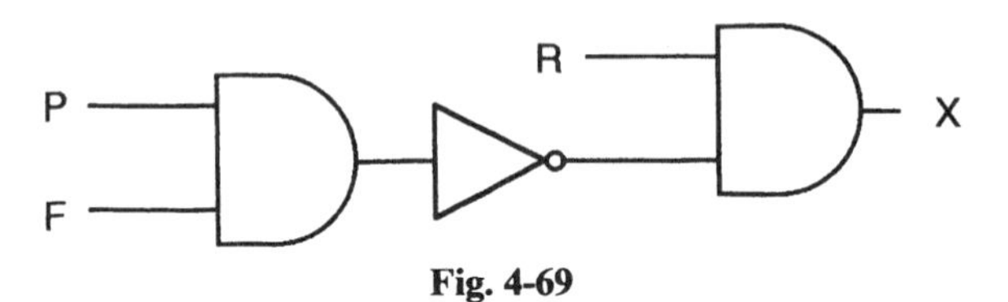

Fig. 4-69

4.40 Repeat Prob. 4.39 for the circuit shown in Fig. 4-70. Comment on the relationship between this and the previous result if Y is designated LT.

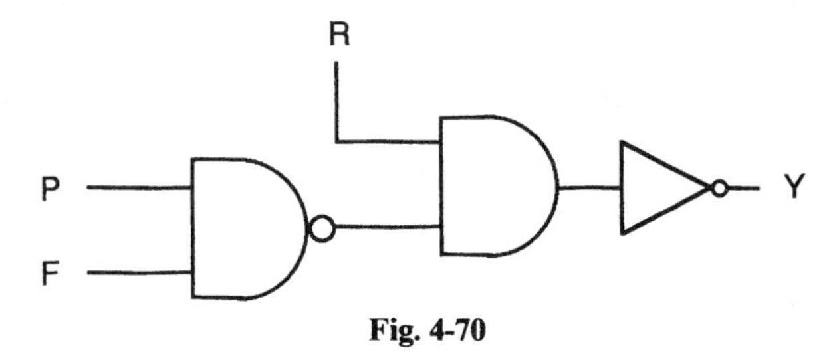

Fig. 4-70

4.41 Draw a mixed-logic diagram for the NAND implementation of the circuit in Prob. 4.40.

Answers to Supplementary Problems

4.18 See Fig. 4-71.

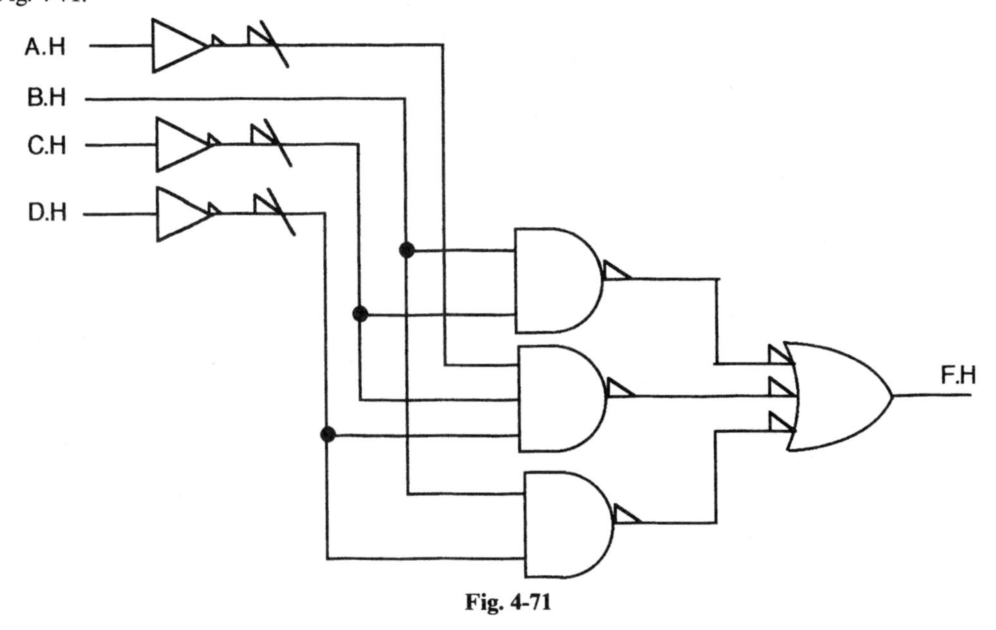

Fig. 4-71

4.19 See Figs. 4-72 and 4-73. The covering shown in Fig. 4-72 plus the four corners yields F = C + B′D′.

	D	0	1	1	0
A	**B** \ **C**	**0**	**0**	**1**	**1**
0	0	1	0	1	X
0	1	0	0	1	X
1	1	X	X	X	X
1	0	1	0	1	X

F

Fig. 4-72

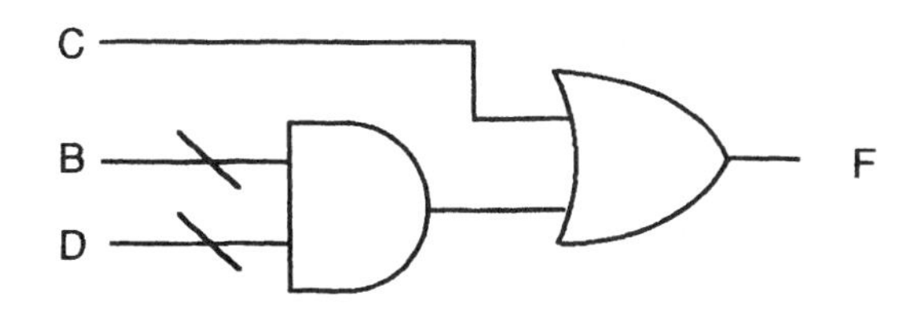

(*a*) Logic diagram.

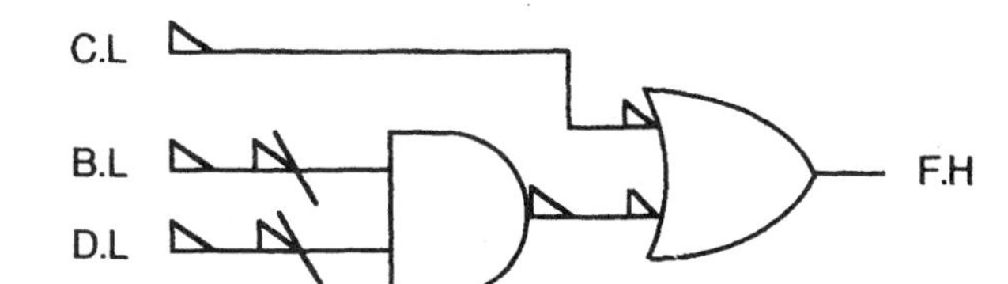

(*b*) Logic implementation with inputs Low-TRUE, outputs High-TRUE and no voltage inverters.

Fig. 4-73

4.20 See Fig. 4-74.

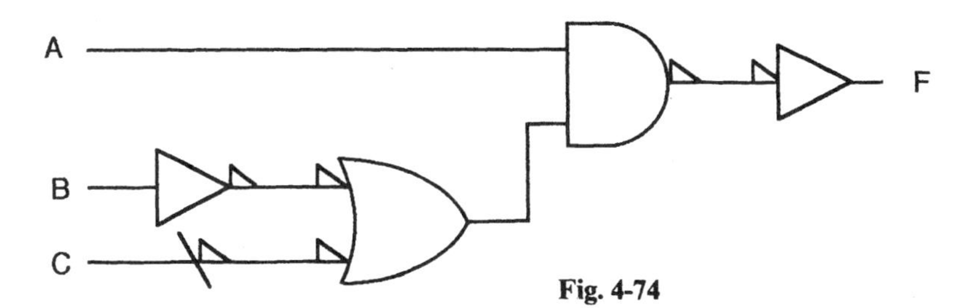

Fig. 4-74

4.21 F = A′ + BC

4.22 F = AB + BC′ + A′D′ and F = AB + BC′ + BD′ are both valid answers. The hardware design for the first case is shown in Fig. 4-75.

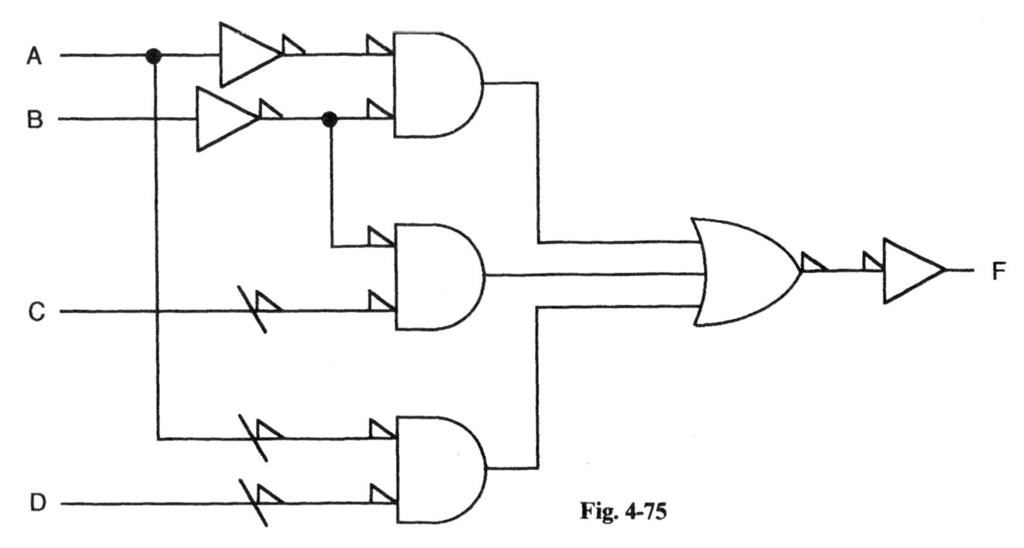

Fig. 4-75

4.23 See Fig. 4-76. Note that the LT bubble convention is being used.

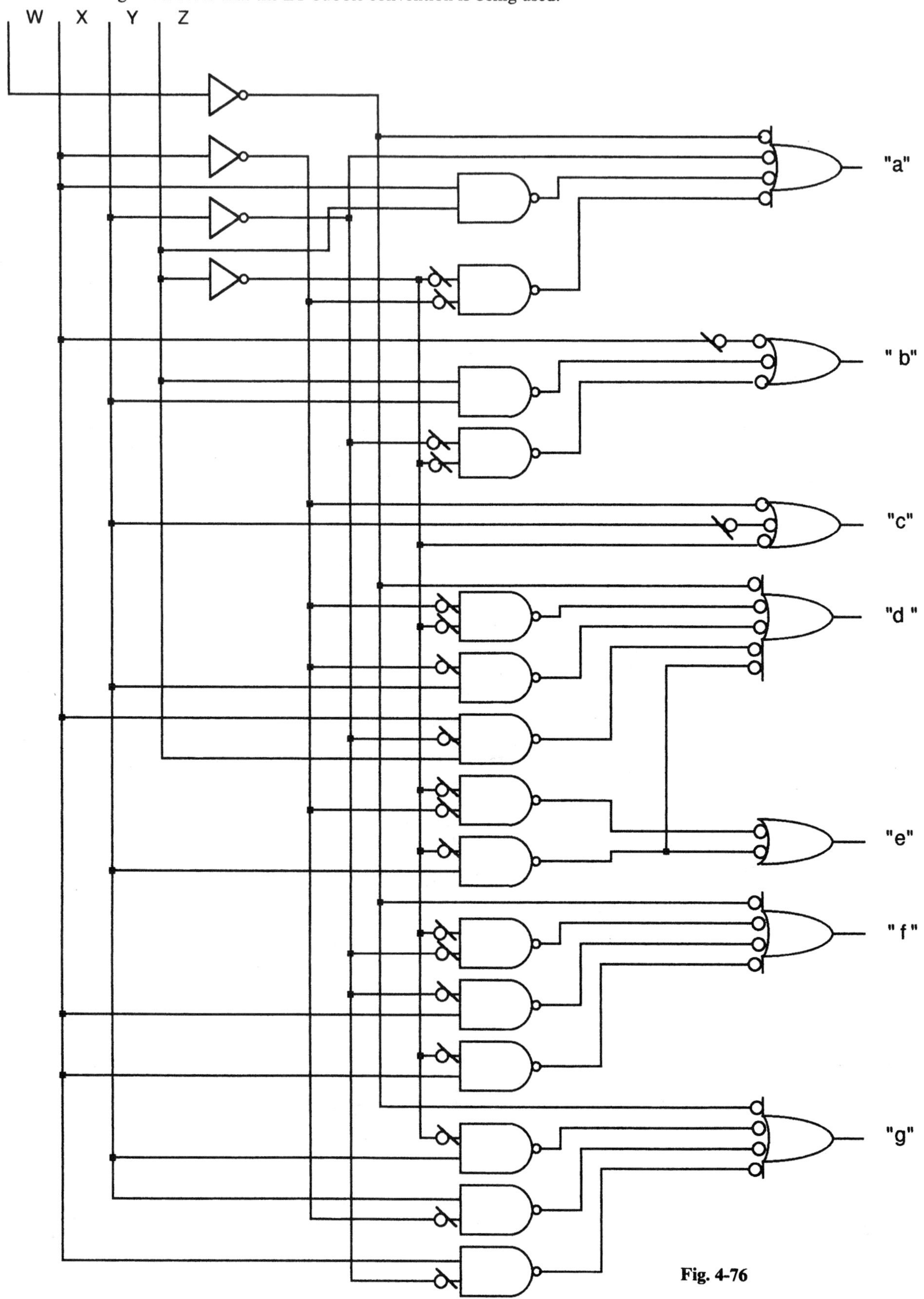

Fig. 4-76

4.24 The specification can be entered directly into a K map, yielding the function $F = A_2' B_2 + A_2' A_1' B_1 + A_1' B_2 B_1$. The required circuit is shown in Fig. 4-77.

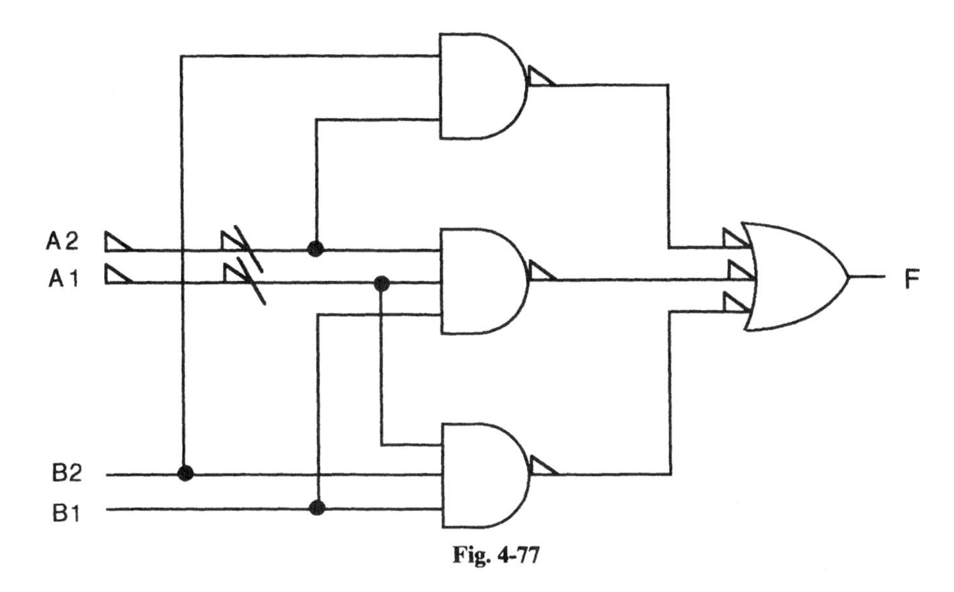

Fig. 4-77

4.25 The simplification, shown in Fig. 4-78, includes a four-corner covering which is omitted from the figure for clarity. The NAND implementation is shown in Fig. 4-79 in which the LT bubble convention is used. Three chips are needed (one dual four-input gate, one quad two-input gate, one 1/3 hex inverter).

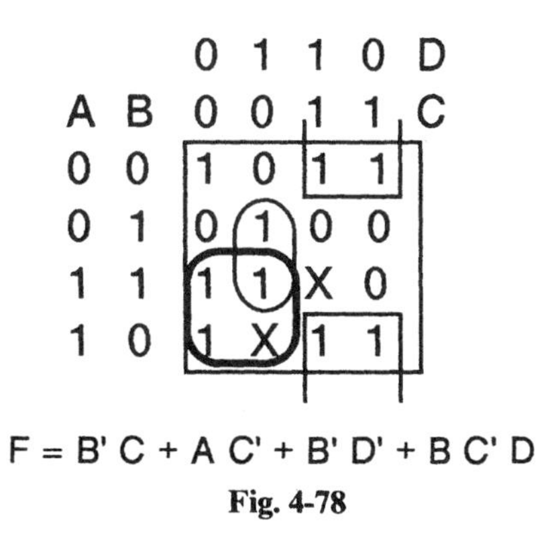

Fig. 4-78

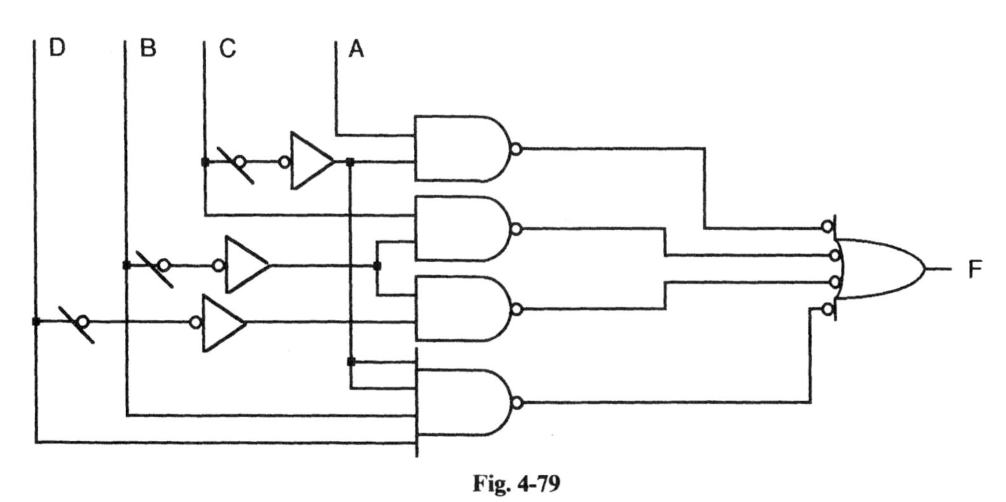

Fig. 4-79

4.26 See Figs. 4-80 and 4-81. Two chips are needed (one triple three-input gate, one quad two-input gate).

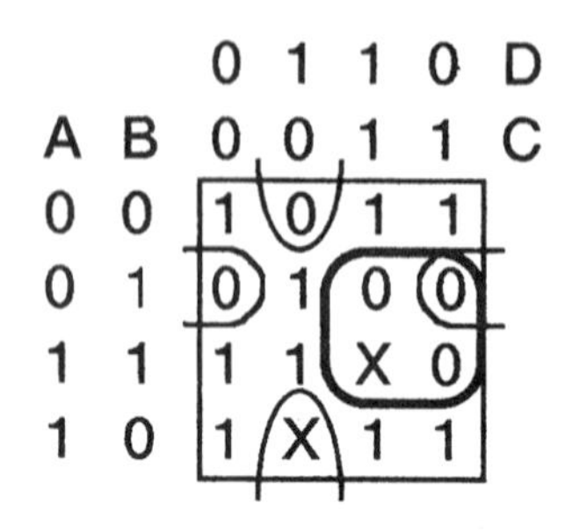

Fig. 4-80

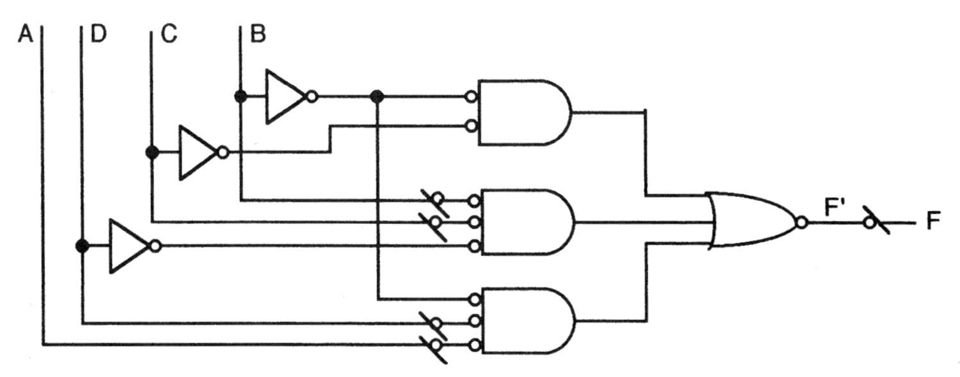

Fig. 4-81 LT bubble convention.

4.27 See Fig. 4-82. The chip count is as follows:

One hex inverter (contains six gates; all used)
One dual four-input gate (contains two gates; all used)
Two triple three-input gates (each contains three gates; all used)
One quad two-input gate (contains four gates; two used)
Total: five chips

4.28 Using $L = AB + A'B' = (A'B + AB')' = (A \oplus B)'$ yields the solutions of Fig. 4-83.

4.29 Using $L' = A'B + AB' = A \oplus B$, the solutions shown in Fig. 4-84 are obtained.

4.30 See Fig. 4-85. The chip count is four: one dual four-input NAND, two triple three-input NAND, one hex inverter.

4.31 From Prob. 3.28, $D = X \oplus Y \oplus B_i$ and $B_o = X'Y + X'B_i + YB_i$. The implementation is shown in Fig. 4-86. The chip count is three: triple three-input NOR, quad two-input NOR, XOR inverters taken from other chips.

4.32 See Fig. 4-87.

4.33 $F = AB + C(A' + B) = AB + A'C + BC$

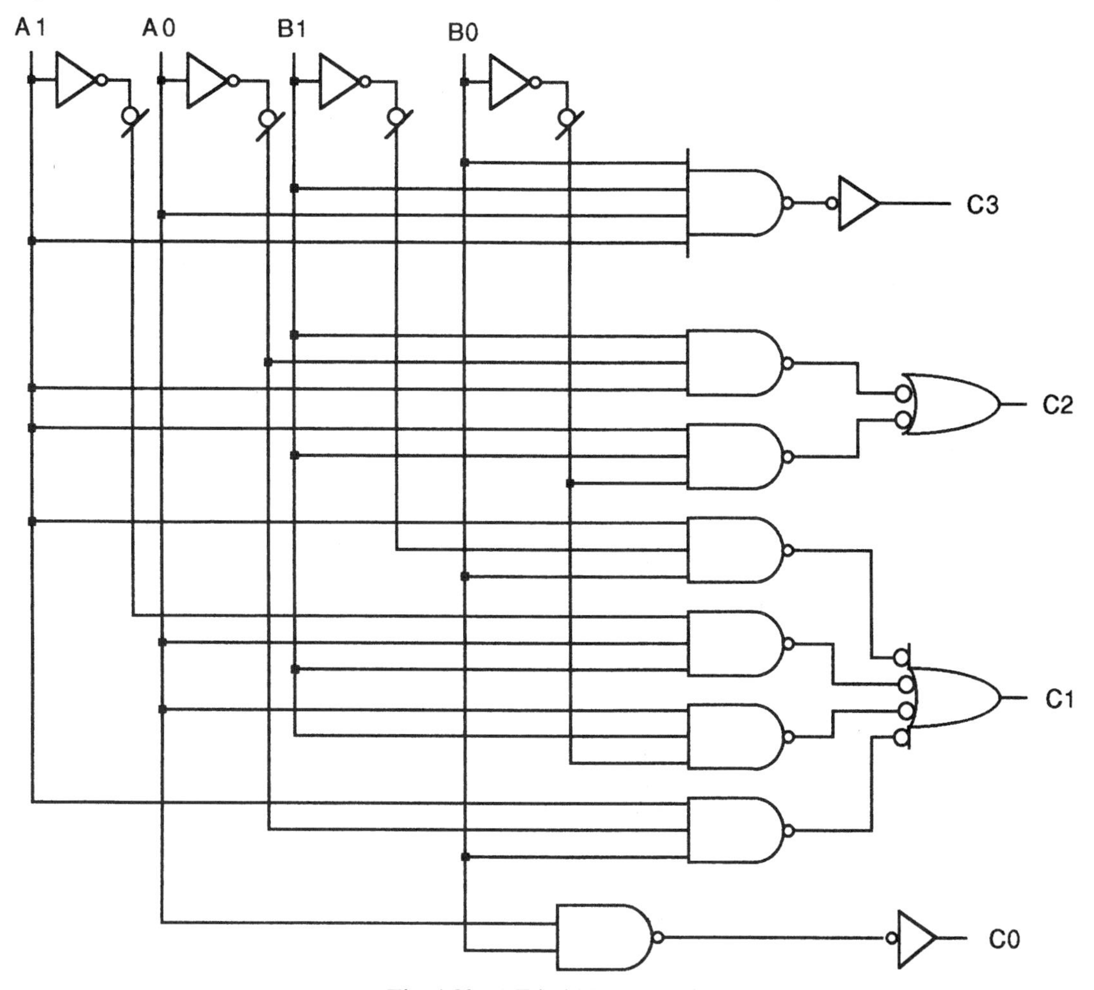

Fig. 4-82 LT bubble convention.

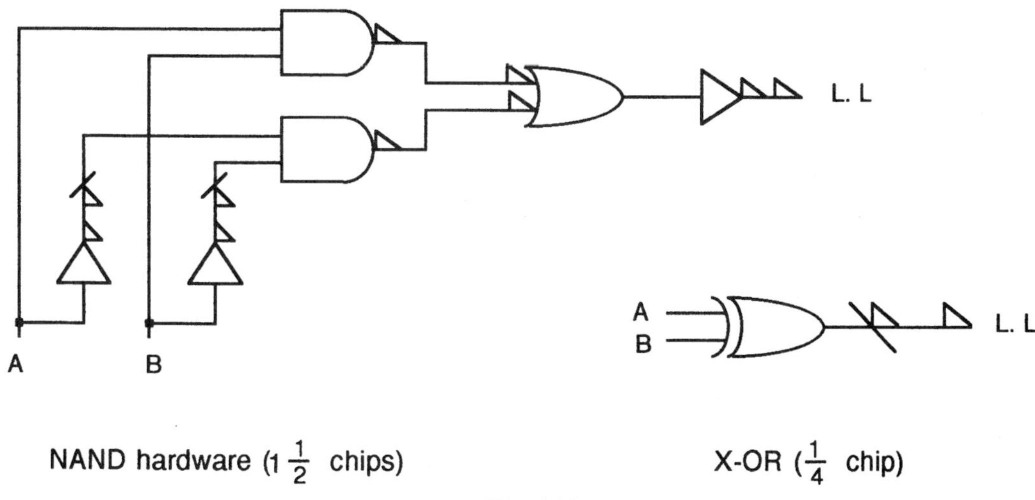

Fig. 4-83

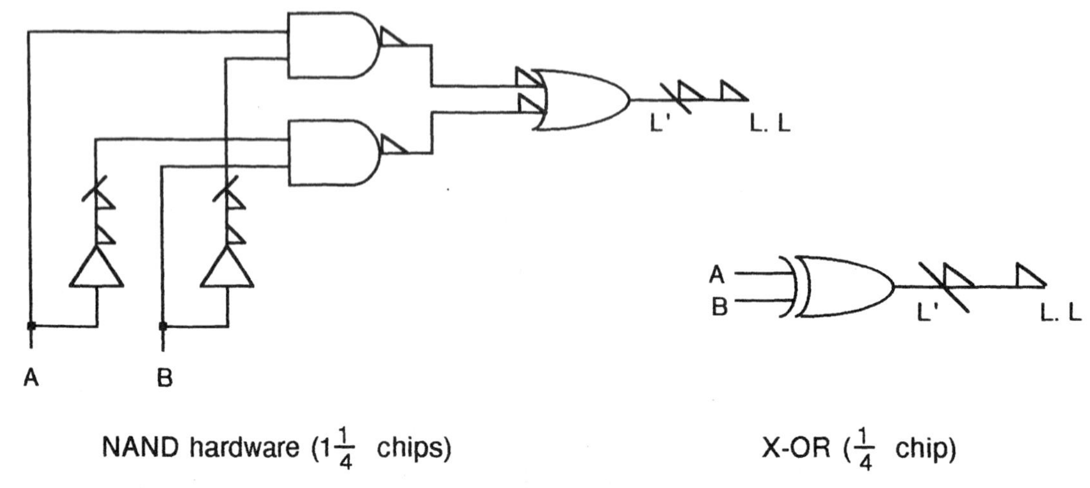

Fig. 4-84

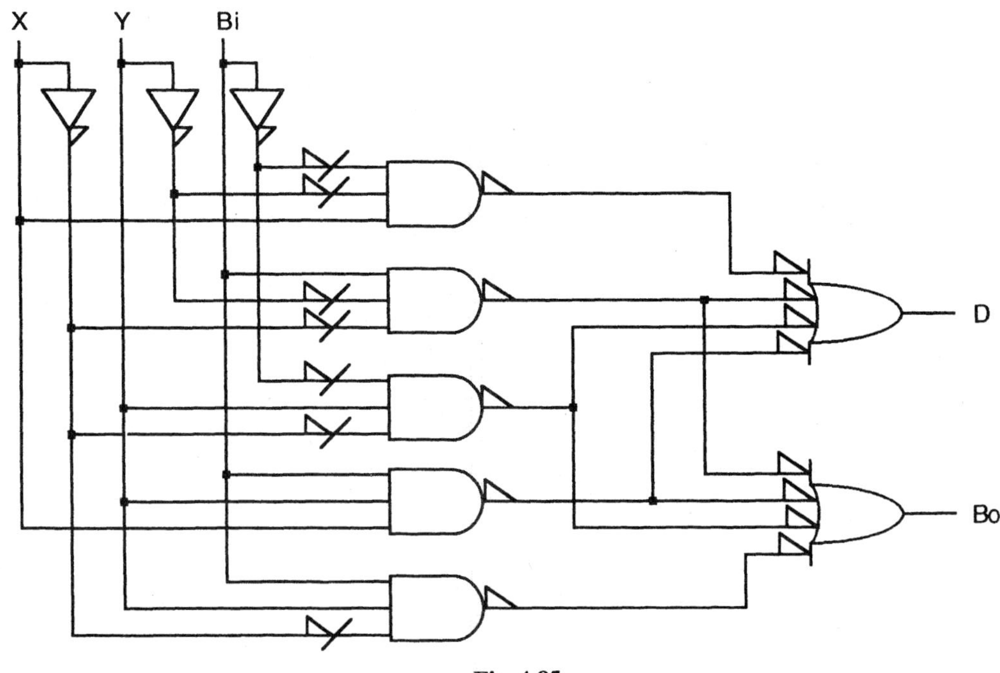

Fig. 4-85

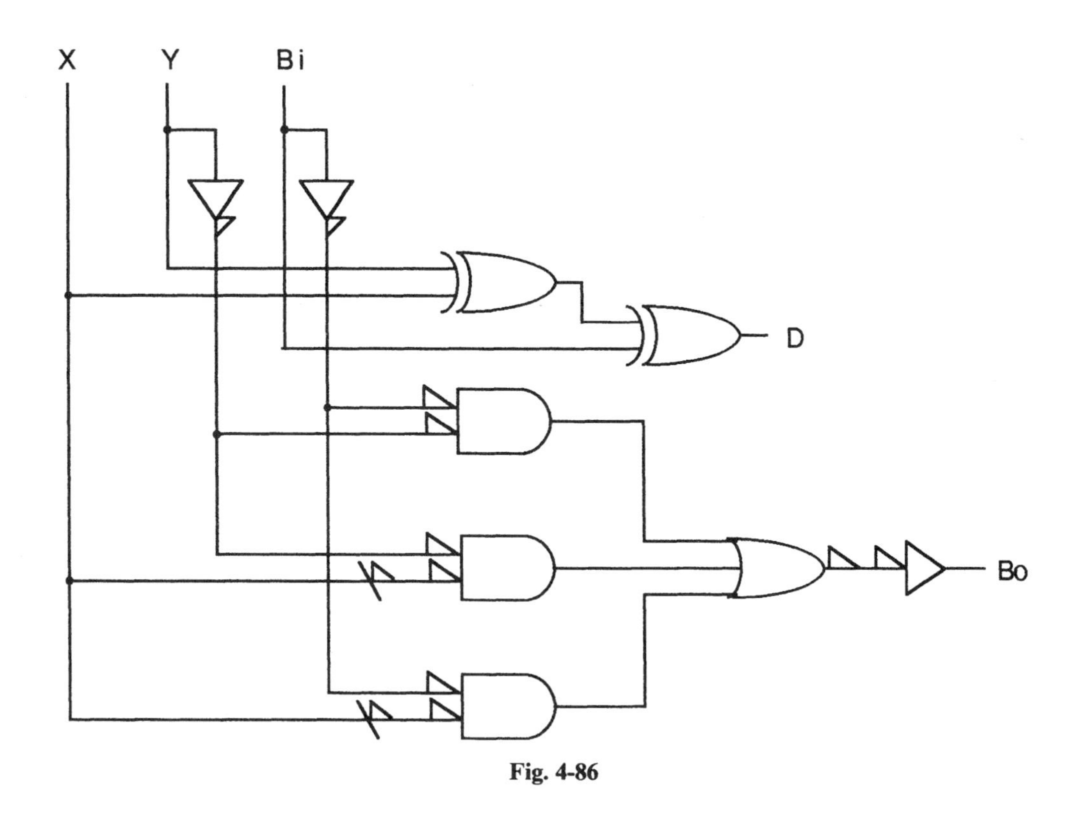

Fig. 4-86

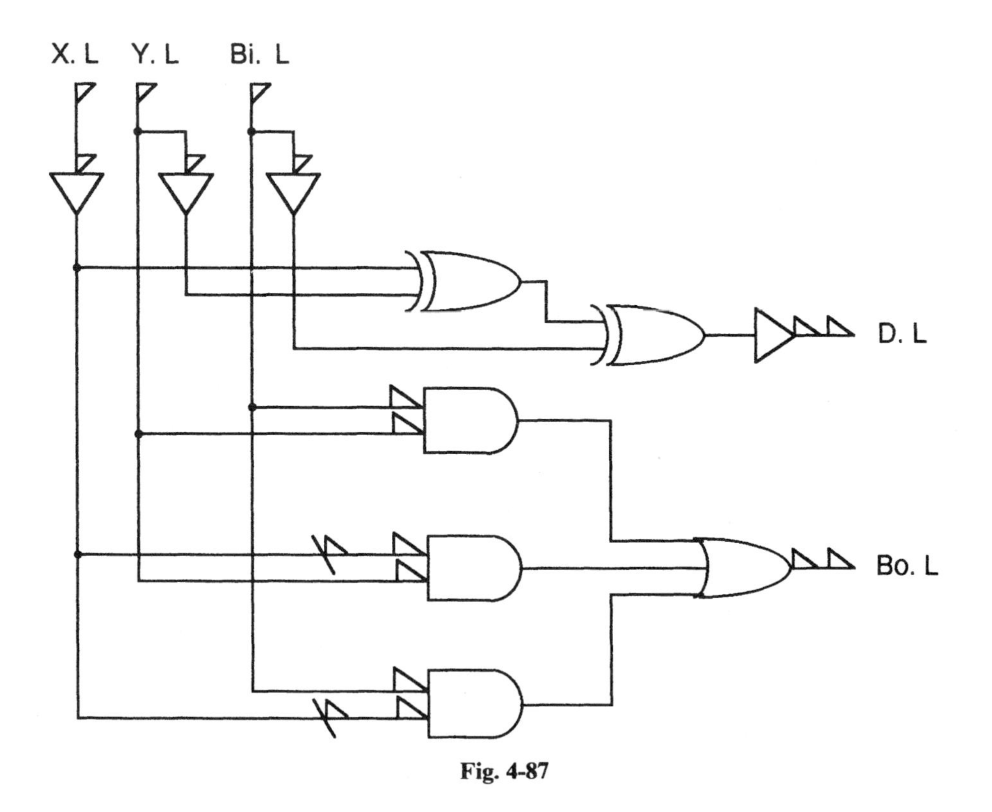

Fig. 4-87

4.34 See Fig. 4-88.

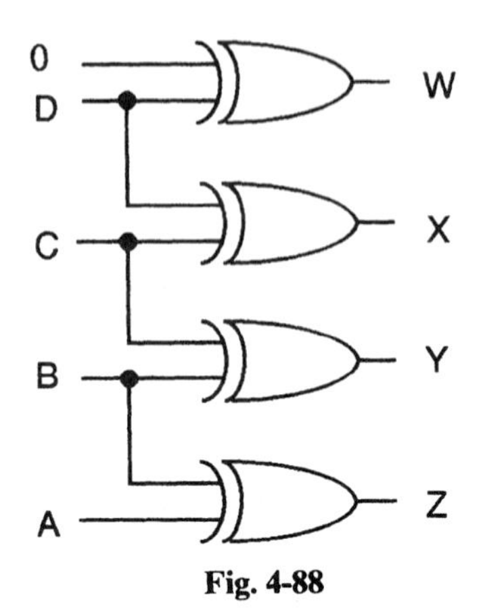

Fig. 4-88

$$W = D = D \oplus 0$$
$$X = CD' + C'D = C \oplus D$$
$$Y = B \oplus C$$
$$Z = A \oplus B$$

Since XORs come four to a chip, only one chip is used here. If, on the other hand, the original equations are used, the following gates are needed:

Four-input AND	8 gates	2 chips
Eight-input OR	1 gate	1 chip
Three-input AND	4 gates	1 chip (use spare gate on four-input OR for one)
Two-input AND/OR	3 gates	1 chip
Four-input OR	1 gate	1 chip

Since all input signals are required in both direct and complemented form, an additonal hex inverter chip may be needed. Thus, six or seven chips will be required for an AND/OR design.

4.35 $F = (B'C + AC + AD)'$; the implementation is shown in Fig. 4-89.

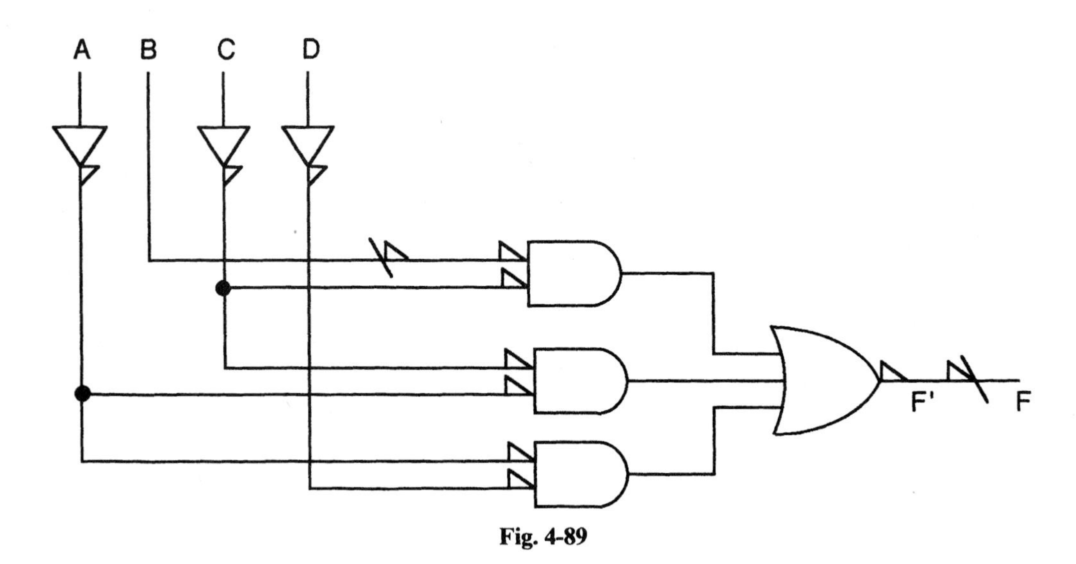

Fig. 4-89

4.36 See Fig. 4-90.

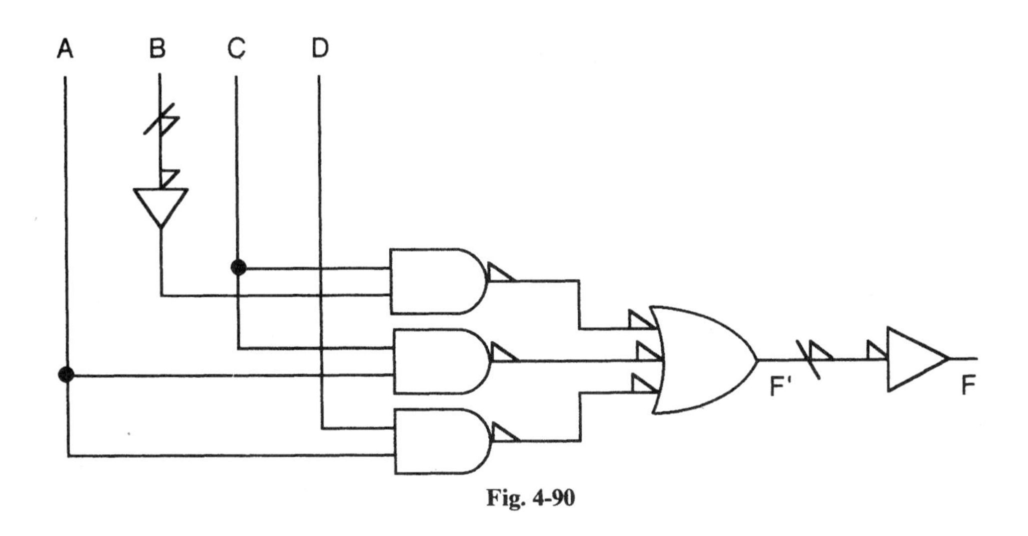

Fig. 4-90

This implementation uses one less inverter. However, the longest signal path passes through four gates, while in Fig. 4-89, the critical timing path is one gate shorter (refer to Probs. 6.11–6.13).

4.37 See Figs. 4-91 and 4-92. Note how the selection of equation form and/or truth values can have a significant impact on design efficiency.

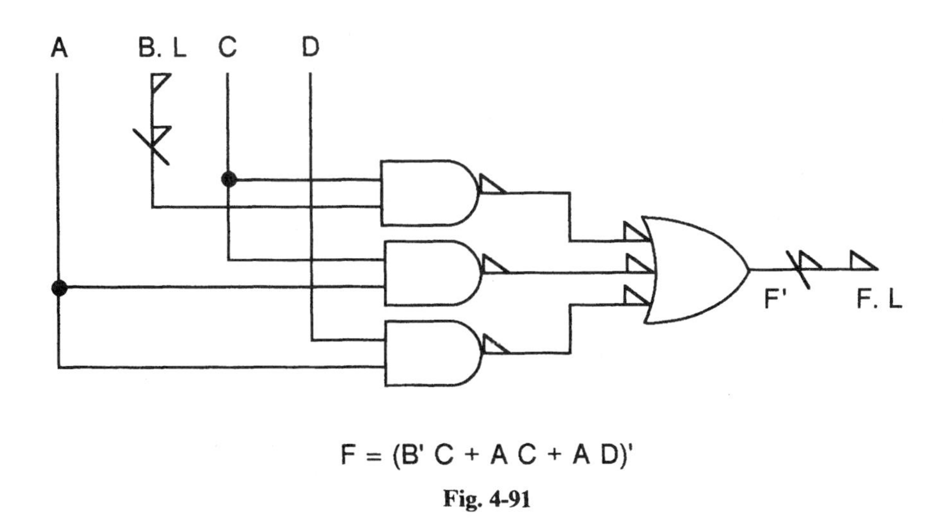

F = (B' C + A C + A D)'

Fig. 4-91

4.38 See Figs. 4-93 and 4-94. We see that the number and placement of required inverters depends on the map covering technique selected and whether NAND or NOR hardware is used.

4.39 $X = R(FP)' = \overline{(R' + FP)}$.

4.40 $Y = (R' + FP)$. Since Y.L = Y'.H, the result becomes identical to that of the previous problem, indicating that the two circuits can be interpreted as being logically equivalent. Refer to Prob. 8.22 for an application of this logic.

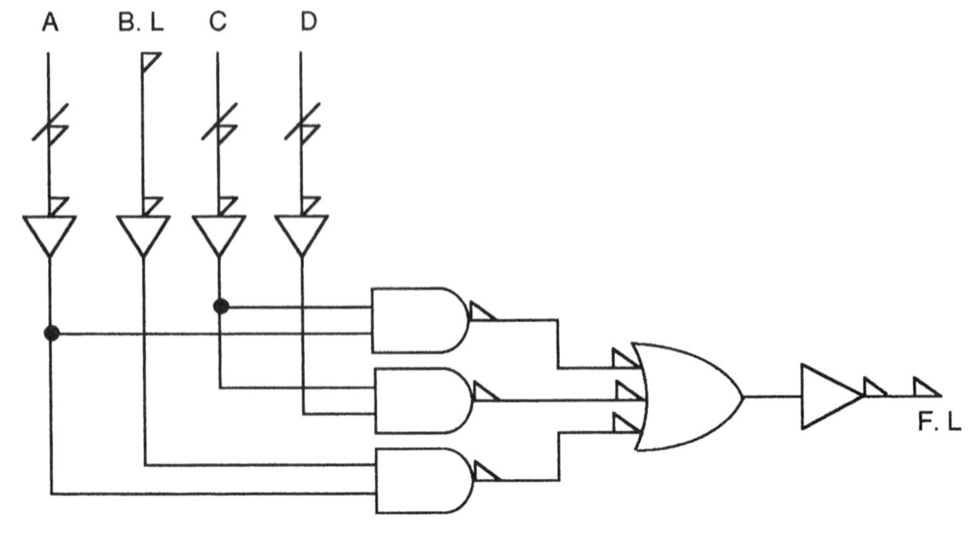

F = A' C' + C' D' + A' B

Fig. 4-92

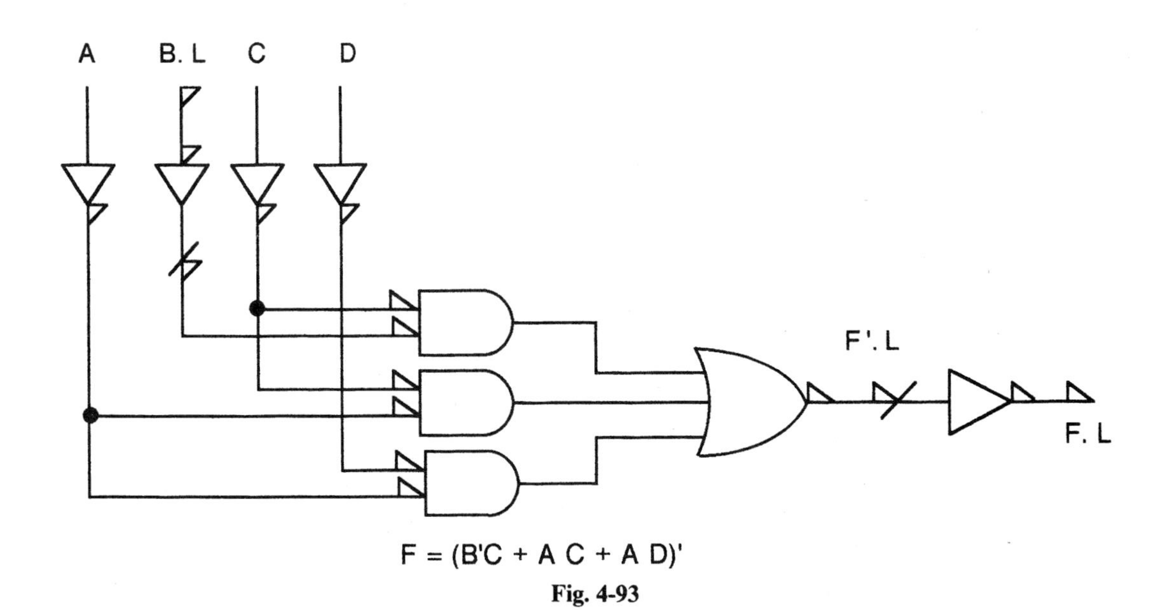

F = (B'C + A C + A D)'

Fig. 4-93

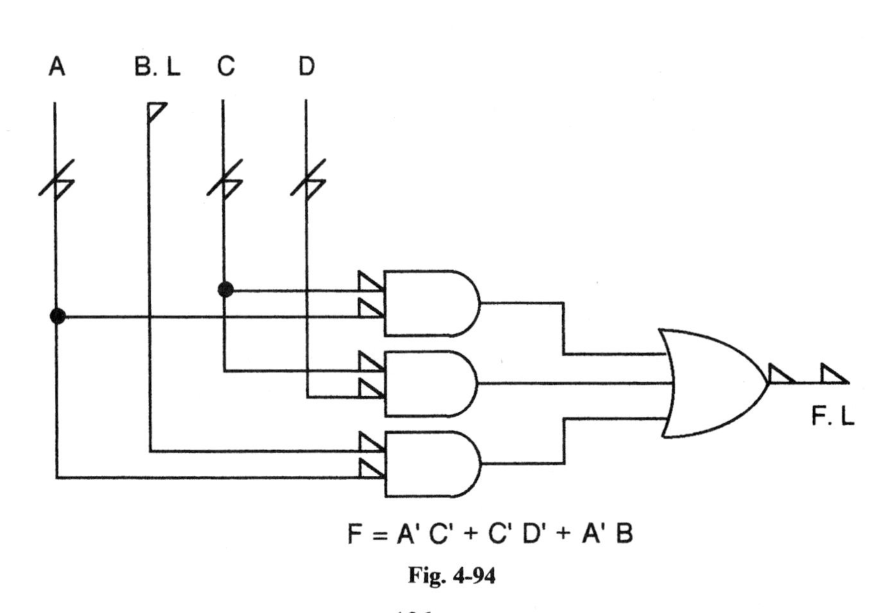

F = A' C' + C' D' + A' B

Fig. 4-94

4.41 See Fig. 4-95. Note how much easier it is to obtain a Boolean equation from this diagram compared to the positive logic forms of the previous two problems.

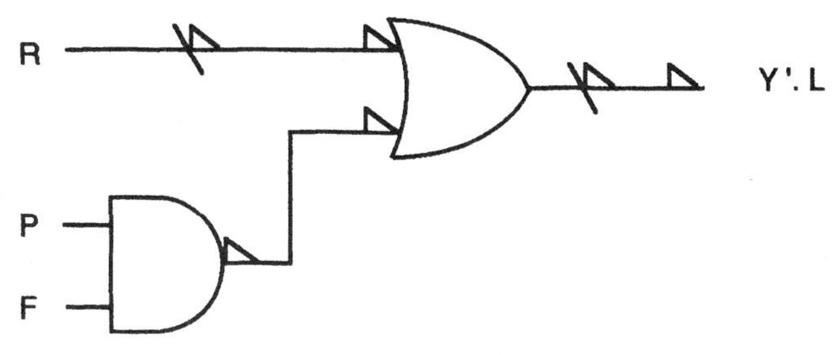

Fig. 4-95

Chapter 5

MSI and LSI Elements

5.1 INTRODUCTION

As IC technology has advanced, it has become possible to fit more and more gates on a single silicon chip and to efficiently create reliable interconnections between them. These advances have permitted the development of single-chip logic circuits designed to accomplish increasingly complex tasks. The resulting chips require much less space than an equivalent circuit realized with individual gates, and the technology has progressed from small-scale integration (SSI) to medium-scale integration (MSI) to large-scale integration (LSI) until, presently, we are producing very large scale integrated (VLSI) devices which are so complex that an entire computer containing up to 1 million gates can be fabricated on a single piece of silicon about the size of a fingernail. In this chapter, three important MSI to LSI devices, the multiplexer, the decoder, and the read-only memory (ROM), will be described in detail along with typical applications.

5.2 MULTIPLEXERS

A multiplexer (MUX) is a device having N select inputs, 2^N data inputs, and a single output. There are several multiplexer symbols presently competing for general use; two of the most common are shown in Fig. 5-1 which depicts an eight-data input, three-select version (refer to App. B). The enable input (EN) is used to turn on the chip. If it is TRUE, the device functions normally for all inputs, whereas if it is FALSE, the chip will not accept data.

The multiplexer may be treated from two distinct viewpoints. First, it can be considered as the digital equivalent of a mechanical rotary switch and, second, as a universal logic module (ULM) which may be used to implement general combinational logic functions, just as interconnected gates do. In either case, it is necessary to examine the device's internal logic in order to understand its operation.

Multiplexer Logic

The logic diagram for a two-select, four-data input multiplexer is shown in Fig. 5-2. In general, each select variable is complemented so that if there are N select variables, there will be 2N lines to which AND gate inputs may be connected. Each AND gate has N + 1 inputs, one of which is a data input as shown.

The circuit is designed so that a data input will be passed to the output of its associated AND gate when the unique combination of select variables which addresses this gate is present on the select inputs. In the two-select case, it can be seen from Table 5.1 that for *each* combination of select variables, there will be a unique AND gate which has its select (nondata) inputs both TRUE at the same time.

Table 5.1

Select Inputs

S_1	S_0	S_1'	S_0'	G1	G2	G3	G4	F
0	0	1	1	0	0	0	D0	D0
0	1	1	0	0	0	D1	0	D1
1	0	0	1	0	D2	0	0	D2
1	1	0	0	D3	0	0	0	D3

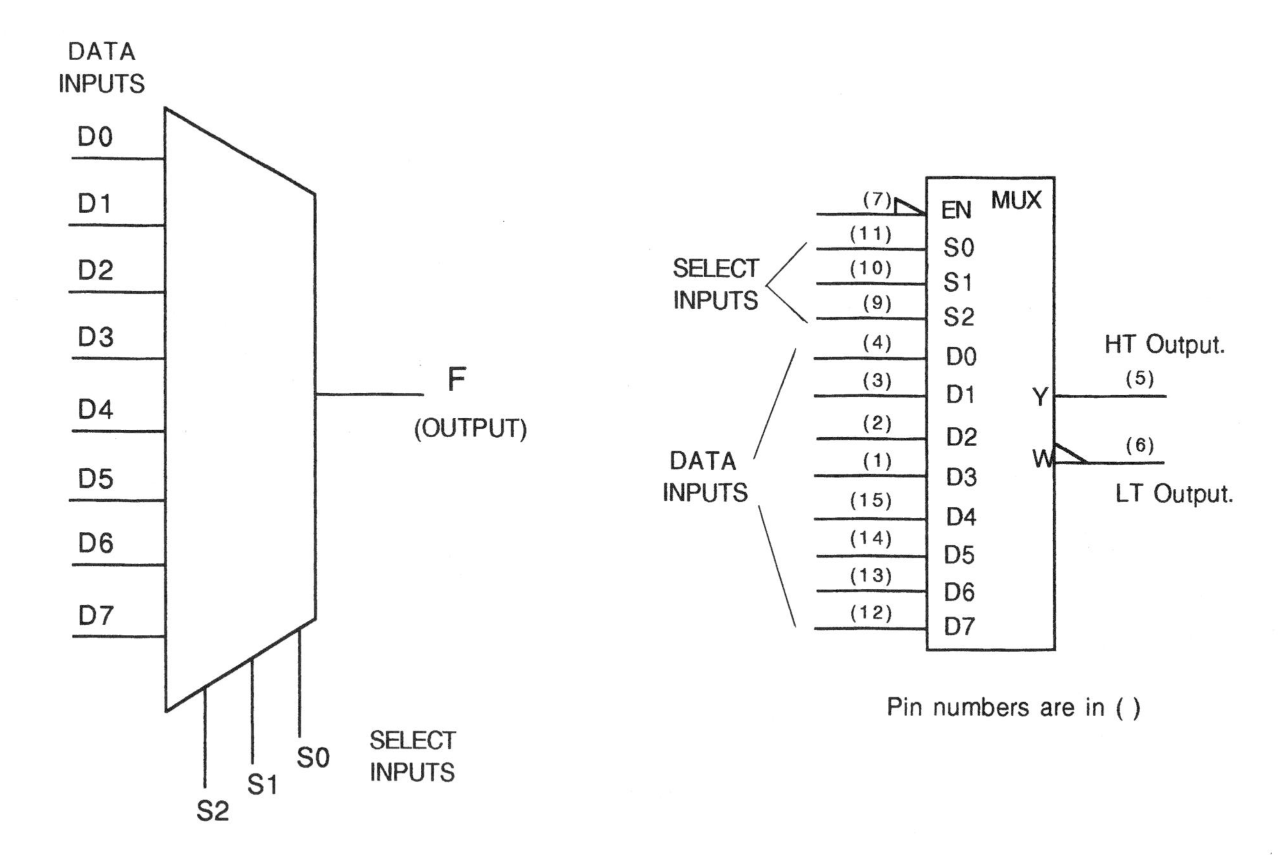

(*a*) Logic symbol.

(*b*) IEEE/IEC symbol.

Fig. 5-1

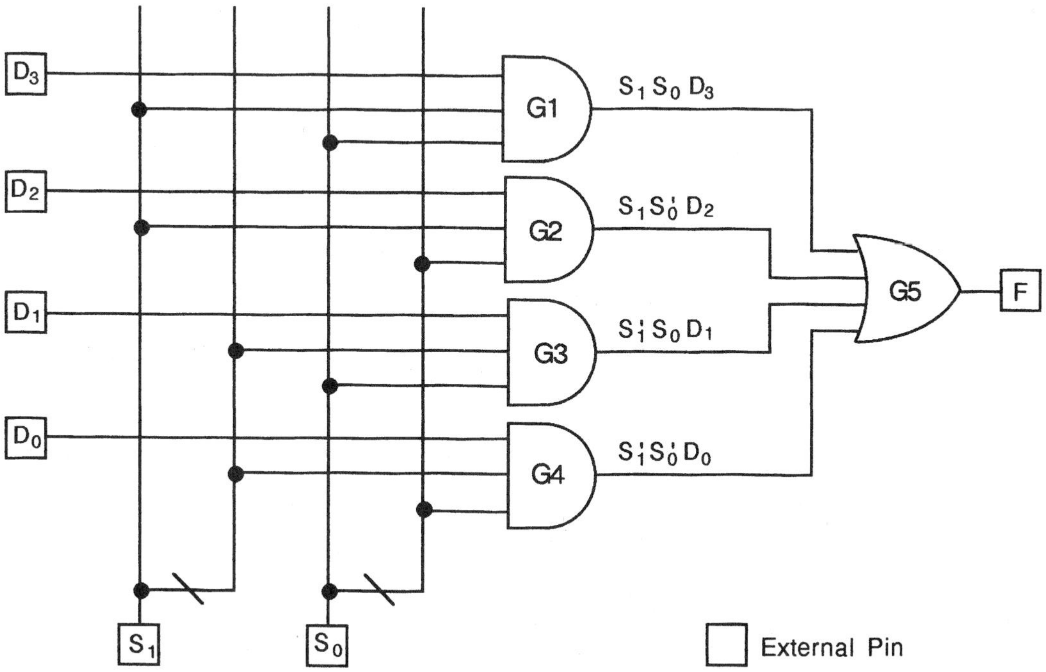

Fig. 5-2

Consider the case $S_1 = 0$, $S_0 = 1$. Since we observe from Fig. 5-2 that $G_3 = S_1' \cdot S_0 \cdot D_1$, the given select input will cause F to equal D_1 since F will be 1 when D_1 is 1 and 0 when D_1 is 0.

The Multiplexer as a Switch

We see that the multiplexer may be viewed as a rotary switch with its select inputs determining which data-input line will be connected to the output, as depicted schematically in Fig. 5-3.

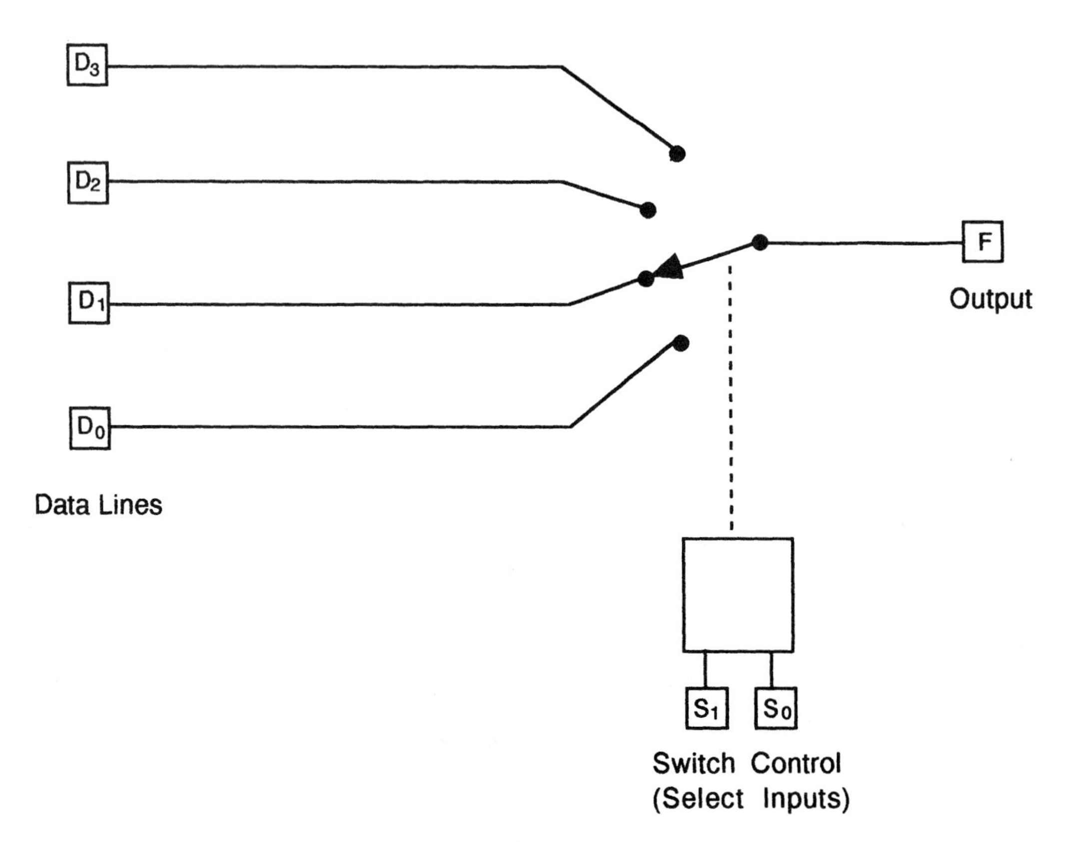

Fig. 5-3

EXAMPLE 5.1 **Multiplexers as data switches.** The circuit shown in Fig. 5-4 has four multiplexers with common select inputs. The outputs of these devices serve as inputs to another multiplexer, so overall the circuit behaves as a single four-select multiplexer having 16 data inputs. Such an arrangement is called a *multiplexer tree.*

If the select inputs are $S_3 S_2 S_1 S_0 = 1101$, determine which data line will be connected to F. $S_1 S_0 = 01$ points to data input 1 of multiplexer 5 through which it is connected to the output of multiplexer 2. Similarly, $S_3 S_2 = 11$ points to input 3 of multiplexer 2 which is connected to signal J. Thus, the unique select input 1101 connects output F to input J.

The Multiplexer as a Combinational Logic Device

There is a direct correspondence between multiplexer logic and a K map which may be best described by examples.

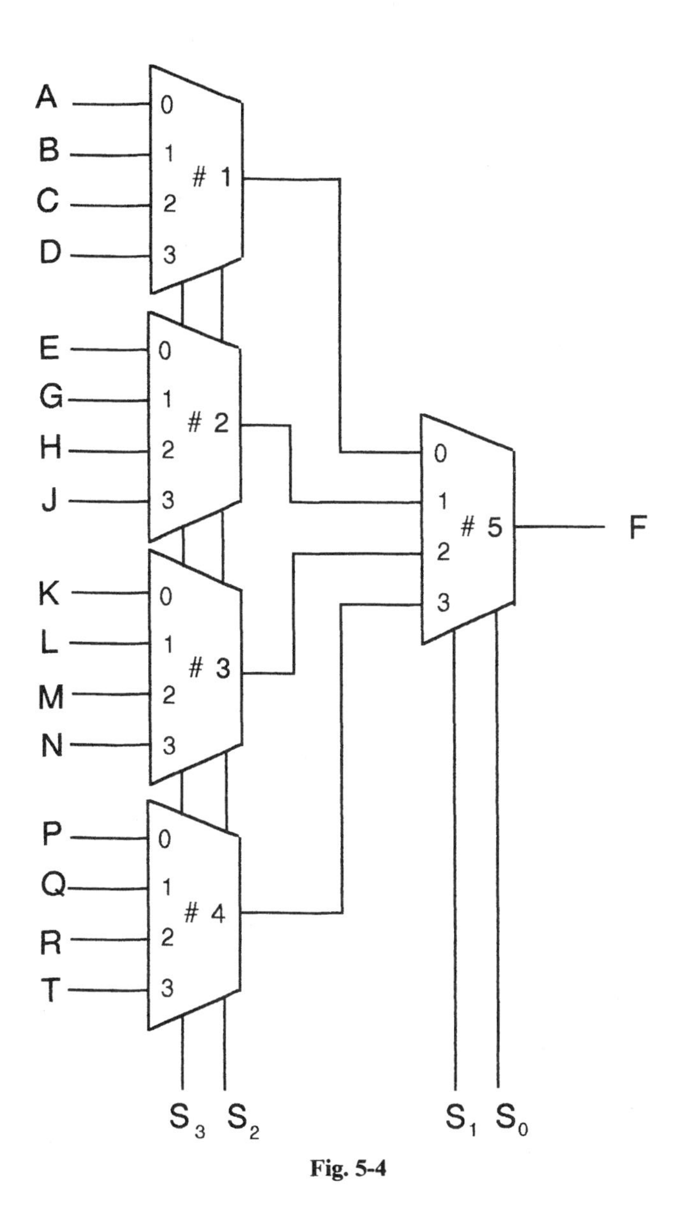

Fig. 5-4

EXAMPLE 5.2 Show the data connections required to cause the output F of a multiplexer to be related to its select inputs A, B, and C by the Boolean expression F = A′B′C′ + A′BC + ABC.

Each K-map square contains the value of logic-function F determined by the unique combination of input variables with which the square is associated. If these variables are thought of as multiplexer select variables, then each K-map square contains the value of the multiplexer output which corresponds to a particular combination of select variables (the address).

In Fig. 5-5, we see the given function displayed on a K map whose squares are numbered with the decimal equivalent of their binary addresses to aid in associating them with corresponding multiplexer data inputs.

On the K map, the square specified by ABC = 011 contains a 1 which indicates that F is TRUE for this input combination. The condition corresponds to the second term of the given equation. In order to obtain a multiplexer output of 1 when select inputs ABC = 011, it is only necessary to connect data input 3 to a voltage corresponding to TRUE or logic 1 as shown because F will be logically connected to this input when select variable combination ABC = 011.

The remaining data connections are determined in a similar fashion. For example, lines 1, 2, 4, 5, and 6 should be connected to logic 0 because the K-map squares specified by the 001, 010, 100, 101, and 110 all contain 0s.

Note that when a multiplexer is used to implement combinational logic, no K-map simplification process is required.

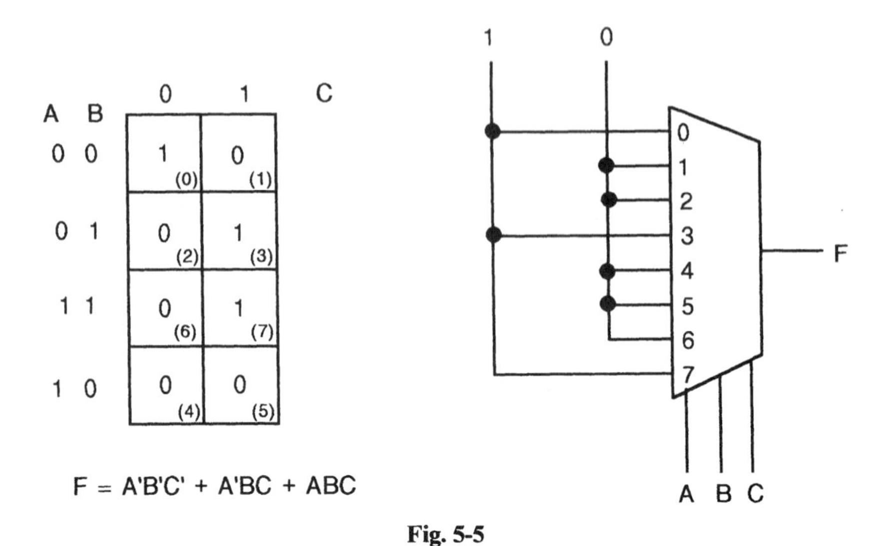

Fig. 5-5

EXAMPLE 5.3 Given the K map shown in Fig. 5-6, show that a more efficient multiplexer logic implementation may be obtained if we *use one less select input than the number of design variables.*

A B \ D C	0 0	0 1	1 1	1 0
0 0	1	1	0	0
1 0	0	1	0	1
1 1	1	1	0	1
0 1	0	1	0	0

Fig. 5-6

Step 1. Select a multiplexer with one less select input than there are design variables. In this example, choose a multiplexer with three select inputs.

Step 2. Choose any three design variables to be used as select inputs. Here, A, B, and D are chosen arbitrarily from the four possible combinations. If the choice is not dictated by some higher-priority design constraint, it should be one that simplifies the analysis as much as possible so as to minimize the possibility of error, as demonstrated in Prob. 5.5.

Step 3. Consider, in sequence, the effects of the various select combinations on the map and multiplexer (refer to Fig. 5-7).

ABD = 000. On the map, the first two upper-left locations in the top row are referenced. Since there is a 1 in both these locations, we see that F = 1 regardless of the value of C. On the multiplexer, the select combination ABD = 000 causes data input D_0 to be connected to F. Thus, the top data line (0) should be hard wired to a voltage corresponding to a logic 1 as shown.

ABD = 001. On the map, the two upper-right locations are referenced. Since both are 0s, F = 0 regardless of the value of C. On the multiplexer, data input D_1 will be connected to F. Thus data line 1 should be connected to a voltage corresponding to logic 0.

ABD = 010. This time, the map-referenced pair differ. Since F = 0 when C = 0 and F = 1 when C = 1, we see that F = C. On the multiplexer, data input D_2 will be connected to F. Since the given logic requires that F = C, data line 2 should be connected to a line carrying the variable C.

ABD = 101. Again, the map-referenced pair differ but this time F = 0 when C = 1 and F = 1 when C = 0, so we conclude that F is the logical inverse of C. On the multiplexer, data line 5 should be connected to C′.

The completely wired multiplexer is shown in Fig. 5-7, and it should be clear, at this point, that data inputs can be only 0, 1, C, or C′. Analysis of the remaining connections is left to the student.

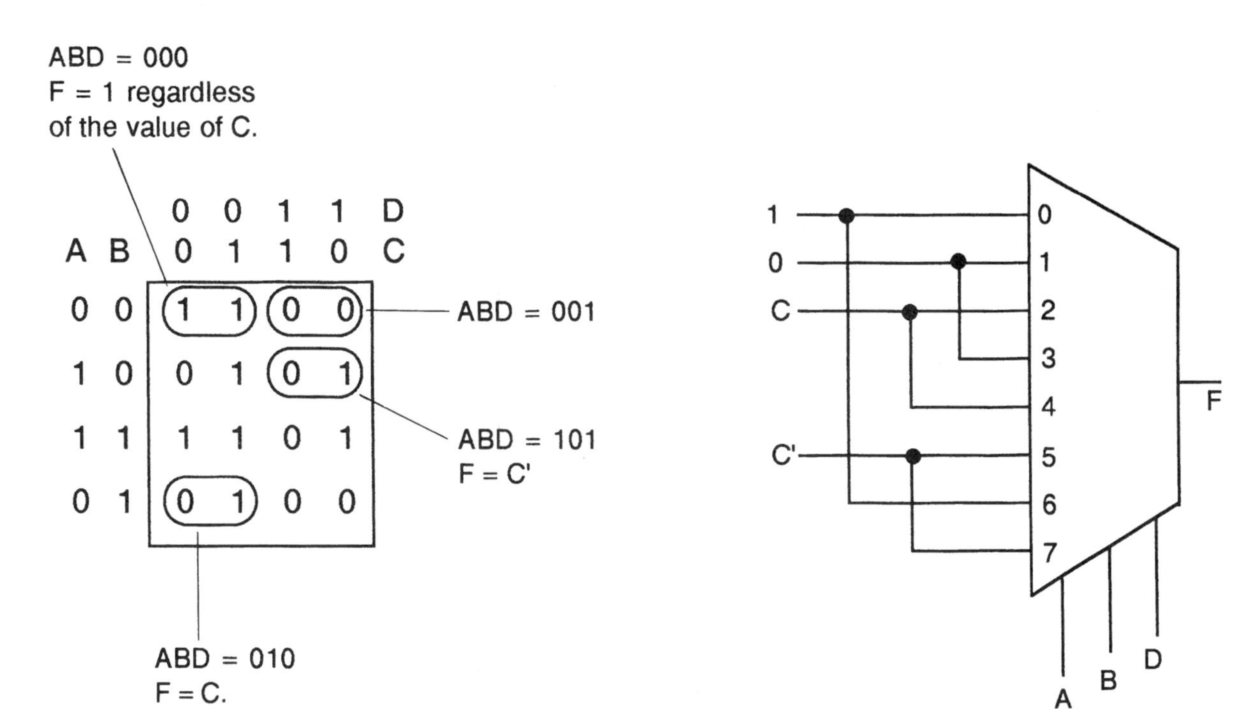

Fig. 5-7

It should be clear that there is a relationship between Boolean expressions, truth tables, K maps, and multiplexers and that, *given a proper size multiplexer for the number of variables, any function can be implemented.*

5.3 DECODERS AND DEMULTIPLEXERS

These two MSI elements are closely related. The *demultiplexer* is an inverse multiplexer in that it takes a single input and distributes it to one of several data outputs according to the state of the select variables. There are 2^N data outputs for N selects. Logical structure is very similar to that shown in Fig. 5-2 and Table 5.1 except that all the data inputs are tied together and there is no OR gate. The logic symbol is shown in Fig. 5-8 which illustrates a multiplexer-demultiplexer combination.

EXAMPLE 5.4 Time division multiplexing. A single transmission facility is to be shared among four subscribers. This may be achieved by using a demultiplexer as shown in Fig. 5-8. On the demultiplexer, L_0 through L_3 represent individual outputs (subscriber lines) while S_{1_d} and S_{0_d} are the select inputs used to determine the route to the output.

Note that the arrangement illustrated may be thought of as a simple telecommunication system in which any one source may be connected to any destination via a single transmission path.

The *decoder* is a special case of a demultiplexer in which *the data input is permanently tied to a logic 1*. Thus, the selected output will be in a logic 1 state while all other outputs are at logic 0. A decoder is termed a "full decoder" if all possible outputs are utilized. The full decoder can be looked upon as a hardware device which represents the rows of a truth table having the select variables as inputs. If, for example, the selects are 101, line 5 of the truth table will be chosen.

If we number the output lines in sequence, it becomes clear that the full decoder may be used as an *address decoder*. Here, the select inputs are treated as a binary number (the address), and a logic 1 will appear on the particular output line selected (addressed) by its unique binary code.

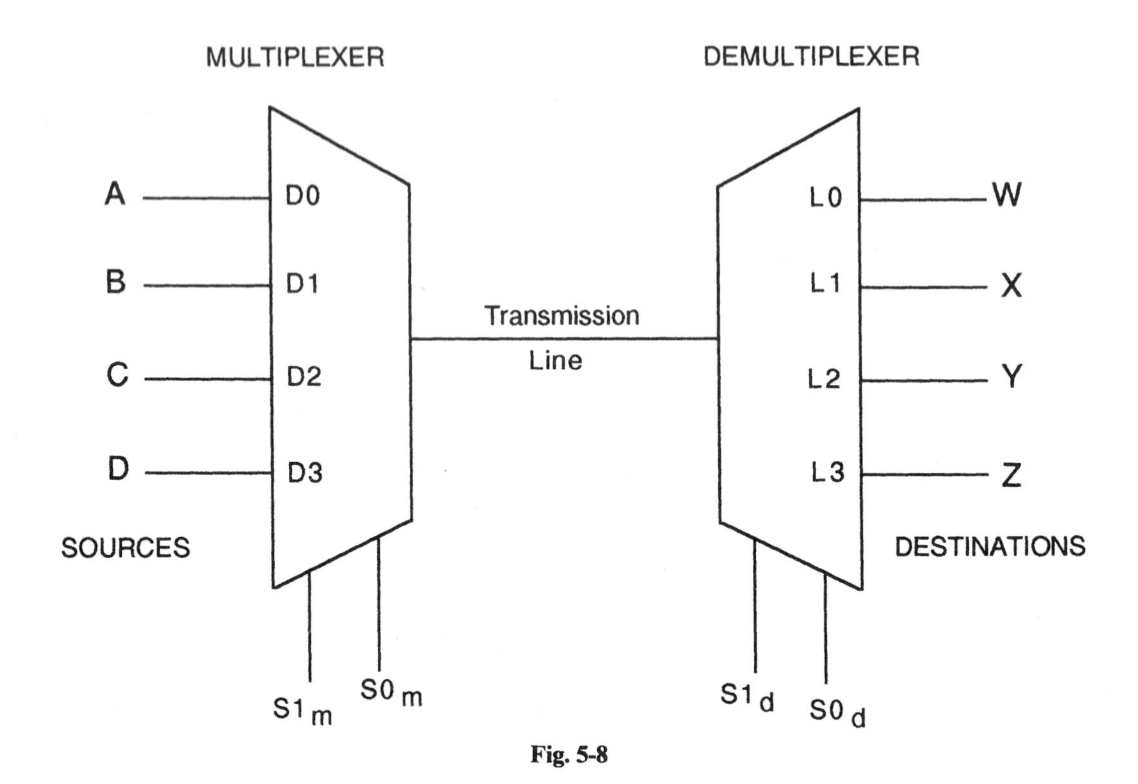

Fig. 5-8

5.4 THE READ-ONLY MEMORY (ROM)

A ROM is shown schematically in Fig. 5-9. It consists of logic which is equivalent to an address decoder (N = 3 in the current example) and a set of OR gates having a very high *fan-in* (large number of inputs per gate). The single vertical line at the input to each OR gate represents, in actuality, all 2^N gate inputs, there being one for each addressed line (decoder output).

A TRUE is produced on a decoder line corresponding to the address input number. This line may be connected to any one of the inputs on any one or more of the OR gates. On the diagram, ■ indicates a connection made and ○ indicates a possible connection not made (optional symbol). In the example of Fig. 5-9, there are eight decoder lines and four OR gates, so the total number of possible connections is $8 \times 4 = 32$.

If we recall that the full decoder is an array of AND gates, each with inputs connected to a unique combination of address bits and which effectively uses the address to "key on" a line in the truth table, then we see that the decoder corresponds to the input side of a truth table. An output variable will be 1 if one or more selected lines corresponding to truth table rows are connected to an OR gate, there being

Table 5.2

	Address						
Line	A	B	C	W	X	Y	Z
0	0	0	0	0	0	0	0
1	0	0	1	1	0	1	0
2	0	1	0	0	0	1	1
3	0	1	1	1	1	0	0
4	1	0	0	0	1	1	0
5	1	0	1	1	0	1	1
6	1	1	0	1	0	0	1
7	1	1	1	0	1	1	1

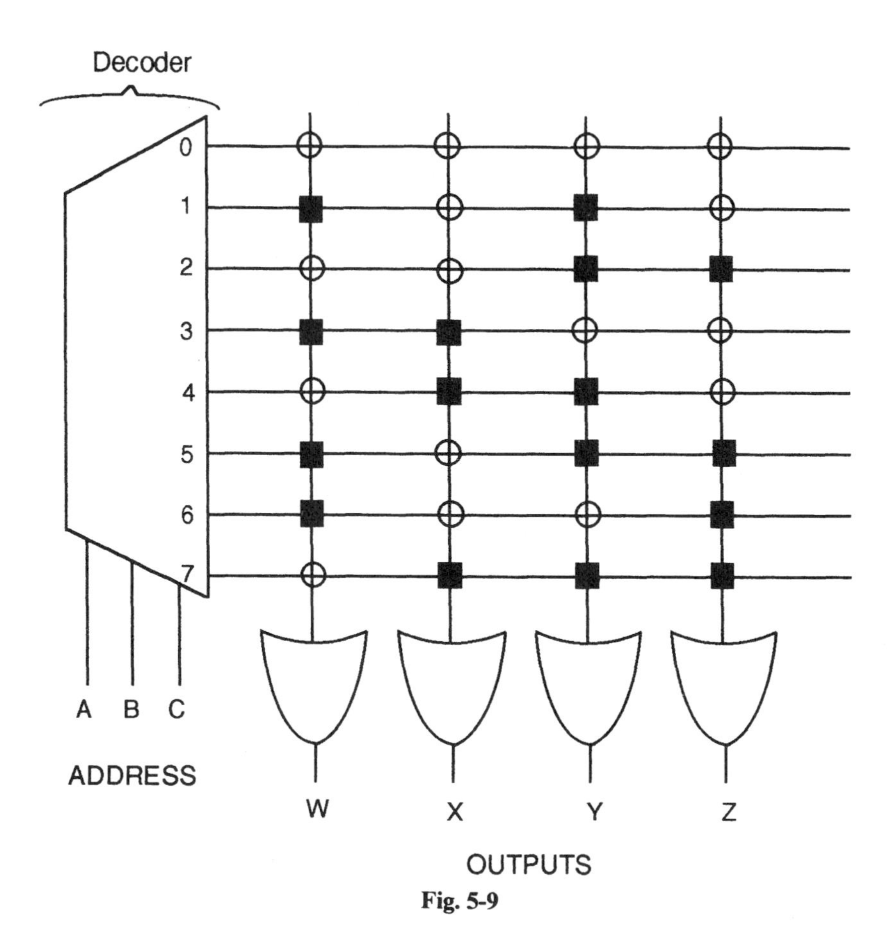

Fig. 5-9

a separate OR gate for each dependent output variable. The list of connections for a given OR gate can be interpreted as a truth table output column. Thus, *a ROM may be interpreted as the hardware representation of a truth table*, and it may be used to implement combinational logic as illustrated by Table 5.2 which represents the connections shown in Fig. 5-9.

Alternately, *the ROM can be considered as a memory device* with data stored as the pattern of connections corresponding to a given address. Customarily, a connection is considered a 1 and the absence of a connection a 0. The pattern of connections along a given address line is termed the "contents" of the address and comprises the data stored there. In Fig. 5-9, if we consider W to be the most significant bit and Z the least, then address 5 is seen to contain binary 1011 or hex B. Data is recovered by setting the address bits appropriately and reading the OR gate outputs.

Programming a ROM

A ROM is manufactured with a full decoder and all the high fan-in OR gates without any connections being made between decoder lines and OR gate inputs. When a customer desires a ROM of a given size, he or she supplies the vendor with information (a *program*) which is usually in the form of a truth table with the contents of addresses represented as hexadecimal numbers. The vendor then makes the appropriate metallic connections, packages the chip, and ships it to the customer. It is possible to create user-programmable ROMs. These components will be discussed in Chap. 8.

EXAMPLE 5.5 Binary to Gray code conversion ROM. The ROM, as a memory device, can be viewed as a lookup table. The comprehensive set of binary inputs to a truth table ($B_3 B_2 B_1 B_0$ in Table 1.1) are used as ROM addresses, and the Gray-coded number corresponding to each binary input is stored at the corresponding address as shown in Fig. 5-10. We see that information in a truth table is effectively looked up by the input selects.

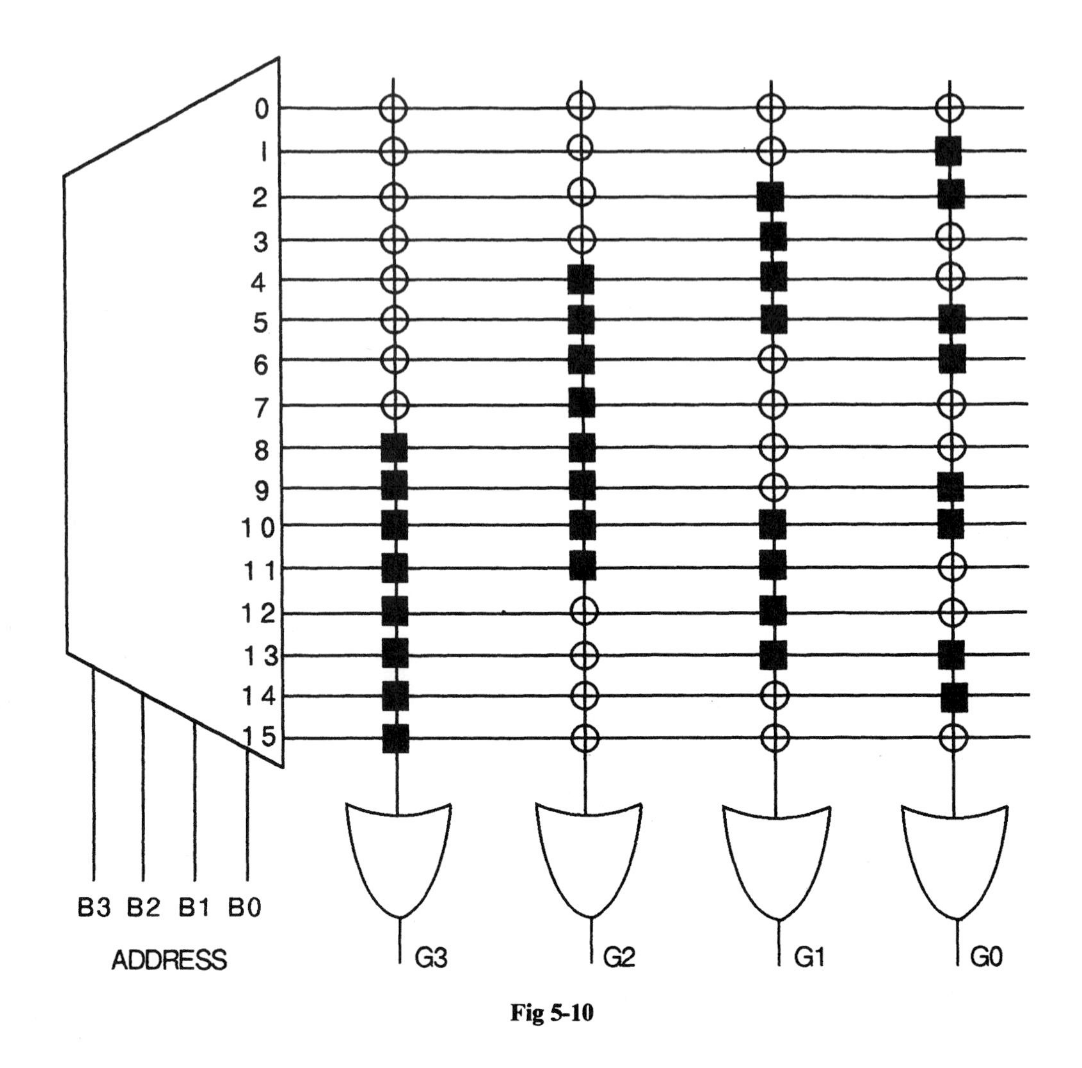

Fig 5-10

Solved Problems

5.1 Draw the logic diagram for a multiplexer having three select inputs.

Determine the AND gate connections by constructing the left side of the truth table like that shown in Table 5.1 (see Table 5.3).

Table 5.3

S_2	S_1	S_0	S'_2	S'_1	S'_0
0	0	0	1	1	1
0	0	1	1	1	0
0	1	0	1	0	1
0	1	1	1	0	0
1	0	0	0	1	1
1	0	1	0	1	0
1	1	0	0	0	1
1	1	1	0	0	0

Note that each row has a unique combination of three 1s. The corresponding connections are shown in Fig. 5-11.

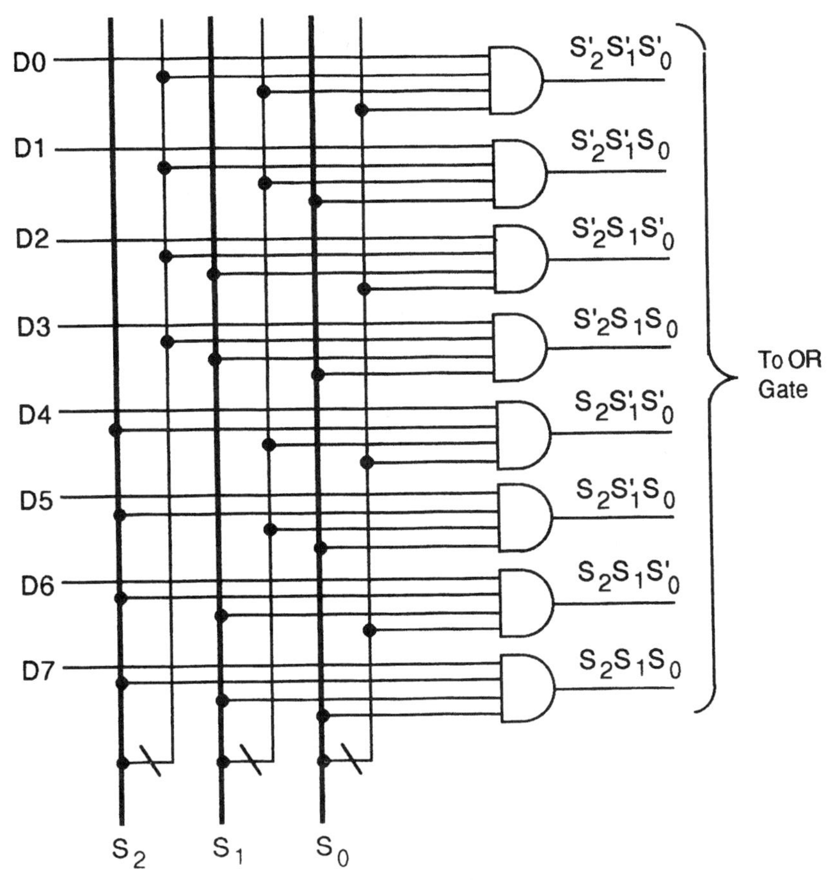

Fig. 5-11

5.2 Implement F = C′D + A′BD + A′B′D′ + ABD′ using a suitably sized multiplexer. Let A, B, and C be the select variables.

First, map the function as shown in Fig. 5-12. Second, group entries into pairs as in Example 5.3. Corresponding decimal equivalents of the inputs addressed by variables A, B, and C are shown in Fig. 5-13 to help illustrate the process. Finally, determine appropriate multiplexer data input values as described in Example 5.3. The resulting connections are shown in Fig. 5-14.

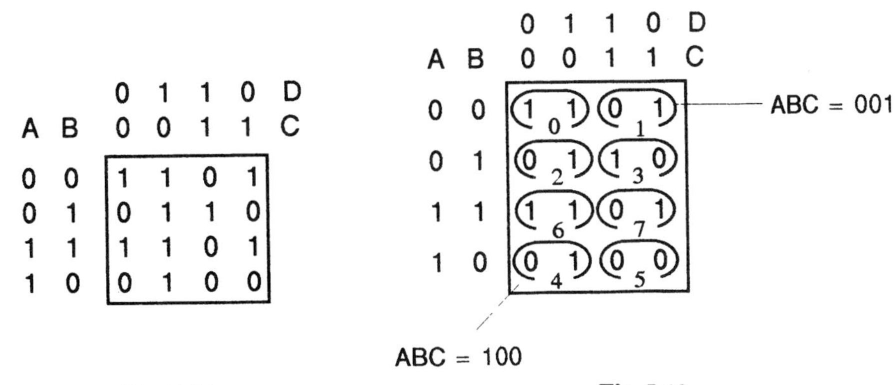

Fig. 5-12 **Fig. 5-13**

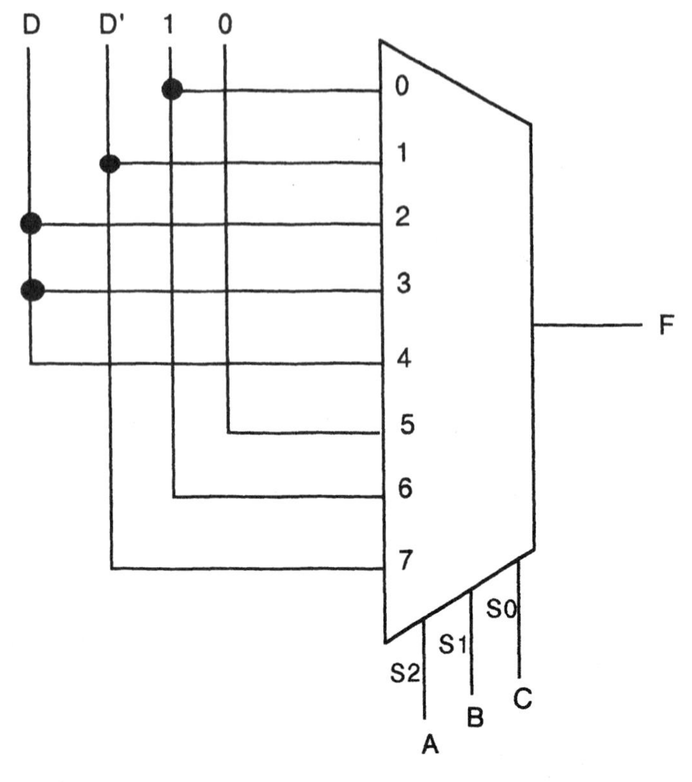

Fig. 5-14

5.3 Repeat Prob. 5.2, this time using A, C, and D as selects.

The map with revised groupings is shown in Fig. 5-15, and the corresponding multiplexer wiring is shown in Fig. 5-16.

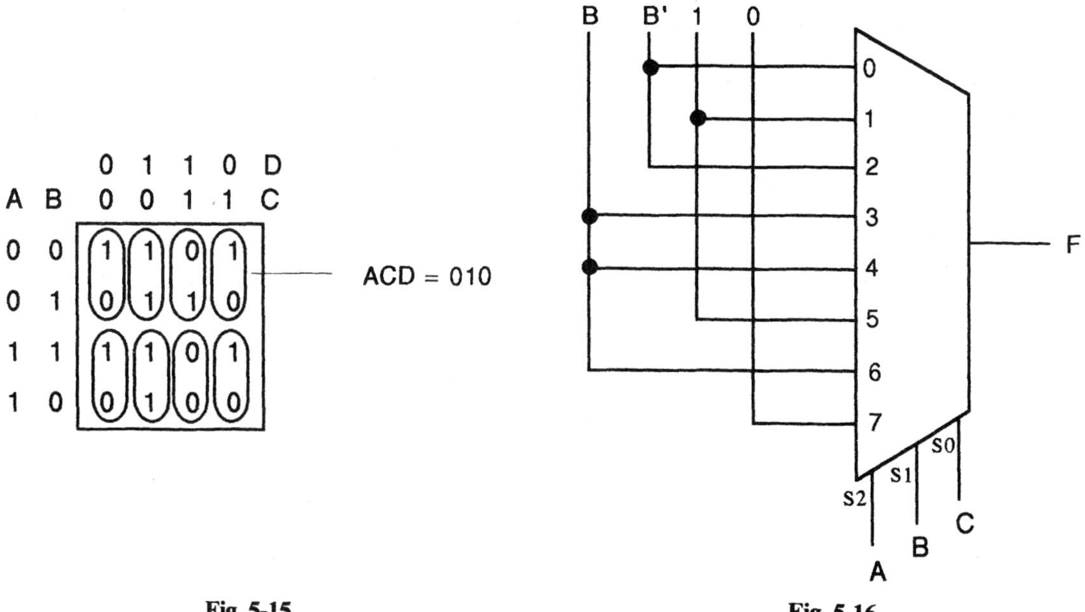

Fig. 5-15

Fig. 5-16

5.4 Repeat Prob. 5.2 using a multiplexer with data inputs consisting only of hard wired 1s or 0s.

In this case, all the variables are used as selects, and each select combination represents a single square on the K map which unambiguously determines a corresponding multiplexer input. Since the multiplexer has four selects, it is twice the size of the multiplexers used in Probs. 5.2 and 5.3. The tabulated results are presented in Table 5.4.

Table 5.4

S_3 A	S_2 B	S_1 C	S_0 D	Data Line	DATA
0	0	0	0	0	1
0	0	0	1	1	1
0	0	1	0	2	1
0	0	1	1	3	0
0	1	0	0	4	0
0	1	0	1	5	1
0	1	1	0	6	0
0	1	1	1	7	1
1	0	0	0	8	0
1	0	0	1	9	1
1	0	1	0	10	0
1	0	1	1	11	0
1	1	0	0	12	1
1	1	0	1	13	1
1	1	1	0	14	1
1	1	1	1	15	0

5.5 Create the K map corresponding to the multiplexer shown in Fig. 5-17 in which Z is the least significant select bit.

For the case where the select inputs are $X = 0$, $Y = 1$, and $Z = 0$, data line 2 will be selected and W′ is routed to F. Thus, F will equal 0 when $W = 1$, and F will be 1 when $W = 0$; a relationship which may be expressed as $f_1 = W'X'YZ'$ is one of the product terms in the total Boolean expression for F. When $XYZ = 001$, line 1 is selected, and we see that the function is equal to 1 regardless of the value of W. This condition is expressed as $f_2 = WX'Y'Z + W'X'Y'Z$ or, equivalently, $F = 1 \cdot X'Y'Z$. The complete logic function F may be constructed by a similar analysis for each input combination, omitting all combinations which yield a 0 output. The resulting Boolean expression is $F = X'Y'Z + W'X'Y'Z' + W'X'YZ' + WX'YZ + WXY'Z' + WXYZ'$ which may now be mapped as shown in Fig. 5-18.

A considerably simpler method is to arrange the K map so that two of the three select variables (X and Y in the present example) are in adjacent rows or columns and the third (Z) has its 1s and 0s in pairs as shown in Fig. 5-19. If this is done, it is easy to group the entries and fill out the map by inspection. Refer to Example 5.3 and Prob. 5.2.

Note that even though the maps of Figs. 5-18 and 5-19 appear different, they actually contain the same information. The lesson to be learned here is that the difficulty of a problem can often be increased or decreased significantly by simple rearrangement of variables or presentation.

5.6 Find an equivalent single multiplexer realization of the logic shown in Fig. 5-20. A larger multiplexer will be required.

The output of M_1 is $F_1 = AC'D' + BC'D + CD'$. The output of M_2 is $F_2 = A'B'C' + A'B + ABD$. Then $F = F_1 + F_2$, which is mapped in Fig. 5-21. Arbitrarily choosing ABC as selects leads to the realization shown in Fig. 5-22.

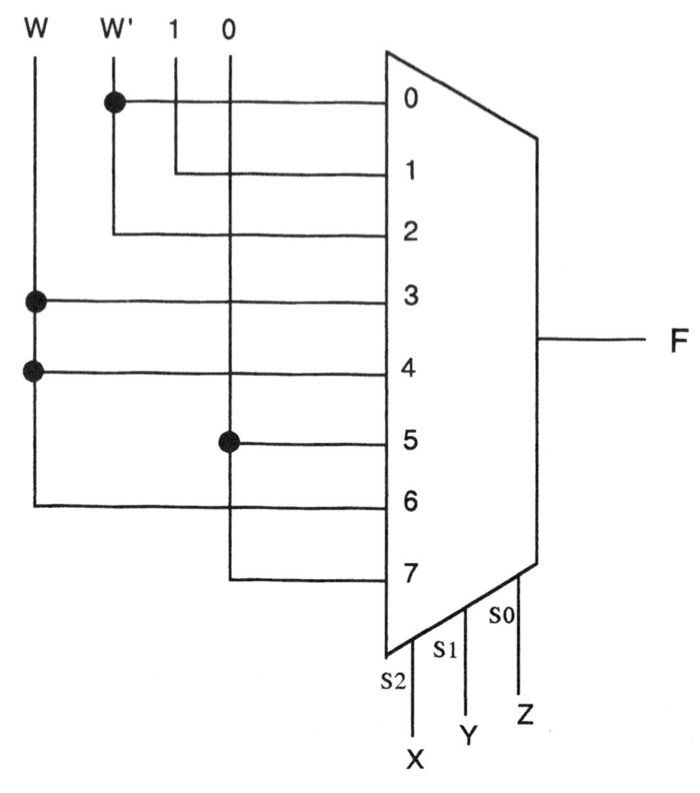

Fig. 5-17

	Z: 0	1	1	0
W X	**Y: 0**	**0**	**1**	**1**
0 0	1	1	0	1
0 1	0	0	0	0
1 1	1	0	0	1
1 0	0	1	1	0

Fig. 5-18

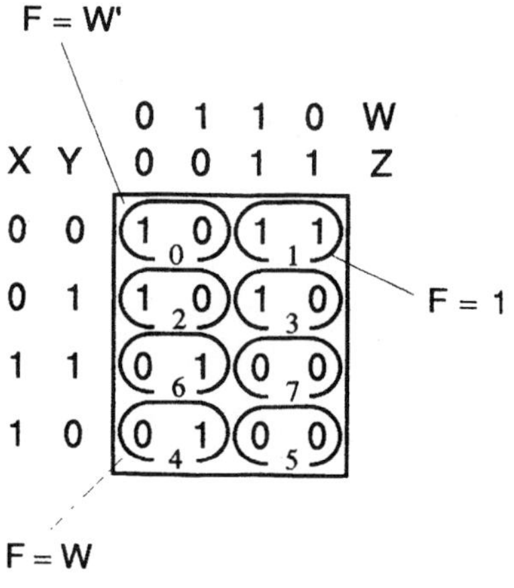

F = W'

	W: 0	1	1	0
X Y	**Z: 0**	**0**	**1**	**1**
0 0	1	$_0$ 0	1	$_1$ 1
0 1	1	$_2$ 0	1	$_3$ 0
1 1	0	$_6$ 1	0	$_7$ 0
1 0	0	$_4$ 1	0	$_5$ 0

F = 1

F = W

Fig. 5-19

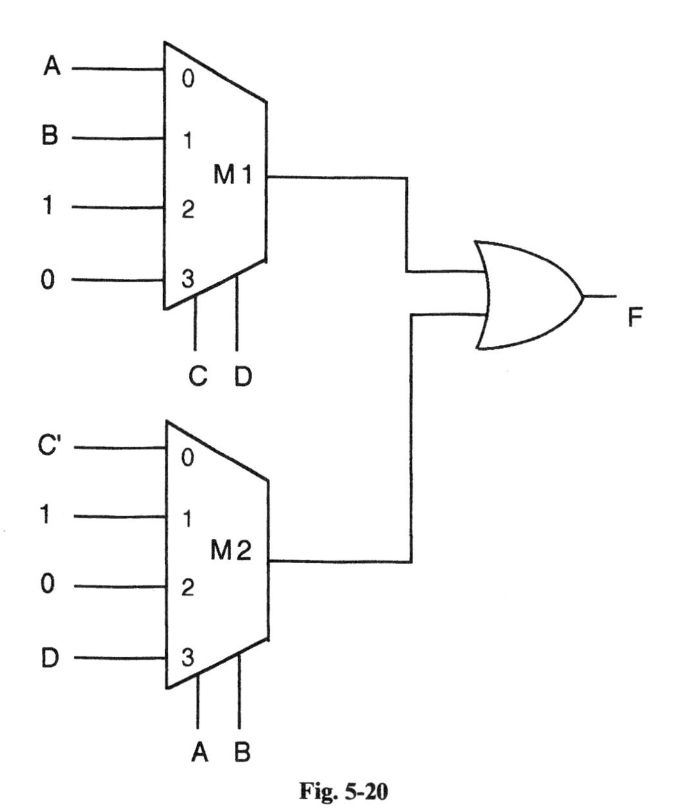

Fig. 5-20

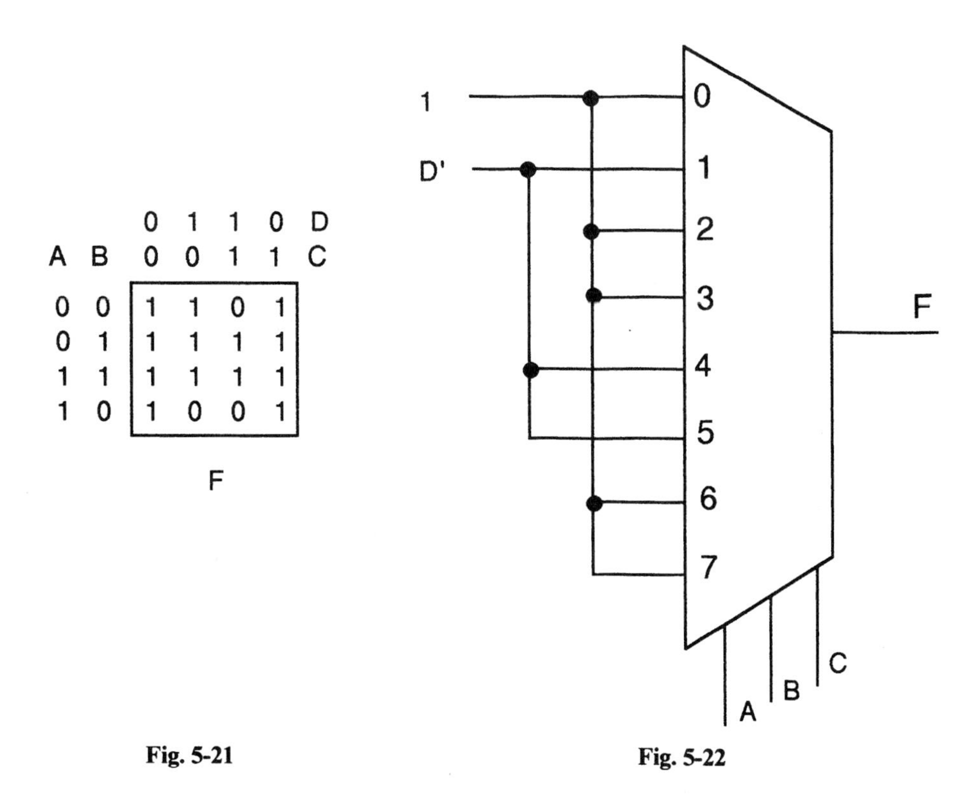

Fig. 5-21

Fig. 5-22

5.7 Create a ROM which will multiply two 2-bit binary numbers.

Call the input numbers $A_1 A_0$ and $B_1 B_0$ and create the truth table necessary to represent the 4-bit product (see Table 5.5). If we treat the input numbers as address variables and assign each output 1 to an OR gate, the required connections may be read directly from the truth table columns as shown in Fig. 5-23. Since circles indicating unmade connections (Sec. 5.4) are redundant, they are omitted.

Table 5.5

A1	A0	B1	B0	F3	F2	F1	F0
0	0	0	0	0	0	0	0
0	0	0	1	0	0	0	0
0	0	1	0	0	0	0	0
0	0	1	1	0	0	0	0
0	1	0	0	0	0	0	0
0	1	0	1	0	0	0	1
0	1	1	0	0	0	1	0
0	1	1	1	0	0	1	1
1	0	0	0	0	0	0	0
1	0	0	1	0	0	1	0
1	0	1	0	0	1	0	0
1	0	1	1	0	1	1	0
1	1	0	0	0	0	0	0
1	1	0	1	0	0	1	1
1	1	1	0	0	1	1	0
1	1	1	1	1	0	0	1

Table 5.6

INPUT (GRAY)				OUTPUT (BINARY)			
G3	G2	G1	G0	B3	B2	B1	B0
0	0	0	0	0	0	0	0
0	0	0	1	0	0	0	1
0	0	1	0	0	0	1	1
0	0	1	1	0	0	1	0
0	1	0	0	0	1	1	1
0	1	0	1	0	1	1	0
0	1	1	0	0	1	0	0
0	1	1	1	0	1	0	1
1	0	0	0	1	1	1	1
1	0	0	1	1	1	1	0
1	0	1	0	1	1	0	0
1	0	1	1	1	1	0	1
1	1	0	0	1	0	0	0
1	1	0	1	1	0	0	1
1	1	1	0	1	0	1	1
1	1	1	1	1	0	1	0

5.8 Create a ROM program which will convert from Gray code to binary.

The Gray code sequence of numbers is rearranged into ascending binary order to serve as the input side of a truth table which has already been discussed in Chap. 2 (refer to Table 2.1). We now know that this sequence may also be interpreted as a list of ROM addresses and that each binary digit corresponding to a particular Gray code input is stored at a unique ROM location and appears as output when its related address is placed on the select lines. The relevant truth table is shown in Table 5.6 and the correctly wired ROM in Fig. 5-24.

5.9 Given the *hardware* diagram shown in Fig. 5-25, create an equivalent ROM.

Notice that there are two instances of unbalanced half arrows. Since we are given a *hardware* diagram, balancing must be achieved without adding any additional hardware. This can be done by the appropriate use of the logical inversion, or slash mark (a nonhardware element), which redefines the truth polarity (LT or HT) at its insertion point. The process may be understood by reviewing the material in Sec. 4.3 in connection with the interpreting of voltage inverters and their relation to the logical inversion process. The resulting balanced diagram is shown in Fig. 5-26 in which the alternative bubble convention is used instead of half arrows.

Once balancing has been achieved, removal of all bubbles and the hardware inverter (*not the slash*) to expose the underlying logic diagram permits reading of the desired Boolean expression by inspection:

$$F = W'X'Z' + W'X(Y \oplus Z) + WX'YZ + WX(YZ)'$$
$$= W'X'Z' + W'XYZ' + W'XY'Z + WX'YZ + WXY' + WXZ'$$

The resulting K map and ROM are shown in Fig. 5-27.

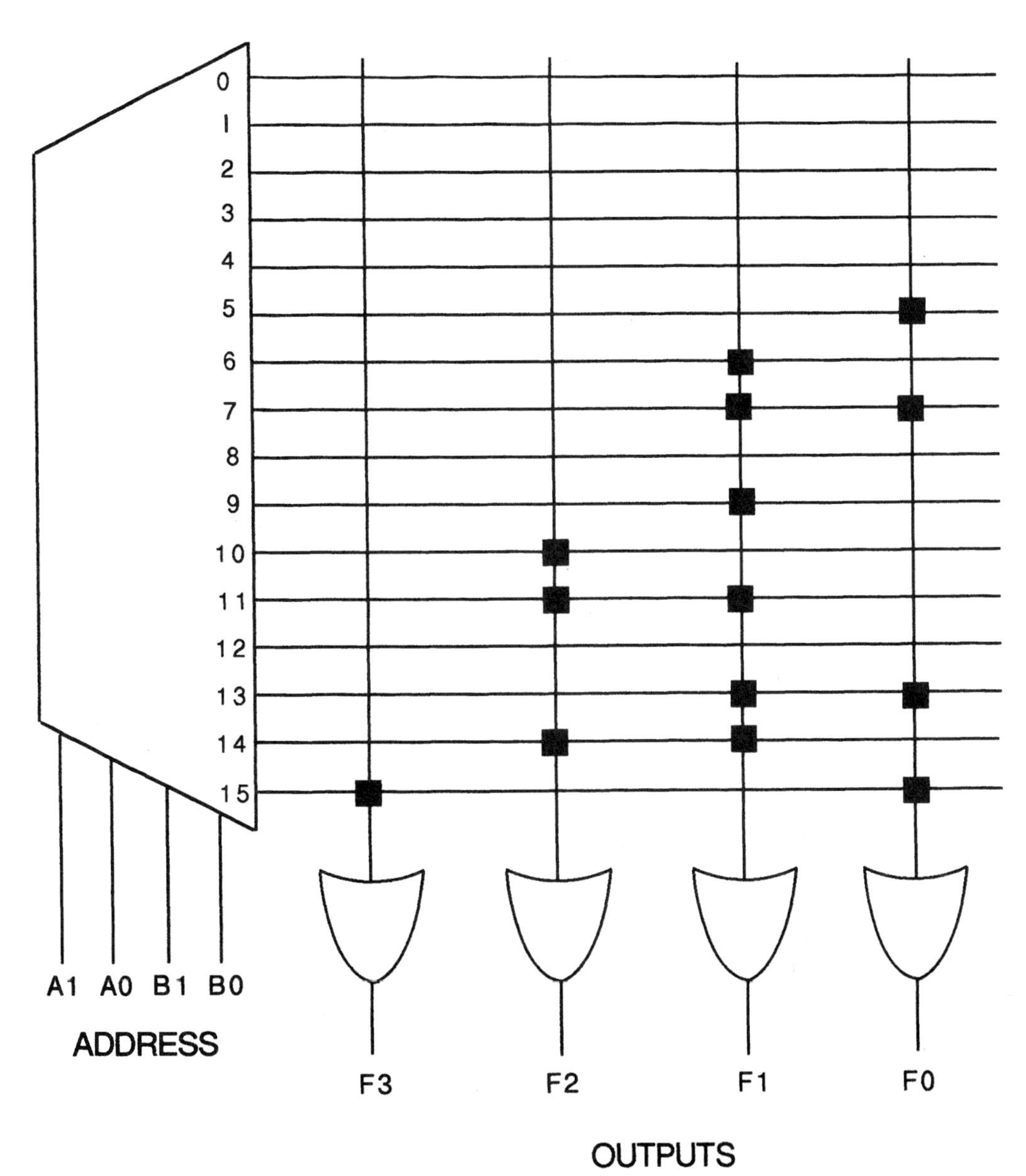
0
1
2
3
4
5
6
7
8
9
10
11
12
13
14
15
A1 A0 B1 B0
ADDRESS
F3
F2
F1
F0
OUTPUTS

Fig. 5-23

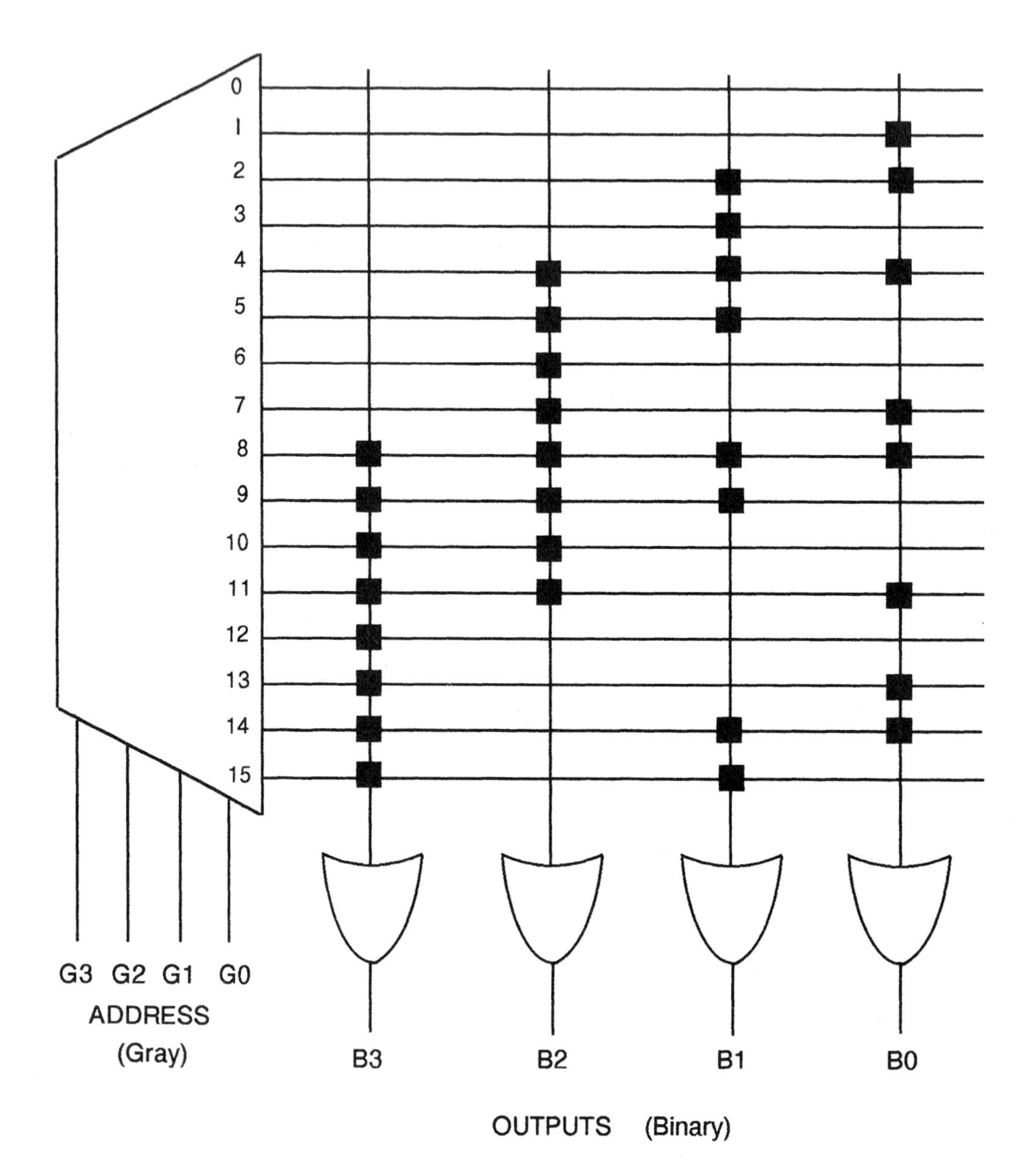
0
1
2
3
4
5
6
7
8
9
10
11
12
13
14
15
G3 G2 G1 G0
ADDRESS
(Gray)
B3
B2
B1
B0
OUTPUTS (Binary)

Fig. 5-24

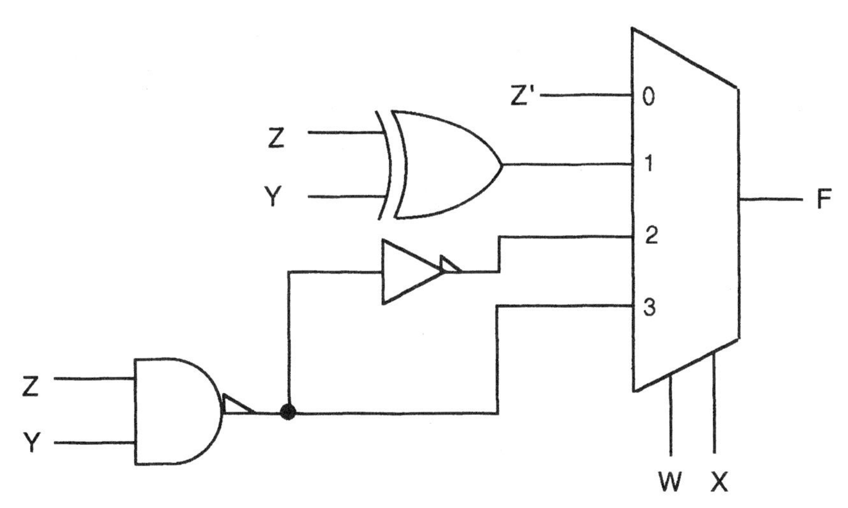

Fig. 5-25

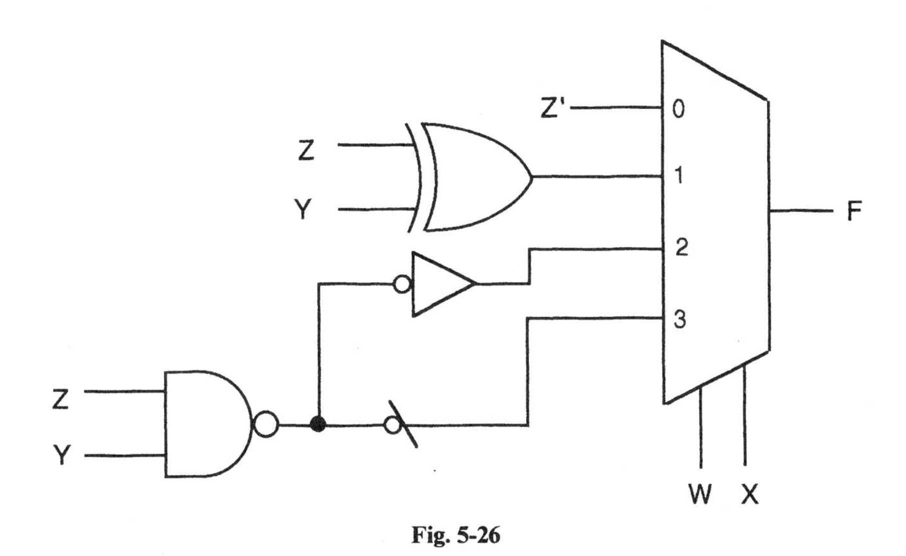

Fig. 5-26

5.10 Implement the logic shown in Fig. 5-28 with a single multiplexer.

From the logic diagram, we see that the Boolean equation is

$$F = XZ' + WYZ + W'X'Z + W'YZ' + WXY$$

Mapping, we get Fig. 5-29. Observe, in Fig. 5-29, that the term WXY is redundant since all 1s are involved in encirclements without it. From the map we get Table 5.7, from which we may now make the multiplexer connections shown in Fig. 5-30.

5.11 Given the multiplexer circuit shown in Fig. 5-31, create an equivalent logic using NAND gates.

By now, the student should be skilled enough to fill out a K map directly from the multiplexer. The process can be aided, however, by constructing a table similar to that of Table 5.7.

Since the goal is a NAND realization, it is reasonable to proceed with a simplification to reduce the number of gates. The K map with an effective 1s covering is shown in Fig. 5-32 and the function is $F = AB + CD + BD'$. The resulting logic diagram for NAND is shown in Fig. 5-33.

		Z: 0	1	1	0
W	X	Y: 0	0	1	1
0	0	1_0	0_1	0_3	1_2
0	1	0_4	1_5	0_7	1_6
1	1	1_{12}	1_{13}	0_{15}	1_{14}
1	0	0_8	0_9	1_{11}	0_{10}

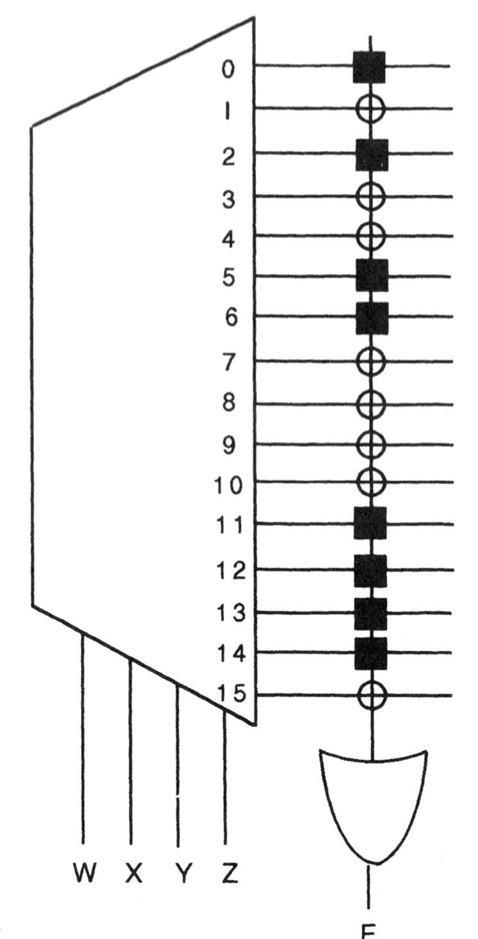

Fig. 5-27

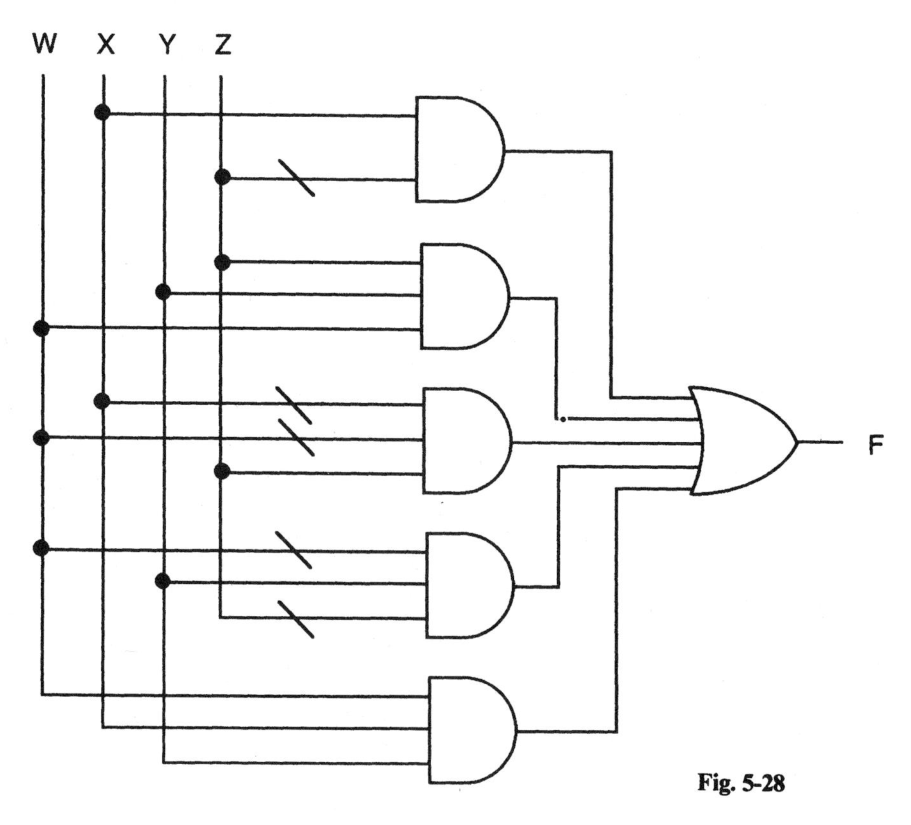

Fig. 5-28

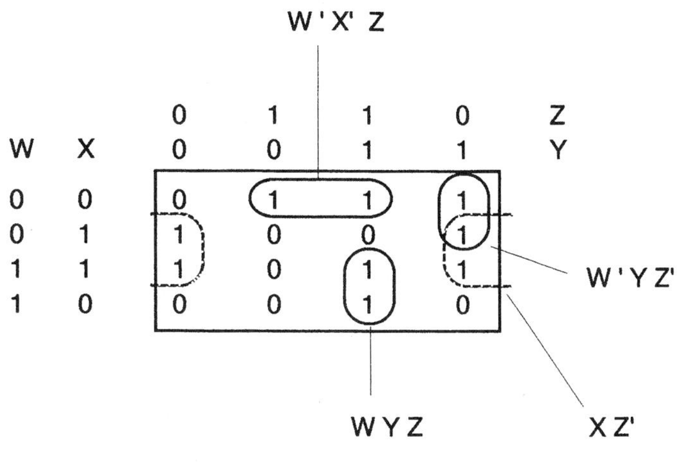

Fig. 5-29

Table 5.7

	W	X	Y	F	
0	0	0	0	Z	F = 0 when Z = 0 F = 1 wnen Z = 1
1	0	0	1	1	
2	0	1	0	Z'	F = 1 when Z = 0 F = 0 when Z = 1
3	0	1	1	Z'	
4	1	0	0	0	
5	1	0	1	Z	
6	1	1	0	Z'	
7	1	1	1	1	F = 1 regardless of Z

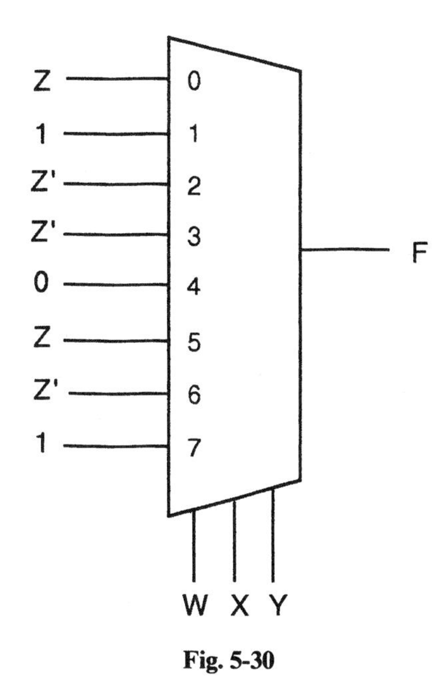

Fig. 5-30

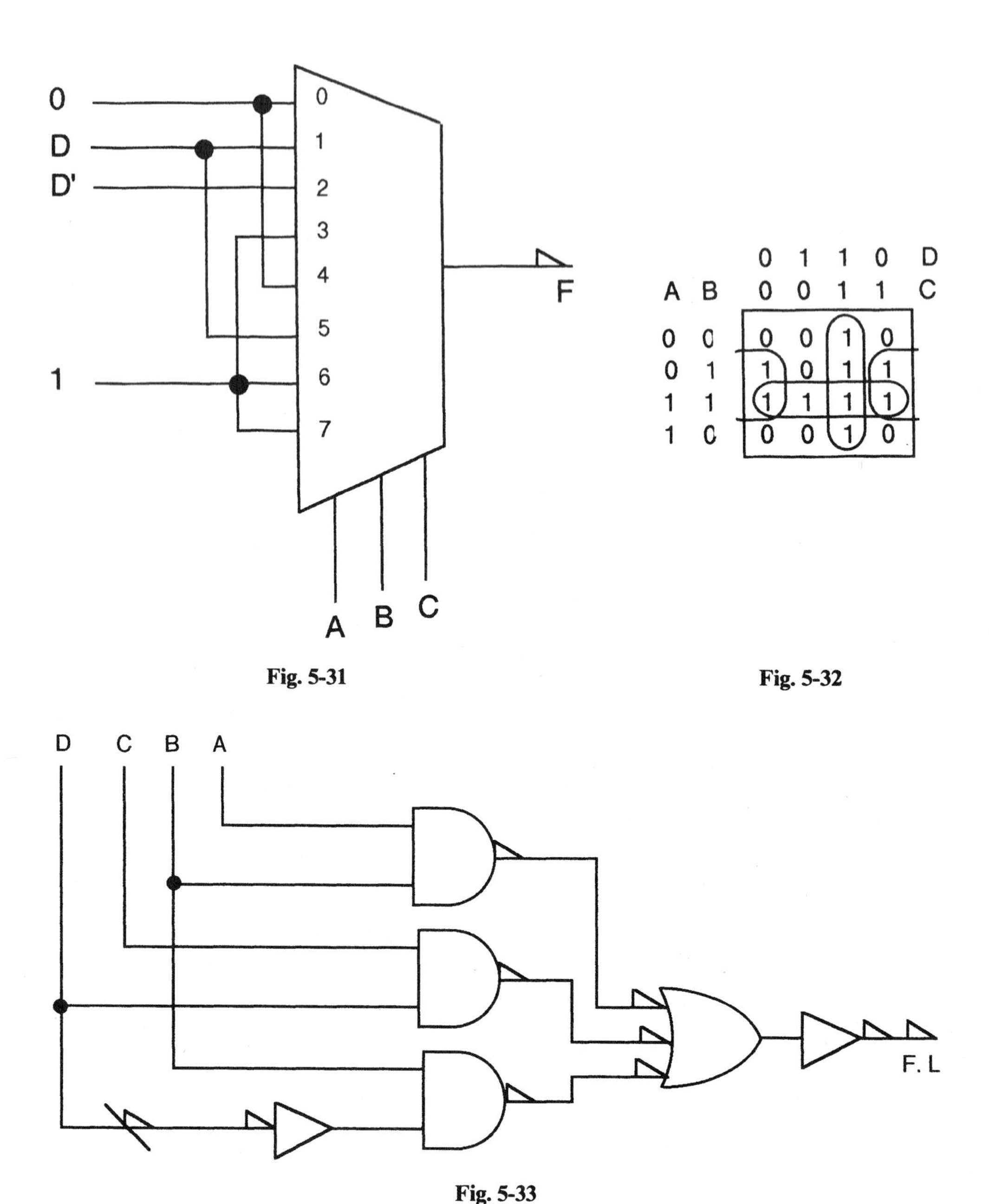

Fig. 5-31

Fig. 5-32

Fig. 5-33

5.12 Design a multiplexer to check parity on a 4-bit digital word; i.e., the circuit should produce a true output if the number of 1s in the word is odd.

There will be four independent variables and one dependent variable (Fig. 5-34). Mapping the function and grouping entries as in Prob. 5.5, we have Fig. 5-35. Then, using A, B, and C as multiplexer selects, we obtain the arrangement shown in Fig. 5-36.

5.13 Given the hardware diagram shown in Fig. 5-37, implement the equivalent logic on a single multiplexer.

The logic may be easily deciphered if we first balance the bubbles by adding logical inversions as shown in Fig. 5-38. After stripping off all bubbles and hardware inverters (leaving the slash marks undisturbed), the Boolean equation may be read directly from the remaining logic diagram as

$$F = XZ' + WYZ + W'X'Z + W'YZ' + WXY$$

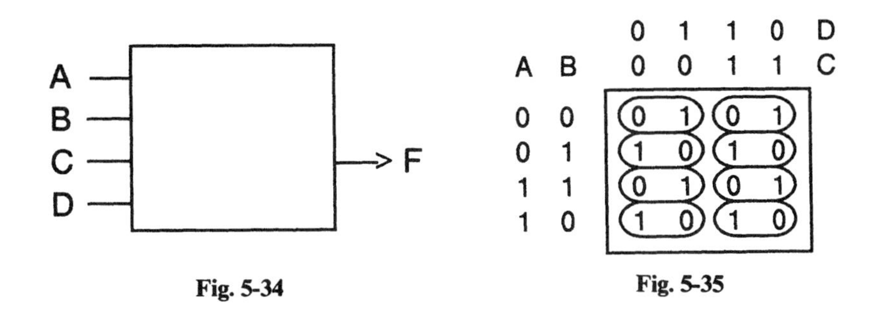

Fig. 5-34

Fig. 5-35

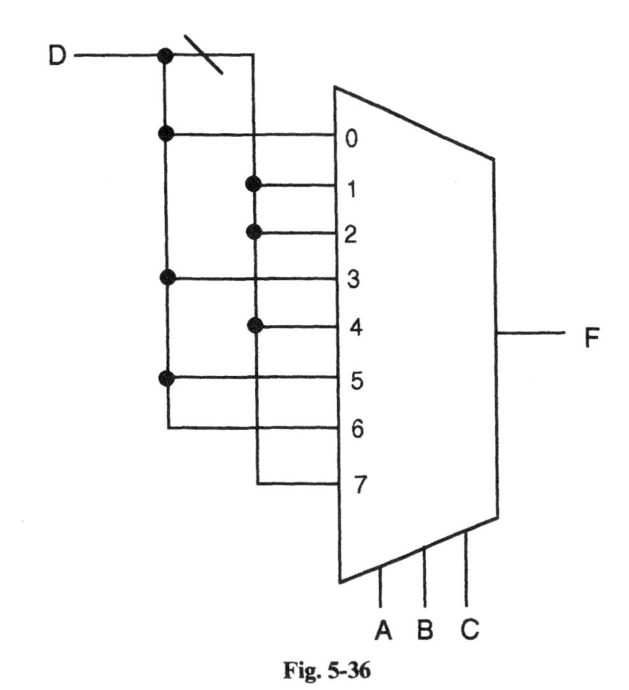

Fig. 5-36

Mapping the equation we get Fig. 5-39.

Finally, after grouping map entries as in Prob. 5.5 and using W, X, and Y as select variables, we obtain the wired multiplexer as shown in Fig. 5-40.

5.14 Determine the Boolean expression corresponding to the hardware diagram shown in Fig. 5-41.

Since there is a hardware inverter between multiplexers M_2 and M_3, the bubbles must be balanced by the addition of a logical inversion as shown in Fig. 5-42. It is assumed that all specified variables are HT. Consider M_3 in Fig. 5-41:

$$A'B' \to F = M_2'$$
$$A'B \to F = M_1$$
$$AB' \to F = M_2'$$
$$AB \to F = M_2$$

Also, $$M_2 = C'D'E + C'DE + CDE'$$

This may be mapped (Fig. 5-43) and simplified by means of a 1s covering to obtain

$$M_2 = C'E + CDE' \tag{5.1}$$

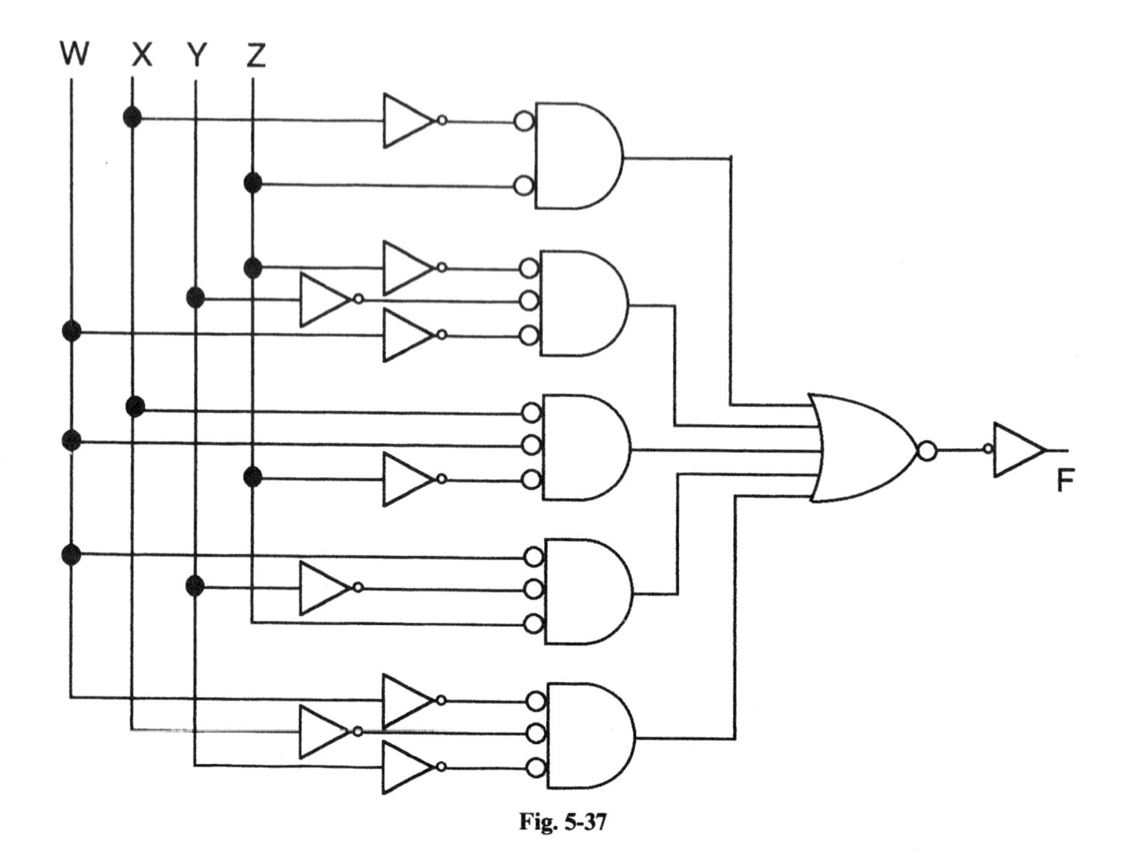

Fig. 5-37

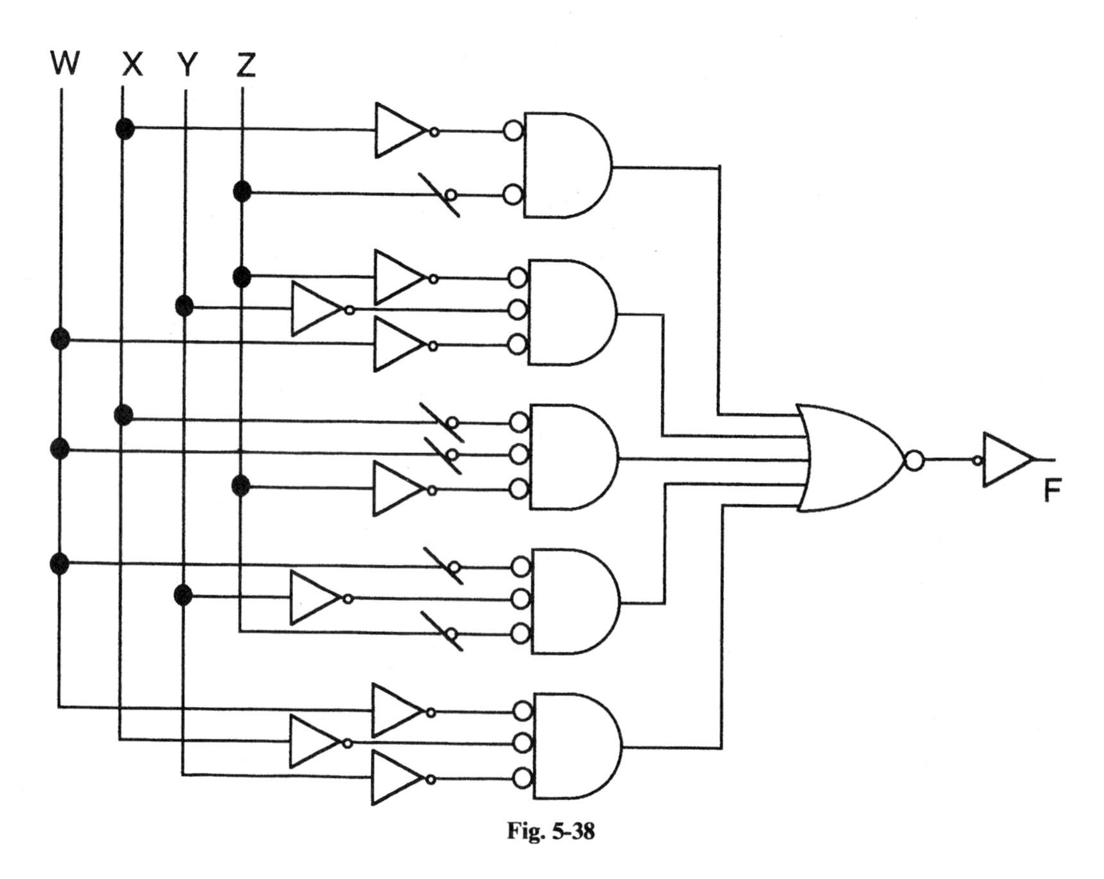

Fig. 5-38

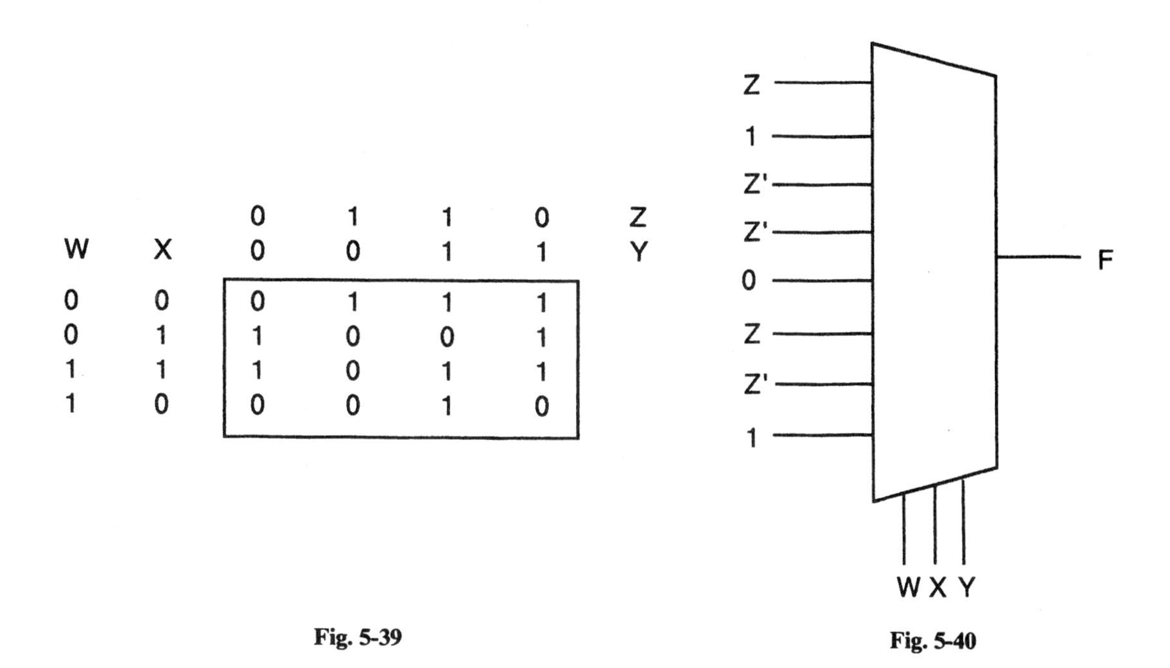

W	X	ZY = 00	ZY = 10	ZY = 11	ZY = 01
0	0	0	1	1	1
0	1	1	0	0	1
1	1	1	0	1	1
1	0	0	0	1	0

Fig. 5-39

Fig. 5-40

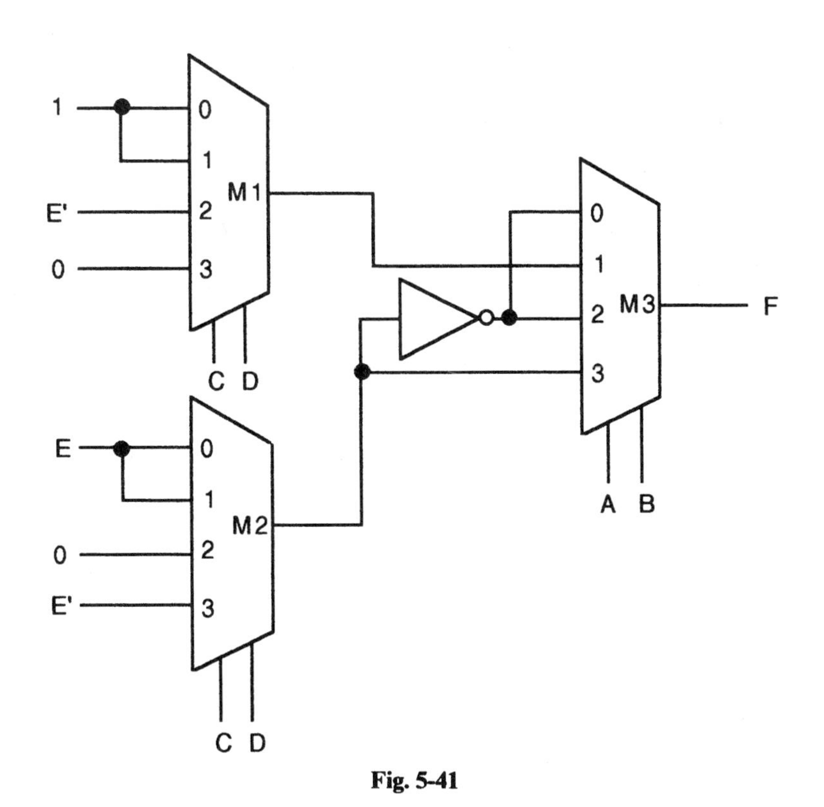

Fig. 5-41

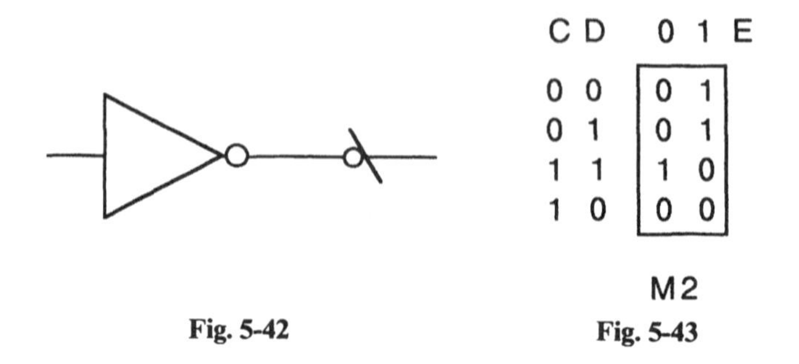

Fig. 5-42

Fig. 5-43

A 0s covering provides us with

$$M_2' = C'E' + CD' + CE \tag{5.2}$$

Similarly,

$$\begin{aligned} M_1 &= C'D' + C'D + CD'E' \\ &= C' + CD'E' \end{aligned} \tag{5.3}$$

We may now construct the function F.

$$\begin{aligned} F &= A'B'M_2' + A'BM_1 + AB'M_2' + ABM_2 \\ &= B'M_2' + A'BM_1 + ABM_2 \end{aligned} \tag{5.4}$$

Substituting Eqs. (5.1), (5.2), and (5.3) into (5.4), we obtain

$$F = B'C'E' + B'CD' + B'CE + A'BC' + A'BCD'E' + ABC'E + ABCDE'$$

which is mapped in Fig. 5-44. Simplification is achieved by means of a 1s covering as shown to yield,

$$F = A'C'E' + CD'E' + B'C'E' + BC'E + B'CE$$

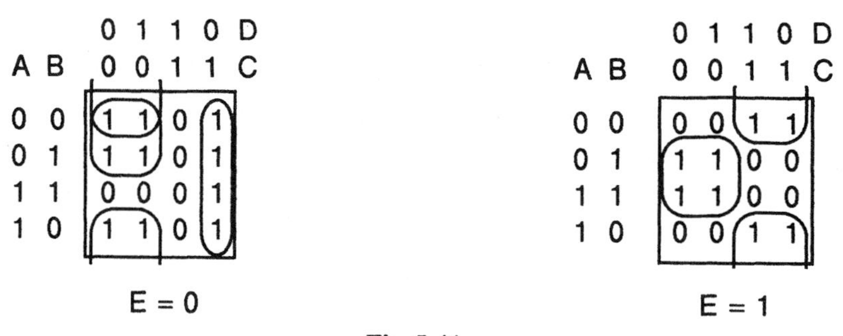

Fig. 5-44

5.15 Find simplified Boolean expressions for each of the outputs of the ROM shown in Fig. 5-45.

The functions may be mapped directly from the ROM connections (Fig. 5-46).

5.16 Given the ROM shown in Fig. 5-47, determine the address at which the hex number E1 is stored. Note the alternative use of solid circles to indicate connections.

Grouping by fours, we obtain \$E1 = (1110)(0001), which is seen to be the data stored on line 6. The binary address is WXYZ = 0110.

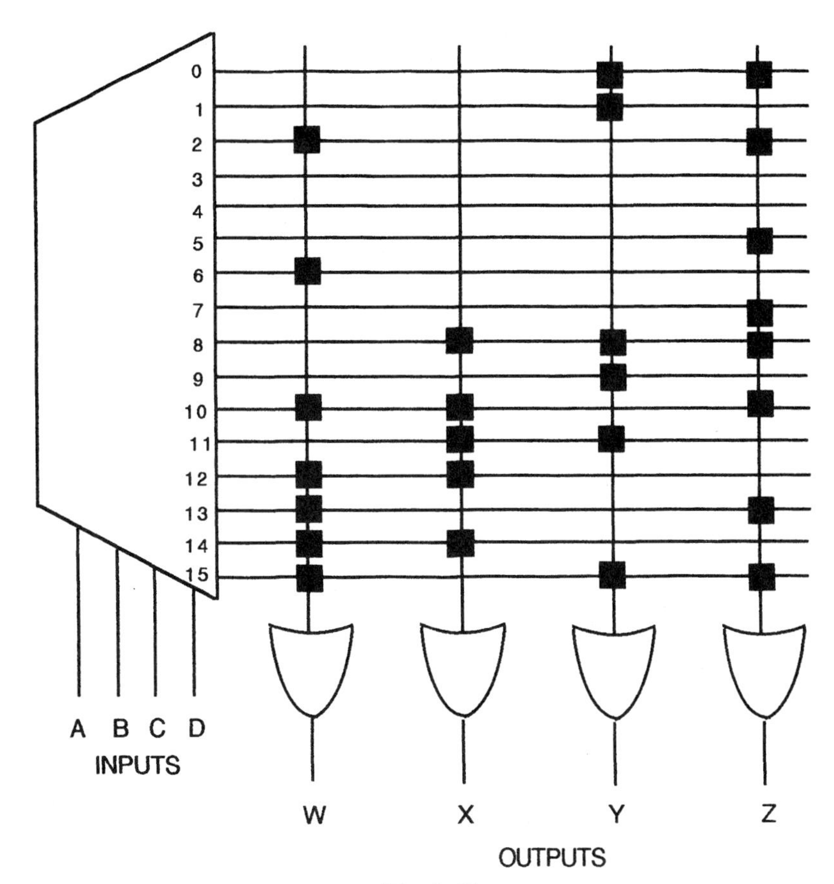

Fig. 5-45

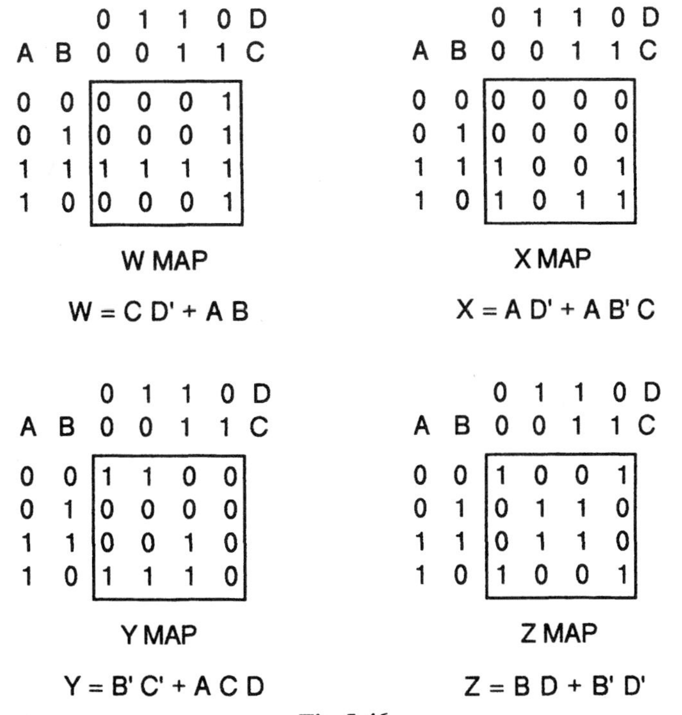

Fig. 5-46

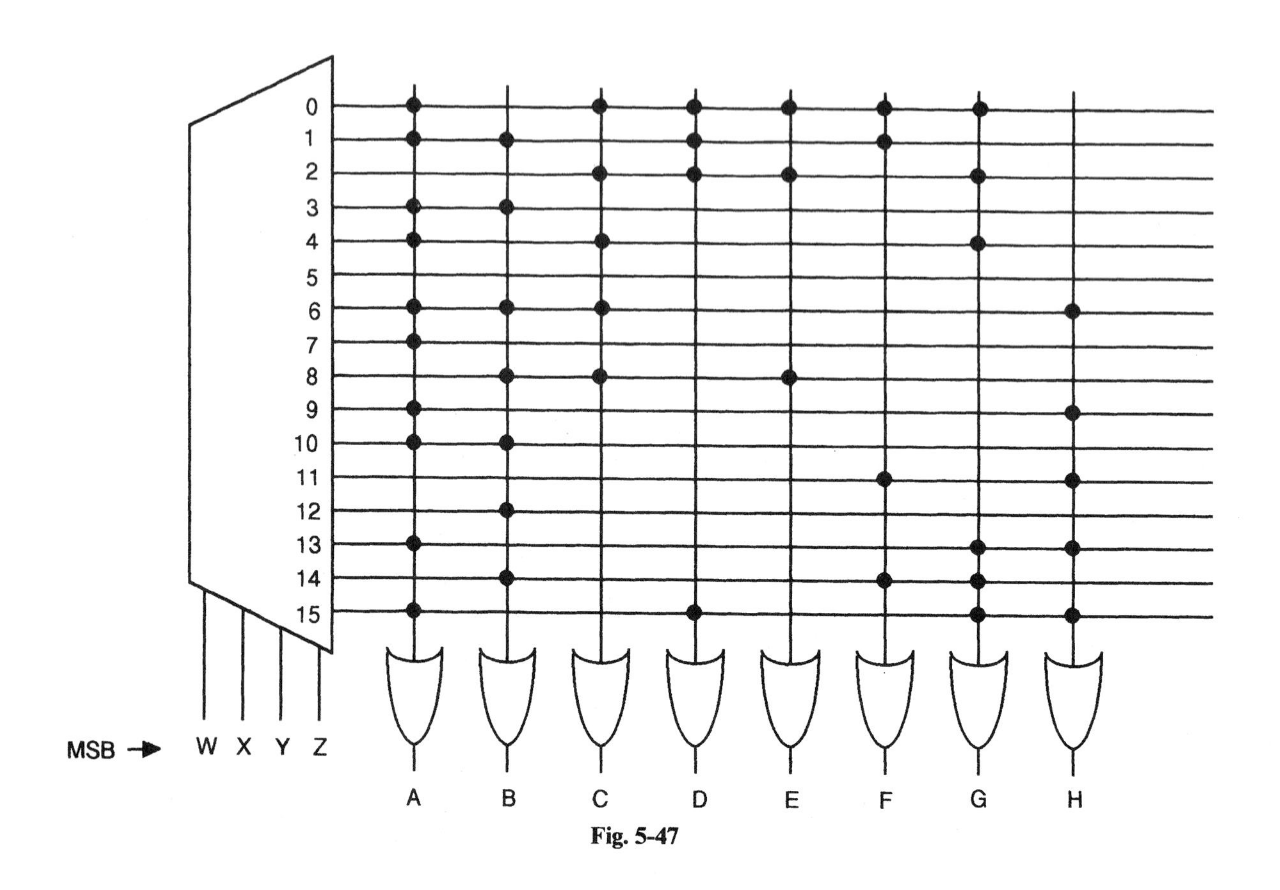

Fig. 5-47

Supplementary Problems

5.17 Given the multiplexer circuit shown in Fig. 5-48, create an equivalent logic circuit using SSI NAND gates.

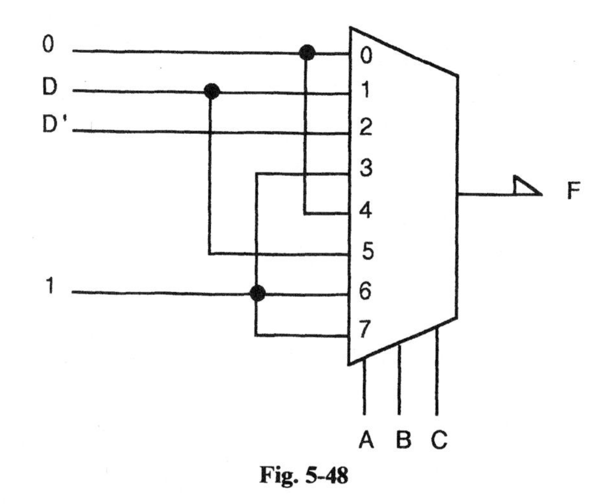

Fig. 5-48

5.18 Implement the logic shown in Fig. 5-49 on a three-select multiplexer.

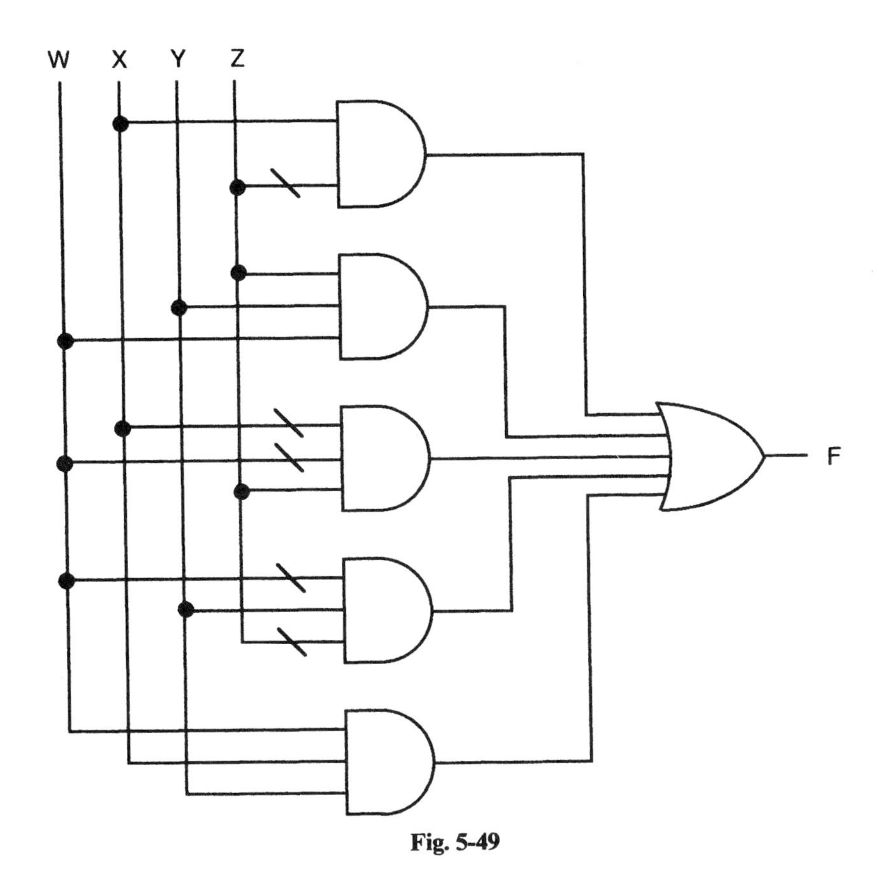

Fig. 5-49

5.19 Implement the multiplexer logic shown in Fig. 5-50 with NAND hardware and minimize the gate count as much as you can. You may assume that all inputs are available in both direct and logically inverted form (HT).

5.20 Implement the parity-checking logic for bit P_3 of the Hamming error-correcting code developed in Prob. 3.26. Use an unsimplified map of the Boolean equation and a three-select multiplexer.

5.21 Repeat Prob. 5.20 using a simplified map and a two-select multiplexer.

5.22 Implement the solution for the 2-bit multiplier of Prob. 3.25 using multiplexers and compare the chip count with the SSI solution of Prob. 4.27.

5.23 Implement the solution of the binary bit subtractor of Prob. 2.27 using multiplexers. Compare the chip count with the SSI solution of Prob. 4.30.

5.24 Implement the binary bit subtractor solution of Prob. 5.20 using three-select multiplexers. Compare your result with the solution to Prob. 5.23.

5.25 In Fig. 5-51, A and F are the most significant bits. Viewing the circuit as a memory device, what are the binary numbers stored at 1011 and 0101?

5.26 Viewing the circuit of Fig. 5-51 as a logic device, draw the K map for variable J and simplify the function.

5.27 Given the wired ROM shown in Fig. 5-52, implement the logic for output F on a three-select multiplexer using W, X, and Z as the select variables. Make W the most significant bit.

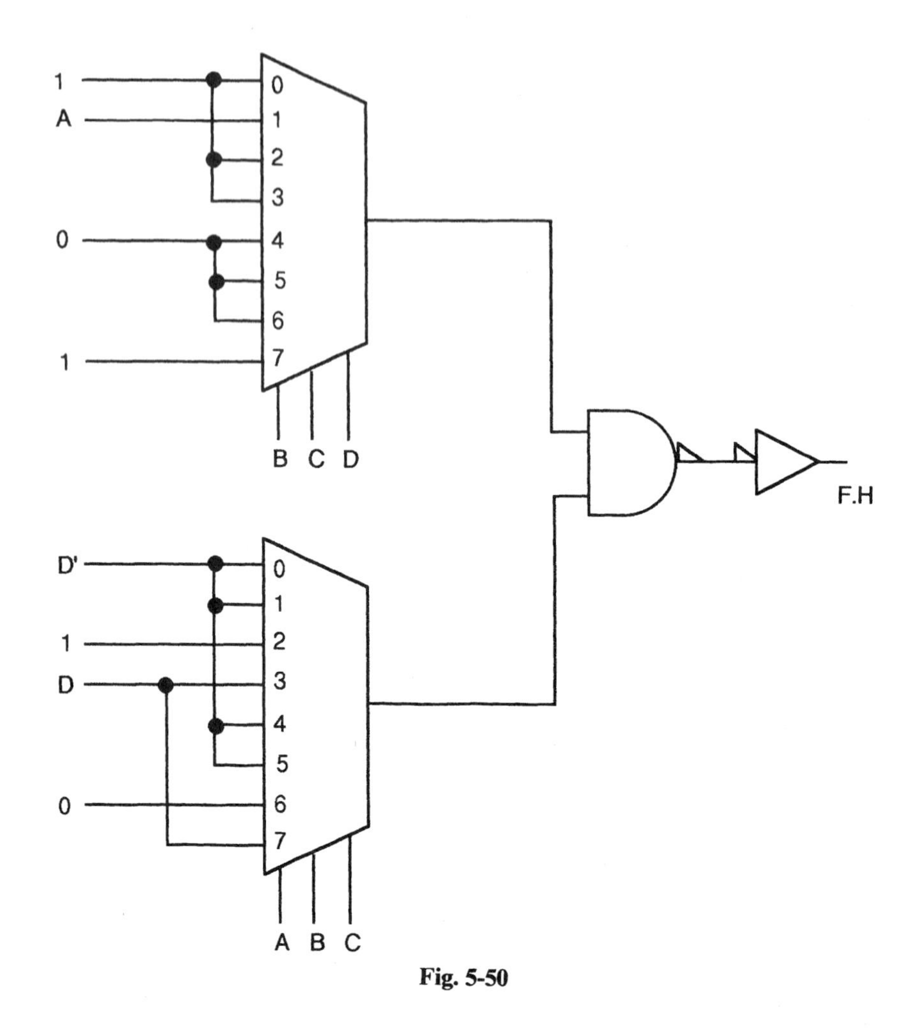

Fig. 5-50

5.28 Design a three-select multiplexer circuit whose output is TRUE if and only if an even number of its four inputs are TRUE.

5.29 Design a ROM to check parity on a 4-bit digital word. That is, the circuit should produce a TRUE output if the number of 1s in the word is odd. Assume binary digits A, B, C, and D, with A being most significant.

5.30 Implement the solution for the seven-segment display driver of Prob. 2.2 using a ROM.

5.31 Treating the ROM shown in Fig. 5-53 as a memory element, determine its contents for all addresses.

5.32 Treating the ROM of Fig. 5-53 as a component, as shown in Fig. 5-54, determinine the system output (Z_1, Z_2) for the input shown.

5.33 With reference to Prob. 2.2, implement the seven-segment decoder driver logic for segment a using a three-select multiplexer.

5.34 Eight-input multiplexers come one to a chip. If a multiplexer costs $0.25, a ROM $1.25, and any chip costs $0.35 to install and test, compare the cost of the multiplexer implementation of the entire seven-segment display driver (Prob. 5.33) with that of a ROM implementation (Prob. 5.30).

5.35 Consider the switching network shown in Fig. 5-55. What signal should be applied to the select inputs to connect b with N?

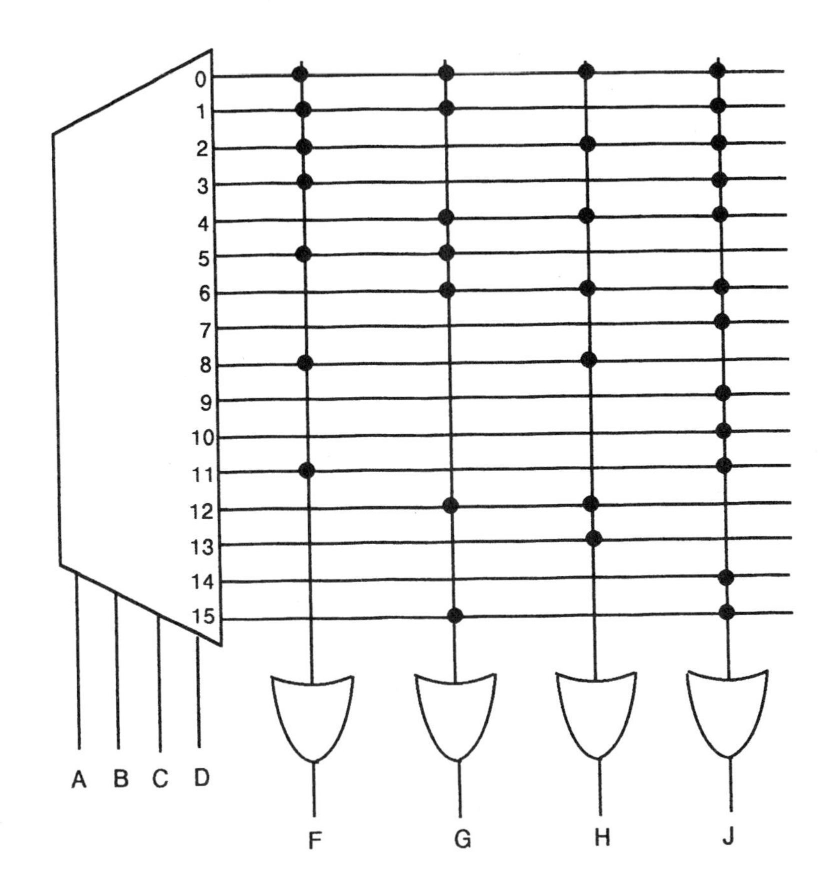

Fig. 5-51

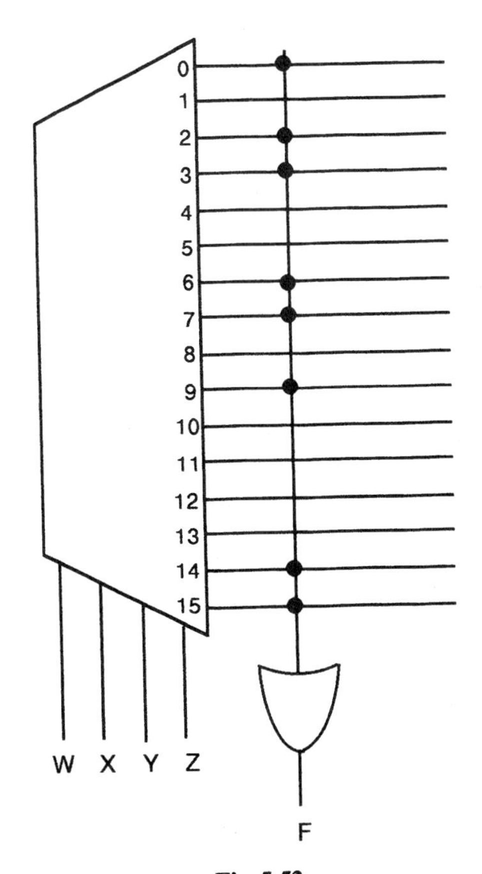

Fig. 5-52

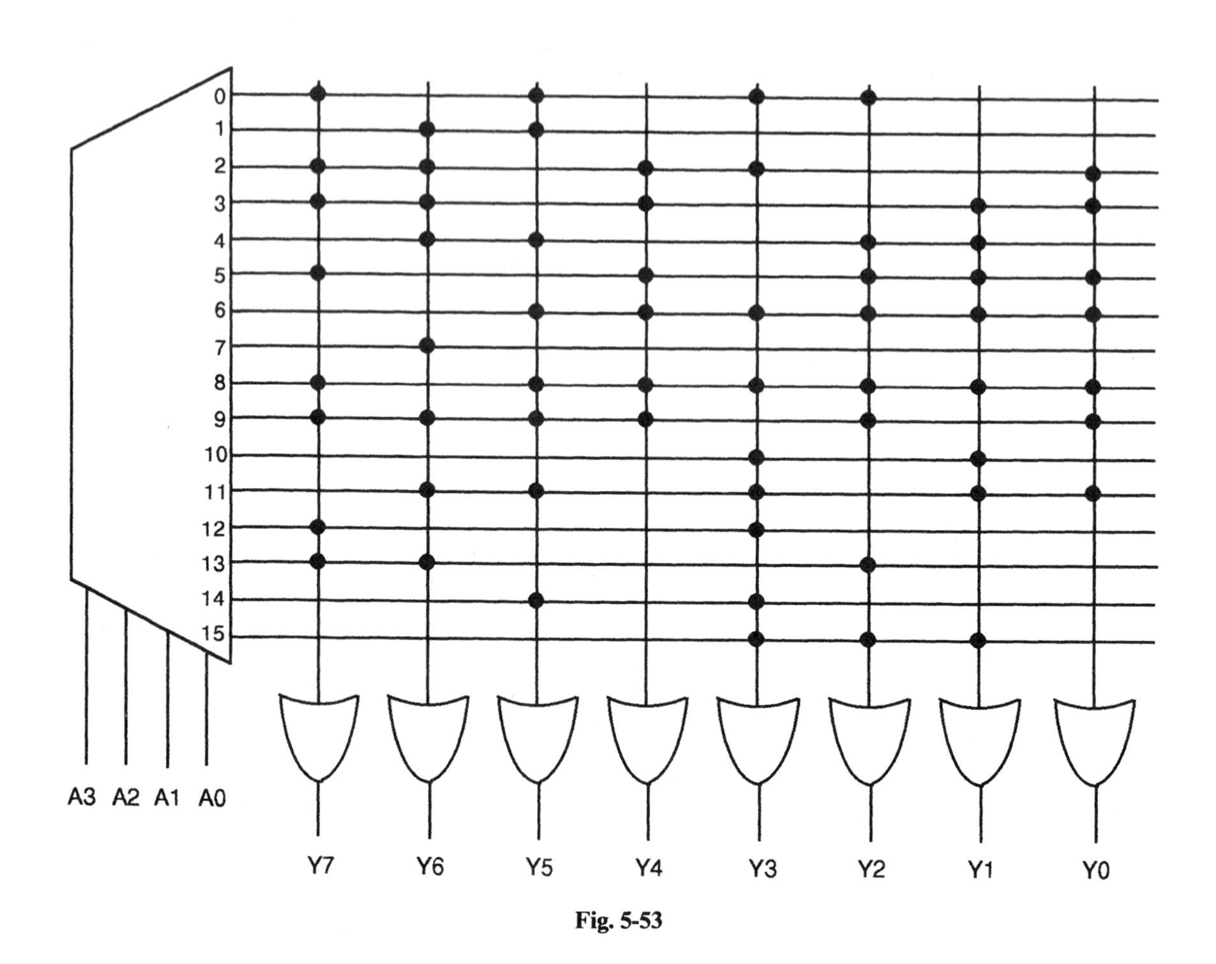

Fig. 5-53

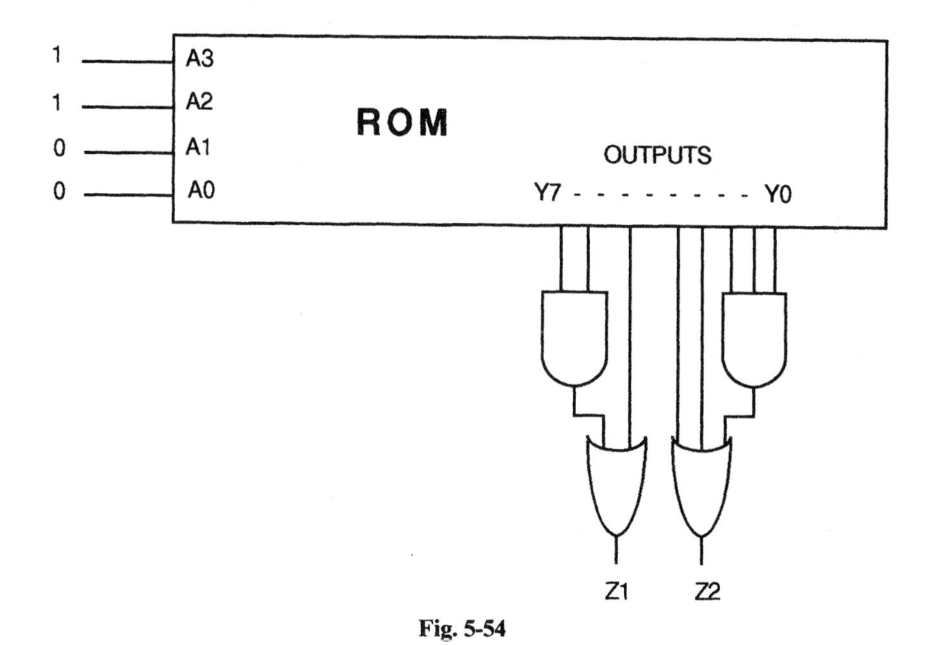

Fig. 5-54

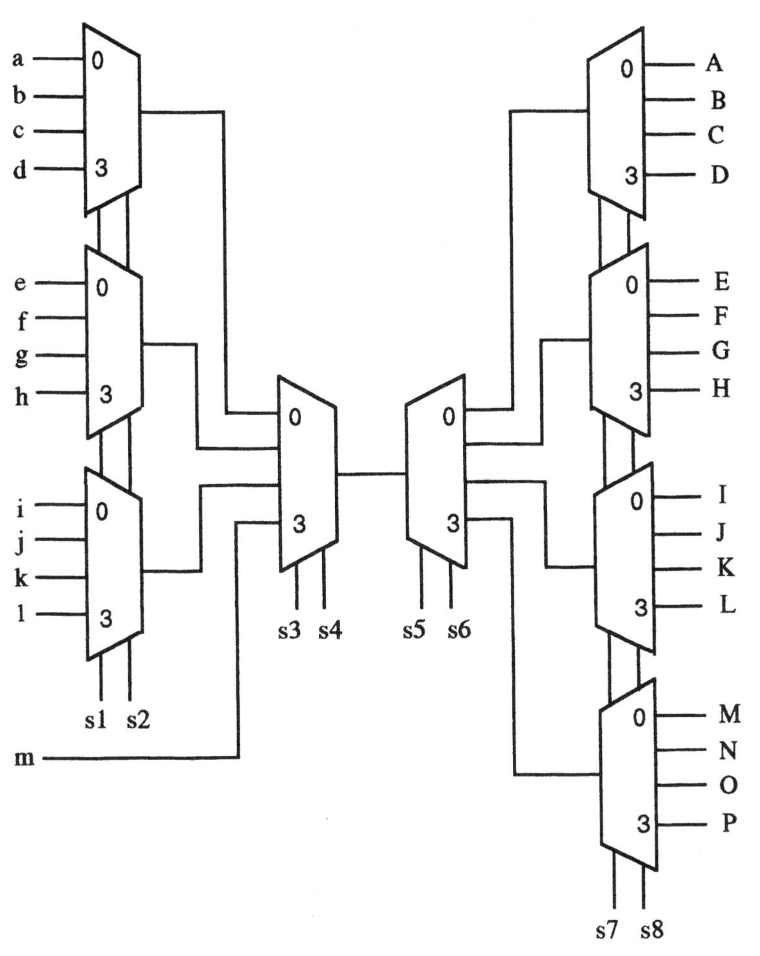

Fig. 5-55

5.36 Consider the switching network shown in Fig. 5-55. What connections are made when the select signals are given by

$$s_1, s_2, s_3, s_4, s_5, s_6, s_7, s_8 = 01110101$$

5.37 For the switching network shown in Fig. 5-55, estimate the number of chips required to implement it with individual SSI gates. Compare this with the multiplexer implementation, noting that four-input multiplexers come two to a chip. Refer to Fig. 5-2 and Sec. 5.3.

5.38 For the two competing designs of Prob. 5.37, each gate chip costs \$0.20 and each dual multiplexer or demultiplexer costs \$0.80. If it costs \$0.35 to install and test any chip and each chip requires 1 in^2 of space, compare the cost and space requirements of each of the two implementations and determine the multiplexer price to break even on cost.

5.39 A designer is considering using a ROM to implement the 2-bit binary multiplier of Prob. 4.27. A gate chip costs \$0.20 and an additional \$0.35 to install and test. If it costs \$0.85 to install and test the ROM, what is the maximum price that can be paid for a ROM to break even on cost relative to the individual gate design?

5.40 A relatively inexpensive commercial ROM is available with eight address lines and eight outputs. How many possible connections are available?

Answers to Supplementary Problems

5.17 See Fig. 5-56.

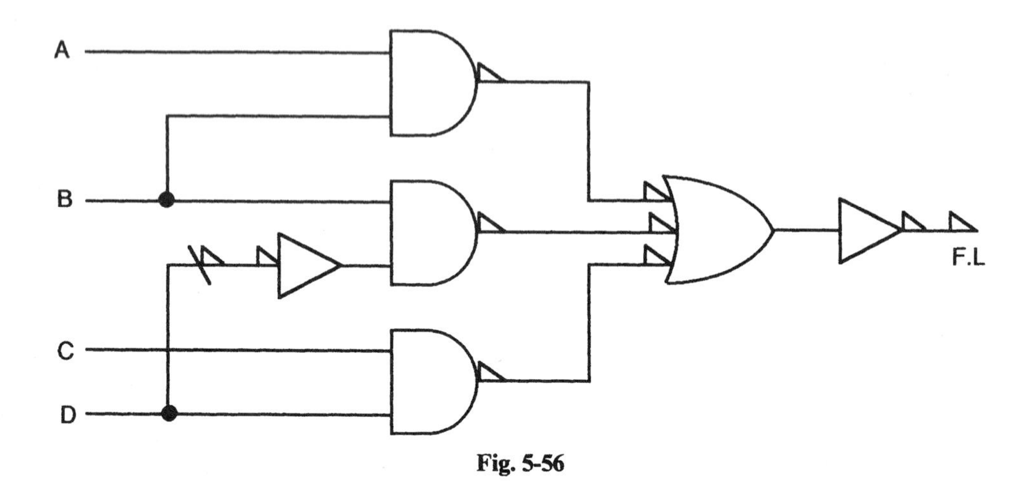

Fig. 5-56

5.18 See Fig. 5-57.

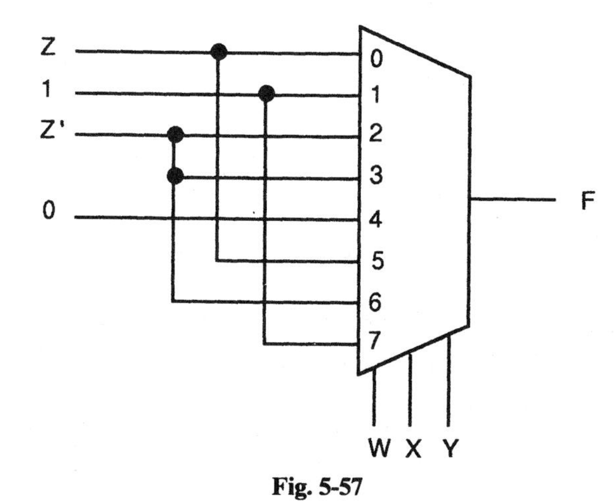

Fig. 5-57

5.19 See Fig. 5-58. Note that the variable A has turned out to have no effect on the output.

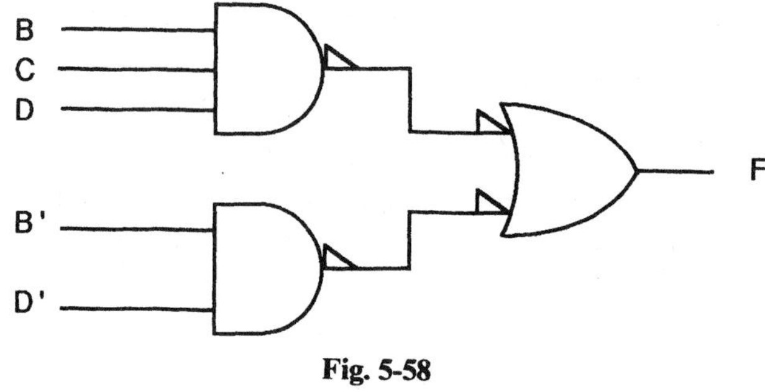

Fig. 5-58

5.20 From Prob. 3.26,

$$\begin{aligned}P_3 = {} & D_4' D_3' D_2' D_1' + D_4' D_3' D_2' D_1 + D_4' D_3 D_2 D_1' \\ & + D_4' D_3 D_2 D_1 + D_4 D_3' D_2 D_1' + D_4 D_3' D_2 D_1 \\ & + D_4 D_3 D_2' D_1' + D_4 D_3 D_2' D_1\end{aligned}$$

See Fig. 5-59.

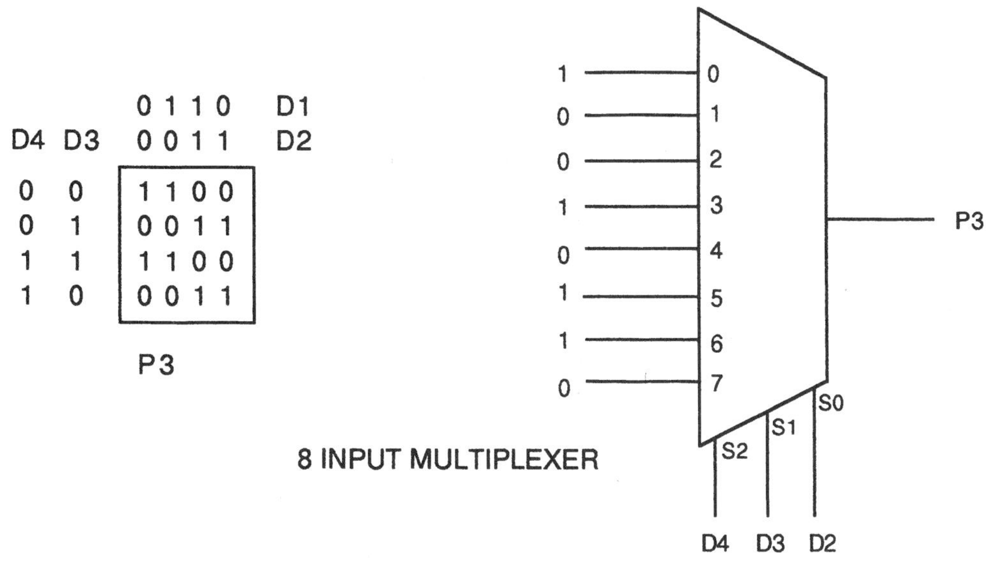

Fig. 5-59

5.21 See Fig. 5-60.

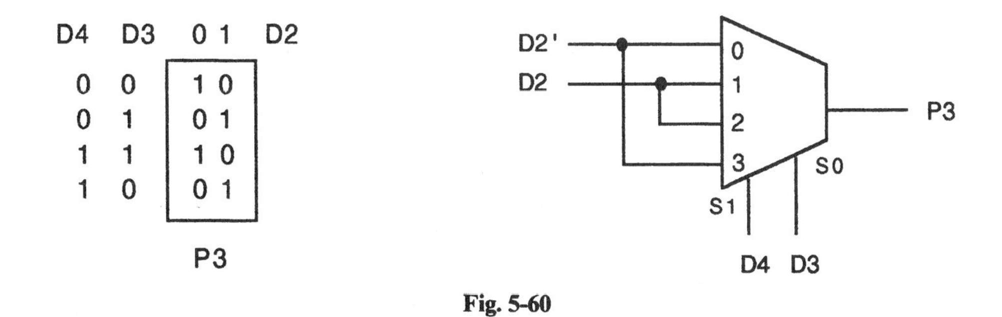

Fig. 5-60

5.22 See Fig. 5-61. Note that only four chips are used compared with five in the optimized SSI case of Prob. 4.27.

5.23 See Fig. 5-62.

5.24 The solution is shown in Fig. 5-63. Since three-select multiplexers are packaged one to a chip, two chips are required. No inverters are needed, however, and the parts layout problem may be eased by this approach.

5.25 1001, 1100.

5.26 See Fig. 5-64. $J = C + B'D + A'D'$.

5.27 See Fig. 5-65.

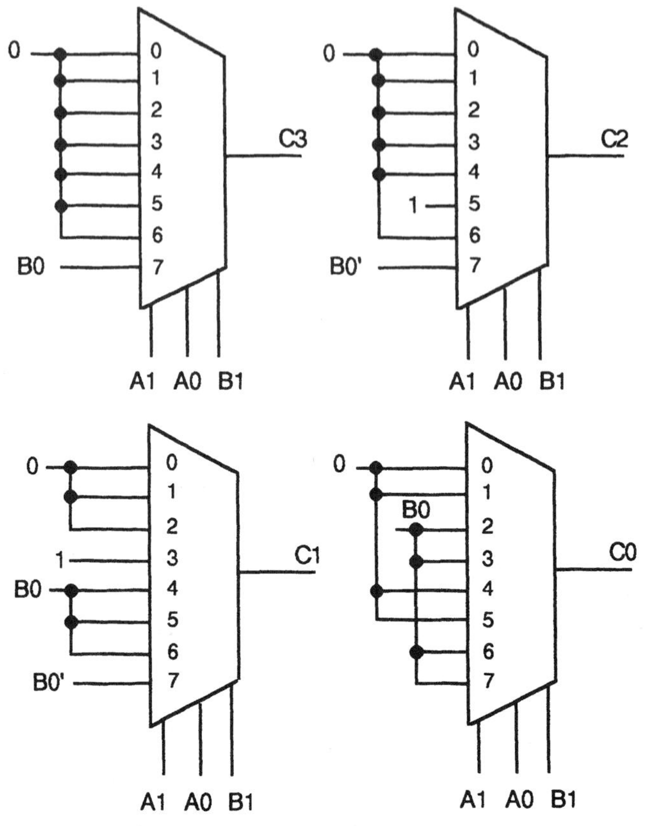

Fig. 5-61

X	Y	Bi	D	Bo
0	0	0	0	0
0	0	1	1	1
0	1	0	1	1
0	1	1	0	1
1	0	0	1	0
1	0	1	0	0
1	1	0	0	0
1	1	1	1	1

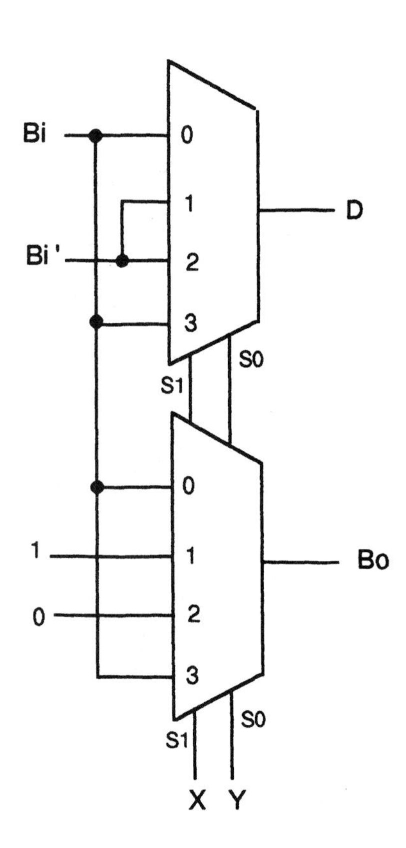

Fig. 5-62

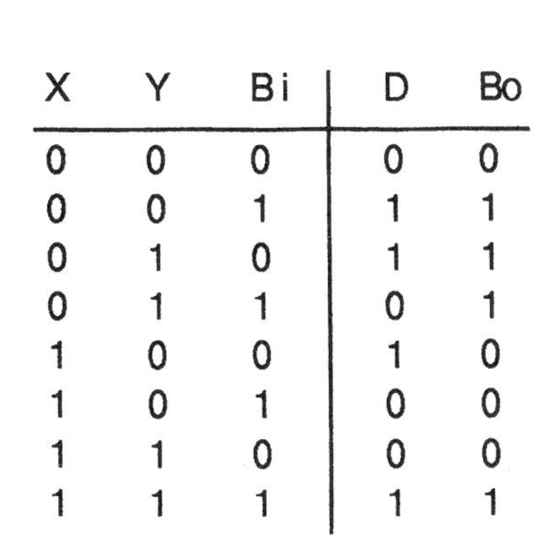

X	Y	Bi	D	Bo
0	0	0	0	0
0	0	1	1	1
0	1	0	1	1
0	1	1	0	1
1	0	0	1	0
1	0	1	0	0
1	1	0	0	0
1	1	1	1	1

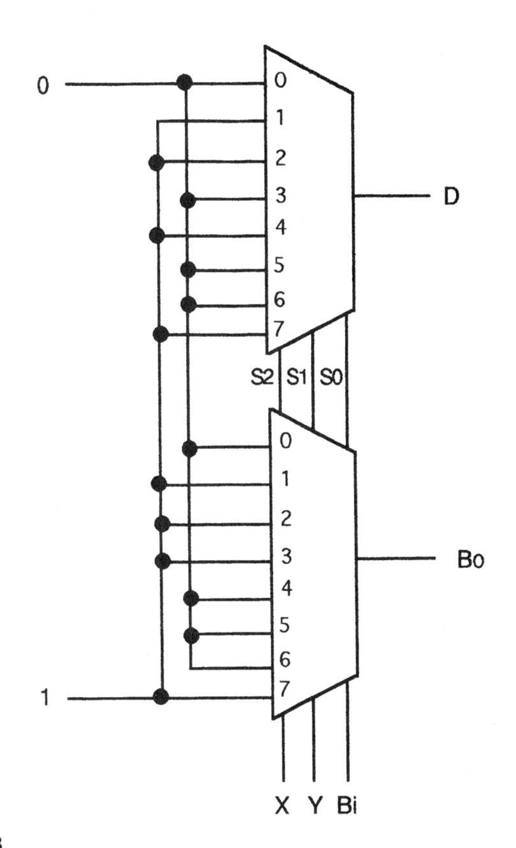

Fig. 5-63

A	B	D: 0, C: 0	D: 1, C: 0	D: 1, C: 1	D: 0, C: 1
0	0	1	1	1	1
0	1	1	0	1	1
1	1	0	0	1	1
1	0	0	1	1	1

Fig. 5-64

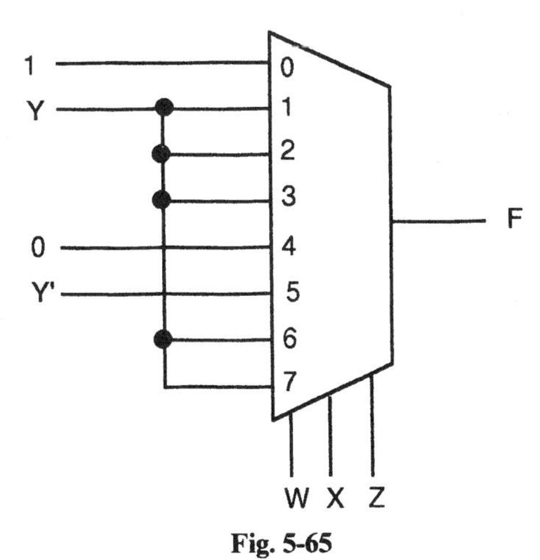

Fig. 5-65

5.28 See Fig. 5-66.

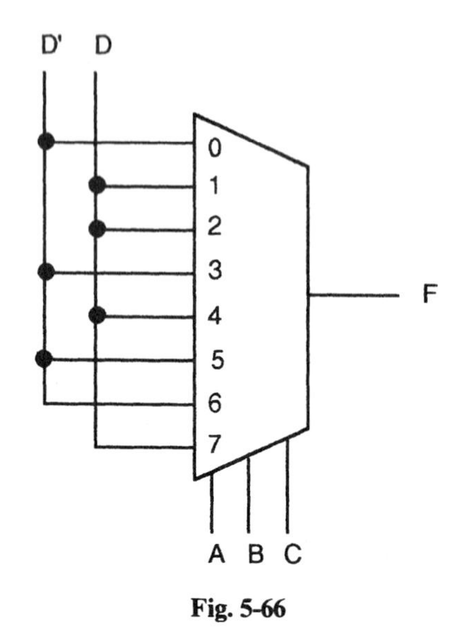

Fig. 5-66

5.29 See Fig. 5-67.

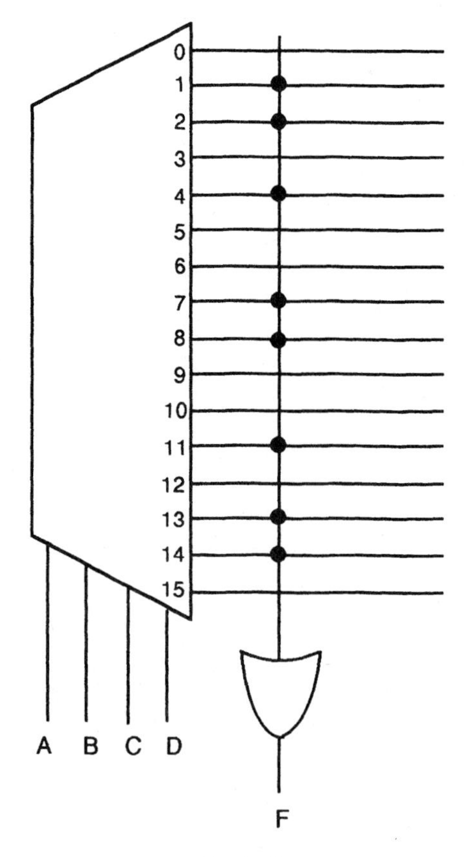

Fig. 5-67

5.30 See Fig. 5-68.

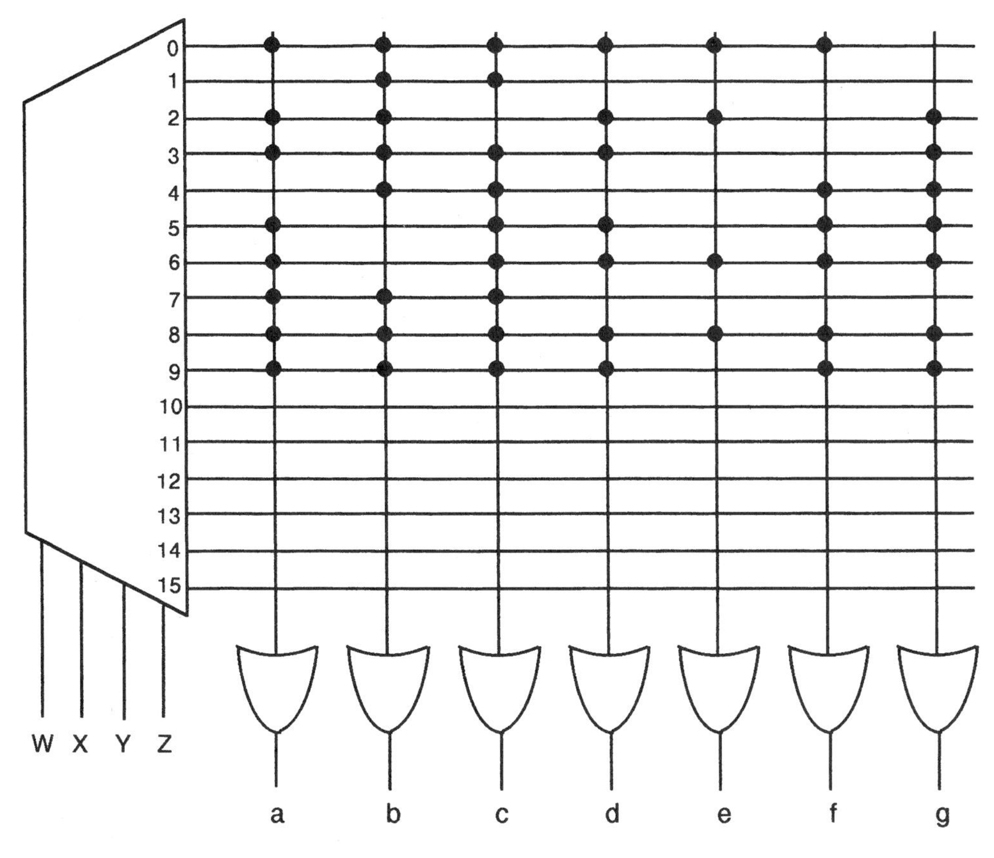

Fig. 5-68

5.31 See Table 5.8.

Table 5.8

Address	Contents	Address	Contents
0	AC	8	BF
1	60	9	F5
2	D9	10	0A
3	D3	11	6B
4	66	12	88
5	97	13	C4
6	3F	14	28
7	40	15	0E

5.32 $Z_1 = 0, Z_2 = 1$.

5.33 See Fig. 5-69.

5.34 Multiplexer implementation cost = 7(\$0.25 + \$0.35) = \$4.20; ROM implementation cost = \$1.25 + \$0.35 = \$1.60.

5.35 $s_1 s_2 = 01, s_3 s_4 = 00, s_5 s_6 = 11, s_7 s_8 = 01$

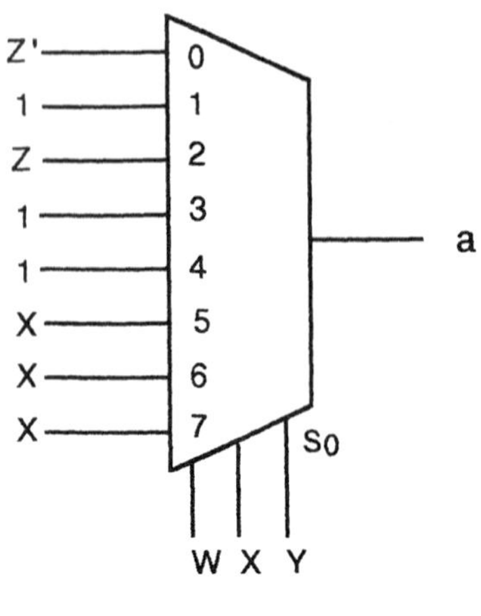

Fig. 5-69

5.36 Point m is connected to point F.

5.37

Gate Type	Number	Chip Contains	Chips Needed
3-input	36	3 gates	12
4-input	4	2 gates	2
Inverters	18	6 gates	3
			Total = 17
Multiplexers	4	2	2
Demultiplexers	5	2	3
			Total = 5

5.38 Seventeen gate chips cost \$3.40 and require \$5.95 for installation; total cost = \$9.35. The space requirement is 17 in^2. Five multiplexer or demultiplexer chips cost \$4.00 and require \$1.75 for installation; total cost = \$5.75. The space requirement is only 5 in^2. The break even point is at a multiplexer cost of \$1.52 (ignoring the space advantage).

5.39 From Prob. 4.27, five chips are needed; thus the cost is \$2.75 for the individual gate approach. This allows a top price of \$1.90 for the ROM.

5.40 $2^8 \times 8 = 2048$. The device is called a 2K ROM since there are approximately 2000 possible connections.

Chapter 6

Timing Diagrams

6.1 INTRODUCTION

The behavior of logic systems is usually described and investigated through the use of timing diagrams. These diagrams display, along a time axis, voltages or logic levels at various points in a digital circuit and can be used to indicate both the functional relationships and time delays which exist between inputs and outputs. The speed with which an input change causes a corresponding change in the output is often a critically important design parameter, and the timing diagram is a valuable aid in the investigation of its impact on system performance. A laboratory instrument widely used in troubleshooting logic hardware is the *logic analyzer*, which is essentially a multichannel storage oscilloscope that displays several timing waveforms simultaneously. Another valuable tool, particularly in the design phase, is the *logic simulator* which consists of software that runs on a computer workstation and is capable of producing families of related timing diagrams from schematic or Boolean inputs. Logic simulators are valuable to designers since they permit the validation of system performance prior to commitment to hardware.

There are two basic types of timing diagrams. First, there is the *microtiming* diagram which is concerned with the propagation delays encountered in individual gates. It is used to detect and display undesirable conditions such as "glitches" and unstable oscillations due to feedback. Second, there is the *macrotiming* diagram which is concerned with the time relationships between signals at various points in the system on a time scale large enough to make the consideration of individual gate delays unnecessary. Various gate input and output waveforms are displayed for a fixed interval, usually with reference to a timing or clock waveform.

6.2 MICROTIMING DIAGRAMS

In digital systems, the hardware implementation of logic gates involves electronic devices which switch between conductive and nonconductive states. It takes a small but finite time for such switching to occur because of electron charge storage and conduction effects which cause a measurable delay between the application of a voltage level change at a gate input and an appropriately recognizable response at the gate's output. This time interval, called the *propagation-delay time* (t_{pd}), is specified by hardware manufacturers on their data sheets and tends to be a constant which is applicable to all gates of the same type.

For the purposes of microtiming logic analysis, every gate in the system is often assumed to have an identical *unit propagation delay*. The microtiming diagram takes these propagation delays into account to show the behavior of a system following a specified change in one or more of its inputs, assumed to occur at $t = 0$.

Microtiming Diagram Preparation

1. The initial values of signals (either logic or voltage) applied to all system inputs must be known and recorded for $t = (0-)$, the time immediately prior to an input change at $t = 0$.
2. Using these values, the state of each internal gate is determined and entered on the timing diagram.
3. A specific input signal is allowed to change its logic state, and this change is entered on the timing diagram at $t = 0$.

4. All connections linking the changed signal to the input(s) of internal gates are identified by examination of the logic diagram.

5. Each of these gates is examined to see if its logic state will be altered by the change(s) in step 4. If it is, the response is entered on the diagram in the next succeeding propagation-delay interval. *Note that since all gates are presumed to have identical delays,* the timing diagram is constructed in discrete steps equal to the unit propagation-delay time t_{pd}.

6. Each changed gate output, if any, is now treated as an input change for any gate(s) to which it is connected, and step 5 is repeated.

7. The process terminates when no more gate outputs are found to change.

EXAMPLE 6.1 Assume that the physical devices (gates and inverters) in the logic circuit of Fig. 6-1 have identical propagation delays. All the inputs and outputs in this example are considered to be HT and, initially, A, B, C, and D are all 1s. At t = 0, C changes to a logic 0. Draw the appropriate logic microtiming diagram, and then convert it to display voltage waveforms.

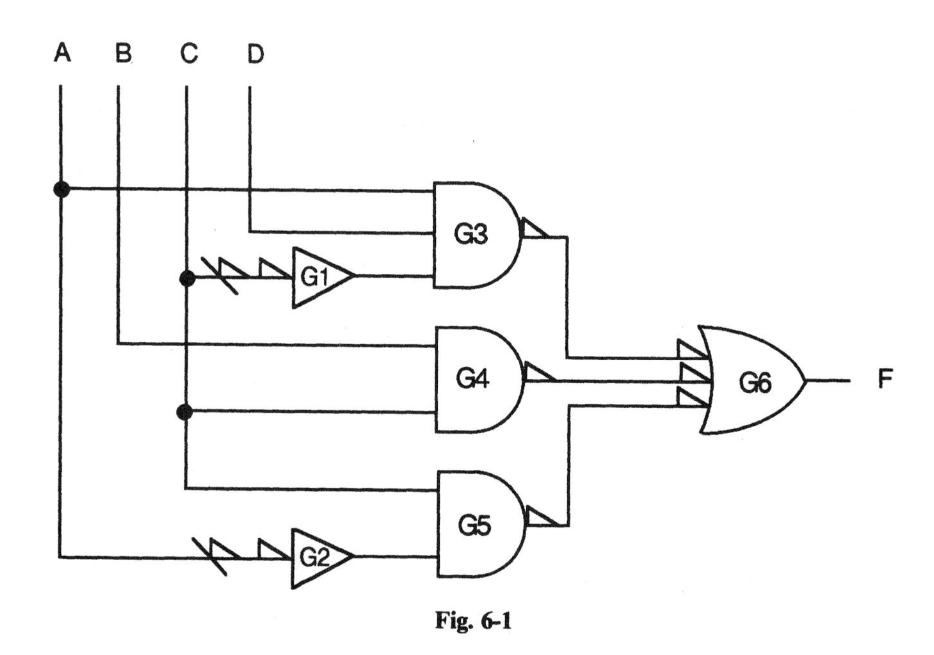

Fig. 6-1

Separation of the various waveforms (particularly if they are hand drawn) is made somewhat less confusing if shading is placed beneath the logic 1 states as shown in Fig. 6-2.

Logic Waveforms

Variable C, which changes from a logic 1 to a logic 0 at t = 0, is seen to connect to gates G_1, G_4, and G_5 as indicated by the notation "XG_1, G_4, G_5" in Fig. 6-2. We examine each of these in turn. Gate G_1 is a voltage inverter. Since it appears in conjunction with a slash mark, its output must change to a logic 1 when C goes to a 0. Note that if an inverter is used solely to match half arrows (as indicated by the absence of an associated slash mark), then *no logical inversion occurs* and the hardware inverter simply passes on a logic change at its input after a delay of one time interval. Gate G_4 is an AND gate, so its output will go to logic 0 if any of its inputs become 0. Thus G_4 will change. Gate G_5 will not be affected by the change to C since one of its inputs is already 0. The changes in the outputs G_1 and G_4 are entered in the diagram at time mark 1, one time-delay interval following the stimulus which caused them (the change in C).

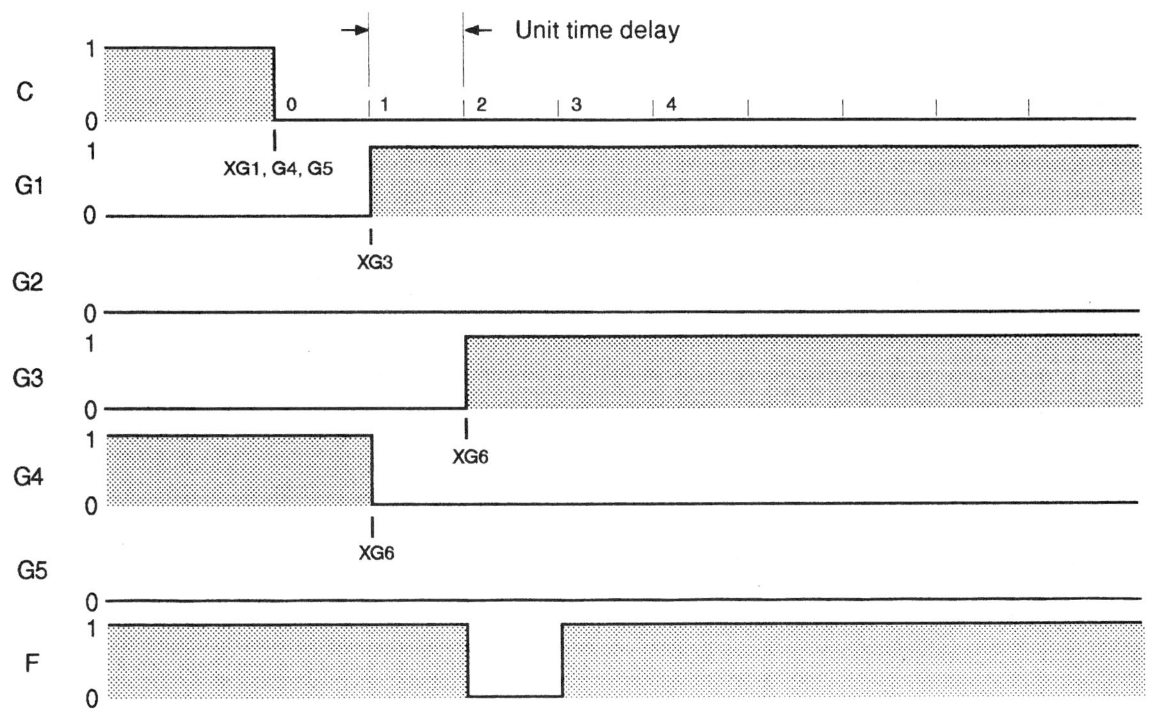

Fig. 6-2 Logic waveforms.

We now look at the gates which have just changed. Gate G_1 is connected only to G_3; G_4 is seen to affect only G_6. Gate G_3 will go to a 1 since all three inputs will be 1s immediately following time interval 1. Gate G_6 will go to a 0 since all three inputs will be 0s immediately following time interval 1. The changes in G_3 and G_6 are entered at time mark 2.

Repeating the preceding process, we note that only the connection between G_3 and G_6 remains to be evaluated: Gate G_6 (F) will be expected to return to a logic 1 since all three inputs will again be 1s. The process terminates since the last gate to change is connected to no others.

Voltage Waveforms

The outputs of G_1, G_2, and G_6 are HT, and, consequently, their logic and voltage waveshapes will be identical. Gates G_3, G_4, and G_5 have LT outputs making it necessary to invert their logic waveforms to obtain voltage equivalents as shown in Fig. 6-3.

6.3 HAZARDS

The logic represented by the diagram in Fig. 6-1 is easily determined to be

$$F = AC'D + BC + A'C.$$

This Boolean expression predicts that when, initially, $A = B = C = D = 1$, F will be 1 and will remain unaltered even though C changes to 0. The waveforms, however (Fig. 6-2), indicate a temporary condition where output F momentarily goes LOW and then returns to a proper HIGH state as required by the logic. This short-term aberration, due to unequal propagation delays in parallel logic paths, is called a *hazard* (sometimes informally referred to as a "glitch").

Since a variable C′ is not available as an input, its requirement by the logic mandates the addition of a hardware voltage inverter which adds an extra unit of delay in its signal path. This, unfortunately,

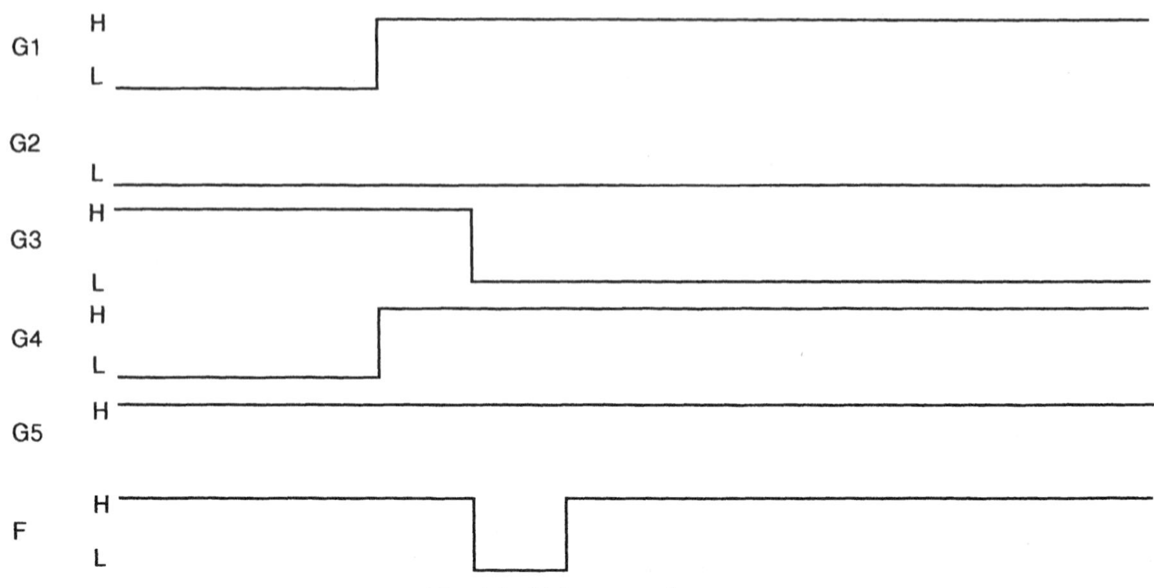

Fig. 6-3 Voltage waveforms.

produces the hazard shown in Fig. 6-2 which arises from the difference in delay between the paths for C and for C′ which converge at G_6.

This type of hazard can be recognized on a K map. Consider the map for the function of Example 6.1 (see Fig. 6-4).

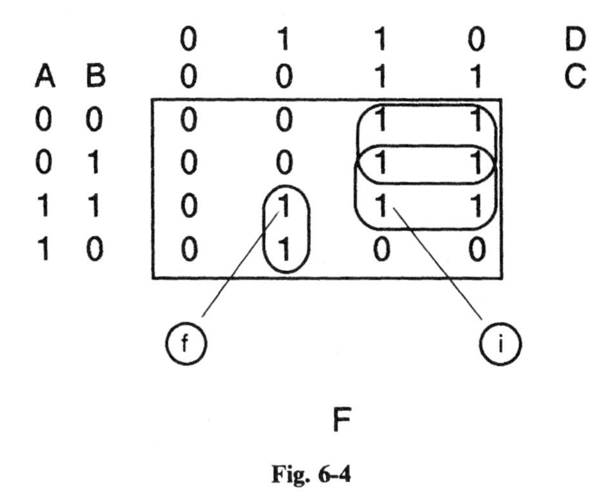

Fig. 6-4

In this map, the transition from initial state ⓘ to the final state ⓕ is seen to occur across the boundary of two coverings which do not overlap. Generally, *in order to determine whether the change in any single variable will result in a hazard condition, one need only look for adjacent initial and final states on the map which are isolated in separate coverings.*

The situation may be fixed with relative ease. A *hazard cover* is added which overlaps source and destination coverings as is shown in Fig. 6-5. This results in an added term in the equation for F and a corresponding increase in the hardware required.

It is important to note that while the map can be used to detect hazards in combinational logic arising from the difference in path lengths between a variable and its complement, there are other classes of timing problems which are not so easily detected (refer to Probs. 6.6 and 6.7). Furthermore, whereas the map may indicate two input states which will be involved in a hazard, it cannot identify

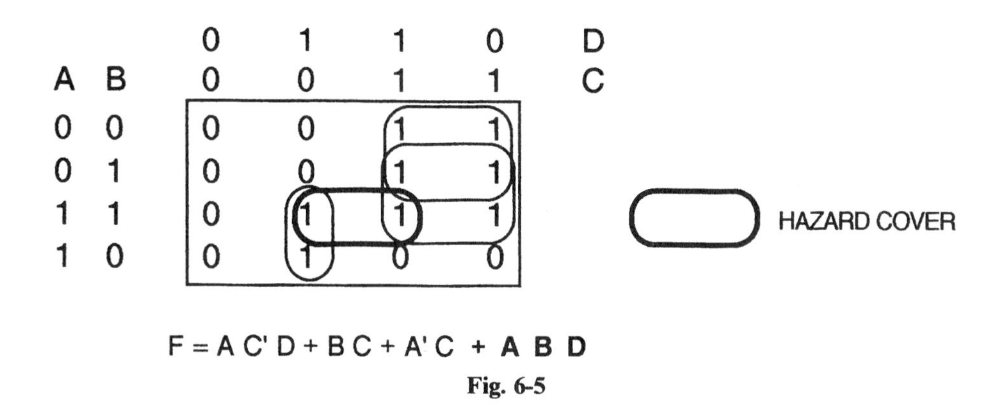

Fig. 6-5

which is the initial one. It is quite possible that a glitch will occur when going from state S_1 to state S_2, but not in the other direction as shown in Prob. 6.1.

Glitches are often unavoidable and, unless countered by proper design, they can cause serious errors in digital circuits. A powerful means of coping with them involves the implementation of strobing or clocking techniques as discussed in Sec. 7.4.

6.4 MACROTIMING DIAGRAMS

Macrotiming diagrams are used to display the outputs of various gates relative to independent inputs or some reference such as a system clock pulse train. They are primarily concerned with circuit behavior viewed on a time scale considerably larger than that used to study propagation-delay phenomena. The effects of individual gate delays are often omitted intentionally or become insignificantly small compared to logic-related sequential transitions shown in the diagram.

EXAMPLE 6.2 The HT inputs A, B, C, and D to the logic shown in Fig. 6-6 vary with time as shown at the top of Fig. 6-7. Assuming that the circuit responds instantaneously to input changes (that is, device propagation delays cannot be distinguished on the time scale used) sketch the output *voltage* waveshapes F and G.

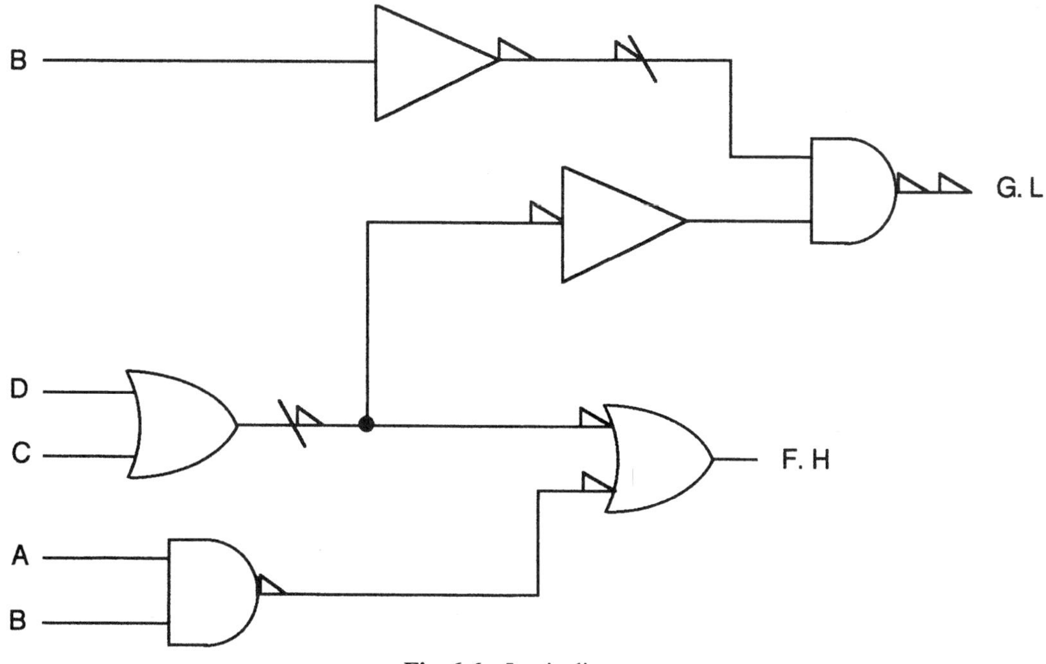

Fig. 6-6 Logic diagram.

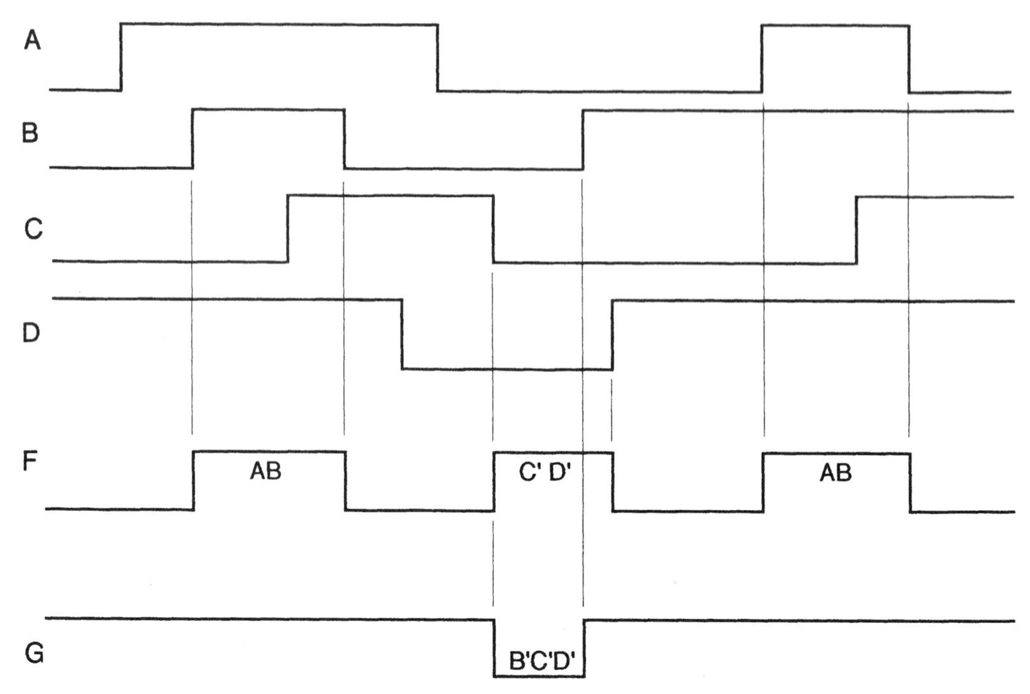

Fig. 6-7 Voltage waveforms.

From the logic diagram, $F = AB + (C + D)'$. Using De Morgan's theorem on the second term yields $F = AB + C'D'$; that is, F will be TRUE if either A and B are both TRUE or C and D are both FALSE. These conditions are met three times for the given inputs and, since F is HT, its corresponding voltage waveform will conform to the logic, as shown in Fig. 6-7.

Output G is seen to equal $B'(C + D)'$ or, equivalently, $G = B'C'D'$. Thus, G will be TRUE only if B, C, and D are all FALSE and, since G is LT, its voltage waveform will be the inverse of that implied by the logic.

6.5 TIMING SIMULATIONS

A major tool for confirming logic design functionality is a computer logic simulator which permits the designer to emulate the breadboarding and testing phases of a design without the need for hardware. Though not a substitute for actual prototyping, the simulator speeds the design process by permitting the early detection of basic flaws and timing problems. Sitting at a workstation monitor, an engineer "assembles" the hardware circuit by a process called *schematic capture* in which the circuit diagram is drawn on the screen by selecting logic components from a software library and interconnecting them by using a mouse or similar pointing tool which permits convenient positioning of objects on the computer screen. A typical screen display from a simulator package called LogicWorks™ from Capilano Computing is shown in Fig. 6-8 which depicts the circuit of Example 6.1.

On the screen shown in Fig. 6-8, the pull-down menu which provides access to the gate library is shown. In the present case, the symbol for NAND has been chosen to emphasize that the analysis will be a hardware simulation. Though most current simulators make use of positive logic conventions, an increasing number are beginning to incorporate mixed-logic symbols, and future versions of LogicWorks™ are expected to do the same.

As a designer creates the circuit schematic, interconnection paths are stored in the computer as a *net list*. The simulation software applies selected signals and/or fixed levels at the network inputs and monitors responses at outputs and any other internal points desired. In Fig. 6-8, inputs A, B, and D are connected to logic 1 via operable switches selected from the I/O menu. Input C is caused to step from 1 to 0 by drawing its waveform (using the mouse) in a timing window, a portion of which is shown in Fig. 6-9. Following a rest command, the simulation is started and the waveforms shown are generated in less

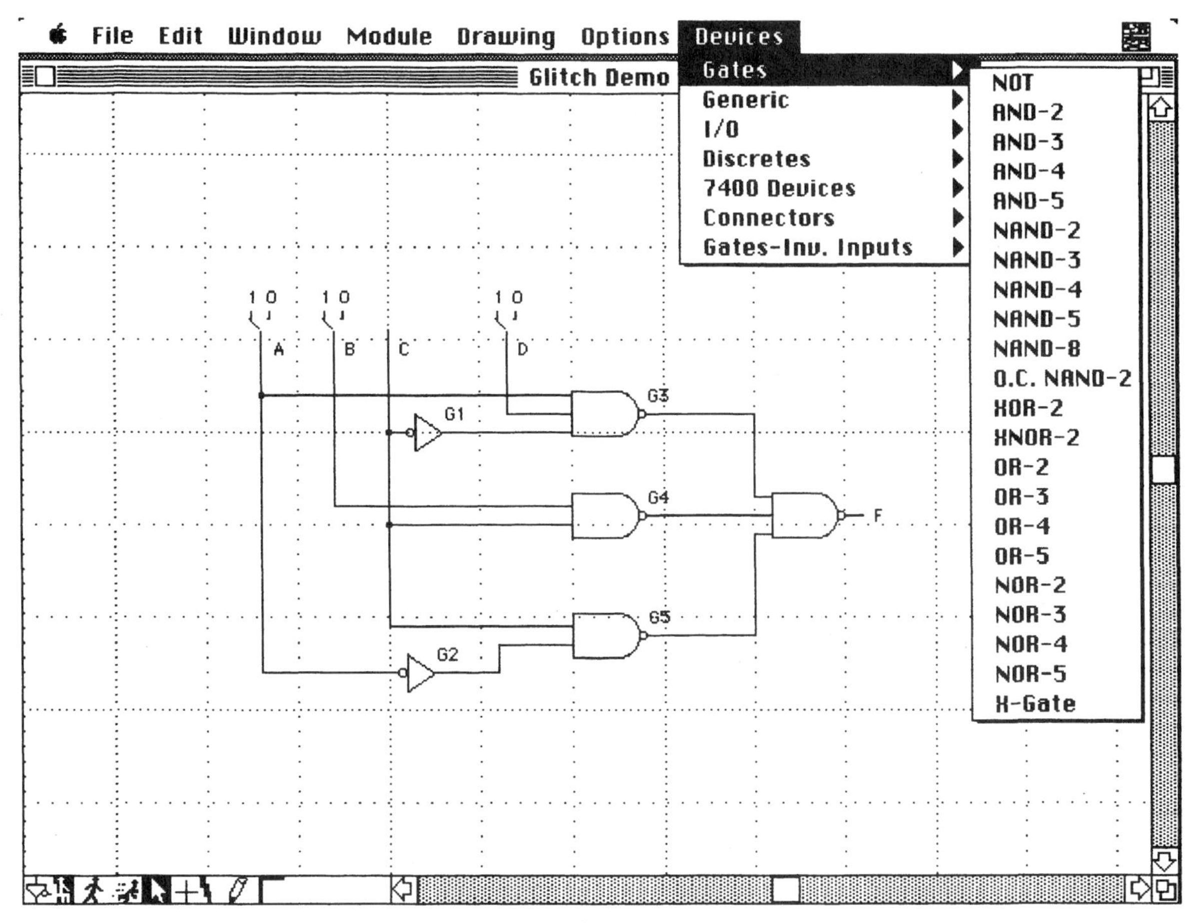

Fig. 6-8

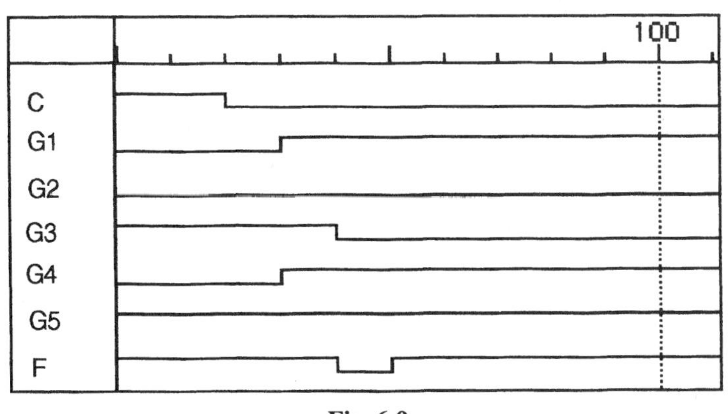

Fig. 6-9

than a second. Compare the results with the manual timing diagram of Fig. 6-3, and note confirmation of the predicted glitch in output F.

In most simulation software, it is possible to vary the delays of selected components and to go glitch hunting. In the current example, all gate delays were set to 10 time units prior to running the simulation. Some software includes glitch-detection routines which predict hazard conditions.

6.6 FEEDBACK IN COMBINATIONAL CIRCUITS

Figure 6-10 shows a *combinational logic circuit which contains feedback.* In this case, two of the output variables (Y_1 and Y_2) are functions of several variables, *including themselves.* In other words, in a circuit with feedback, the value of a given variable is affected by the variable itself.

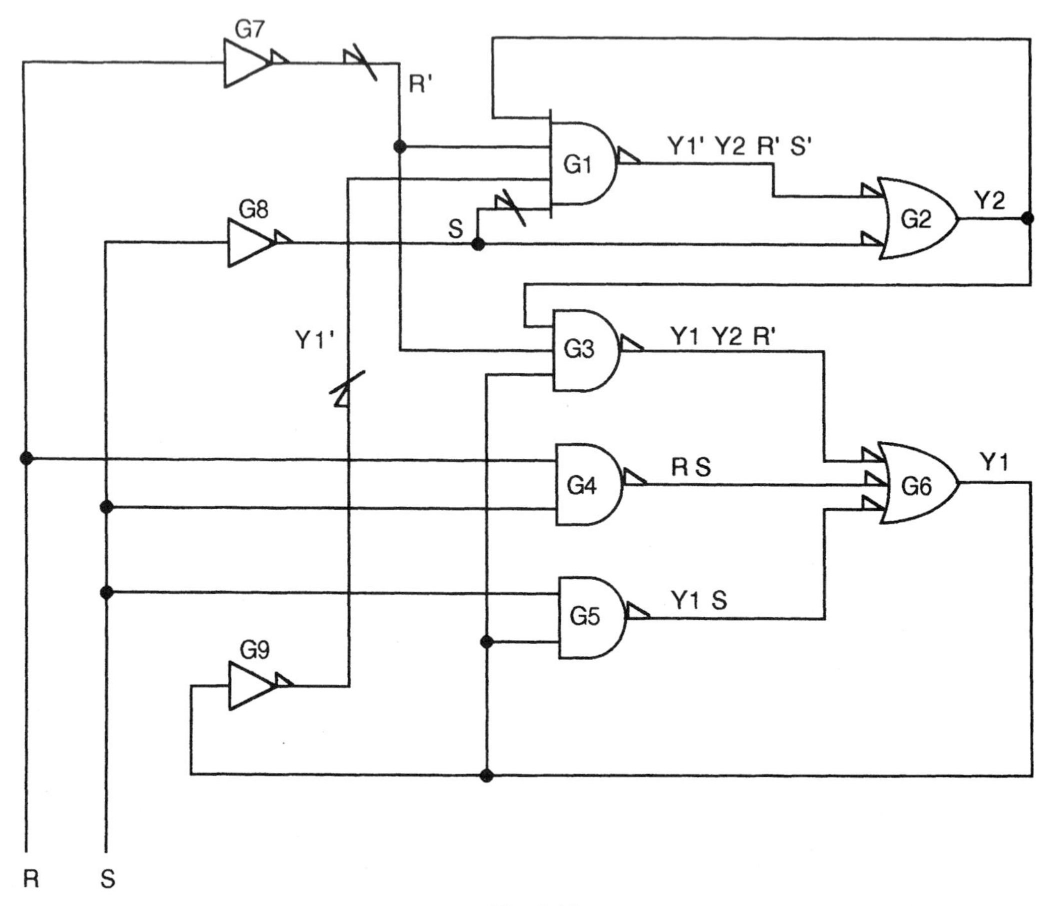

Fig. 6-10

Under certain conditions, such circuits can be unstable since the arrival of a feedback signal can cause the output to change which, in turn, will modify the input and cause the output to change again, and so on. If a K-map hazard analysis predicts a glitch, we might expect the *probability of continuous oscillation* between two states, the rate being dependent upon circuit propagation delays.

We see from Fig. 6-10 that

$$Y_2 = Y'_1 Y_2 R'S' + S$$
$$Y_1 = Y_1 Y_2 R' + RS + Y_1 S$$

K maps for Y_1 and Y_2 are shown in Fig. 6-11 where we observe, in the Y_2 map, that there are nonoverlapping coverings involving adjacent 1s. This indicates that a glitch is to be expected in Y_2 when $R = 0$ and S changes, warning us that we would be well advised to analyze this circuit further to check for oscillatory behavior.

It is obvious that, as circuit designs become more complex and incorporate feedback, the construction of microtiming diagrams by hand becomes a burdensome task. The computer simulator makes the job vastly easier, permitting designers to check out circuits much more completely before investing in

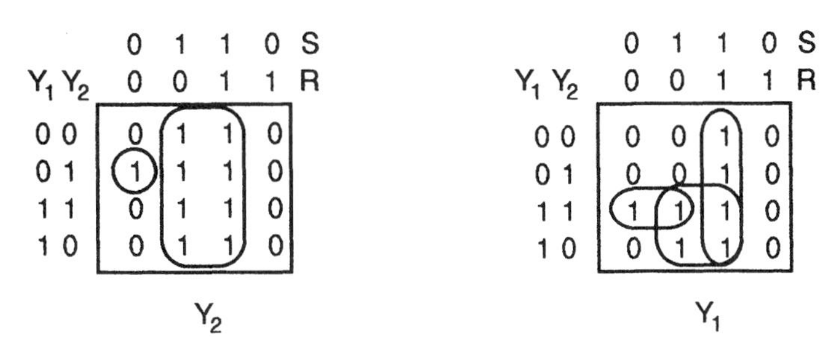

Fig. 6-11

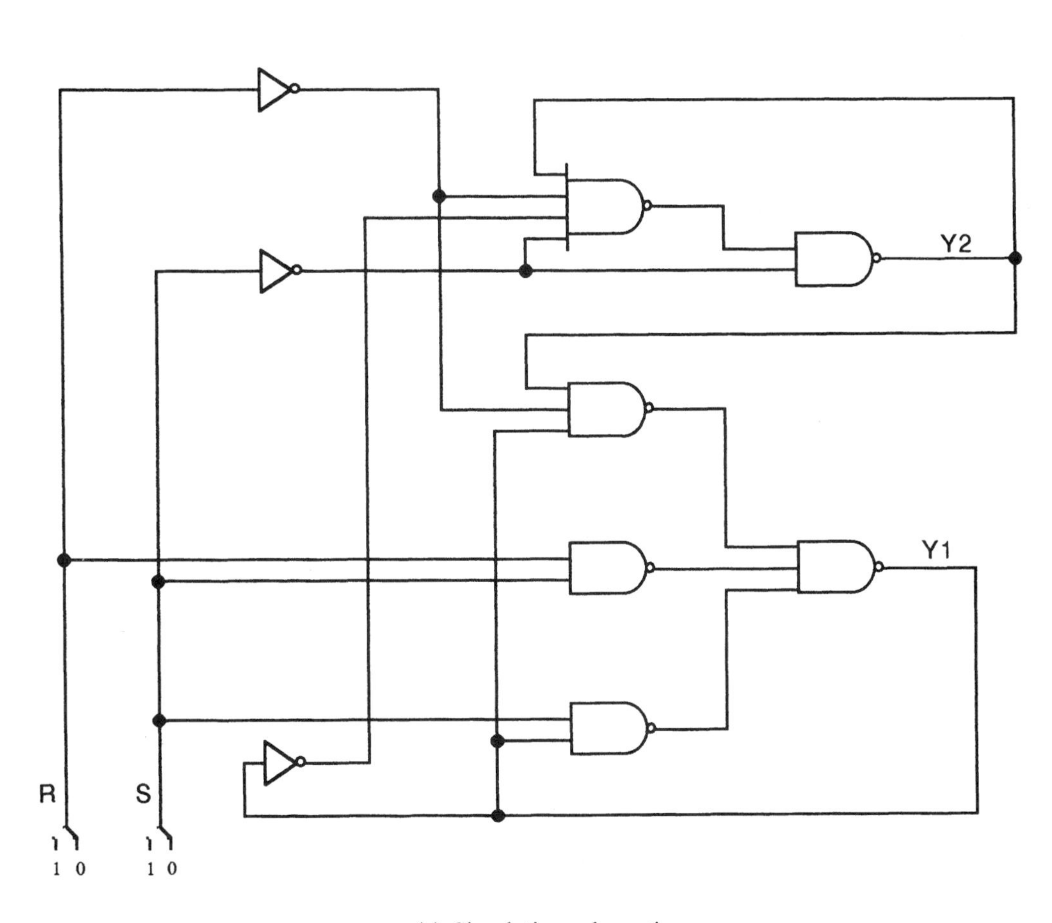

(*a*) Simulation schematic

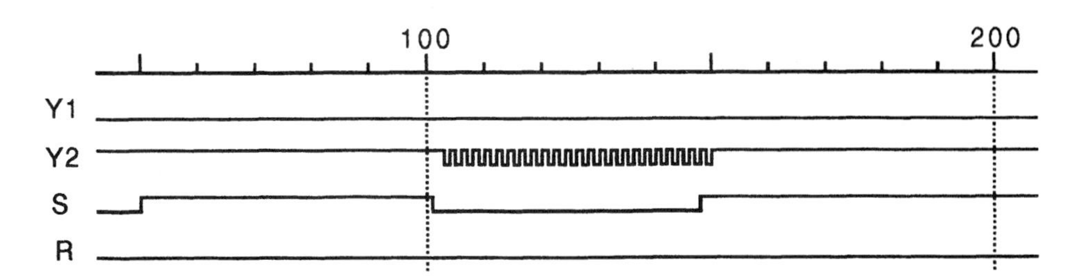

(*b*) Predicted waveforms (microtiming diagram)

Fig. 6-12

hardware. As an example, consider the case where the feedback circuit of Fig. 6-10 is constructed with NAND hardware. The predicted instability is easily demonstrated by a LogicWorks™ simulation as shown in Fig. 6-12.*

Note that, as predicted by the K maps, instability occurs in Y_2 when R = 0 and S changes.

Solved Problems

6.1 Consider the NAND implementation of the function F = (AB)′ + AD shown in Fig. 6-13. Assuming that all gates have the same time delays, draw a logic microtiming diagram for the case where A changes while B = D = 1.

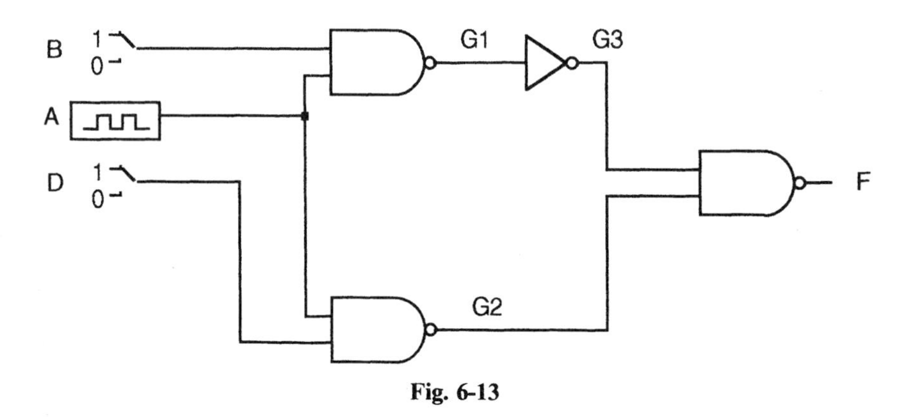

Fig. 6-13

Prior to beginning a step-by-step signal trace, it is advisable to determine the underlying logic. This may be easily done by applying the conversion process discussed in Prob. 4.10 which provides us with the required mixed-logic equivalent reproduced in Fig. 6-14.

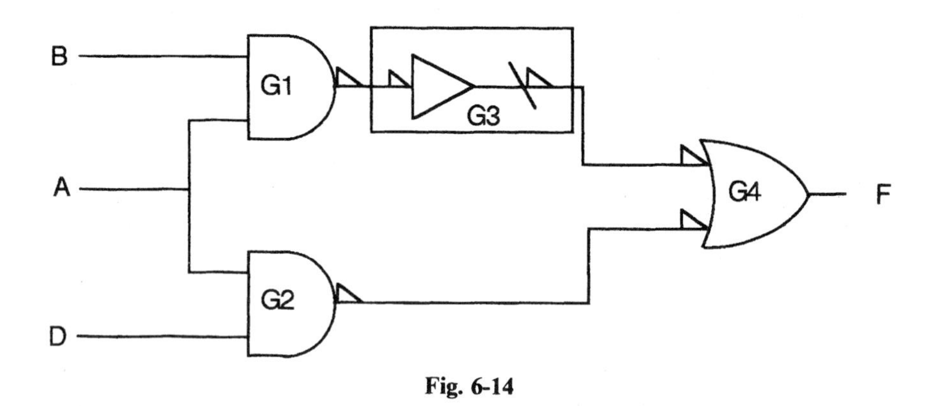

Fig. 6-14

Microtiming diagram generation is described in Example 6.1. The connections radiating from each variable change are traced, the affected gate or gates are identified, and the resulting logic level changes (if any) are entered in the diagram one time-delay interval following the stimulus. The resulting timing

* The simulation of circuits with feedback can sometimes fail to function properly because of the lack of unambiguous initial conditions. A brief discussion of how this problem is handled when using LogicWorks™ can be found in App. C.

diagram is shown in Fig. 6-15. The given Boolean expression leads us to believe that F will remain equal to 1 regardless of the state of A, and, as expected, there is no response to a 0-to-1 change in input A. Contrary to Boolean prediction, however, a 1-to-0 change in A produces a glitch two time-delay intervals following the input transition. This result may be compared with simulation waveforms if all LT outputs are inverted (see Fig. 6-16).

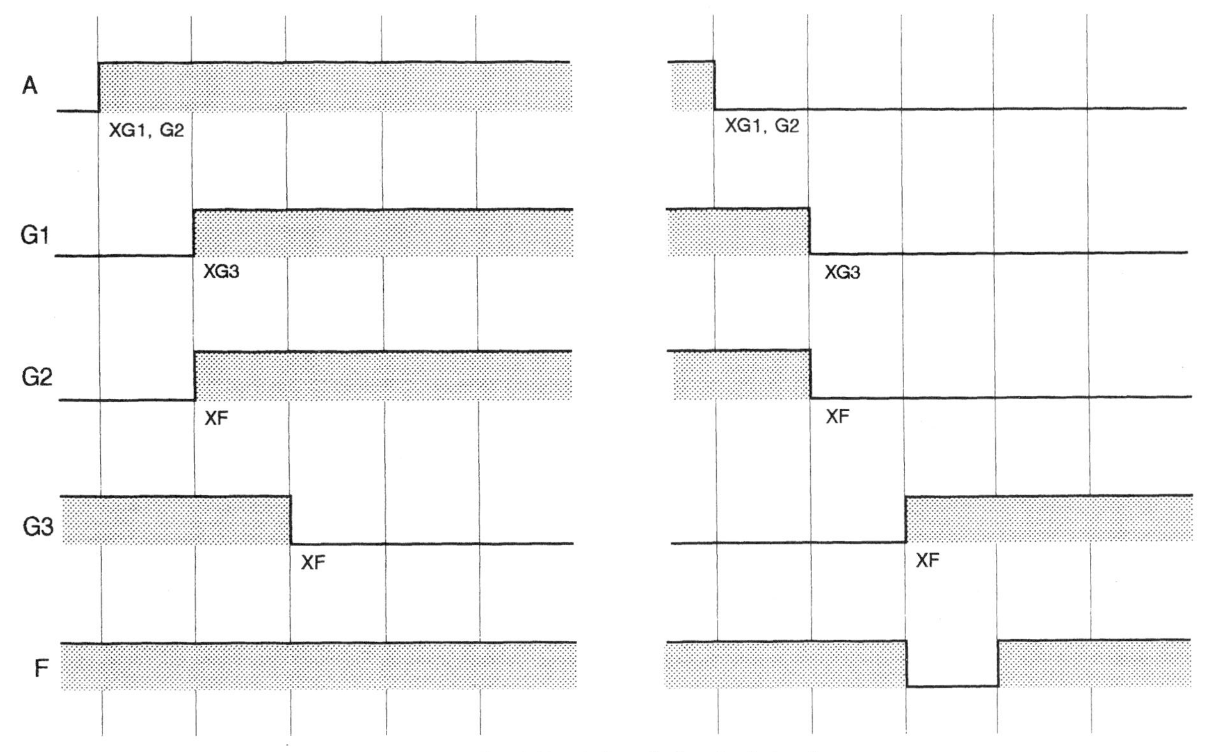

Fig. 6-15 Microtiming logic levels.

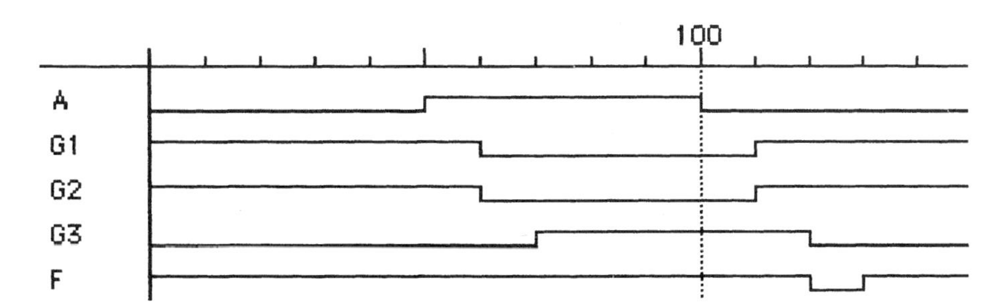

Fig. 6-16 Timing simulation. Note the glitch in F following a transition in A from 1 to 0.

6.2 For the circuit in Prob. 6.1, draw the K map, add a hazard covering to eliminate the glitch, and discuss its impact on the hardware.

Application of De Morgan's theorem to the given function yields F = A′ + B′ + AD. The K map with encirclements corresponding to this Boolean form is shown in Fig. 6-17 where the glitch may be readily identified as adjacent 1s in isolated coverings as indicated. A hazard covering can be added over the two indicated 1s, or it may be extended over the entire right-hand column. The former arrangement requires an additional AND gate. The latter yields a simple Boolean term, but it requires an inverter to balance the half arrows, so, in the present case, there is probably not much reason to choose one method over the other. If some other specification or design rule doesn't tip the balance, a professionally engineered coin toss is permissible.

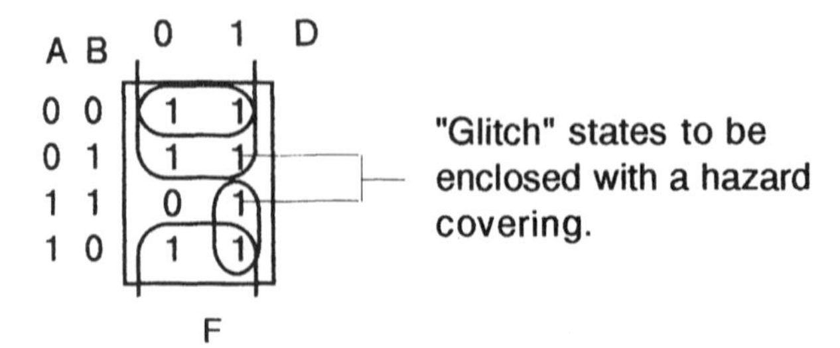

Fig. 6-17

6.3 Assuming that all hardware (gates and inverters) in Fig. 6-18 have the same time delay, sketch the complete logic *micro*timing diagram. Initially, all input variables are 1s. At t = 0, C changes to a 0. All inputs and outputs are HT.

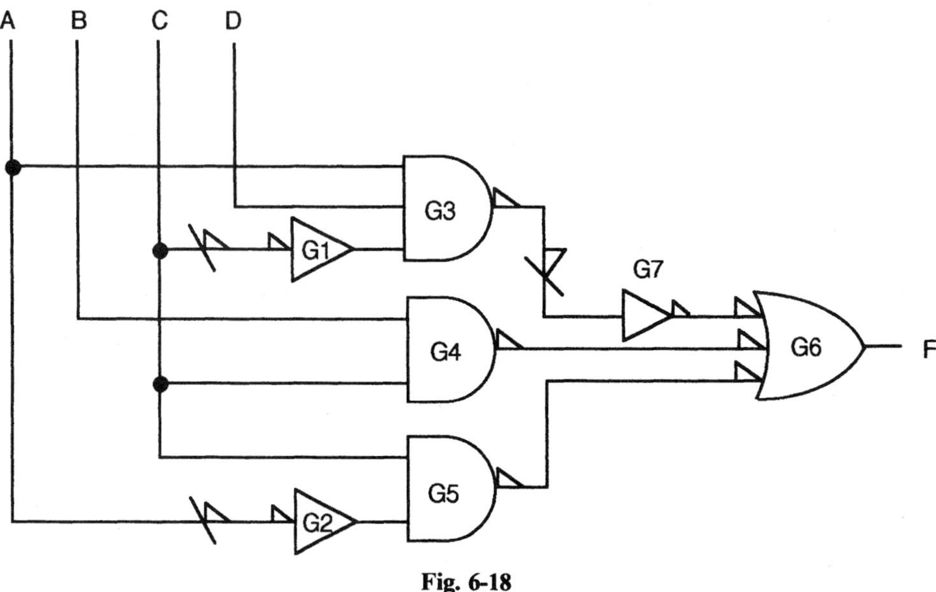

Fig. 6-18

Since A, B, and D remain 1s, it is relatively easy to trace the logic. When C changes, it directly affects gates G_1, G_4, and G_5 which, if they respond, will do so after one time-delay interval. Gate G_5 does not change state since its lower input remains at logic 0. When gate G_4 changes, it can't affect G_6 because G_7 is at logic 1. However, inverter G_7 goes to logic 0 in the third time interval in response to the change in gate G_3, causing G_6 (and F) to go to 0 in interval 4. The logic microtiming diagram is shown in Fig. 6-19.

It is interesting to compare this circuit to that of Fig. 6-1 which differs only in the absence of inverter G_7. Though, in Fig. 6-18, the reconvergent paths for variable C have three time delays in the upper branch and only one in the lower, no glitch occurs because of the logic structure.

6.4 The circuit shown in Fig. 6-20 is a common interconnection of gates called a latch which will be described in detail in Chap. 7. Note the cross-coupled feedback arrangement where the output of one gate is used as an input for the other. Assume that the output states are as shown and both inputs are initially at logic 0. At time t = 0, input R changes to a 1. Given that both gates have equal delay times, sketch the microtiming logic diagram.

Because of the logical inversions, the feedback inputs to gates G_1 and G_2 are 1 and 0, respectively. With both external inputs at logic 0, the given outputs are seen to be consistent. When **R** changes, it affects G_2 which goes to a logic 1 after one time-delay interval. This causes G_1's feedback input to become 0, and,

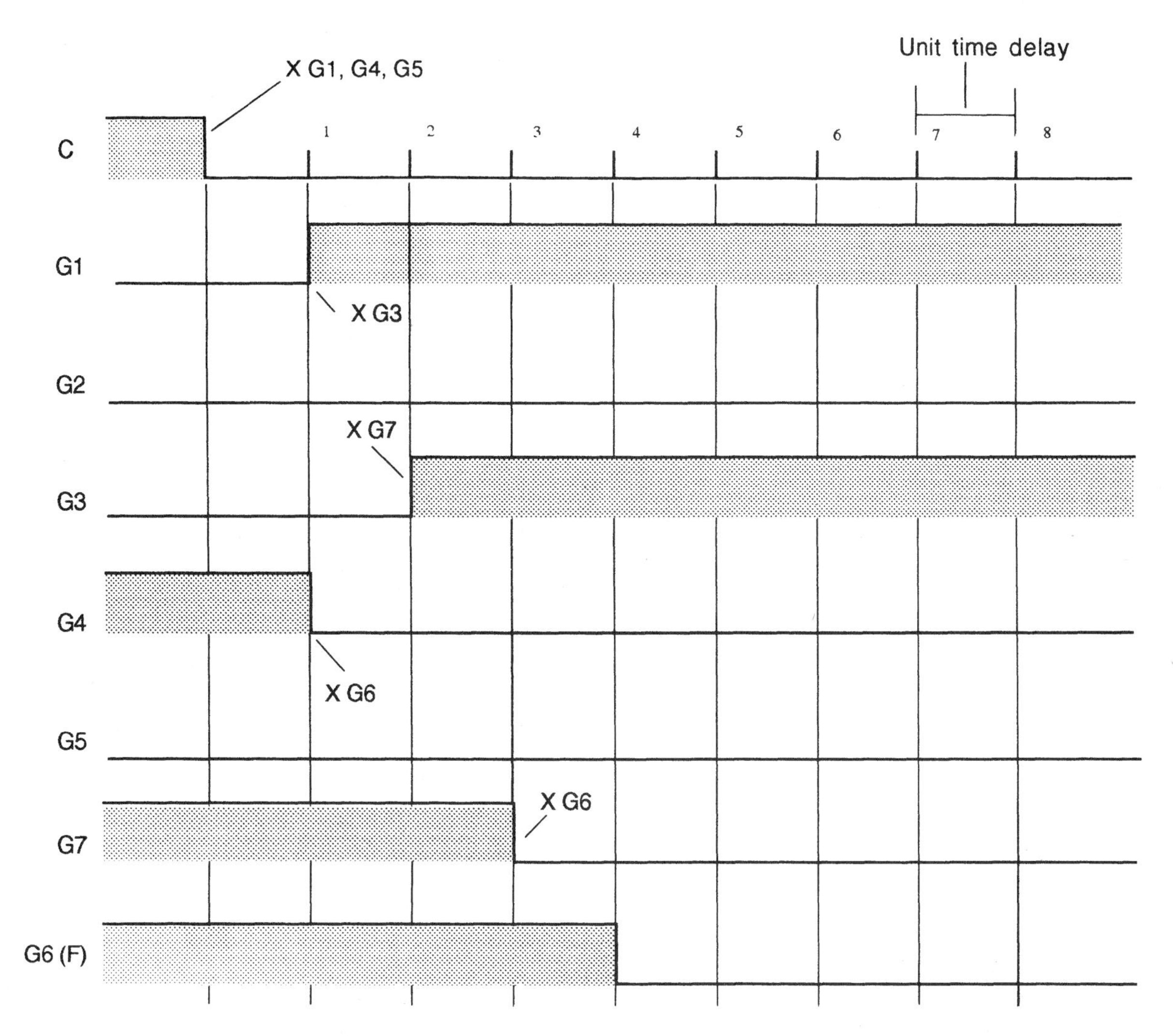

Fig. 6-19 Logic waveforms.

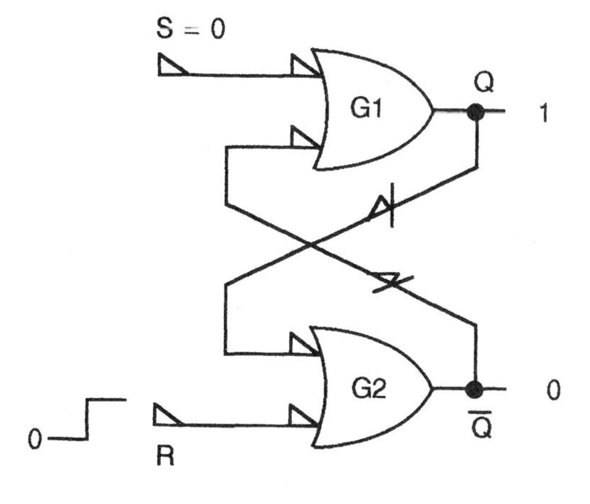

Fig. 6-20

since this gate now has zeros at both inputs, its output Q will transfer from 1 to 0 after one more delay interval (refer to Fig. 6-21).

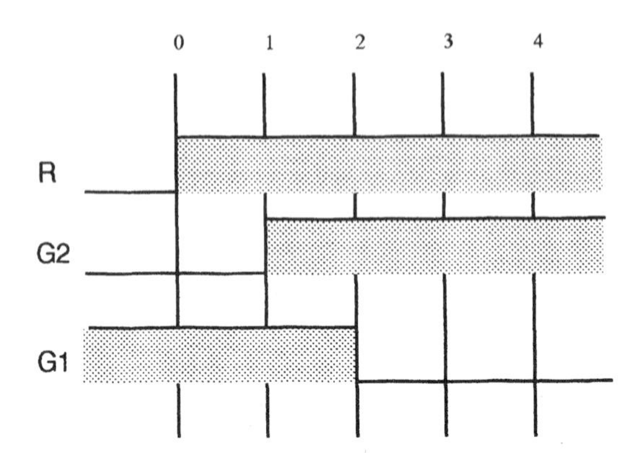

Fig. 6-21 Logic waveforms for the latch.

We see that G_2 will not respond to the change in G_1 because it is already in the 1 state and the process terminates. The new input 1 arriving via inversion from G_1 will, however, "latch" G_2 in the 1 state so that it will remain there even though input R returns to logic 0. More on this in Chap. 7. Note that it takes two time-delay intervals for the circuit to settle into its final state.

6.5 In Fig. 6-22, the initial conditions are Q = 0, A = 0, B = 1, and C = 0. At time t = 0, input C changes to a 1 and, six time-delay intervals later, it returns to a 0. Draw the microtiming diagram for 12 time-delay intervals, assuming that all gate delays are the same.

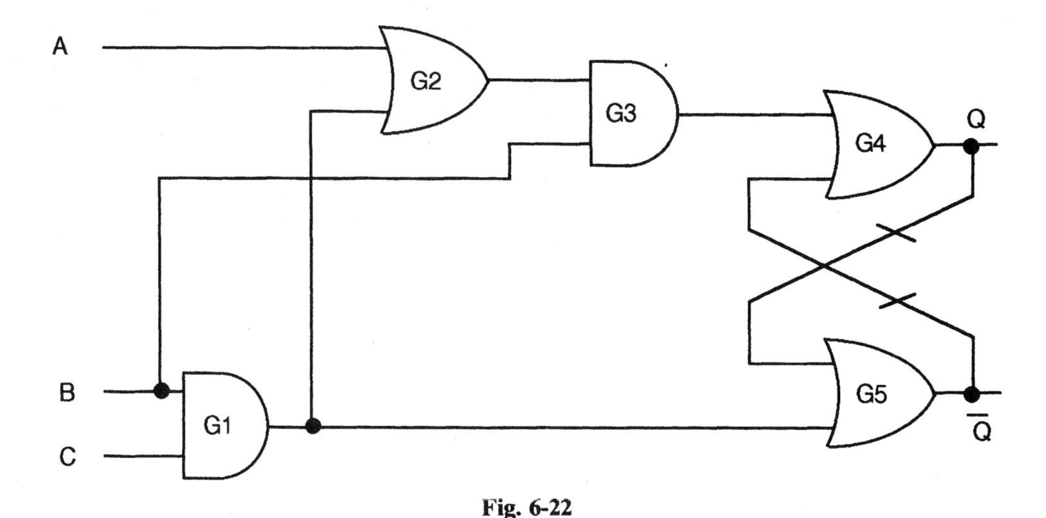

Fig. 6-22

From the given terminal conditions, we may deduce the logic states of all gates in the circuit by making use of simple Boolean relations. These states are entered in the timing diagram prior to t = 0, at which time input C is caused to step. After noting those gates potentially affected by each change, the appropriate responses are plotted one time-delay interval later as shown in Fig. 6-23.

The G_5 transition at time division 8 has no effect since gate G_4 is already in the 1 state via its input from gate G_3. The process therefore terminates until another input change occurs.

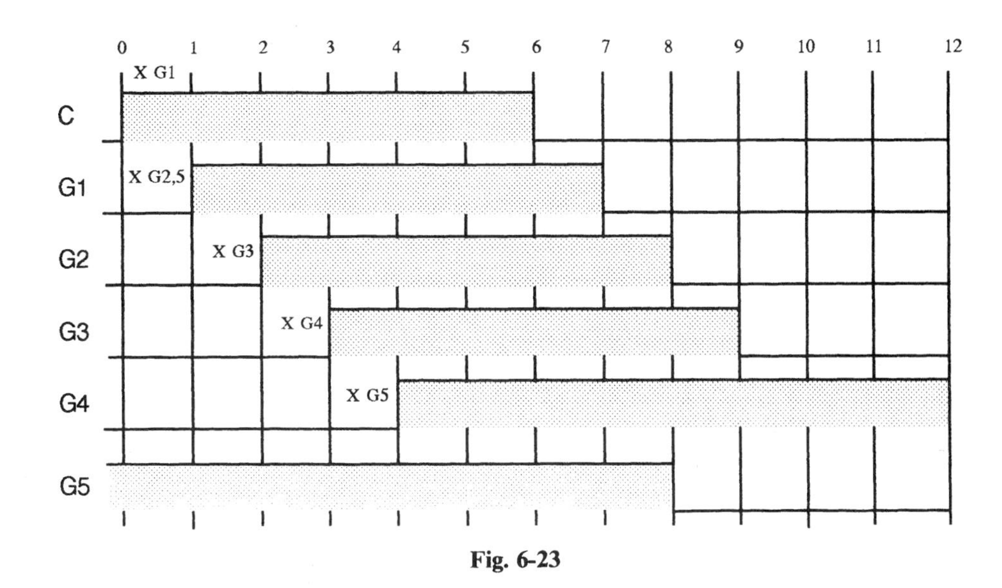

Fig. 6-23

6.6 Show that if a hardware inverter is inserted into one of the feedback paths of a latch as shown in Fig. 6-24, the circuit can become unstable and oscillate continuously.

The analysis is most easily carried out if we don't have to worry about HT and LT designations. Therefore, we will first convert the hardware diagram of Fig. 6-24 to a logic diagram by removing half arrows (Fig. 6-25). Inverters are usually removed also, since they serve only to convert LT to HT or vice versa and do no logic. In the present case, however, we intend to investigate the effect of propagation delays and must consider the delay in hardware inverter G_3 even though we treat it logically as a piece of wire.

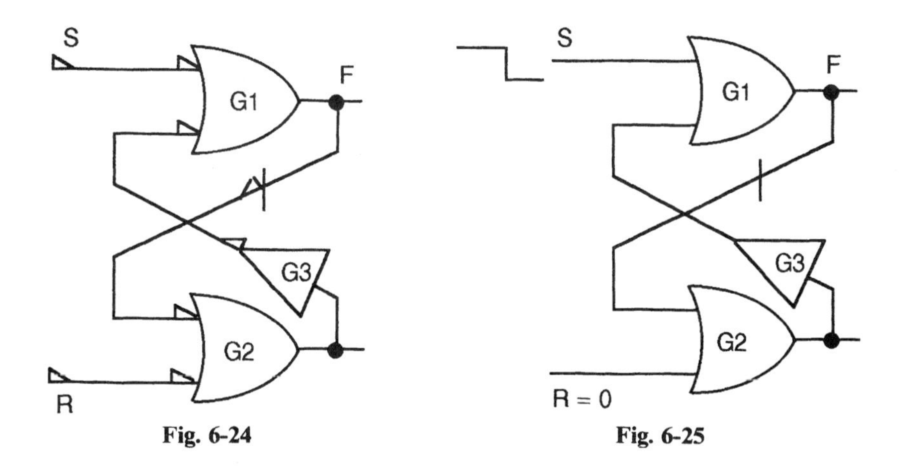

Fig. 6-24 Fig. 6-25

It should be clear, from Fig. 6-25, that a logical asymmetry has been introduced into the latch circuit. A change in the output of G_1 is logically inverted before passing to G_2, whereas a change in G_2's output passes unaltered to the input of G_1. If $S = 1$, the output of G_1 will be 1, the output of G_2 will be 0, and a change in R should have no effect since OR gate G_1 only requires a single TRUE input to have its output TRUE. The circuit is stable. If, on the other hand, S changes from 1 to 0 when $R = 0$ as shown, then we see that gate G_1 will have two logic 0 inputs and F will go to 0 after one time-delay interval. This is interpreted as a 1 at the input to G_2 (note the slash) whose output goes to a 1 after one more delay interval, and this 1 is passed on unaltered by G_3 which adds an additional delay increment. Thus, three propagation-delay intervals after the change in S occurs, G_1 receives a countermanding signal to return to the 1 state. This change, of course, starts the chain all over again and a continuous oscillation should be expected with a repeating period of six propagation-delay intervals. An applicable microtiming diagram is shown in Fig. 6-26.

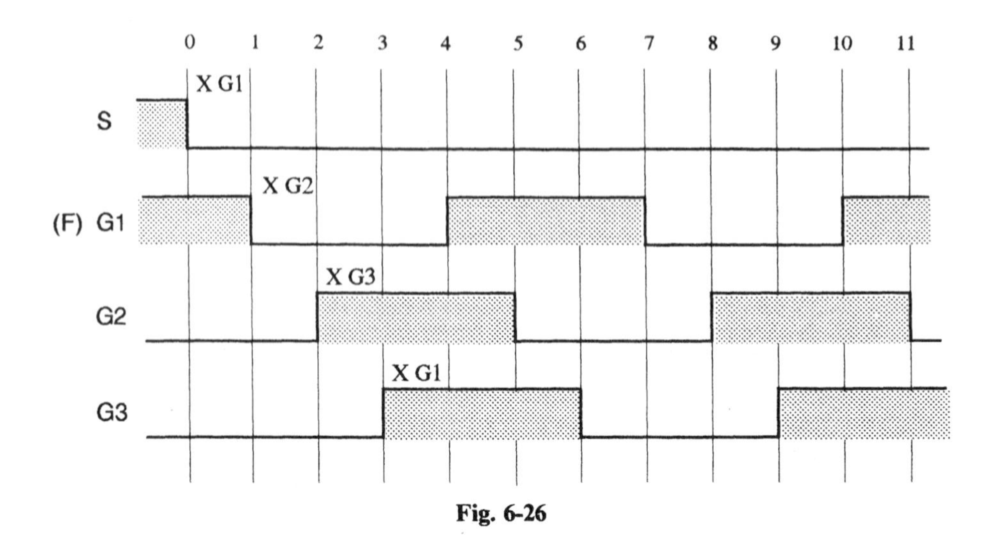

Fig. 6-26

Since the microtiming diagram is constructed one step at a time, it makes the sequential analysis easier to trace and comprehend than trying to reason through the process with words alone (eloquent as they may be).

6.7 Demonstrate that instability in the latch circuit of Fig. 6-20 can also result if the input change is reduced in duration to a single time-delay interval.

As in Prob. 6.4, we assume that input R goes from 0 to 1 when Q = 1 and Q′ = 0. If, following this step, R returns to 0 before the signal it initiated propagates through G_1 and back to the input of G_2 with a 1 to latch it, then G_2 will return to the 0 state instead of remaining in the 1 state as expected. Further analysis shows that a continuous oscillation will occur until either R or S becomes a logic 1. The logic waveforms are shown in Fig. 6-27 where arrows indicate the cause-and-effect relations involved. This problem illustrates the fact that designers must consider the duration of inputs as well as connection paths in order to assure that their circuits will be stable.

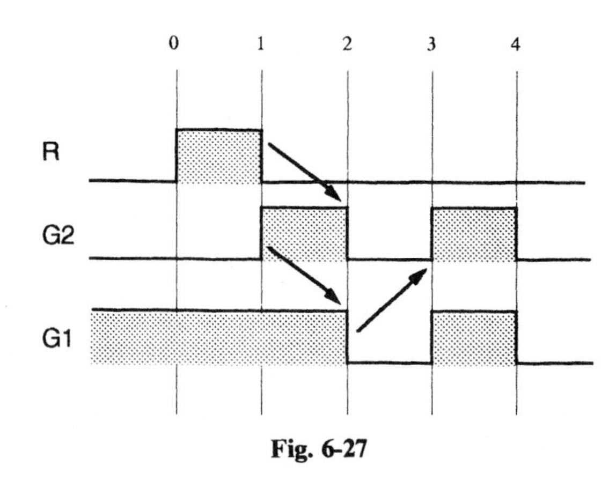

Fig. 6-27

6.8 Draw the macrotiming diagram for the circuit and logic waveforms shown in Fig. 6-28.

$G = D \oplus S_0$ and $H = (XS_1)'$. The output of the multiplexer will be $F = XS_1$ when H = 0 and F = GH when H = 1.
These relationships are shown in Fig. 6-29 in time relation to the given input variables.

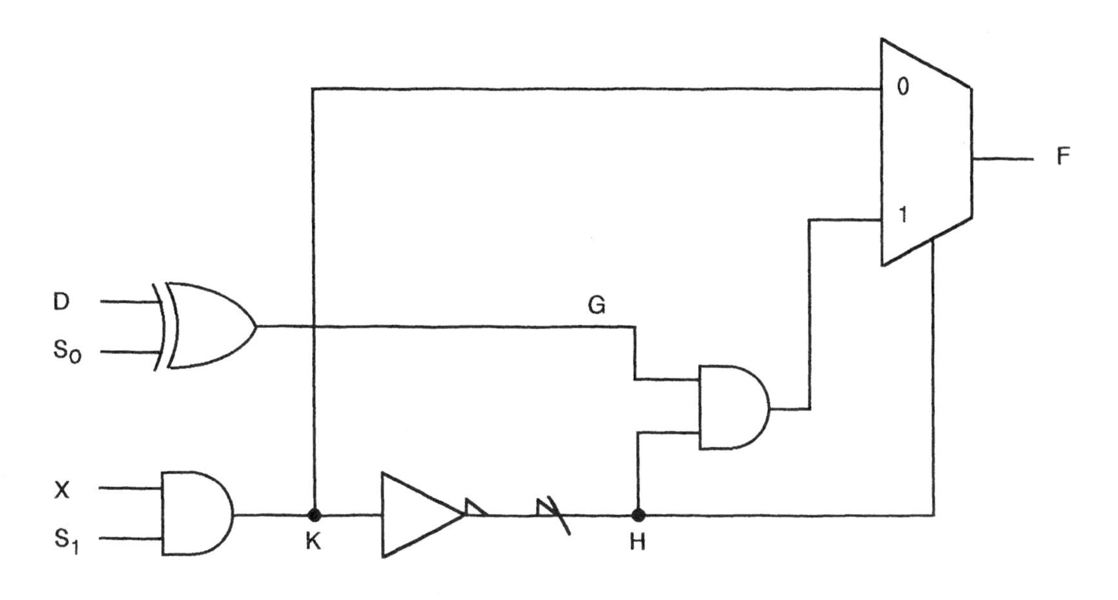

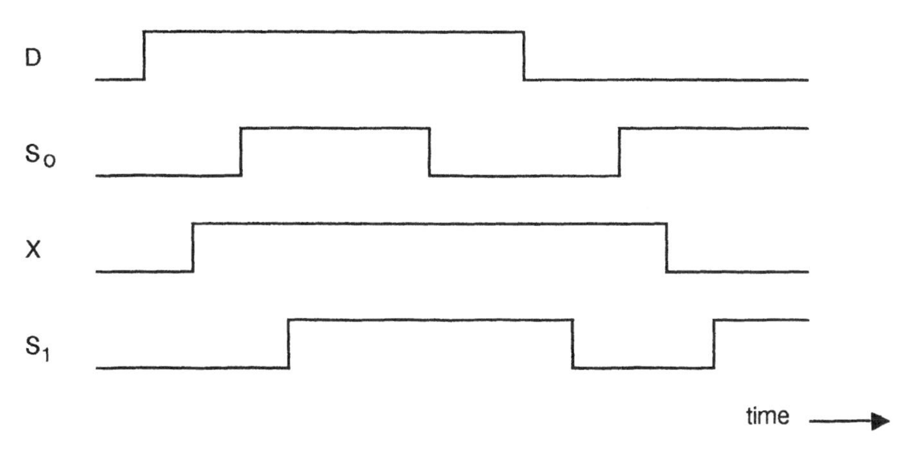

Fig. 6-28

6.9 The logic circuit shown in Fig. 6-30 represents a master-slave JK flip-flop (which will be described in detail in Chap. 7). The initial conditions present on its external pins (represented by boxes) are as follows: J = 1, K = 1, CLK = 0, CLR = 0, PR = 0, Q = 1, Q′ = 0. Sketch a micro-timing diagram to show what happens when the CLR input steps from 0 to 1 at time t = 0.

Before the timing diagram can be started, it is necessary to determine the state of every gate at t = 0. This may be accomplished by reasoning from the given initial conditions as follows:

- Since Q′ = 0, all three inputs to G_8 must be 0. Thus, $G_6 = 0$.
- The state of G_5 can't be determined yet because the lower input to G_7 is 1, and this is sufficient to account for the given Q = 1.
- $G_{10} = 0$ because the other input to G_6 is 1.
- $G_4 = 0$ because the other input to G_{10} is 1.

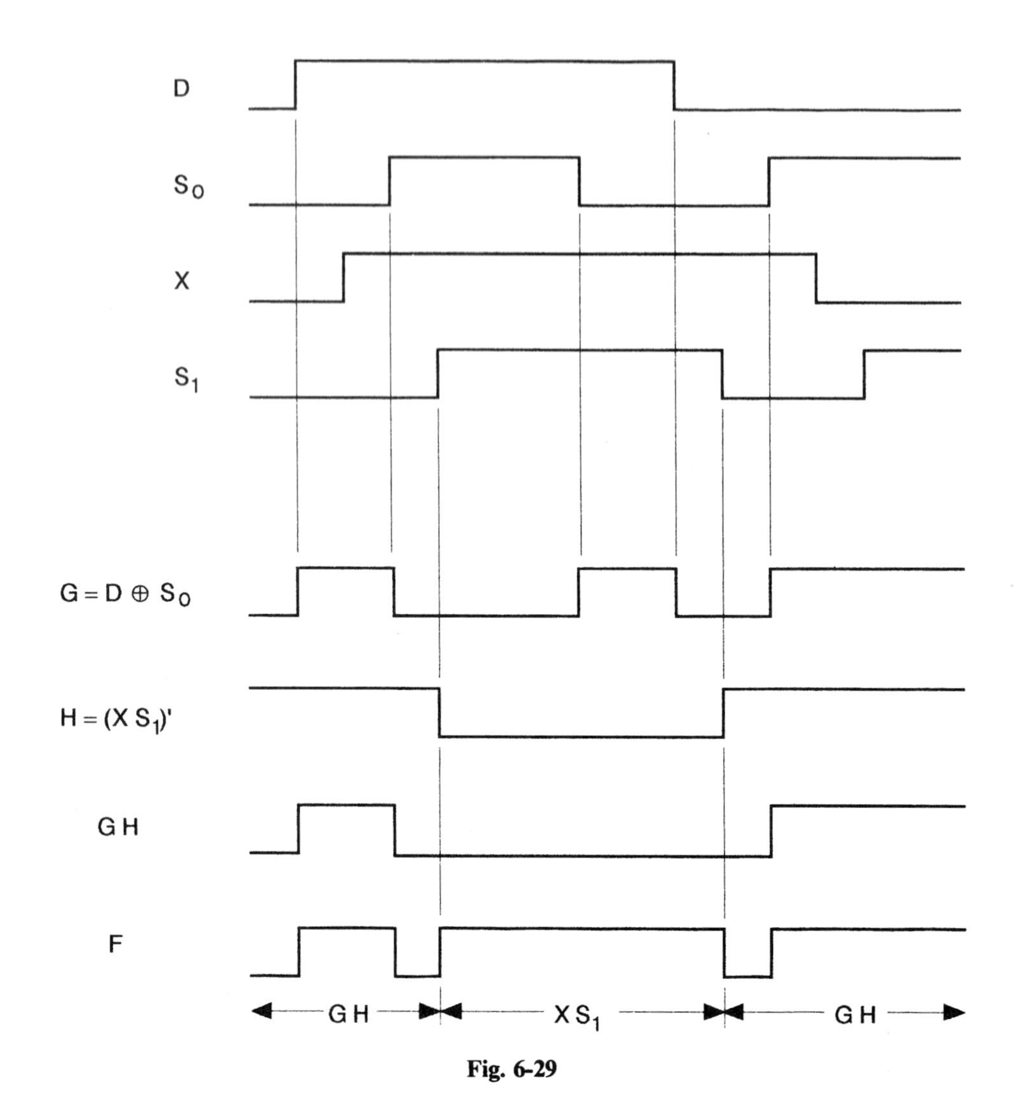

Fig. 6-29

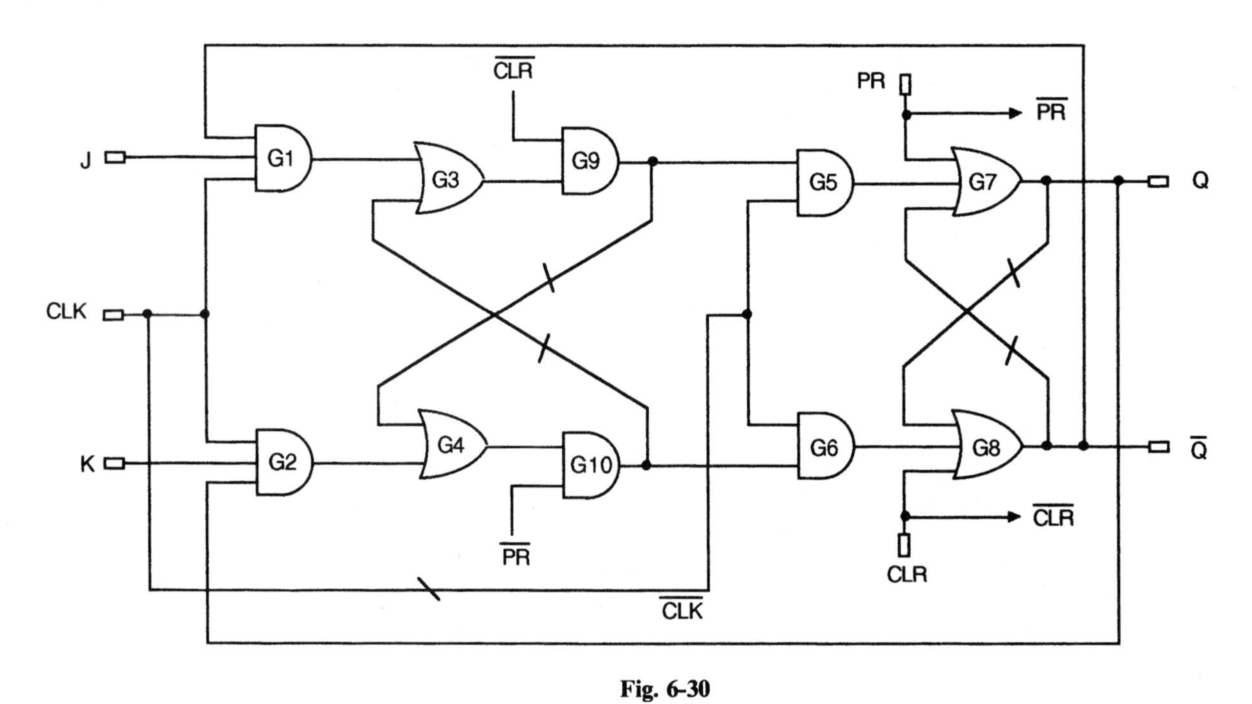

Fig. 6-30

- $G_1 = 0$ and $G_2 = 0$ because CLK $= 0$.
- $G_3 = 1$ via the inversion of output G_{10}.
- $G_9 = 1$ because both its inputs are 1s. This may also be deduced from the input requirements for G_4.
- $G_5 = 1$ because both its inputs are 1s.

At t = 0, the CLR input goes from 0 to 1. One propagation-delay unit later, G_8 and G_9 respond in time interval 1, as whown in Fig. 6-31.

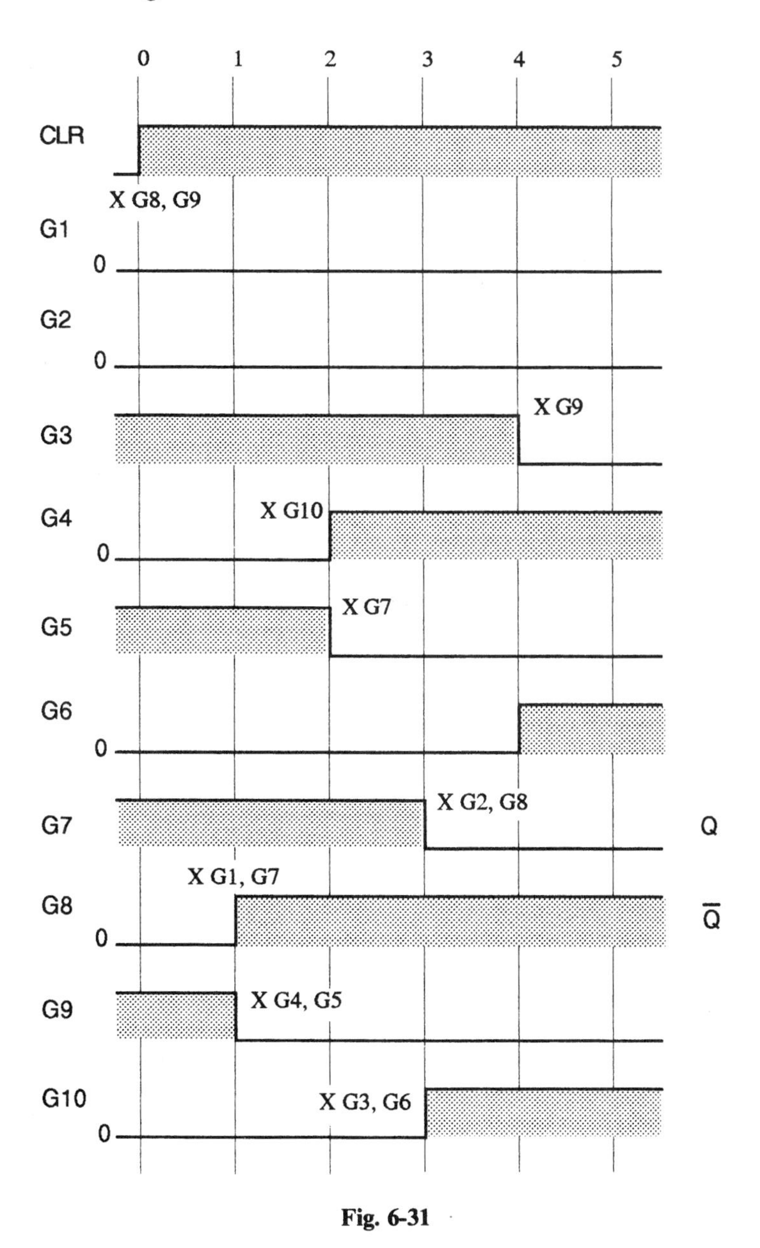

Fig. 6-31

The response continues to propagate.

- G_8 is connected to G_1 and G_7. G_1 can't change because CLK remains at 0; G_7 won't change because of the 1 input from G_5.
- G_9 is connected to G_4 and G_5 and both of these gates respond in time interval 2.

- G_4 affects G_{10} and G_5 affects G_7, both of which respond in interval 3.
- G_7 is connected to G_2 and G_8, neither of which will respond.
- G_{10} is connected to G_3 and G_6, both of which respond in interval 4.
- G_3 connects to G_9 and G_6 connects to G_8. Neither of these gates will respond, and the process terminates.

Notice that this circuit requires four time-delay intervals for the response to proceed to completion. Applying a CLR input of lesser duration could result in some unpleasantness, as demonstrated in Prob. 6.7.

6.10 Consider a synthesis problem in which a circuit is required that will produce a particular output in response to a given set of inputs. The specification is presented in Fig. 6-32 where it is assumed that the output F responds to inputs A and B after one propagation-delay interval has elapsed. Design the circuit.

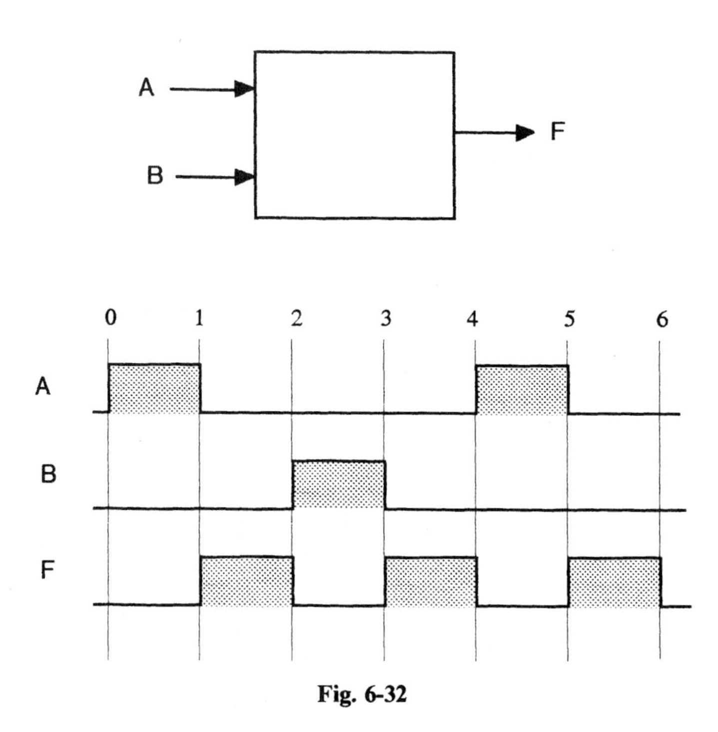

Fig. 6-32

To view the functional relationship between inputs and output, we redraw the waveforms as a macro-timing diagram by removing the delay as shown in Fig. 6-33. A truth table may now be prepared by reading down each time-interval column:

Interval	A	B	F
0 - 1	1	0	1
1 - 2	0	0	0
2 - 3	0	1	1
3 - 4	0	0	0
4 - 5	1	0	1
5 - 6	0	0	0

(Rows 3 - 4 through 5 - 6: Redundant)

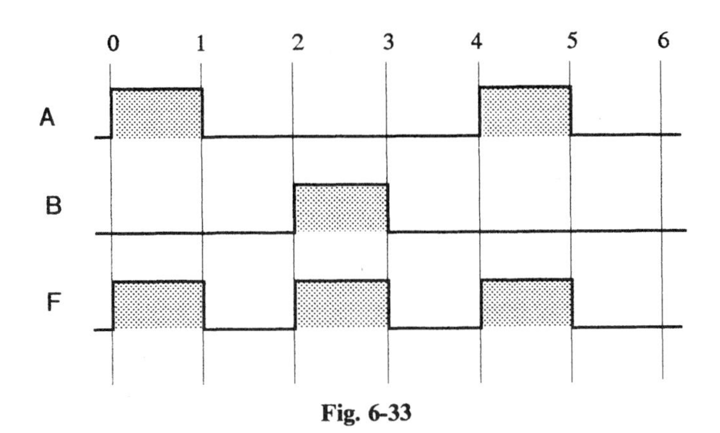

Fig. 6-33

We next remove redundant entries and, noting that the 11 state does not occur, treat it as a "don't care".

A	B	F
0	0	0
0	1	1
1	0	1
1	1	X

Synthesis, or design, problems rarely have unique answers, and the designer must select from a number of possible approaches and solutions. For example, in this exercise, one could write F = A′B + B′A and use five hardware elements as shown in Fig. 6-34; or a somewhat more experienced designer might recognize the exclusive-OR function. An inspired designer, making use of the knowledge that the 11 state never occurs, would simply use a single OR gate.

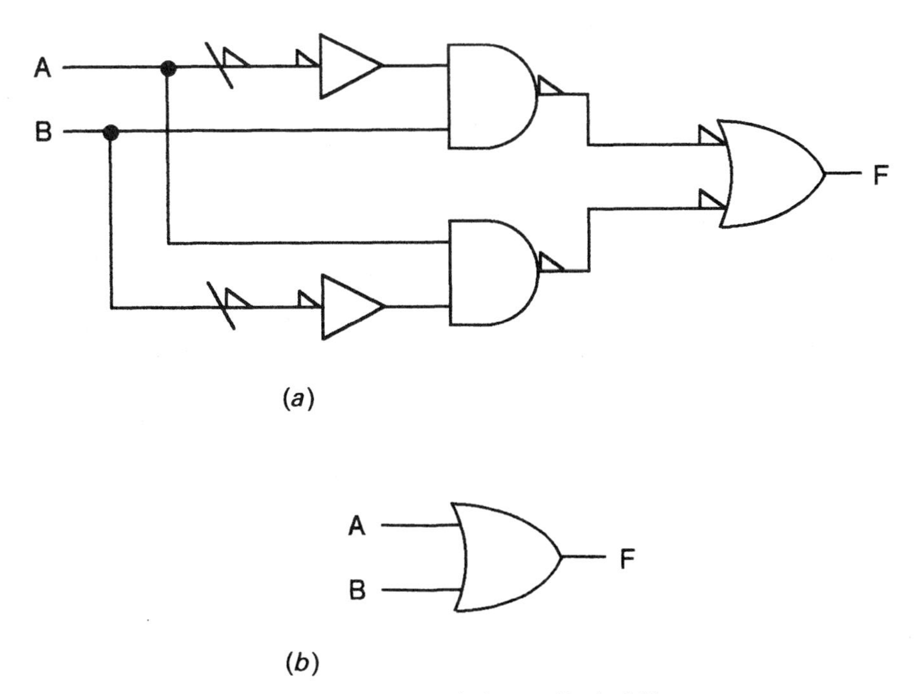

A

B

F

(*b*)

Fig. 6-34 Two solutions to Prob. 6.10.

It should be pointed out that synthesis problems are usually complex and require a more sophisticated approach than that required by this simple example. Suppose that F were to be 1 in interval 3–4. In this case, a conflict is observed with the requirements of interval 1–2, and some additional design power must be applied (see the material on state machines in Chap. 10).

Supplementary Problems

Note: Problems 6.11 through 6.13 illustrate how hardware and truth assignments affect circuit timing and economy.

6.11 The circuit shown in Fig. 6-35 was developed in Prob. 4.36. If, for all gates, the shortest propagation delay is 5 ns and the longest delay is 15 ns, find the longest and shortest path delays in the circuit.

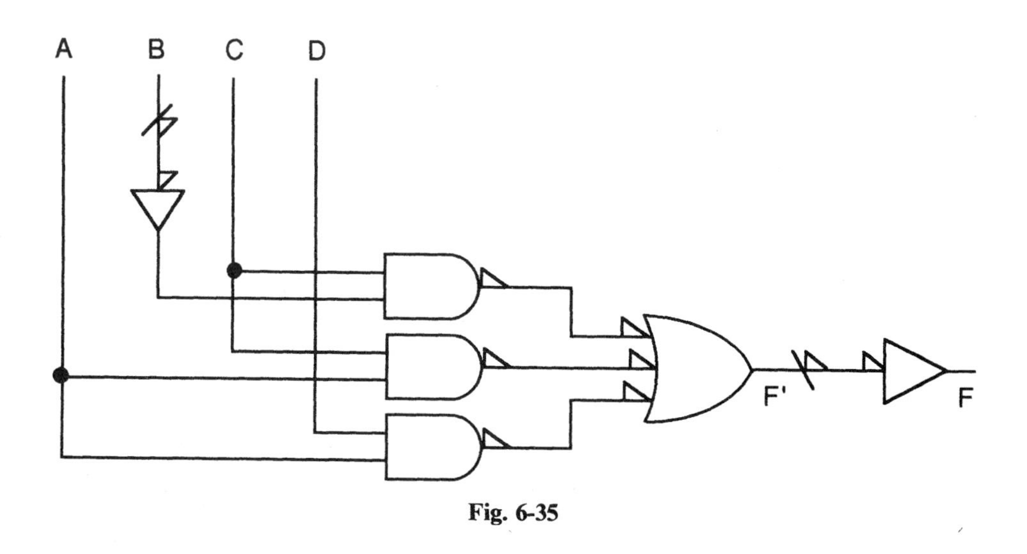

Fig. 6-35

6.12 Two other solutions for F were obtained in Prob. 4.38. Using the same propagation delays specified in Prob. 6.11, find the longest and shortest path delays (refer to Fig. 6-36).

6.13 The fabrication of a gate or inverter on a custom chip requires the following allocation of resources:

Each gate	2 units
Each input, after the first	0.5 unit
Each interconnecting wire segment	0.25 unit

If each silicon resource unit costs \$0.02 and each unit of path delay over 15 ns cost \$0.10, compare costs of the two implementations in Prob. 6.12, assuming worst-gate propagation delays of 15 ns.

6.14 In Fig. 6-37, all hardware (gates and inverter) have the same time delay. Sketch the complete microtiming logic diagram assuming that initially inputs A and D are 1s and, at time t = 0, L cycles as shown in Fig. 6-38. All inputs and outputs are HT.

6.15 The multiplexer in Fig. 6-39 responds to a change in its select input after two gate-delay intervals. Assume that immediately prior to t = 0, inputs R and S are 0, Q = 0, and that multiplexer inputs A and B are 0 and 1, respectively. Sketch the microtiming diagram for the case where S switches to 1 at t = 0.

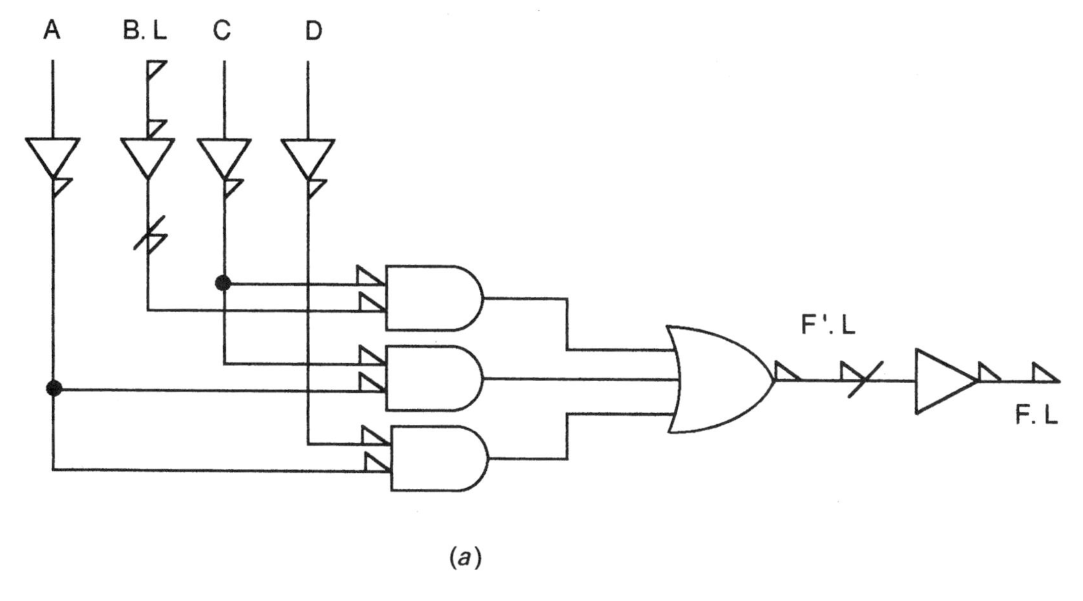

(a)

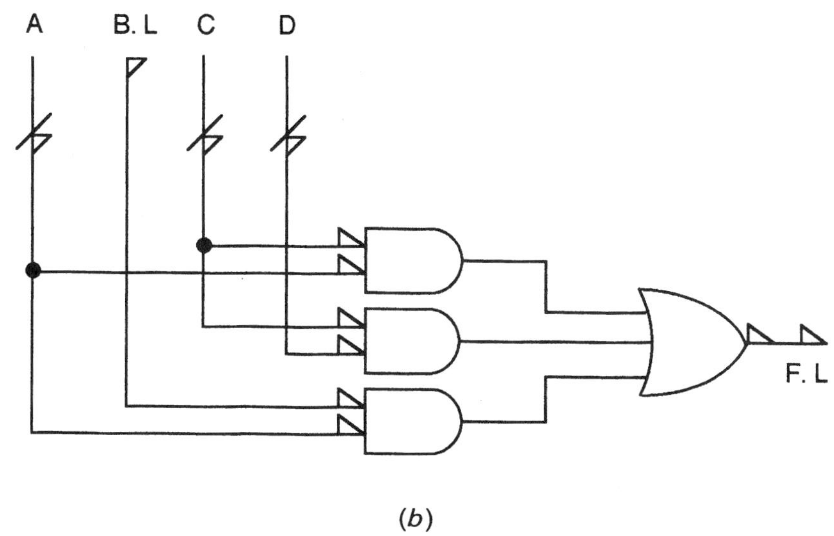

(b)

Fig. 6-36

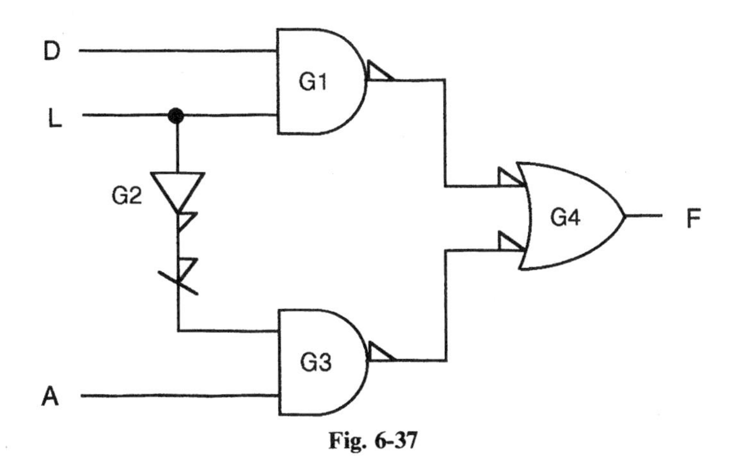

Fig. 6-37

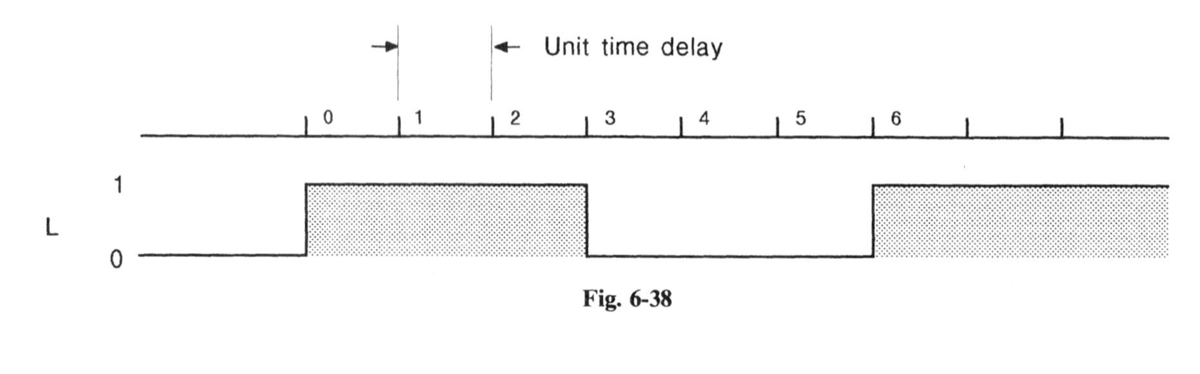

Fig. 6-38

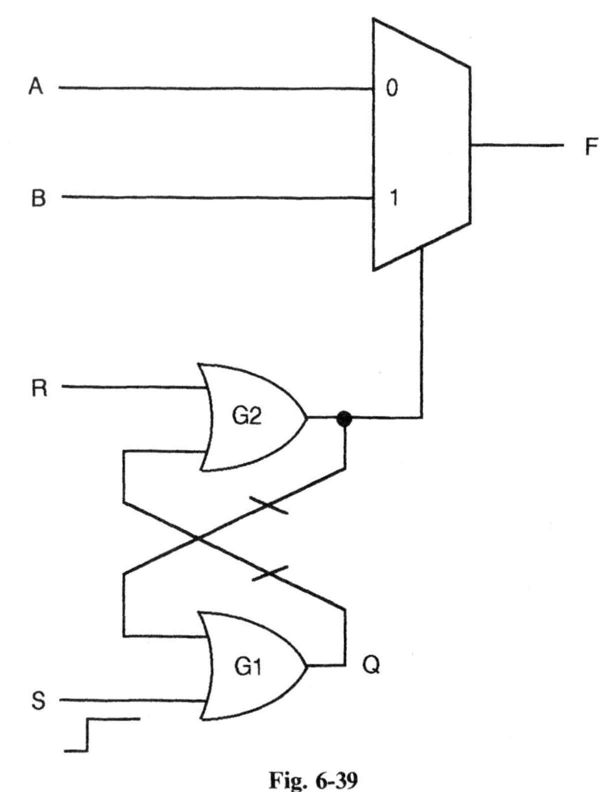

Fig. 6-39

6.16 For the circuit shown in Fig. 6-40, determine whether there are any hazards, and, if so, determine the transitions which cause them.

6.17 For the circuit of Prob. 6.16, create a voltage microtiming diagram for the transition involving G2 and G3.

6.18 With reference to the circuit of Prob. 6.16, create a voltage microtiming diagram for the transition involving G1 and G2.

6.19 What modification can be made to the circuit of Fig. 6-40 to prevent the glitch of Prob. 6.17 from occurring?

6.20 In the circuit of Example 6.1 (Fig. 6-1), determine all possible hazards. First determine potential hazards by inspection for reconvergent paths and then investigate completely by mapping.

6.21 The loop of inverters shown in Fig. 6-41 is a circuit of the sort used to test the speed of microcircuits. Show that if the number of inverters is odd, an oscillation will develop.

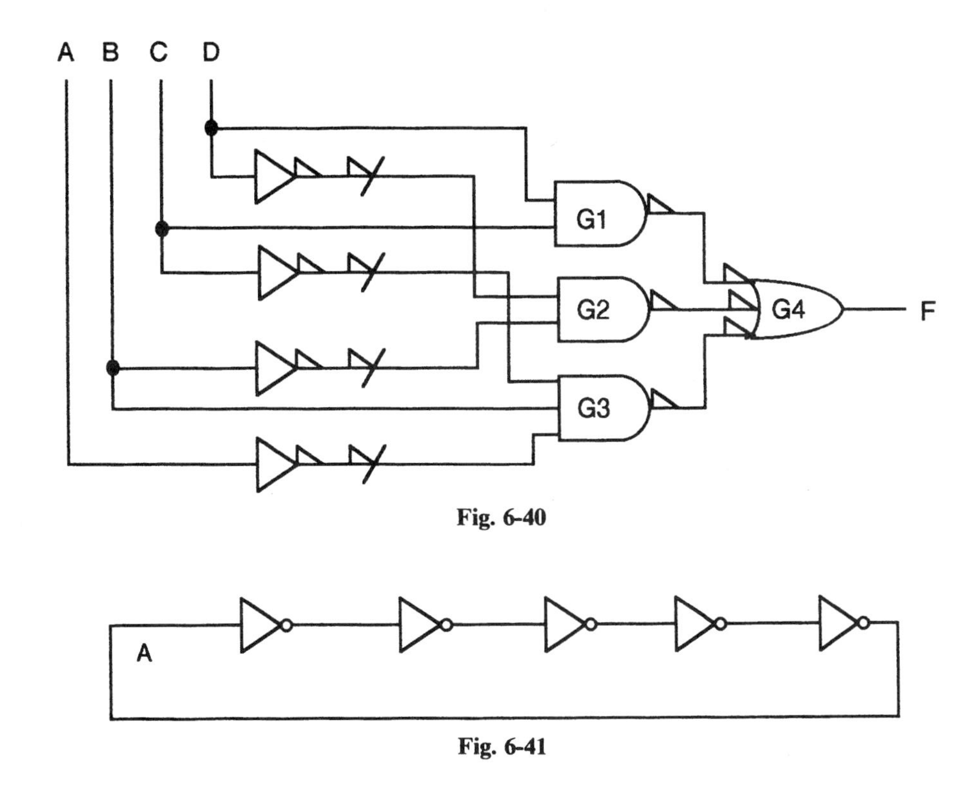

Fig. 6-40

Fig. 6-41

Answers to Supplementary Problems

6.11 An input change can take anywhere between 15 and 60 ns to appear at the output.

6.12 The longest and shortest path lengths are equal and time through the network will range between 10 and 30 ns. Note how the selection of hardware to match truth values can affect performance.

6.13

	(*a*)		(*b*)	
Resources	24.75	\$0.50	13.5	\$0.27
Excess time	45 ns	\$4.50	15 ns	\$1.50
Total		\$5.00		\$1.77

6.14 See Fig. 6-42. Note that a glitch occurs when L goes from 1 to 0 and not the reverse.

6.15 See Fig. 6-43.

6.16 Mapping the logic F = B′D′ + CD + A′BC′ shows three cases of touching but nonoverlapping coverings. Hazards occur for the following transitions.

A B C D	to	A B C D
0 1 0 0		0 0 0 0
x 0 1 1		x 0 1 0
0 1 1 1		0 1 0 1

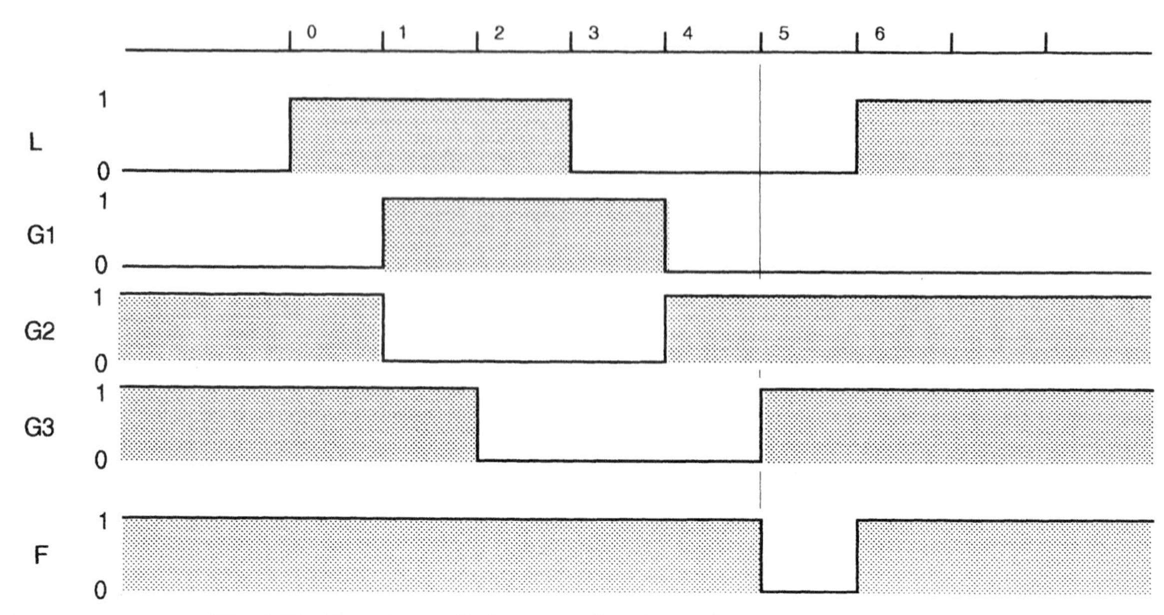

Fig. 6-42 Note that a glitch occurs when L goes from 1 to 0 and not the reverse.

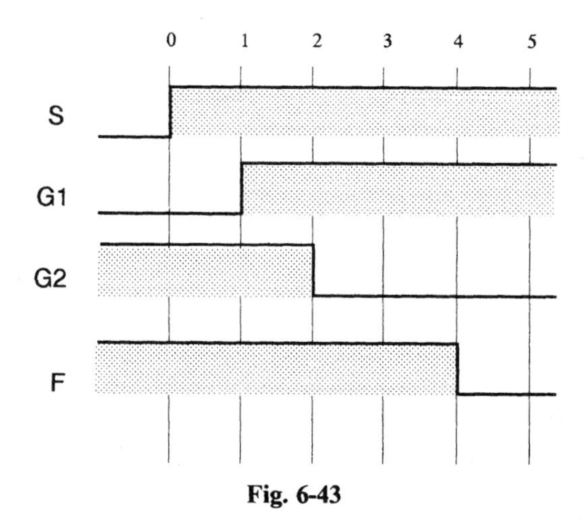

Fig. 6-43

6.17 See Fig. 6-44. Gate-propagation delays have been set at 5 time units to emphasize the glitch.

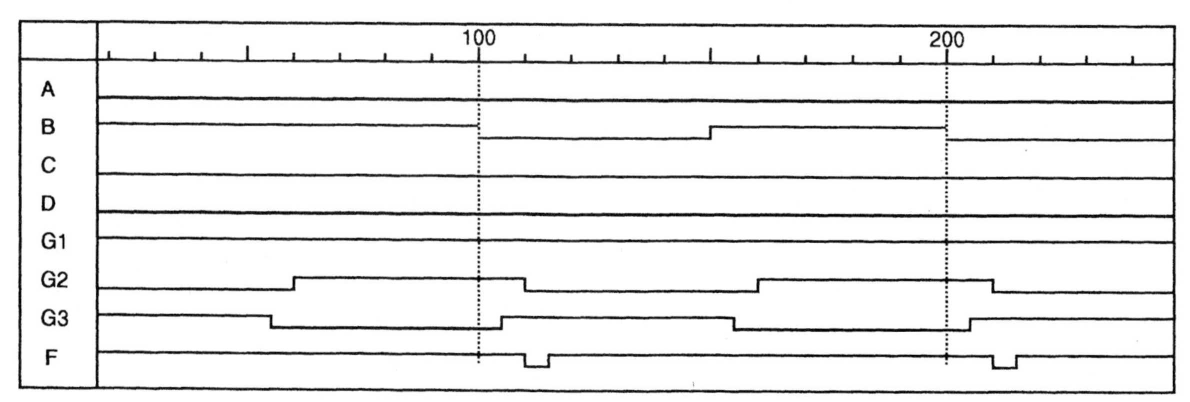

Fig. 6-44 Note the glitch at the appropriate transition.

6.18 See Fig. 6-45. All gate-propagation delays are 1 time unit.

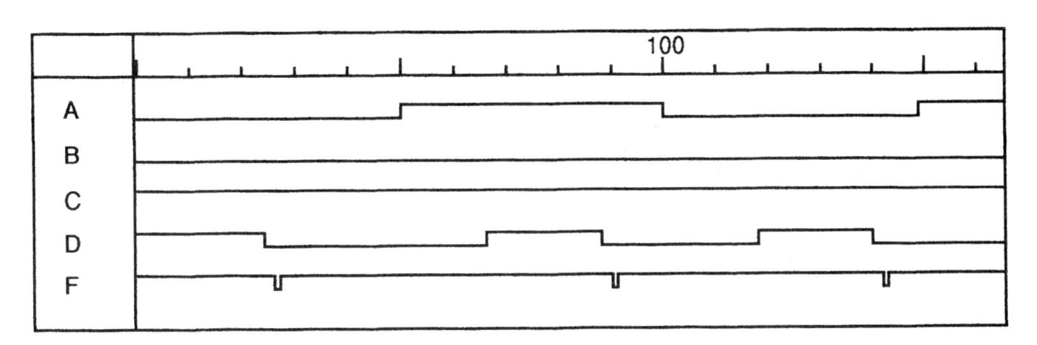

Fig. 6-45 Note that the glitch is independent of A.

6.19 Add the hazard covering shown in Fig. 6-46 by converting G_4 to a four-input gate and wiring the term A′C′D′ using an additional NAND gate.

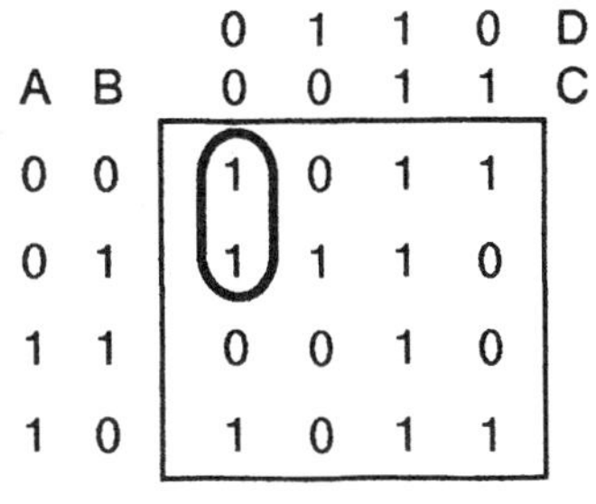

Fig. 6-46

6.20 There are two reconvergent paths from an input; one from A and one from C. The Boolean equation for the circuit is F = AC′D + BC + A′C which, when mapped, shows a hazard only exists for C (the ABCD to ABC′D transition). The potential hazard on A is covered.

6.21 Assume that point A is 0. If there are n integers, a 1 will appear at the end of the chain if n is odd. Thus, the signal A will change to 1 after an elapsed time delay of $n(t_{pd})$. The 1 will, in turn, become a 0, and so on. The simulation is shown in Fig. 6-47.

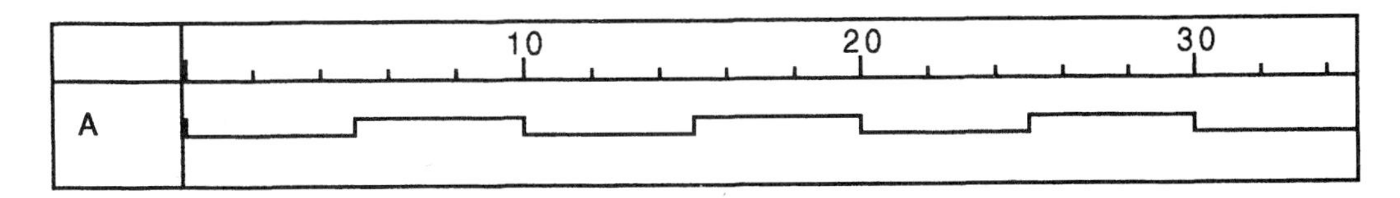

Fig. 6-47

Chapter 7

The Flip-Flop

7.1 INTRODUCTION

In addition to performing combinational logic, most modern digital systems also store commands and data which is waiting to be processed or which is the result of computation. This chapter deals with various embodiments of the basic digital memory element, the bistable circuit or *flip-flop*. Its most primitive form is called a *latch* which consists of a pair of logic gates with their inputs and outputs interconnected in a feedback arrangement which permits a single bit to be stored. Adding components to the basic latch and combining latches in various ways produces several different flip-flop types, each of which has characteristics designed to optimize performance in specific applications. This chapter is concerned with RS, JK, D, and T flip-flops and their operating modes and triggering characteristics.

7.2 THE BASIC LATCH

Figure 7-1 shows the logic for a simple memory element called a latch.

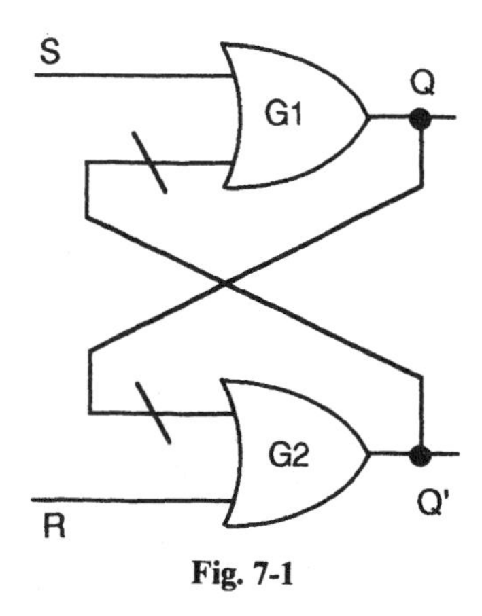

Fig. 7-1

This circuit has two inputs (S and R, corresponding to "set" and "reset," respectively) and at least one output (Q). Sometimes a second output (Q′) is included as shown. It is easy to trace the closed feedback loop: Signal Q is logically inverted and connected to gate G_2 which produces Q′. This signal, in turn, is logically inverted and coupled back to the input of gate G_1 which generates Q, thus completing the loop. This arrangement can be used to store a single bit of information when both R and S are logical 0s.

Assume that output Q is a 1 when inputs R and S are both 0s. This output is logically inverted to a 0 and fed back to serve as one input to G_2. Both G_2 inputs are now 0s, and, consequently, a 0 is produced at its output Q′. This signal is inverted and fed back to the input of G_1 whose output Q must consequently be a 1 thus validating the original assumption $Q = 1$. Note that Q′ is a 0 under these circumstances. This situation will be maintained until an input change occurs.

A similar analysis can be made to demonstrate that $Q = 0$ is also a stable state. In this case, Q′ will be 1, and we see that output Q′, as expected, is always the logical inverse of Q when R and S are both 0.

Suppose, now, that S is changed to a 1 so that the input state is R = 0, S = 1. This places a 1 on at least one of G_1's inputs which causes Q to be 1. This 1 is logically inverted and appears as a 0 at an input to G_2 and, since R = 0, a 0 output from G_2 is assured. This 0 is inverted and fed back to the other input of G_1 thereby reinforcing its output Q = 1. The logic 1 on S which initiated the process can be removed any time after the feedback process has had time to complete its operation and the output remains at Q = 1. Thus, a 1 on S when R = 0 ensures that Q = 1, and this state will be maintained when S returns to 0. We say that the circuit has been latched. In a similar manner, it is easily shown that inputs of R = 1 and S = 0 result in Q = 0 and that this state is maintained (latched) when R returns to 0.

Thus, if the condition S = 0, R = 0 is regarded as a normal or rest state, the circuit's outputs tell us which input was last connected to a logic 1; in other words, *we may interpret the latching behavior as a memory function.*

The simple latch may be implemented in either NOR or NAND hardware as shown in Fig. 7-2 (see Prob. 7.17 for a positive logic form). Note that the logic is the same in both cases and only the truth values of inputs and outputs are different.

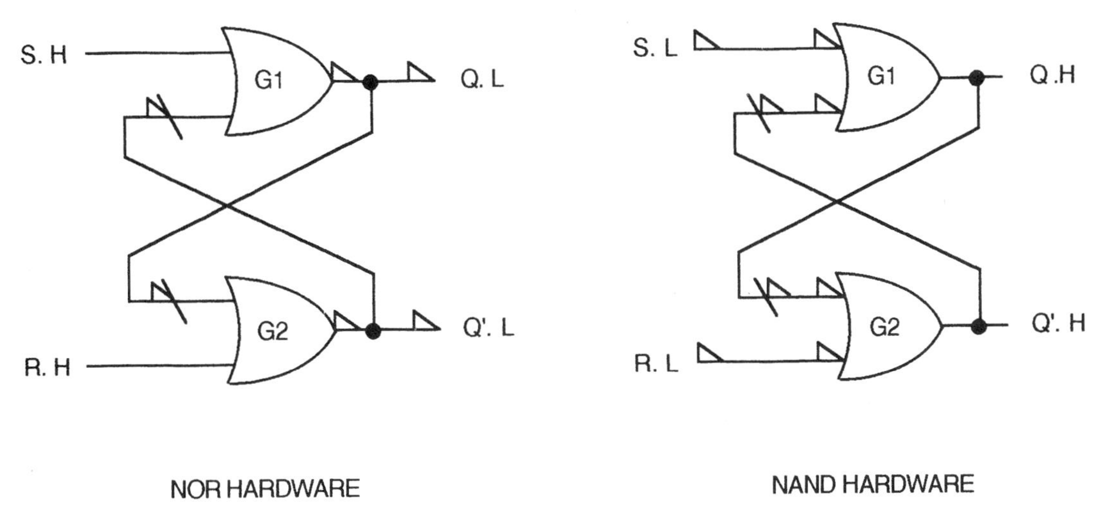

Fig. 7-2

7.3 THE CHATTERFREE SWITCH

Data is often entered into a digital system by means of manually operated switches. These switches can have many forms ranging from the spring-loaded contacts under the individual caps of computer keyboards to push-button switches on digitally controlled machine tools. A common characteristic of all these mechanical switches is that their spring contacts have a tendency to bounce when actuated, causing a short series of repetitive make-and-break connections lasting for several milliseconds. If, in a given system, a closed electrical contact represents a logic 1 and an open contact a 0, it should be clear that contact bounce can be a serious problem since multiple 1s and 0s might be generated when only a single 1 was intended.

The latch can serve as a very handy "debouncer" circuit as shown in Fig. 7-3. When the switch SW is in the upper position as shown, the voltage V (which corresponds to a logic 1) is applied to the S input. No current flows in the lower resistor, and input R will be at ground potential which corresponds to a logic 0. Thus, the upper switch position corresponds to S = 1, R = 0. When the switch is moved to the lower position, voltage V is applied to R and the input condition is S = 0, R = 1. Mechanically, switch bounces are very small. The moving part of the switch will bounce on and off its contact point (R or S) and never back and forth between these terminal points. At the top of a bounce, the latch input is S = R = 0, a condition where we have seen that no output change occurs. Thus, the latch will change

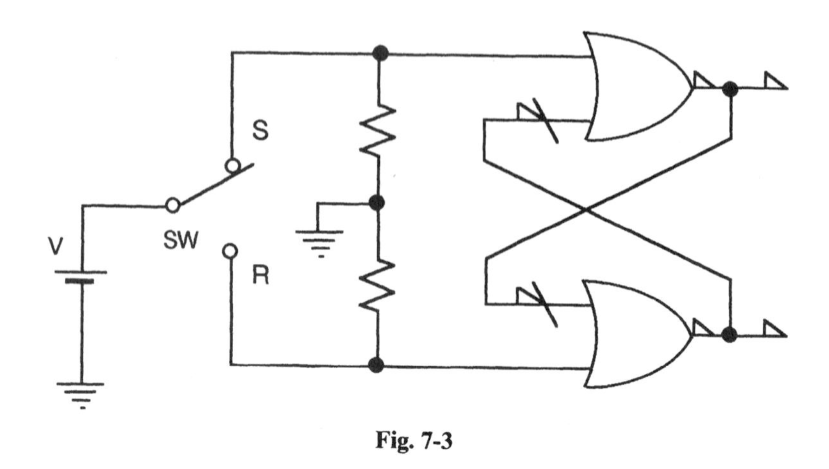

Fig. 7-3

state only when the switch is manually transferred from one position to another, and it will respond only to the initial contact, ignoring any bounces which occur.

7.4 CLOCKED RS FLIP-FLOP

As was pointed out in Chap. 1, the data stored in a digital system is vulnerable to noise. The simple latches shown in Fig. 7-2 are termed "1s catching" since a 1 appearing on either the R or the S input can be caught (latched) by the flip-flop which reacts the same whether this 1 was caused by true data or noise.

If the R and S inputs are ANDed with an enabling signal (often called a clock) as shown in Fig. 7-4, then the input to the latch will only respond to noise on the R and/or S inputs during the clock pulse duration. Many designers have adopted the approach of allowing the data (R, S) lines to be vulnerable to noise while devoting a great deal of resource to make the clock signal relatively noisefree. In this case, the latch is vulnerable to spurious signals only during the time that the clock pulse is true. It follows that a narrow clock pulse will result in increased noise immunity for the circuit, which is often called a *clocked*, *gated*, or *strobed latch*.

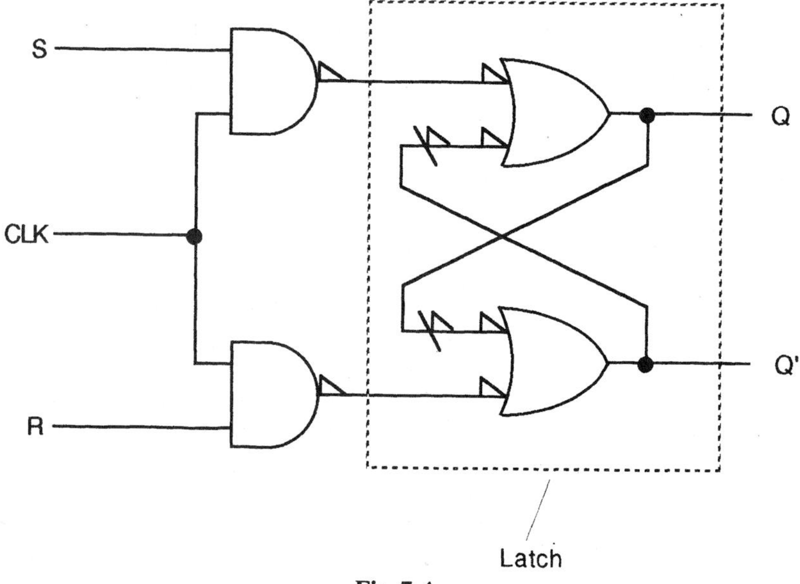

Fig. 7-4

The clocked RS flip-flop is often characterized by a *state table* (Table 7.1). Here Q_n is the value of Q prior to the clock pulse, while Q_{n+1} is the value assumed by output Q when a TRUE clock level is applied. The first row states that if S and R are both 0, the value of Q after the clock is equal to the value before the clock; i.e., no change will occur. The next two rows indicate that the Q output matches S if R and S are complements. The input combination R = 1, S = 1 is forbidden. This is because a condition exists where the output state can be indeterminate. If the R, S inputs are are both 1 and the clock is 1, then both outputs will be 1. The problem occurs if, after the clock returns to 0, the inputs change to R = S = 0. In this case, the output state will depend upon which AND gate output reaches 0 first, the outcome being determined by the winner of a race and not upon logic.

Table 7.1

S	R	Qn+1
0	0	Qn
0	1	0
1	0	1
1	1	forbidden

The clock serves another useful function besides improving noise immunity. We may think of it as an *enable* input which, when TRUE, connects the latch to the RS data lines. This is important since, as we have seen in Secs. 6.2 and 6.3, propagation delays can cause expected logic levels to appear at gate inputs at slightly different times and can even cause momentary false signals (glitches). The enable or clock signal delays the operation of a latch or other digital circuit until its inputs have time to become established at their final values. The periodic application of an enabling signal forms the basis for the *synchronous operation of digital systems.* Each enabling pulse, in effect, advances digital processing by one step and, since these pulses occur at regular intervals like clockwork, the interchangeability of the terms *enable* and *clock* seems appropriate.

The Strobed Data Latch

If an inverter is added between the R and S inputs of Fig. 7-4 (as shown in Fig. 7-5), then reference to Table 7.1 indicates that output Q will follow data input D when the enable input is TRUE and data will be latched into the flip-flop (stored) when the enable input is FALSE.

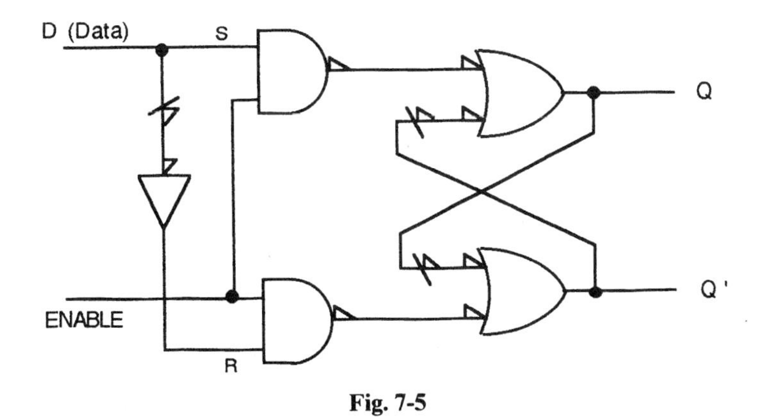

Fig. 7-5

EXAMPLE 7.1 In the circuit of Fig. 7-5, assume that Q is LOW at $t = t_0$ and an enabling strobe input is applied at $t = t_1$ and removed at $t = t_2$. The output, in response to a given data waveform, is shown in Fig. 7-6 in which propagation delays are assumed negligibly small for the time scale chosen.

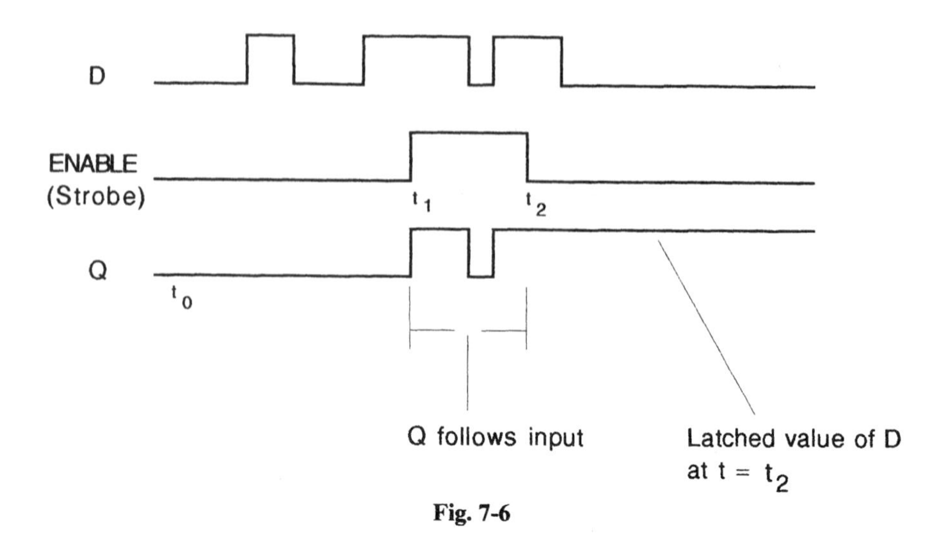

Fig. 7-6

As long as the strobe is 0, the latch inputs RS = 00 and no change will occur. When ENABLE = 1, Q will follow D (as per Table 7.1); we say that *Q is transparent to D* during this interval. When ENABLE returns to 0 at $t = t_2$, the flip-flop will store the value at the D input existing immediately prior to time t_2 when the strobed latch is disabled. It is important to bear in mind that all hardware has propagation delays and *timing considerations must be taken into account if input signal pulse widths are sufficiently small* (refer to Sec. 7.9).

An array of strobed latches, each connected to one line of a multiline data bus, is capable of capturing an n-bit word if n latches are used and all are strobed simultaneously by a common enable signal as shown in Fig. 7-7.

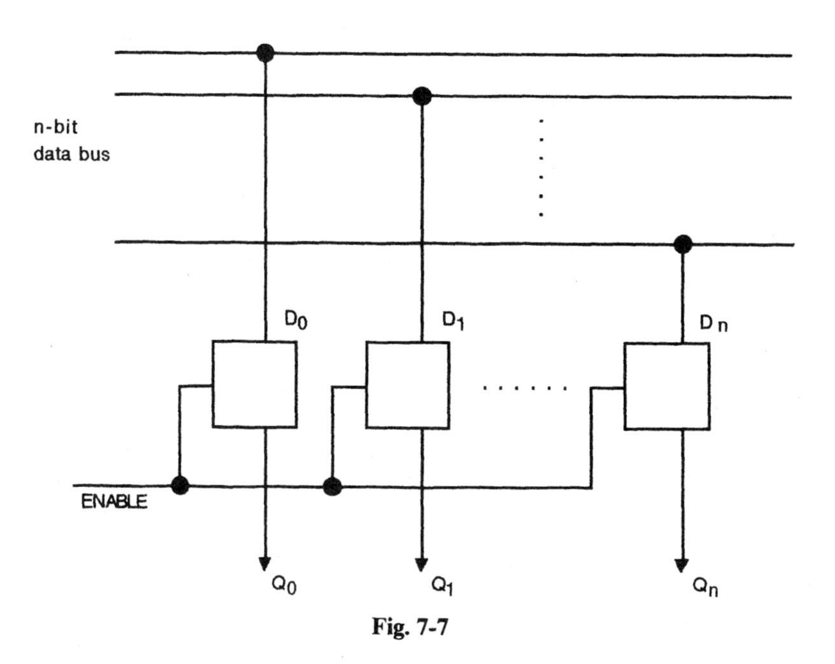

Fig. 7-7

7.5 THE JK FLIP-FLOP

The RS flip-flop's forbidden state represents opportunity. By adding components, designers have modified the basic clocked RS flip-flop so that the circuit will toggle (i.e., invert its output state) when clocked. This design modification presents some challenge since propagation time as well as logic connections must be considered. An initial attempt is shown in Fig. 7-8. The designer has fed the flip-flop's outputs back to the input gates in an effort to steer clock pulses to the proper side of the latch to cause toggling when a 11 input condition occurs. If Q is 1, the AND gate with input K is enabled and, if Q is 0, the gate with input J is enabled. Analysis of the circuit shows that response to JK = 00, 01, and 10 is identical to that of the clocked RS flip-flop. In the 11 state, however, as soon as toggling occurs, the feedback reverses, which means, unfortunately, that as long as the condition J = K = CLK = 1, we can expect an oscillatory condition to exist as shown in the simulation waveforms of Fig. 7-9.*

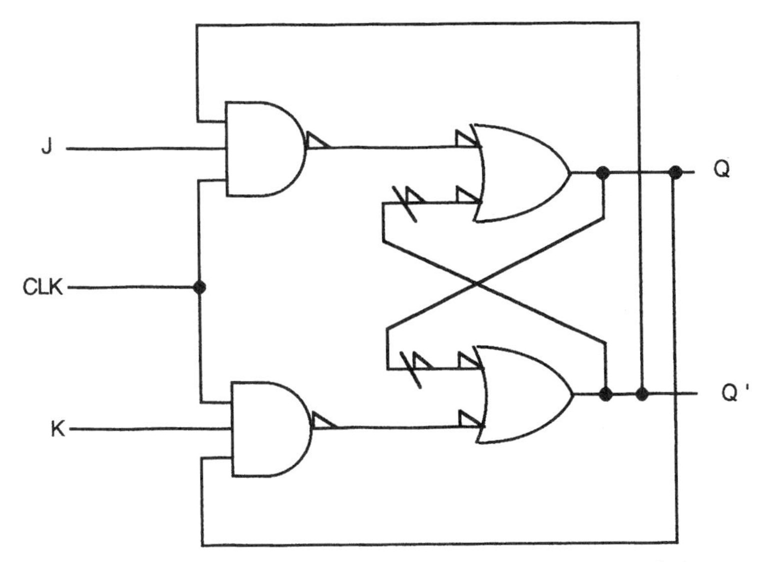

Fig. 7-8 A "flawed" gated latch design. Inputs are J and K.

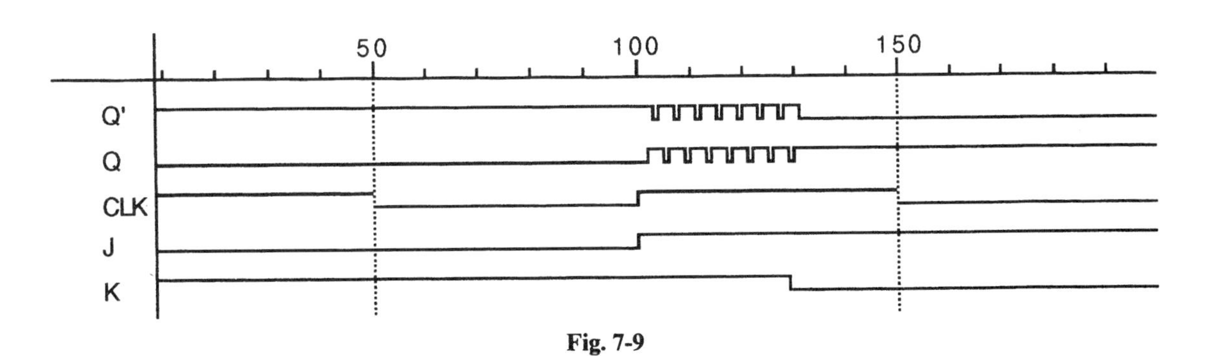

Fig. 7-9

To make the circuit operate properly, we needed either to provide a very narrow clock pulse or a means of delaying the feedback until the clock pulse is over. The second approach is easier to accomplish reliably, and its implementation is illustrated in the master-slave configuration shown in Fig. 7-10.

The master-slave flip-flop comprises two clocked RS flip-flops connected in series. Note that the clock pulse applied to the slave is inverted relative to the clock applied to the master, creating, in effect, a feedback delay equal to one clock pulse width.

Suppose that inputs J and K are TRUE and the applied clock input is also TRUE. The inverted clock CLK′ will be FALSE and will block any signal flow through gates G_5 and G_6 so that the Q_s and Q'_s values cannot change during this time. The CLK signal itself is steered through either G_1 or G_2

* As mentioned previously in a Chap. 6 footnote, simulation of circuits with feedback (such as the cross-coupled latch in Fig. 7-8) can sometimes experience problems. Refer to App. C.

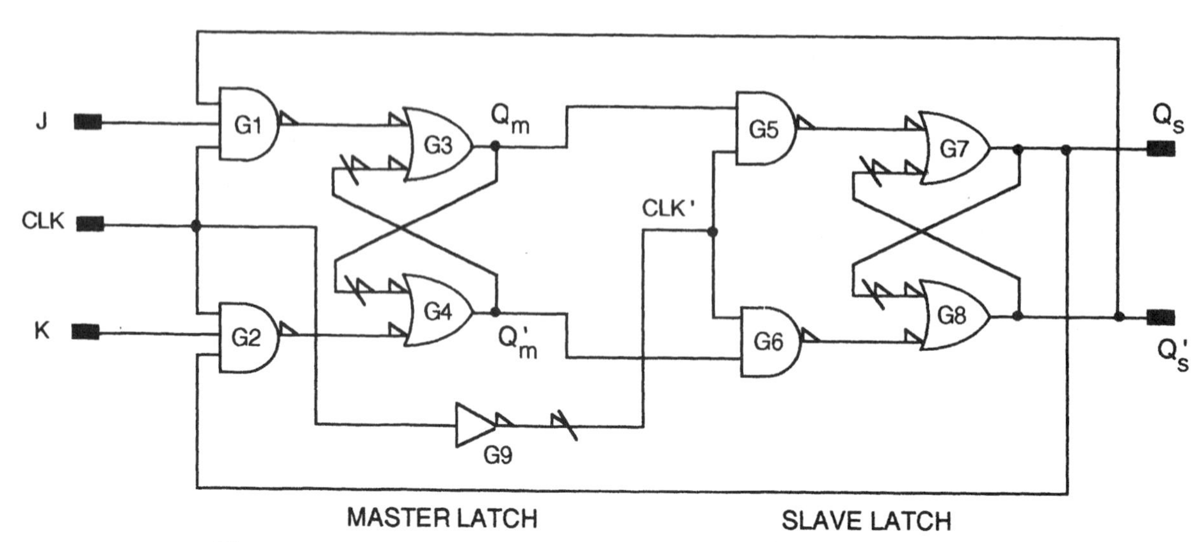

Fig. 7-10 Master-slave JK flip-flop. Black rectangles represent external pins.

(depending on the state of Q_s) and toggles the master. Since the passage of signal from master to slave is blocked, the slave outputs (and hence the feedback signals) do not change.

When the CLK returns to FALSE (and, accordingly, CLK′ changes to TRUE), the input gates G_1 and G_2 are blocked and will not respond to changes in Q_s. Gates G_5 and G_6 are now enabled, and the master outputs are transferred to the slave. Note that the slave latch outputs alone are connected to external pins on the IC chip so that the JK flip-flop of Fig. 7-10 will change output state *only* when a TRUE to FALSE *transition* of the CLK input occurs. This type of synchronous clocking is a form of *edge triggering* in which the circuit responds only to input data which is present *immediately prior* to the triggering transition of the clock waveform. Depending upon design, either increasing (positive) or decreasing (negative) clock transitions can be used to initiate triggering.

A summary of the clocked JK flip-flop's synchronous operation is given in Table 7.2. Here, Q_{n+1} represents the Q output state *following the triggering edge of an applied clock pulse.* We see from the table that Q mimics J when J and K are complementary, that Q remains unchanged when JK = 00, and that the flip-flop toggles when JK = 11.

Table 7.2 JK Synchronous (Clocked) Operation

J	K	Qn+1	Mode
0	0	Qn	Inhibit
0	1	0	Clear
1	0	1	Set
1	1	Q'n	Toggle

7.6 JK FLIP-FLOP WITH PRESET AND CLEAR

It is often desirable to add asynchronous (unclocked) features to a synchronous circuit so that, for example, it may be reset to a particular state at any time. Consider the modified JK flip-flop, shown in Fig. 7-11. This circuit is operated in one of two modes.

1. Asynchronous: The clock is disabled (CLK = 0). In this case, the preset (PR) and clear (CLR) inputs act as the set and reset inputs of an unclocked RS flip-flop.
2. Synchronous: PR and CLR are disabled (PR = CLR = 0), and the clock is activated.

The modes are not intended to be mixed, and the designer does so only at his or her own peril.

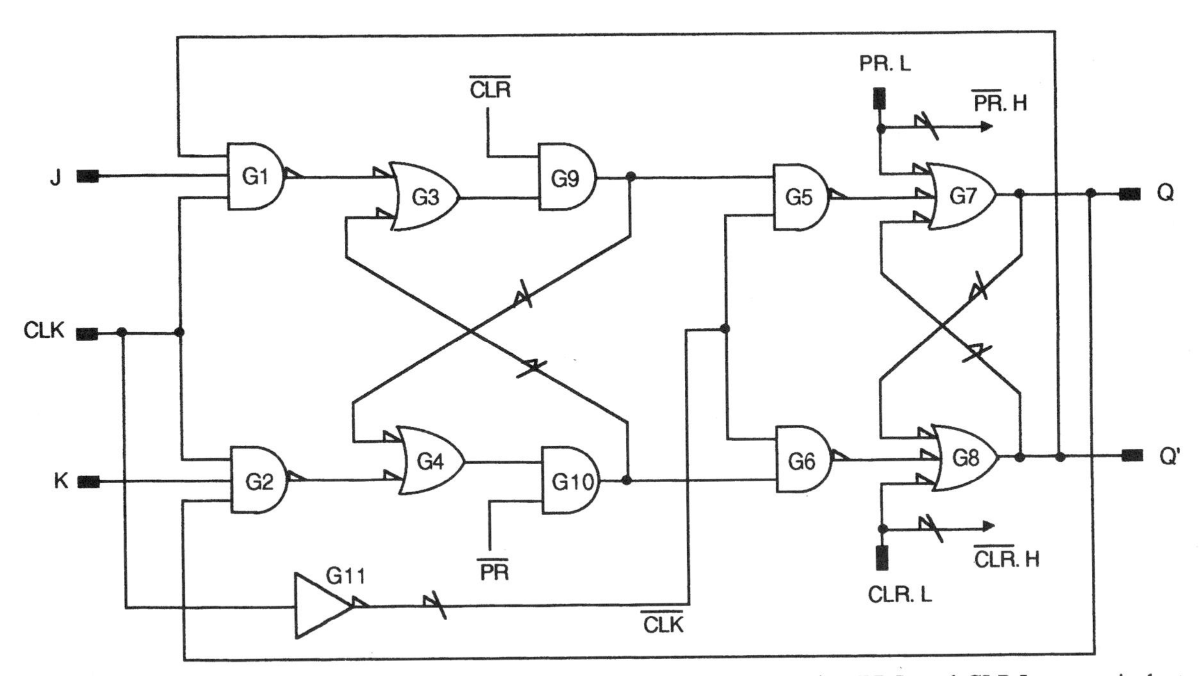

Fig. 7-11 JK flip-flop with preset and clear (mixed-logic description). Note that PR.L and CLR.L are equivalent to PR′.H and CLR′.H, respectively.

A state table for the master-slave JK flip-flop with preset and clear inputs is shown in Table 7.3. Manufacturers often provide such tables with H and L entries representing HIGH and LOW voltages, respectively. Their relationship to the logic values 0 and 1 depend upon whether the inputs and/or outputs of the hardware being specified are HT or LT. Note that when the flip-flop is operated asynchronously with preset and clear inputs, it is expected that the clock will remain at logic 0.

Table 7.3 JK Asynchronous Operation

PR	CLR	Q
0	0	Q
0	1	0
1	0	1
1	1	forbidden

7.7 SIGNAL PROPAGATION WITHIN THE FLIP-FLOP

Further insight into the functioning of the master-slave JK flip-flop may be gained by studying its microtimining diagram. Construction of such a diagram has been illustrated in Prob. 6.9, where the process was described in detail. It should be clear that as circuits grow in complexity, the use of simulation to study timing waveforms becomes increasingly advantageous.

EXAMPLE 7.2 Response of the JK flip-flop to a clock transition. Assume, in Fig. 7-11, that J = K = 1, PR = CLR = 0, and that Q_m (output of G_9) and Q are both 1. Initially, the clock input CLK is 0. At some arbitrary time, the clock changes to 1, remains there for 50 time-delay units, and then returns to 0. Assume also, that in the given hardware implementation, all gates have a propagation delay of 5 time units.

The initial logic values for all gates can now be determined. Consider G_1 before the clock transition occurs. Its inputs are J = 1, CLK = 0, and Q′ = 0; since the logic is AND, its output must be 0 (which, with this hardware, corresponds to a high voltage). At G_2, the inputs are K = 1, CLK = 0, and Q = 1, and its output is also 0. The analysis can be extended to each gate in turn until all internal logic levels are ascertained.

When CLK goes to 1, gates G_1, G_2, and G_{11} are affected. One or more of these gates will change state, as dictated by the logic, after one propagation-delay interval has elapsed. The process continues, as described in Prob. 6.9, until more gates respond (unless, of course, an instability exists). The complete LogicWorks™ simulation is shown in Fig. 7-12.

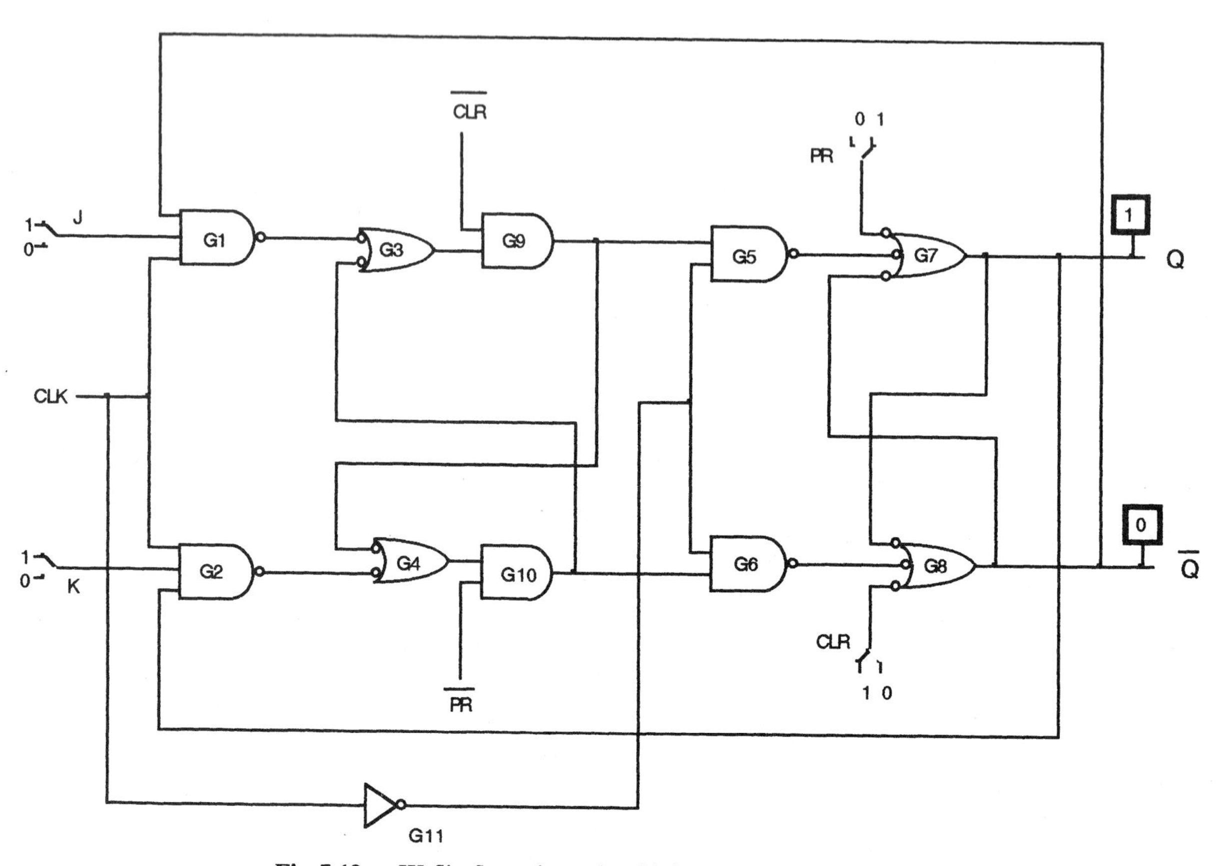

Fig. 7-12*a* JK flip-flop schematic with inputs set for toggle mode.

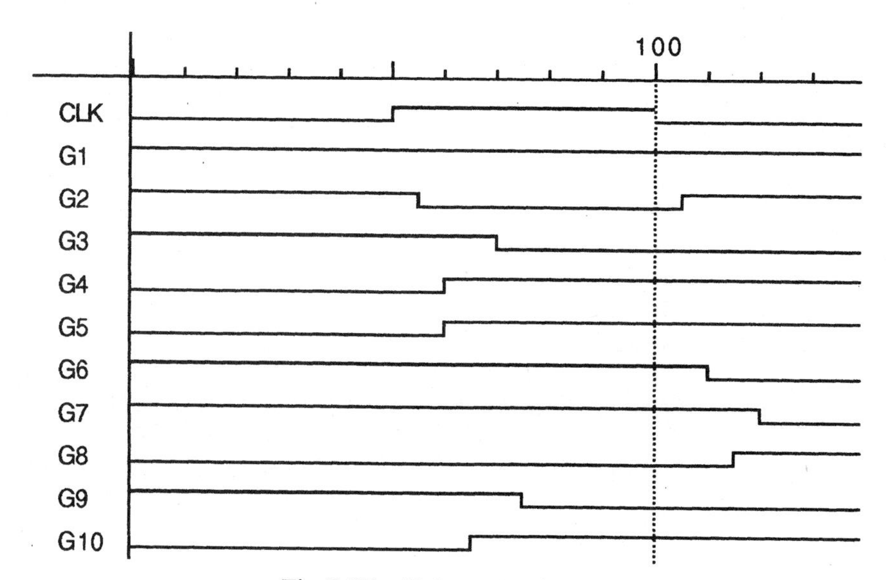

Fig. 7-12*b* Voltage waveforms.

Note that *the output changes only on the falling edge of the clock pulse* and that G_8 responds one time-delay interval ahead of G_7. It is also interesting to observe that the output of G_1 remains high (at logic 0) throughout the process since, at no time, are all three of its inputs TRUE.

7.8 OTHER FLIP-FLOP TYPES

The state of the master-slave JK flip-flop described in Sec. 7.6 is determined by the signals on its J and K inputs at the rising (0-to-1) transition of the clock pulse though the outputs (Q and Q′ of the slave unit) which do not change until the falling edge of the clock pulse. Flip-flops of this type are called *negative-edge-triggered. Positive-edge-triggered* devices are also available. Although the JK flip-flop is widely used in digital electronics, there are other types, the D and T being the most common.

The D Flip-Flop

This flip-flop (D for delay) is similar to the strobed data latch of Sec. 7.4 except that it is edge-triggered (nontransparent). It is defined as a device with a status (Q output) which, after being clocked, is identical to the signal on its D input prior to the clock pulse. Positive logic state tables for the device are shown in Table 7.4. Table 7.4*a* is for a synchronous (PR = CLR = 0) D flip-flop, while Table 7.4*b* is for an asynchronous (CK = 0) D flip-flop.

Table 7.4*a*

D	Qn+1
0	0
1	1

Table 7.4*b*

PR	CLR	Q
0	0	Q
0	1	0
1	0	1
1	1	forbidden

The D flip-flop can be modeled as a master-slave device having a logical inversion between its J and K inputs. Like the JK, it may be equipped with asynchronous inputs, PR and CLR (or S and R).

EXAMPLE 7.3 Consider the circuit composed of D flip-flops shown in Fig. 7-13. Assume that, initially, all Q outputs = 1 and that transitions occur on the *leading* edge of each clock pulse. Sketch the timing diagram for three clock pulses following the application of a brief CLR pulse.

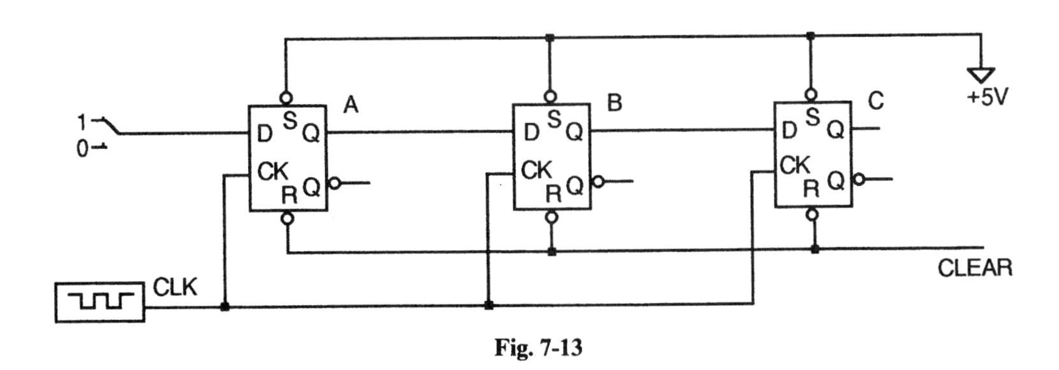

Fig. 7-13

The D flip-flops shown have LT preset (S) and clear (R) inputs so that the 5-V connection to the common set line is interpreted as a logic 0, or FALSE (0 V would be TRUE). Assuming that the clear line is initially at a high voltage also (logic 0), we see from Table 7.4 that if CLR goes to 0 V (logic 1), the outputs will go to 0 and remain there when CLR returns to logic 0 (until a 0-to-1 clock transition occurs). Note that since the outputs are HT, logic 0 is represented by 0 V.

The first rising CLK transition following the clear signal will cause flip-flop A to go to a 1 since its input D is 1. The other two flip-flops will not respond since their D inputs are 0. The second clock pulse will cause B to go to 1 because its D input, which is connected to A, is now 1. Similarly, output C will go to 1 on the third clock pulse and no further transitions will occur.

A LogicWorks™ simulation of the expected waveforms is shown in Fig. 7-14. Note that the effects of small propagation delays are included.

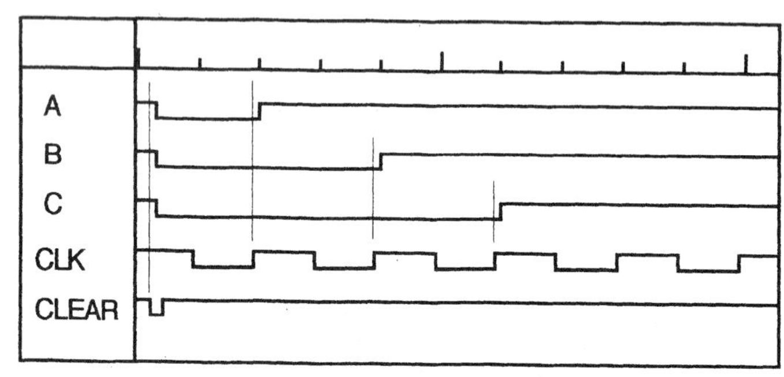

Fig. 7-14 Voltage waveforms.

The T Flip-Flop

This flip-flop (T for toggle) operates such that when its single input T is 1, its output Q will change with every clock pulse. When T is 0, clocking has no effect. This device can be modeled as a JK flip-flop with its J and K inputs both permanently tied to logic 1. As in other flip-flop types, the T may also contain asynchronous clear and preset inputs. The state table for the synchronous case is shown in Table 7.5.

Table 7.5

T	Qn+1
0	Qn
1	Qn'

EXAMPLE 7.4 Draw the macrotiming diagram for the group of T flip-flops shown in Fig. 7-15. Assume that all flip-flops have been initially cleared to 0 and that triggering occurs on the *trailing edge* of each clock pulse.

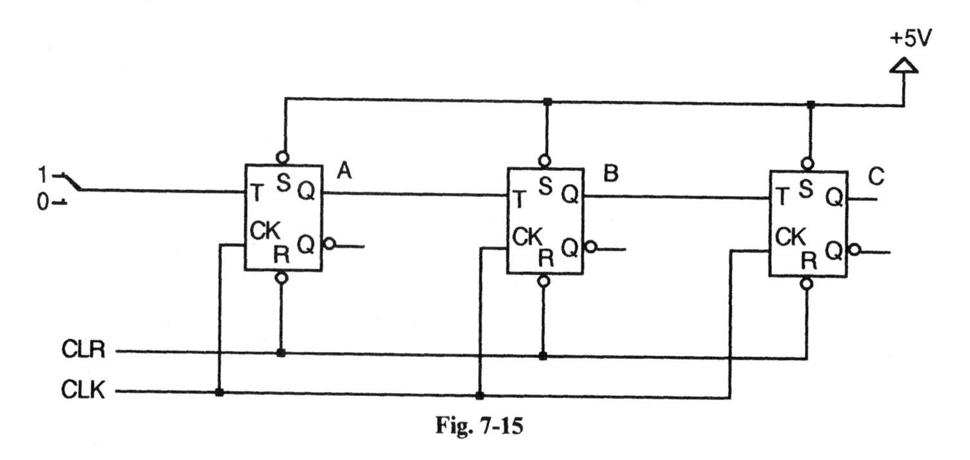

Fig. 7-15

Since flip-flop A has its input fixed at a logic 1, it will change states (toggle) in synchronization with every negative-going clock transition. Flip-flops B and C, on the other hand, will only toggle if their inputs are at logic 1 immediately prior to the negative-going clock pulse. Thus, we see in Fig. 7-16, that upon receipt of the second clock pulse, flip-flop B will toggle because its input from A is 1 just prior to the trigger, while flip-flop C will not change because its input from B is 0 at this time. The circuit generates a repetitive sequence ABC = 000, 100, 010, 111, 000, etc.

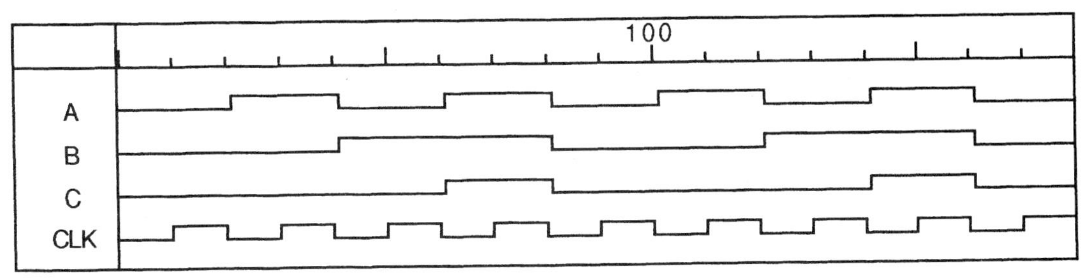

Fig. 7-16

7.9 FLIP-FLOP TRIGGERING AND TIMING

Regardless of their triggering modes, there are two particular times of interest for all flip-flop types. One, the *setup time* (t_{su}), is the minimum time that the inputs (J, K, etc.) must be stable prior to the arrival of the clock pulse edge which initiates a transition. The other is the *hold time* (t_h), which is the time that the inputs must remain stable after the initiating edge of the clock pulse arrives. The setup and hold times together establish an interval, relative to the clock pulse, wherein the inputs must remain invariant. If the inputs change during this time window (shown shaded in Fig. 7-17), the flip-flop manufacturer will not guarantee the final state of the device after clocking.

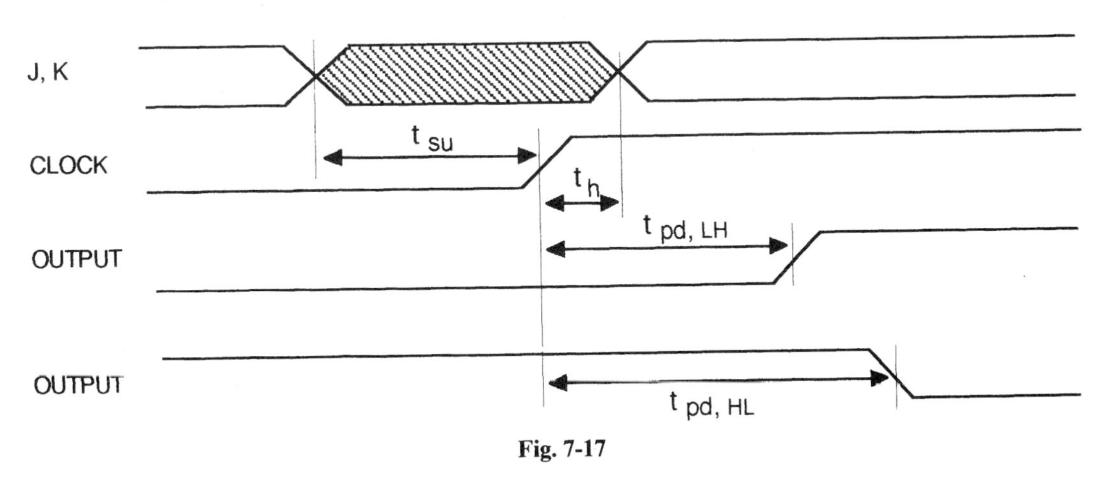

Fig. 7-17

In addition to the setup and hold times, two propagation-delay times are also specified: $t_{pd,\,HL}$ (the delay, relative to the clock pulse, required for an output HIGH-to-LOW transition) and $t_{pd,\,LH}$ (the delay, relative to the clock pulse, for an output LOW-to-HIGH transition). These delays establish the minimum time which must be allowed after clocking before the output can be assumed stable. Figure 7-17 shows the relationship between the various flip-flop timing parameters.

In currently available logic devices, the various delay times discussed above range from nanoseconds to tens of nanoseconds (1 ns = 10^{-9} s). Note that in real circuits, transitions do not occur as steps; they have finite rise and fall times as shown. For simplicity, these are usually depicted as being linear, and the various timing specifications are measured midway between the minimum and maximum levels.

EXAMPLE 7.5 Often, in synchronous logic circuits, the outputs of an array of flip-flops (called a register) are fed into a network of AND-OR combinational logic that is used to produce the signals which determine what the register's new state will be following the next clock pulse. Figure 7-18 shows such a system. Determine the maximum clock rate for the given worst-case timing.

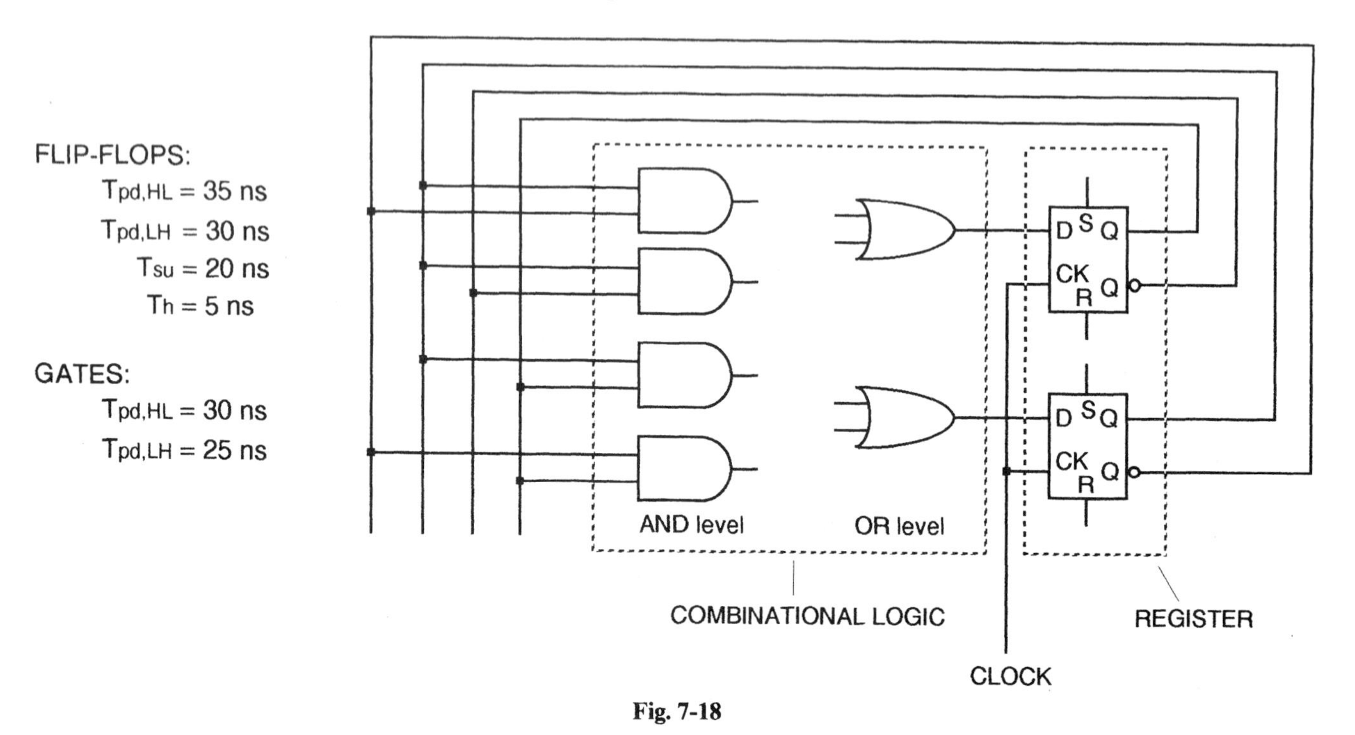

Fig. 7-18

Consider timing as starting with the arrival of a clock pulse:

1. Since it is not known which type of transition will take place at the flip-flop outputs, choose the worst case (that is, $t_{pd} = t_{pd,\,HL} = 35$ ns). After this time has elapsed, the flip-flop outputs are guaranteed stable.
2. Similarly, after a subsequent 30 ns, we are assured that the flip-flop outputs will have propagated through the AND level.
3. Another 30 ns takes us through the OR level.
4. The signal must remain stable at the D inputs for an additional 20 ns before another clock pulse can be applied.
5. Since a signal cannot propagate through the network in less than 5 ns, the hold time requirement is fulfilled.

The various delays are shown schematically in Fig. 7-19. The minimum clock period is the sum of the individual delays or 115 ns. The maximum clock rate is $1/115 \times 10^{-9} = 8.7$ MHz.

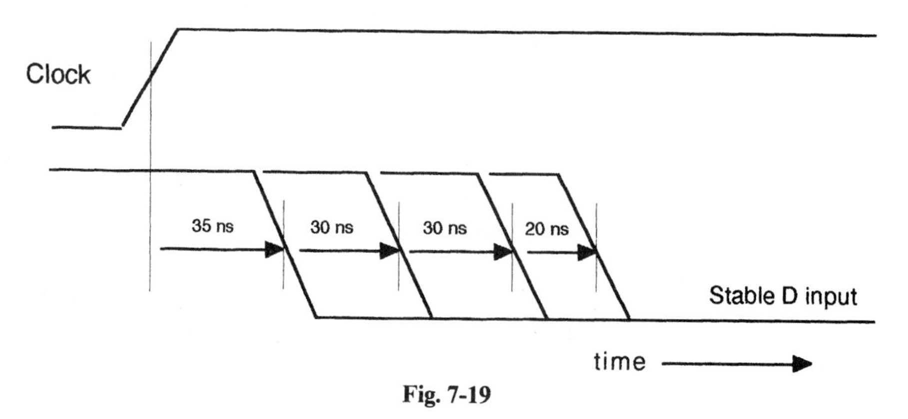

Fig. 7-19

7.10 METASTABILITY

If a flip-flop is subjected to changing input signals during its setup or hold times, a condition known as metastability can develop where the output is neither HIGH nor LOW but somewhere in between. The probability that a metastable condition will persist for a given time duration decreases with the time duration, meaning that of all possible metastable durations, the shorter ones are most likely to occur. A probability density function expressing this relationship is observed to be approximately exponential as shown in Fig. 7-20.

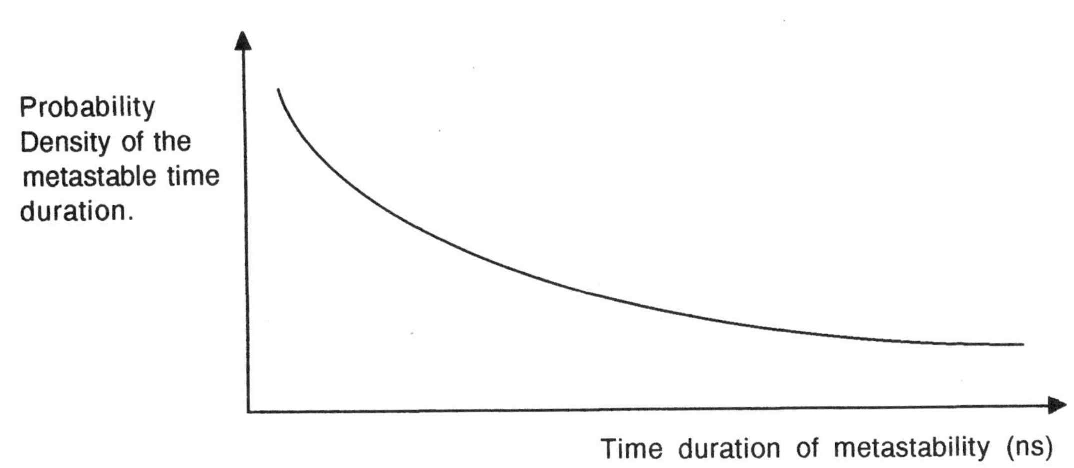

Fig. 7-20

Errors due to metastable behavior can occur in high-speed circuits where one flip-flop's output is expected to trigger another soon after being triggered itself or where data applied to a flip-flop is asynchronous so that there is a finite probability that an input will change too close to a clock pulse. A failure occurs if a flip-flop is in a metastable state when its outputs are required to be stable.

Metastable behavior is often specified in terms of mean time between failure (MTBF):

$$\mathrm{MTBF} = \frac{1}{F_{clk} \times F_{data} \times T_W \times \exp(-T_R \Delta t)} \tag{7.1}$$

F_{clk} is the flip-flop clock frequency, F_{data} is the clocking frequency of the data input signal (D, J, K, ...), T_W and T_R are constants which depend upon the technology used in constructing the flip-flop, and Δt is the time delay between the triggering edge of the clock pulse and the strobe which enables (or clocks) a succeeding device making use of the flip-flop's output. In most cases, flip-flop triggering and data strobing are provided by successive pulses of a single clock source so that Δt is equal to the period $1/F_{clk}$.

EXAMPLE 7.6 A D flip-flop is clocked at 10 MHz and has unsynchronized input data that changes at a rate of approximately 10 MHz. Determine MTBF when, for the technology used, $T_W = 0.5$ s and $T_R = 0.75 \times 10^9\ \mathrm{s}^{-1}$. Repeat the calculation for a 20-MHz clock rate.

The assumption here is that a clock edge triggers the flip-flop, which is expected to have a stable output by the time the next triggering clock edge arrives. Thus, $\Delta t = (F_{clk})^{-1}$.

$$\begin{aligned} F_{clk} &= 10^7\ \mathrm{s}^{-1} \\ F_{data} &= 10^6\ \mathrm{s}^{-1} \\ T_W &= 0.5\ \mathrm{s} \\ T_R &= 0.75 \times 10^9\ \mathrm{s}^{-1} \\ \Delta t &= 10^{-7}\ \mathrm{s} \end{aligned}$$

From Eq. (7.1), compute MTBF $= 7.47 \times 10^{19}$ s. Since this is so large, we are comforted by the fact that we will surely not live long enough to observe a failure.

Suppose the clock frequency is now doubled to $2 \times 10^7\ s^{-1}$. This affects Δt which has an exponential, and thus, a significant effect. Recalculation yields MTBF = 1932 (32 min) s, which is significant indeed! Clock pulses are too close to allow proper operation, and the circuit is essentially useless.

The MTBF equation implies that if the chip manufacturing process can be altered to double the value of constant T_R, then the deleterious effect of the increased clock frequency can be almost entirely overcome. With $T_R = 1.5 \times 10^9$ and $F_{clk} = 2 \times 10^7\ s^{-1}$, MTBF $= 3.73 \times 10^{19}$ s.

Because T_W and the input data rate do not appear in an exponent, we see that they are of lesser importance relative to T_R and the clock frequency as far as metastable errors are concerned.

Solved Problems

7.1 A manufacturer has created the NAND hardware circuit shown in Fig. 7-21 to implement a clocked RS flip-flop. It is specified that the propagation delay for a HIGH-to-LOW transition ($t_{pd,\,HL}$) at the outputs Q and Q′ is longer than that for a LOW-to-HIGH transition ($t_{pd,\,LH}$). Construct a microtiming diagram and verify that the specification is correct.

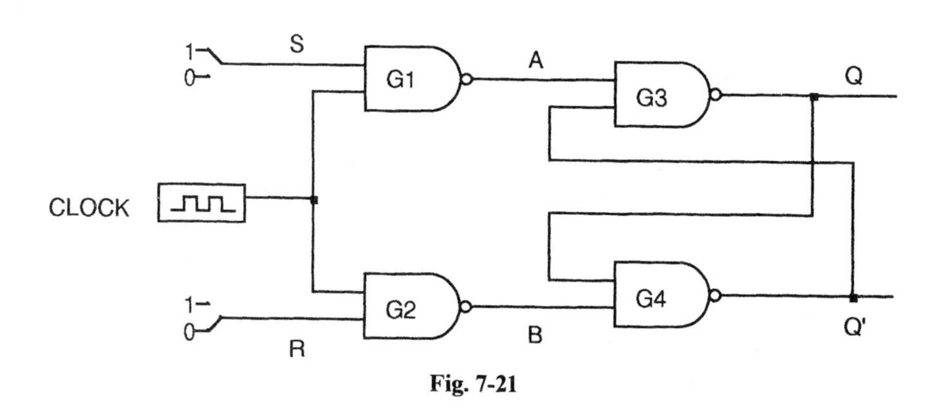

Fig. 7-21

Figure 7-21 is a positive logic *hardware* diagram which may be conveniently analyzed by converting it to a pure logic form. In circuits comprising interconnected NAND gates, each NAND symbol may be regarded as an AND gate followed by a slash (logical inversion) as demonstrated in the lower branch of Fig. 4-43 (part 2 of Prob. 4.10).

Consider the set action of this circuit (switches positioned as shown). Initially, the flip-flop is assumed to be reset; that is, Q = 0 (LOW) and Q′ = 1 (HIGH). Upon the appearance of a positive clock pulse, gate G_1 has 1s at both inputs. One propagation-delay interval later, its output goes to logic 0 (LOW). This changes the inputs to gate G_3 from 11 to 01, so after an additional propagation delay, Q changes from LOW to HIGH. This transition causes the inputs to G_4 to go from 01 to 11 which, after yet another propagation delay, produces a HIGH-to-LOW change in Q′.

A similar analysis of the reset action shows that, once again, an output LOW-to-HIGH transition initiates the complementary HIGH-to-LOW transition. Microtiming waveshapes are shown in Fig. 7-22 where, it is clear that for both set and reset, $t_{pd,\,LH}$ is shorter than $t_{pd,\,HL}$ and the manufacturer is correct.

7.2 An early attempt to design the circuit of Prob. 7.1 is shown in Fig. 7-23 in which gates G_1 and G_2 are AND rather than NAND. The circuit does not work. Determine what is wrong and draw the microtiming diagram.

In this case, whenever the clock is LOW, the inputs to the Q and Q′ gates both include a 0 because the outputs of gates G_1 and G_2 are both LOW, and, consequently, Q and Q′ will both be HIGH. This is not an

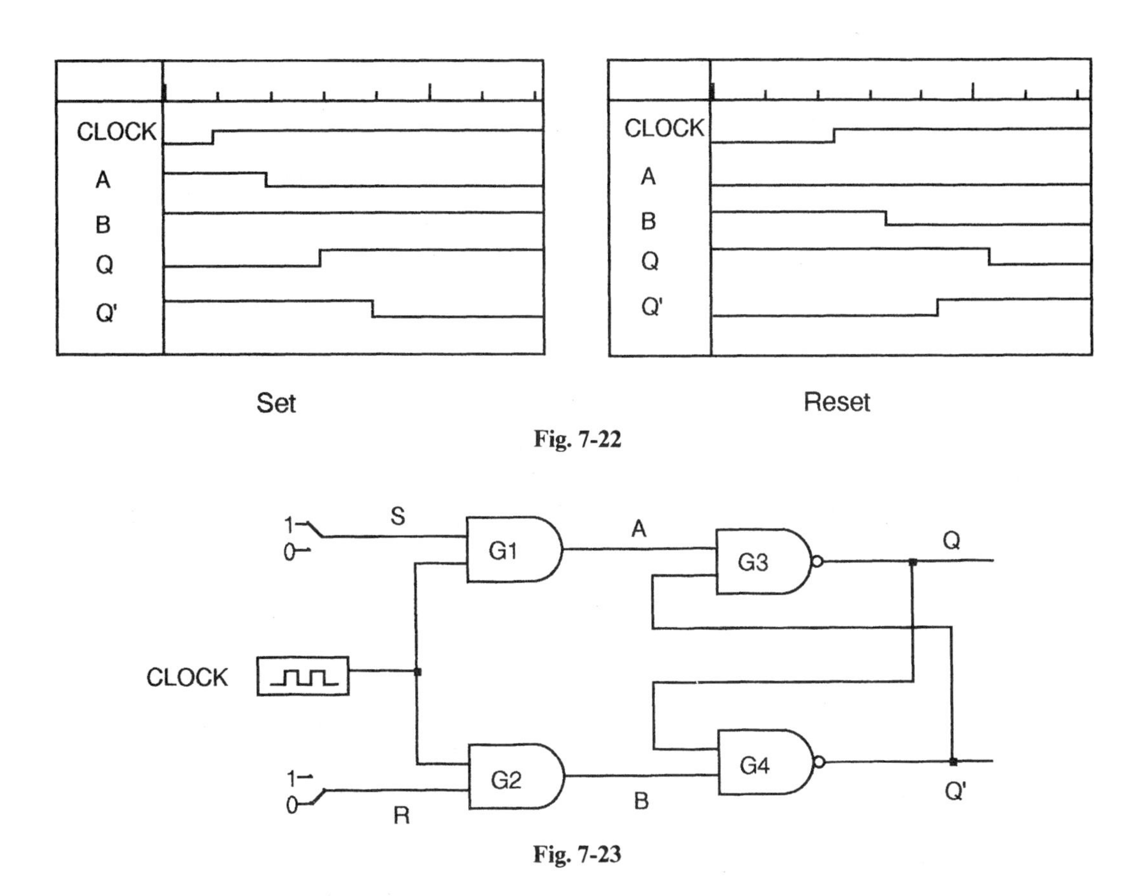

Fig. 7-22

Fig. 7-23

allowed state for an RS flip-flop. Furthermore, when the clock goes HIGH, the circuit does not operate properly. The Q output will go LOW after two time-delay intervals, and Q′ will remain unchanged because it is held by the LOW output of gate G_2. Thus, during a positive clock pulse, the output state will be just the opposite of that desired and it will not latch. The microtiming diagram is shown in Fig. 7-24.

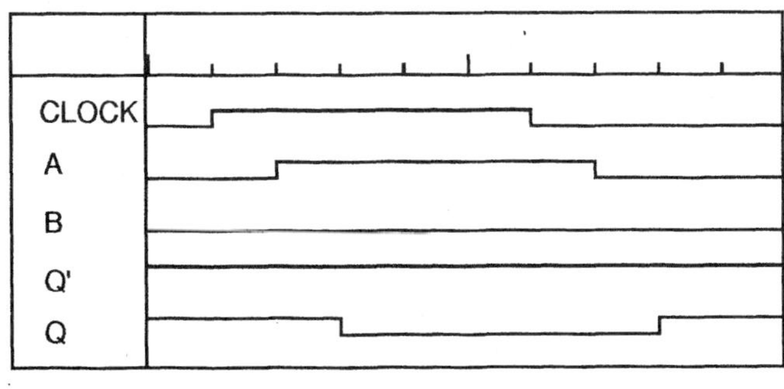

Fig. 7-24

7.3 A student has designed the circuit shown in Fig. 7-25 to implement a clocked RS flip-flop with hopes that it will function as indicated in Table 7.1. Determine whether the circuit will perform properly and if the relation between propagation delays stated in Prob. 7.1 are applicable.

With the circuit arrangement shown, a positive-going clock pulse will pass through the gate G_1, producing a logic 1 at the upper input to G_3. This change will only have an effect if both inputs to gate G_3 were previously LOW, implying that Q′ must be LOW prior to the clock pulse in order for Q to change. Thus, a clock pulse will *reset* the flip-flop (cause Q to go to 0) when S = 1 and R = 0. If the switch settings are reversed, we see that clocking will *set* the flip-flop (cause Q to go to 1).

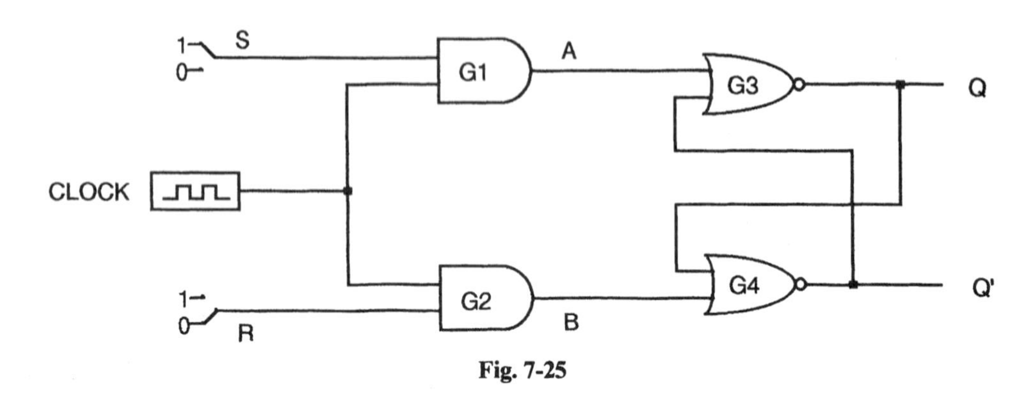

Fig. 7-25

Note that in both cases, a 00 input is compatible with either a 01 or 10 output condition, meaning that the circuit can hold a set or reset state; i.e., it can latch. There is only one small problem: in order to be classified as an RS flip-flop, the Q output must follow S rather than be its complement as it is in the given circuit. This situation is easily corrected if the S and R or Q and Q′ terminal designations are interchanged.

Microtiming diagrams for both set and reset input conditions are shown in Fig. 7-26. We see that output HIGH-to-LOW transitions precede LOW-to-HIGH transitions, so *relative propagation-delay times are contrary to those stated in Prob. 7.1.*

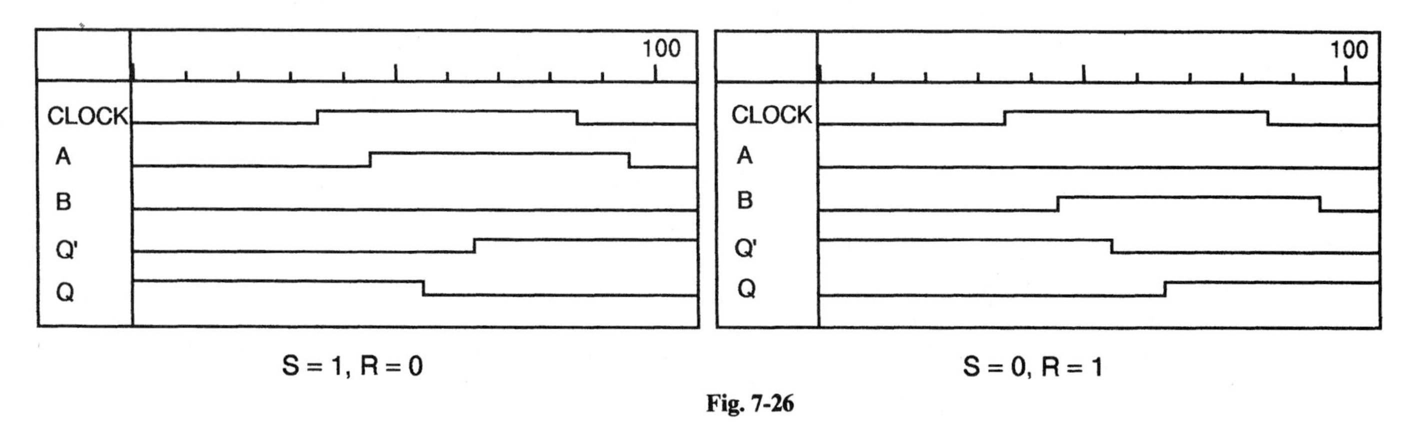

Fig. 7-26

7.4 Draw a NOR hardware implementation of the RS flip-flop of Prob. 7.1.

The logic diagram is shown in Fig. 7-27 from which a corresponding NOR implementation may be easily derived (Fig. 7-28). Note that both the inputs and outputs must be LT if additional hardware is to be avoided.

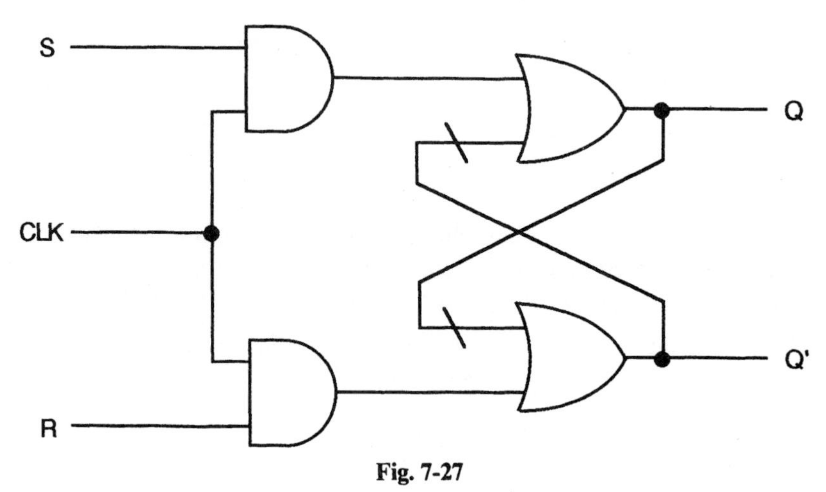

Fig. 7-27

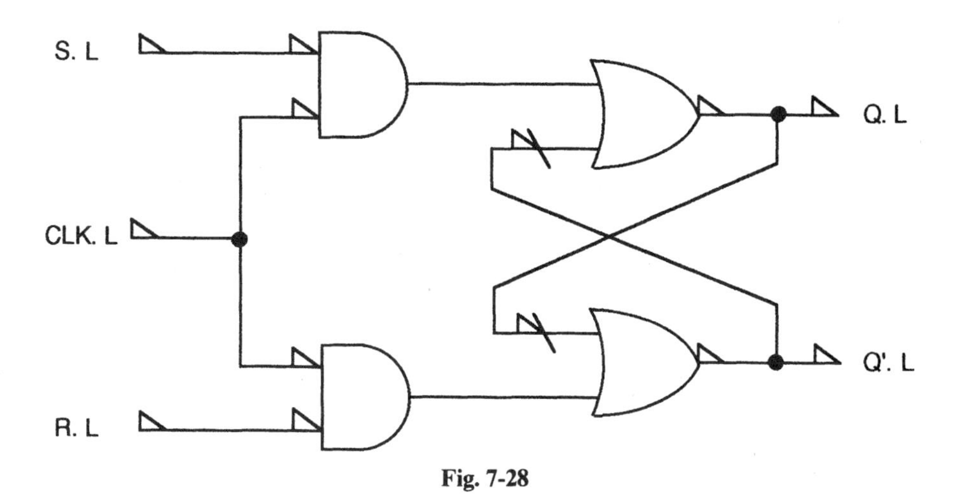

Fig. 7-28

7.5 The circuit of Fig. 7-29 is the NAND hardware implementation of logic for a set dominant latch. It functions such that when R and S are simultaneously TRUE, output Q will be TRUE. Demonstrate that this is the case.

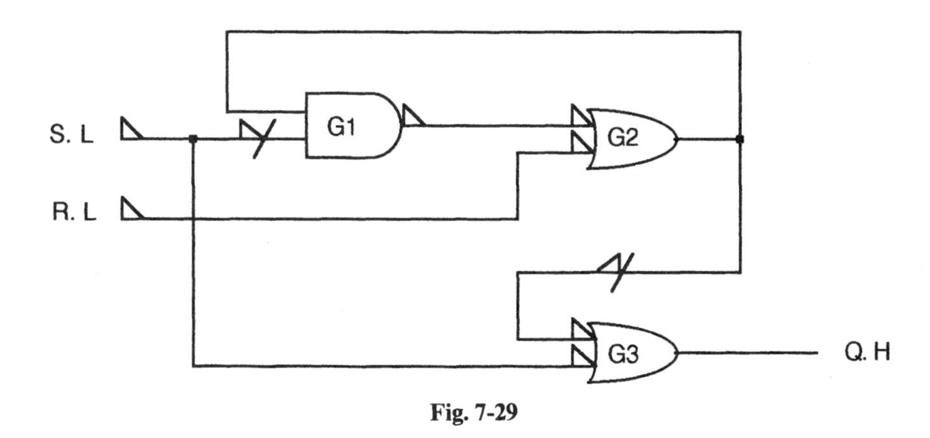

Fig. 7-29

The circuit is most conveniently analyzed by removing all half arrows and viewing it as a logic diagram.

Case 1. S = 1, R = 0.

Q = 1 because one of G_3's inputs is connected to S, which is 1. R will have no effect on Q, but it causes G_2 to be 0 since the output of G_1 is 0.

Case 2. R remains at 0; S goes to 0.

The output of G_2 remains at 0 because it is latched via G_1. Thus, when S becomes 0, output Q remains at logic 1.

Case 3. R = 1, S = 0.

G_2 = 1, which becomes latched via G_1 so long as S = 0. The logical inversion following G_2 and the direct connection to S cause both inputs to G_3 to be 0s; consequently, Q = 0.

Case 4. S remains at 0; R goes to 0.

The output of G_2 remains at 1 and R cannot affect G_3. Thus, Q remains at logic 1.

Case 5. R = 1, S = 1.

Q = 1 because of the direct connection from S to G_3. This time, G_2 is held at logic 1 by R so that if S returns to 0, Q will follow it. Note, however, that if R returns to 0 ahead of S, G_1 will stay at 0, G_2 will go to 0 and latch there, and, because of the logical inversion, Q will remain at 1.

A simulation output, shown in Fig. 7-30, illustrates typical behavior.

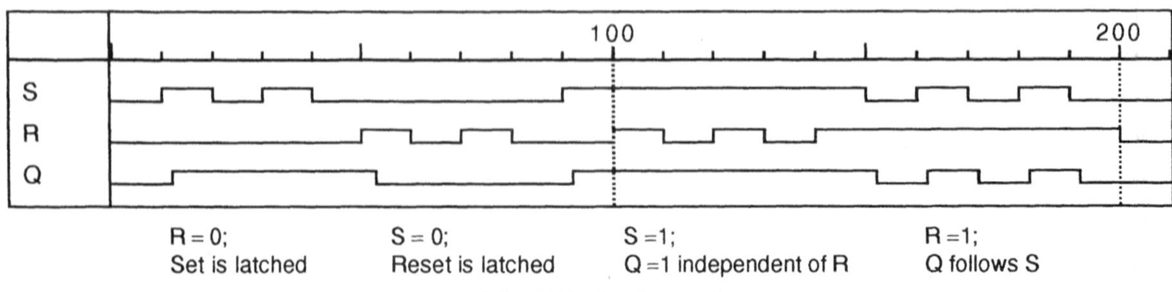

Fig. 7-30 Logic waveforms.

If a Q′ output is desired, it can be obtained by adding an inverter. Though the resulting RS flip-flop is more complicated than the latch in Fig. 7-2, it does not exhibit the simpler circuit's ambiguous behavior where it is possible for the Q and Q′ outputs to be simultaneously equal to logic 1. As is often the case, improved performance is achieved at the cost of simplicity.

7.6 Figure 7-31 shows the NAND logic implementation of an edge-triggered JK flip-flop whose proper function depends upon the difference in propagation delays between two signal paths rather than upon the master-slave action described in Sec. 7.5. Assume that Q = 0, J = 1, and K = 0 and draw a logic microtiming diagram to illustrate the sequence of events which occur when a positive-going clock pulse is applied.

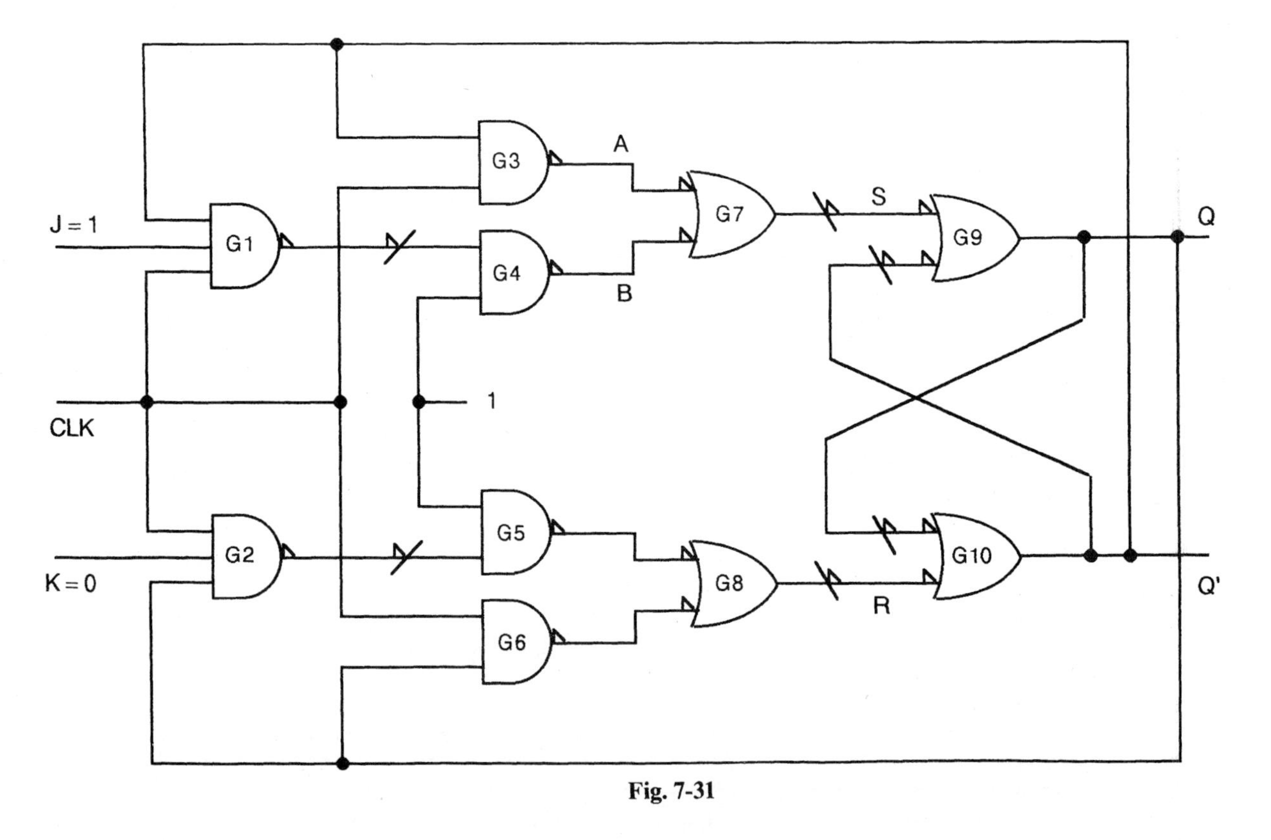

Fig. 7-31

We are primarily concerned with the upper half of the circuit since gate G_2, having a fixed 0 input from K, will not change.

Determine the relevant initial states as follows:

- $G_1 = 0$ since CLK = 0.

- A = 0 since CLK = 0.
- B = 1 since the output of G_1 is inverted.
- $G_7 = 1$ and, because of the inversion, the set input S to latch G_9–G_{10} is 0.
- Similar analysis shows that the reset input R to latch G_9–G_{10} is also 0.

Assume that the clock input rises (CLK → 1). Then A → 1 and B → 0. Note that A changes one time-delay interval ahead of B because the CLK signal passes through only one gate instead of two.

The sequence at the input to G_7 is AB = 01, 11, 10. Since one of its inputs is always TRUE in this sequence, G_7 remains at logic 1, S remains at 0, and Q does not change.

Assume that the clock input falls (CLK → 0). Again, A changes before B and the sequence is AB = 10, 00, 01. In this case, the input to G_7 will be at 00 for one propagation-delay interval, momentarily causing G_7 to go to 0 and S to 1. Output Q will go to 1 and latch there via G_{10}.

Note that in order for latching to occur, the pulse signal S must last at least as long as the combined propagation delays of gates G_9 and G_{10}. In practical circuits, this can be achieved by alterations in appropriate gate structures.

The microtiming diagram is shown in Fig. 7-32.

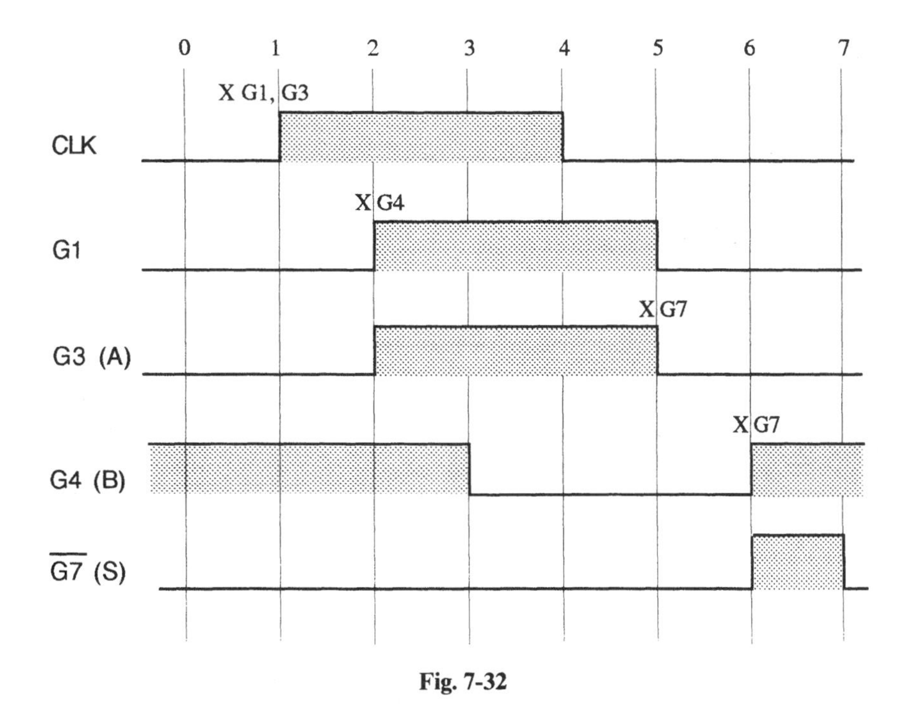

Fig. 7-32

7.7 Consider the edge-triggered flip-flop of the preceding problem. If an input changes (J in this case), even as precipitously close to the last possible moment as a single propagation-delay interval before the trailing edge of the clock pulse, no output response will occur. Demonstrate that this is the case with reference to the microtiming diagram of Fig. 7-32.

Refer to Figs. 7-31 and 7-32 and assume that J goes to 0 in an interval preceding the falling edge of the clock pulse. G_1 will follow in the next time interval since all its inputs must be at logic 1 for it to remain TRUE. Gate G_3 is unaffected by J, but G_4 will respond to the change in G_1, causing signal B to return to logic 1 earlier. It can be seen that if this happens, the 00 input to G_7 will not occur and, consequently, S and Q will remain unchanged.

An exactly similar analysis can be applied to the K input. Thus, it is clear that an instruction to set or reset the flip-flop will only be executed if it remains in place at least until the clock pulse falls. Note, however, that in the real world, where there are finite rise and fall times, a metastable condition is possible if an input changes too close to the active edge of a clock pulse (refer to Sec. 7.10).

7.8 The circuit in Fig. 7-33 has two T flip-flops which trigger on the trailing edge of the clock pulse and a D flip-flop which triggers on the rising edge. The input, initially at logic 1, switches its values on the trailing edge of every fourth clock pulse. Assuming that all flip-flops are cleared at t = 0, draw a timing diagram showing the waveforms at outputs A, B, and C.

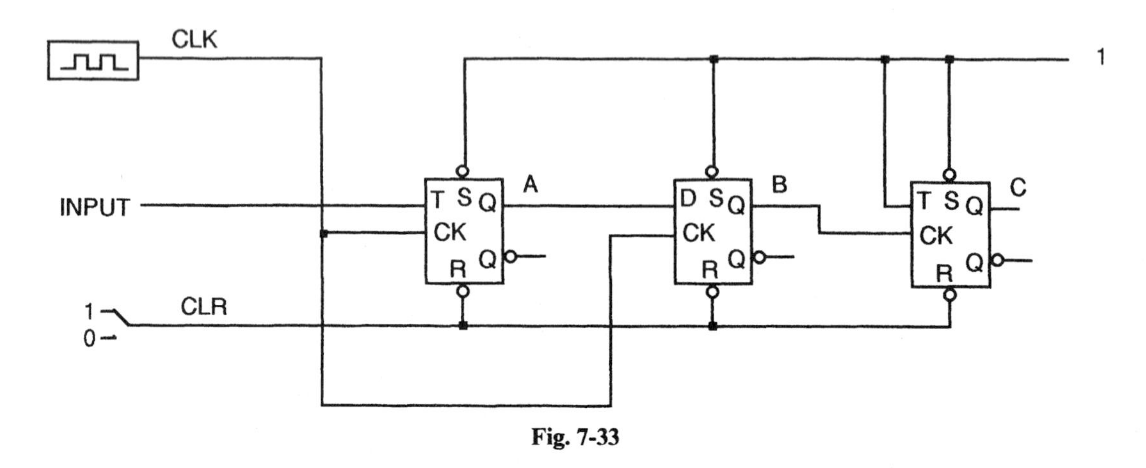

Fig. 7-33

Flip-flop A toggles when input = 1; otherwise it doesn't change. Flip-flop B assumes the value of A when the clock rises. In effect, the D flip-flop delays A by an interval equal to the spacing between clock pulses. Flip-flop C toggles whenever B changes from 1 to 0. Waveforms are shown in Fig. 7-34, which includes a small propagation delay of 1 time unit.

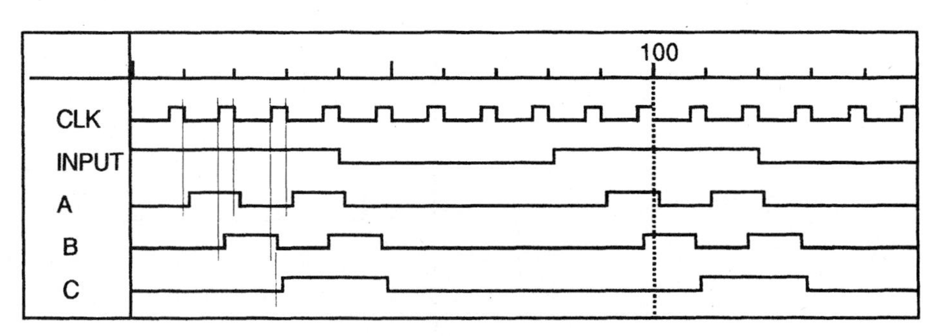

Fig. 7-34

Note that input changes coincident with a falling clock pulse edge do not affect flip-flop A until the succeeding clock pulse. The simulation, quite properly, accounts for the fact that a flip-flop will not respond when input signal setup and hold times are zero. Comments on the simulation of this problem may be found in App. C.

7.9 Two negative-going edge-triggered JK flip-flops are interconnected as shown in Fig. 7-35. Assume that both have been preset to 1, following which, a clock signal is applied. Neglecting propagation delays, draw the waveshapes at A′ and B for five clock pulses.

A sequential problem of this type may be approached in an organized fashion by creating a listing of the states following each clock pulse (see Table 7.6).

The JK inputs to each flip-flop act as instructions that tell the flip-flop which state to assume when the next clock pulse arrives (refer to Table 7.2). Since $J_A = B'$ and $K_A = 1$, the initial instructions for flip-flop A are $J_A K_A = 01$, which indicates that the flip-flop will be cleared (A = 0) at the falling edge of the next clock pulse. Similarly, the JK inputs to flip-flop B are 11 which instruct it to toggle (change state). These instructions are circled in the first row of Table 7.6 and their results are indicated by pointers to entries in the next row which corresponds to the situation immediately following the trailing edge of the first clock pulse.

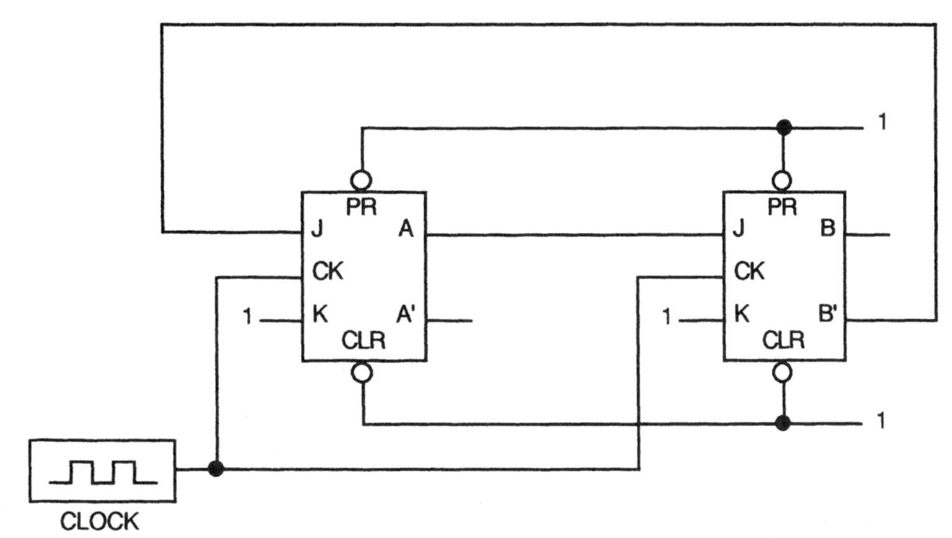

Fig. 7-35

Table 7.6

PULSE	A	B	$J_A = B'$	K_A	$J_B = A$	K_B
0	1	1	0	1	1	1
1	0	0	1	1	0	1
2	1	0	1	1	1	1
3	0	1	0	1	0	1
4	0	0	1	1	0	1
5	1	0	1	1	1	1

When the second clock pulse arrives, we should expect flip-flop A to toggle ($J_A = K_A = 1$) and flip-flop B to remain unchanged since the input $J_B K_B = 01$ calls for B to go to 0, a state that it is already in.

The table is filled out one row at a time, the entries in columns A and B (corresponding to flip-flop outputs) being determined by appropriate JK states in the preceding row. When the process is completed, waveforms may be inferred by simply reading down the columns which correspond to variables of interest. The result is shown in Fig. 7-36.

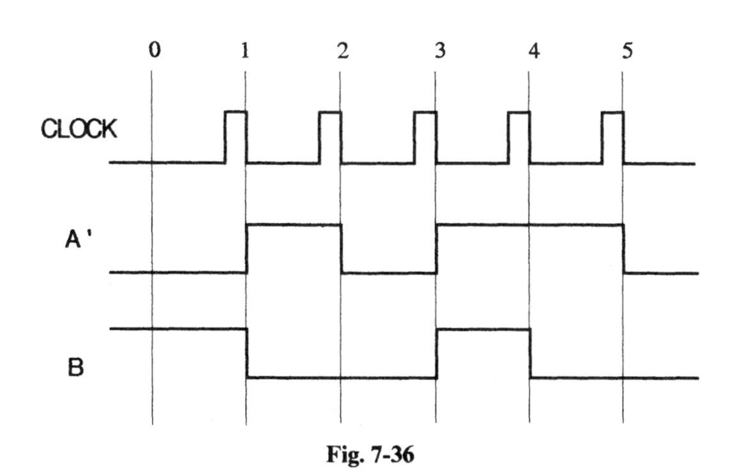

Fig. 7-36

7.10 Figure 7-37 indicates the time variation of inputs to a JK flip-flop. Assuming that the device is initially cleared, sketch output Q in relation to the clock pulses.

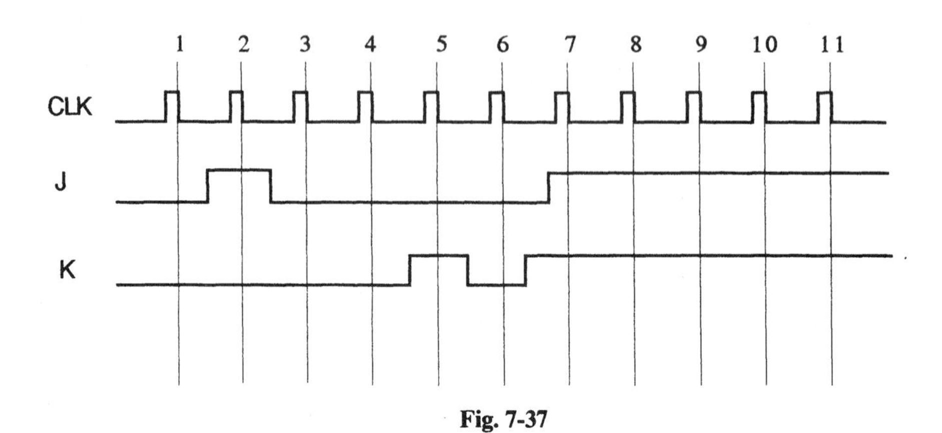

Fig. 7-37

As in the preceding problem, an ordered listing of successive states is helpful. In this case, because the J and K inputs are independent of output Q, the form of our tabulation will be essentially the same as that given in Table 7.2. Table 7.7 should be self-explanatory.

Table 7.7

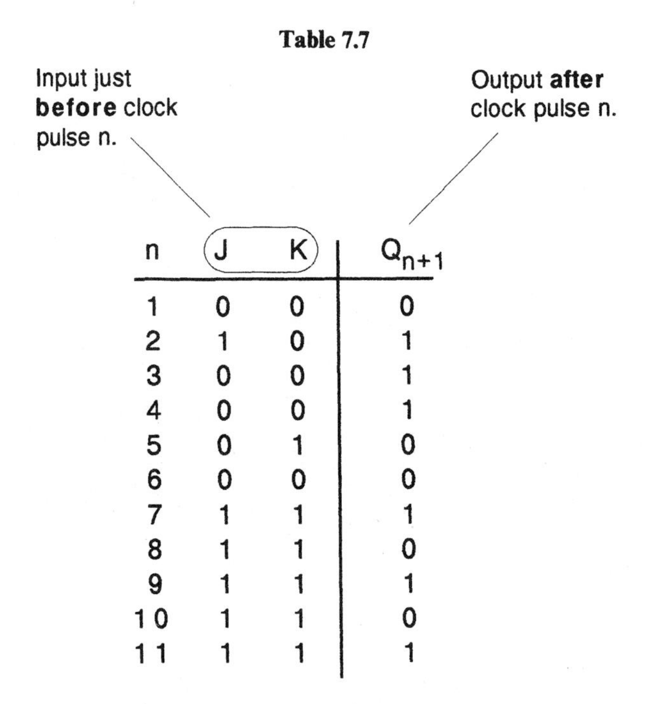

n	J	K	Q_{n+1}
1	0	0	0
2	1	0	1
3	0	0	1
4	0	0	1
5	0	1	0
6	0	0	0
7	1	1	1
8	1	1	0
9	1	1	1
10	1	1	0
11	1	1	1

The output, as a function of time, shown in Fig. 7-38, is obtained by reading down the right-hand column of the state table. Note that the exact time between clock pulses at which the inputs change has no effect on the resulting output. This is an important characteristic of edge triggering.

7.11 Inputs D, S_o, X, and S_1 are given with the circuit of Fig. 7-39. Assuming that the flip-flop responds to negative-edge triggering and is initially preset (Q = 1), sketch the output Q, as a function of time, for eight clock pulses.

The inversion between the J and K inputs constrains the circuit to function as a D flip-flop, transferring input data to the output after each clock pulse.

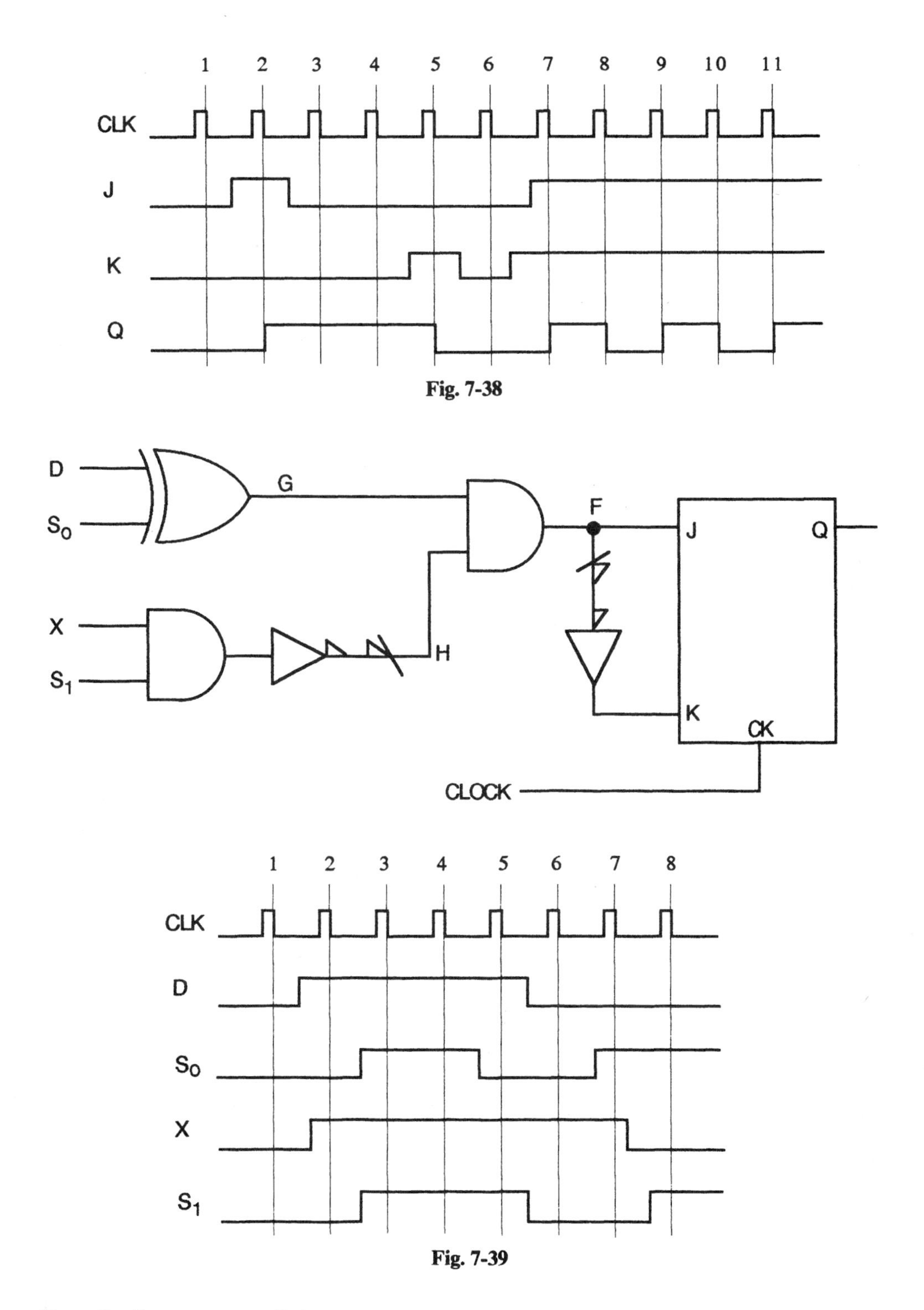

Fig. 7-38

Fig. 7-39

From the diagram, we see that

$$G = D \oplus S_0$$
$$H = \overline{XS_1}$$
$$F = GH$$

The values of variables G, H, and F may be determined for any instant of time by simply drawing a vertical line at any point in the timing diagram of Fig. 7-39, noting the values of the inputs, and applying the above logic equations. Note that for the combinational logic portion of the circuit, the variables are not affected by or related to clock pulses. Output Q, on the other hand, will be equal to the value of F immediately prior to each negative-going clock pulse edge and will remain constant between clock pulses.

For this problem, we will draw intermediate waveforms instead of tabulating the logic states (see Fig. 7-40).

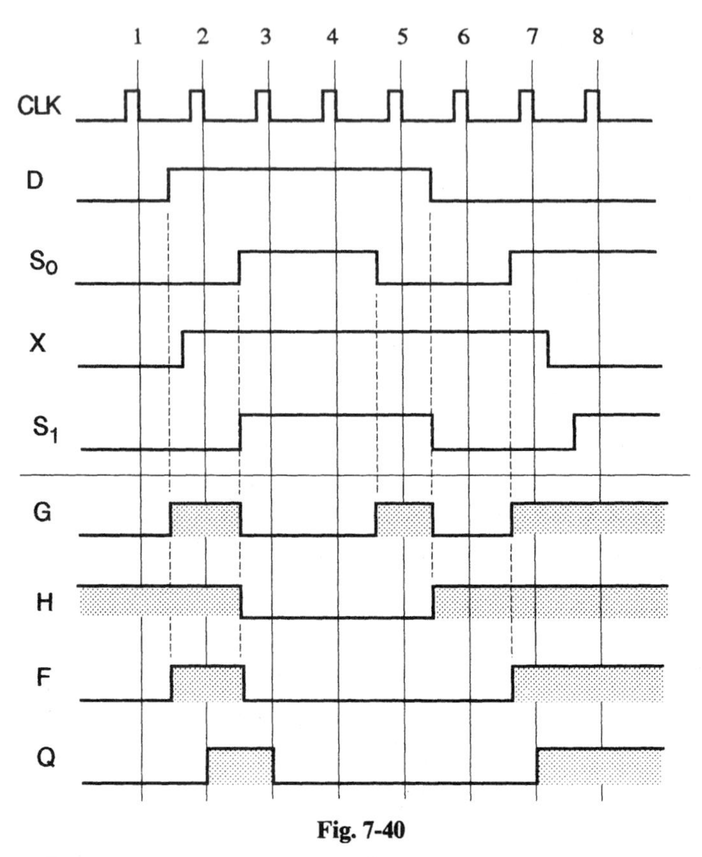

Fig. 7-40

7.12 Determine waveshapes at the flip-flop outputs for the circuit shown in Fig. 7-41 assuming that the flip-flops are initially cleared prior to the application of the first clock pulse. The given flip-flops are triggered by leading (LOW to HIGH) edges of the clock pulses and propagation delays are negligibly small.

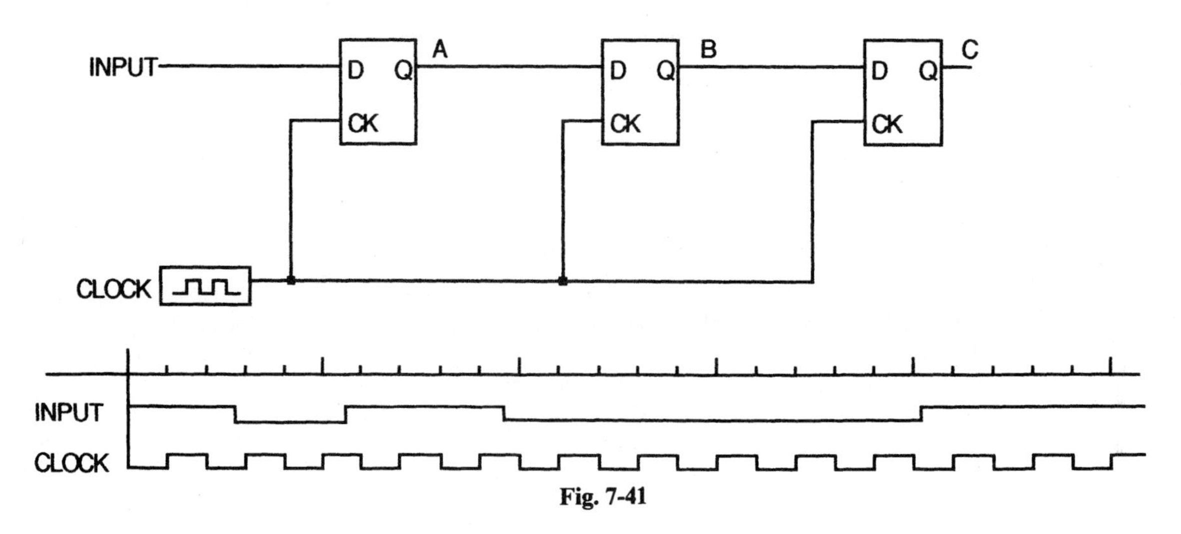

Fig. 7-41

The flip-flops are D (delay) types which are constructed so that whatever is on a D input prior to the triggering edge of a clock pulse will be latched at the Q output immediately following triggering. This behavior should be obvious in the LogicWorks™ simulation shown in Fig. 7-42 where an observation of the relative waveform positions clearly indicates the delay feature of the circuit.

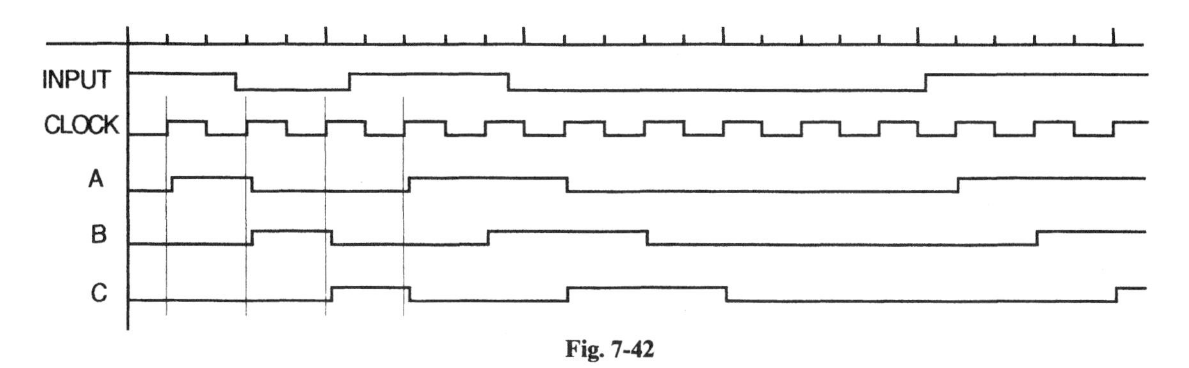

Fig. 7-42

7.13 Substitute T flip-flops for the D flip-flops in Prob. 7.12, and sketch the waveshapes at A, B, and C. Again, assume that the flip-flops are initially cleared prior to the application of the first clock pulse and that they are triggered by the leading (LOW to HIGH) edge of the clock pulse. Use the same input and clock waveshapes shown in Fig. 7-41.

The T flip-flop toggles (changes state) when its input T is 1 and remains unchanged when T is 0. Consider the response to the first clock pulse in the waveforms of Fig. 7-43. Flip-flop A toggles because its input is 1 prior to the leading edge of the clock pulse. Flip-flop B does not change because its input A, being slightly delayed relative to the clock, is LOW at the moment the clock goes HIGH. Flip-flop C remains unchanged because B is LOW when the clock pulse occurs. Note that transitions in waveforms A, B, and C are aligned relative to each other since all are toggled by the same clock signal.

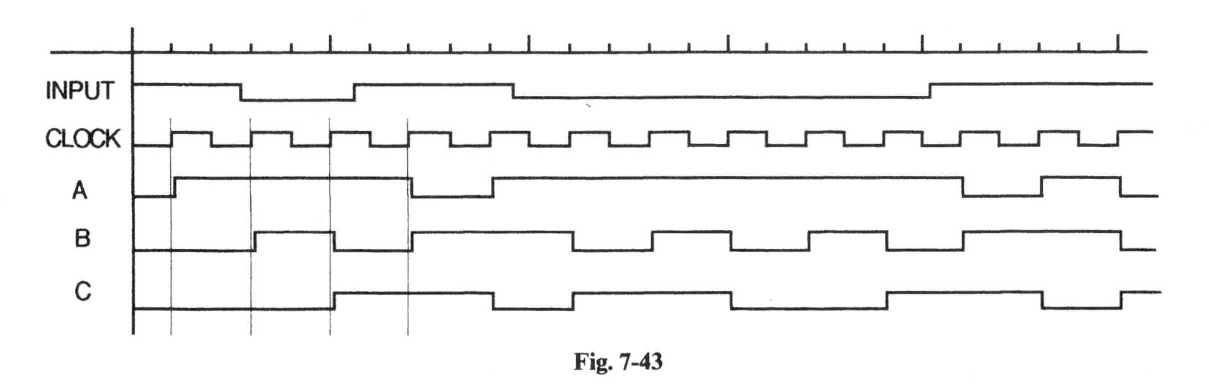

Fig. 7-43

7.14 In Fig. 7-44, the flip-flops are initially cleared ($A = B = C = 0$). Assuming that A is the most significant multiplexer select bit and that the flip-flops are negative-edge-triggered, sketch the logic waveform at F for seven clock pulses, ignoring propagation delays.

Note that flip-flop C is triggered from B and, with $J = K = 1$, all flip-flops will toggle when clocked. Prior to the first clock pulse, $ABC = 000$ which selects line 0. Thus, initially, $F = 1$. The first clock pulse causes flip-flops A and B to toggle; C will not respond to a positive transition so that $ABC = 110$ which selects line 6, connecting F to $D = 1$. The second clock pulse toggles A and B, the latter causing C to toggle. At this point, $ABC = 001$ selecting line 1 and $F = 0$. The third clock pulse toggles A and B; C remains unchanged and $ABC = 111$ (line 7). This connects F to D so that, in the interval between the third and fourth pulses, output F will follow D (which is seen to change). Continuing the analysis yields Table 7.8 and the waveforms of Fig. 7-45.

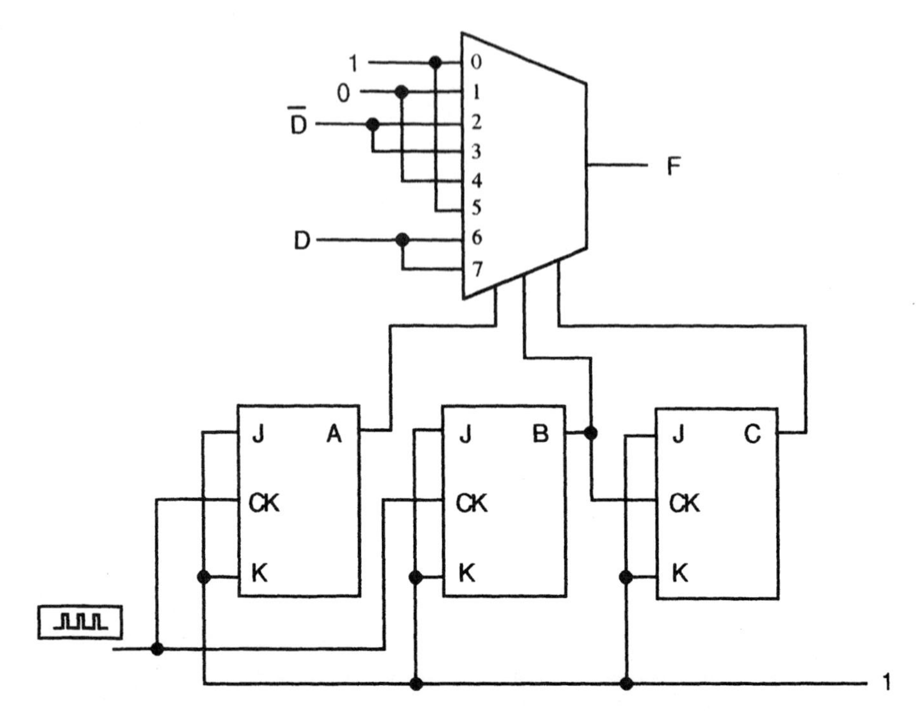

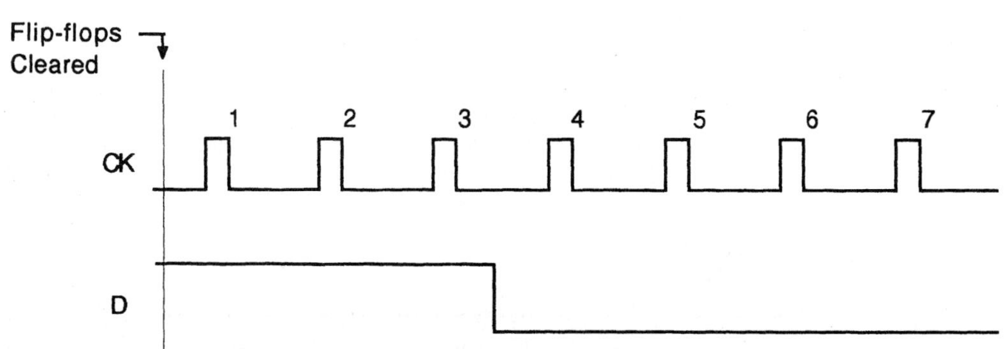

Fig. 7-44

Table 7.8

PULSE	A	B	C	Line	F
	0	0	0	0	1
1	1	1	0	6	D
2	0	0	1	1	0
3	1	1	1	7	D
4	0	0	0	0	1
5	1	1	0	6	D
6	0	0	1	1	0
7	1	1	1	7	D

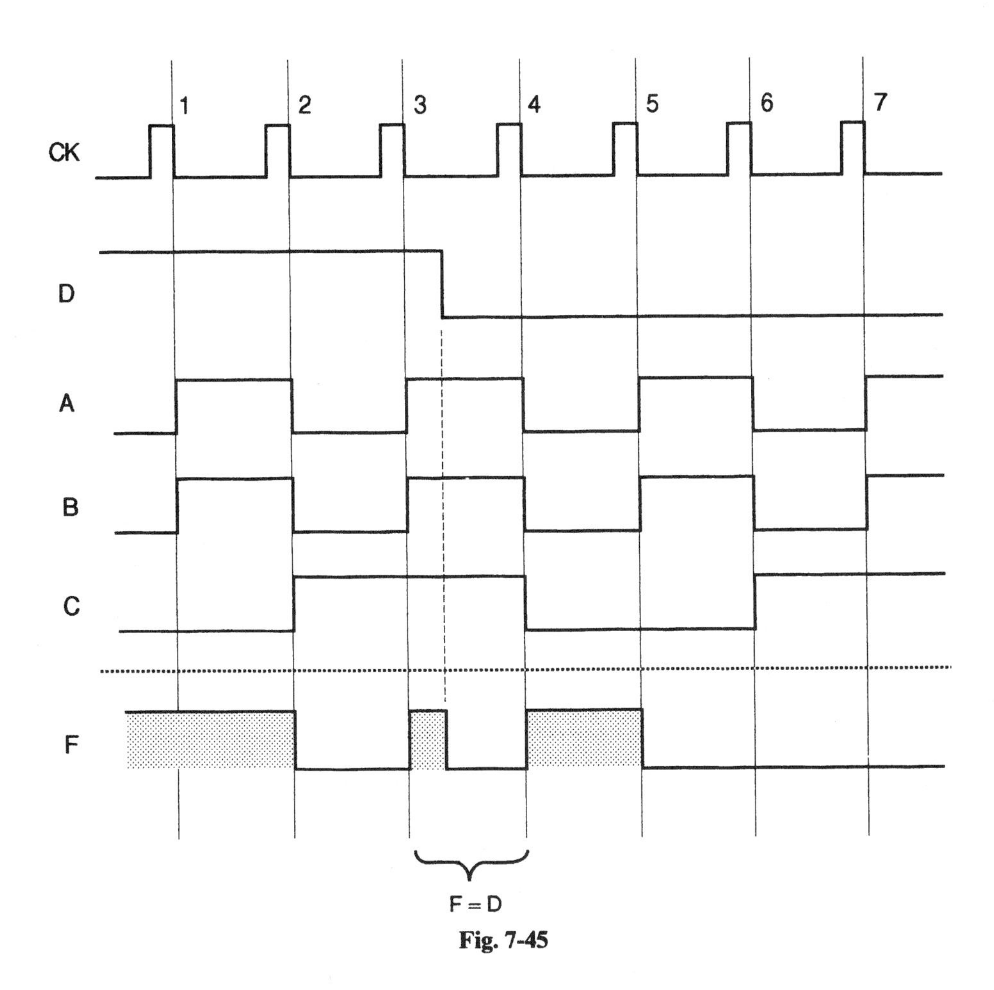

Fig. 7-45

7.15 Given the JK flip-flop circuit and input waveforms shown in Fig. 7-46, sketch the output waveshape Q. Assume that the flip-flop has been preset (Q = 1) before clock pulses are applied and that it changes state on the 1-to-0 transition of a clock pulse. Ignore propagation delays.

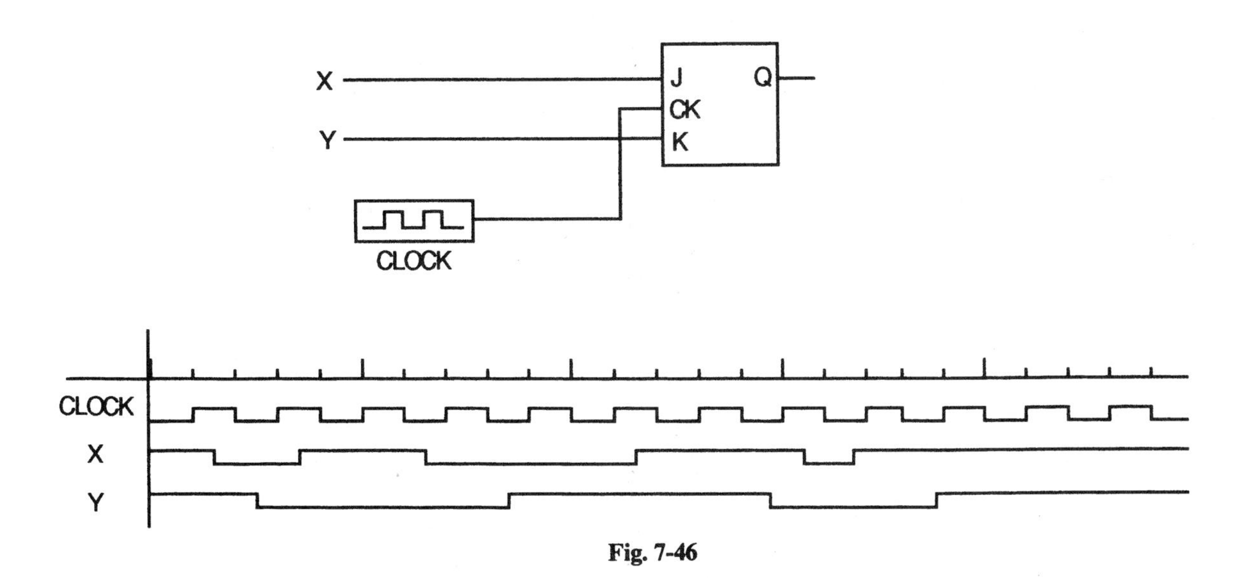

Fig. 7-46

Recall the truth table of the JK flip-flop in synchronous operation (Table 7.2).

J	K	Q_{n+1}
0	0	Q_n
0	1	0
1	0	1
1	1	Q'_n

Apply these rules at every appropriate clock pulse edge to yield the result shown in Fig. 7-47.

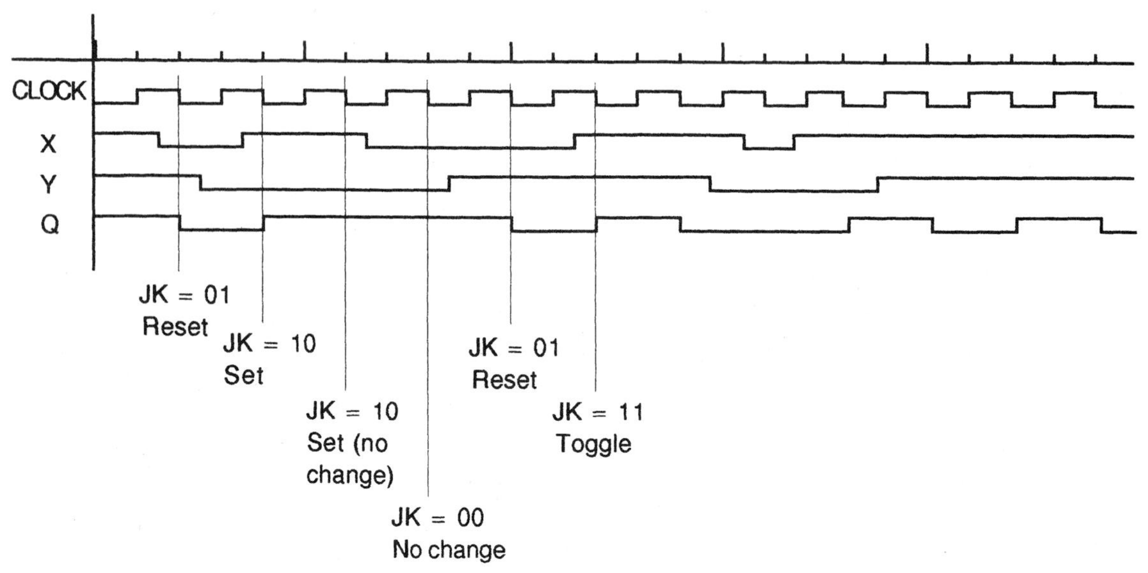

Fig. 7-47

Supplementary Problems

7.16 It is desired to make the clocked latch of Fig. 7-4 produce LT outputs Q and Q′ and respond to LT signals on R, S, and CLK. Determine a hardware configuration which uses a minimal number of gates.

7.17 A positive logic NAND implementation of the RS latch is shown in Fig. 7-48. Draw the corresponding pure logic diagram and state table.

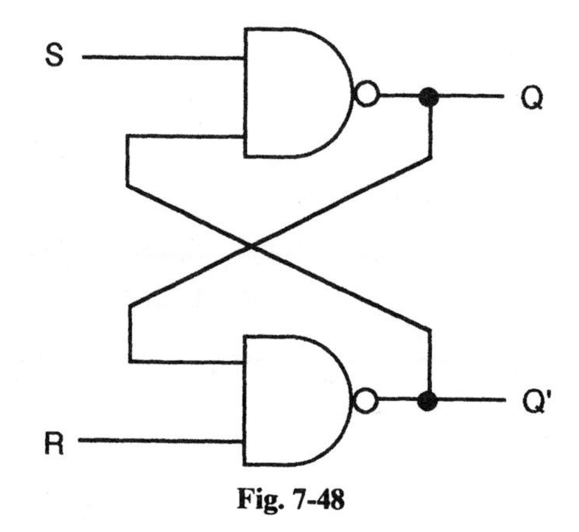

Fig. 7-48

7.18 A circuit is to be designed which combines the characteristics of a D and T flip-flop in a single device. The specifications are as follows: When the D and T inputs are simultaneously TRUE, the device will not change state; for all other input combinations, the device is to behave as a normal D or T flip-flop. Create the applicable synchronous state table, using Table 7.2 as a guide.

7.19 Consider the circuit shown in Fig. 7-49. Initially, the master-slave flip-flops are both cleared. For the input X waveform shown, sketch waveforms for variables A and Z, assuming that triggering occurs when the clock pulse falls.

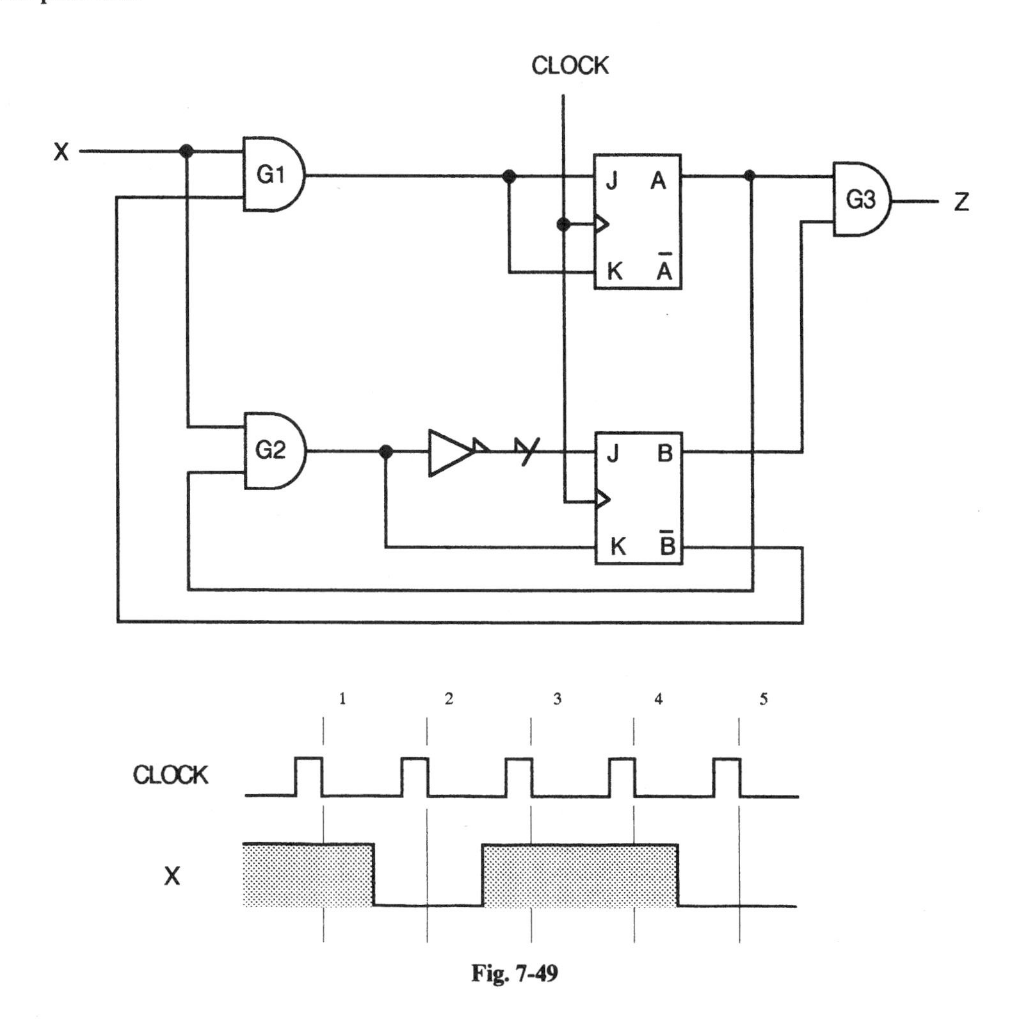

Fig. 7-49

7.20 An engineer, in designing an edge-triggered JK flip-flop of the type shown in Fig. 7-31 (Prob. 7.6), attempts to reduce gate count by omitting the latch G_9–G_{10}, leading to the circuit shown in Fig. 7-50.

The designer's reasoning is as follows:

Initial conditions: Q = 0, J = 1, K = 0.
Clock rises.
Point A → 1, B → 0.
The change at A occurs before the change at B.
The sequence at the input of G_7 is AB = 01, 11, 10 and Q remains unchanged.
Clock falls.
Again, A changes before B, the sequence being AB = 10, 00, 01.
During the brief period when AB = 00, Q → 1 which is latched via G_5, G_8, and G_4.

The above analysis is correct except for one little incorrect assumption! Develop a microtiming diagram and find the flaw.

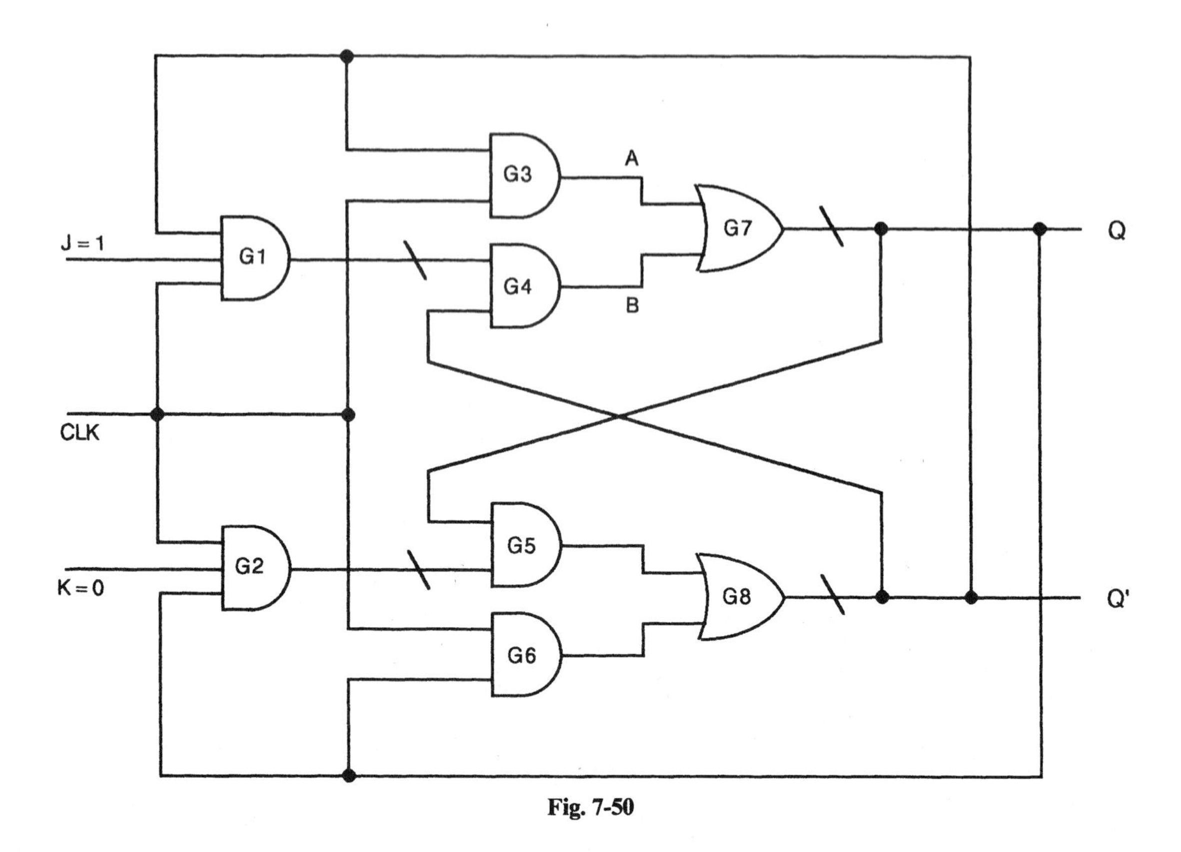

Fig. 7-50

7.21 Consider the circuit of Fig. 7-51. With reference to Prob. 7.17, determine outputs Q and Q′ when the clock is LOW.

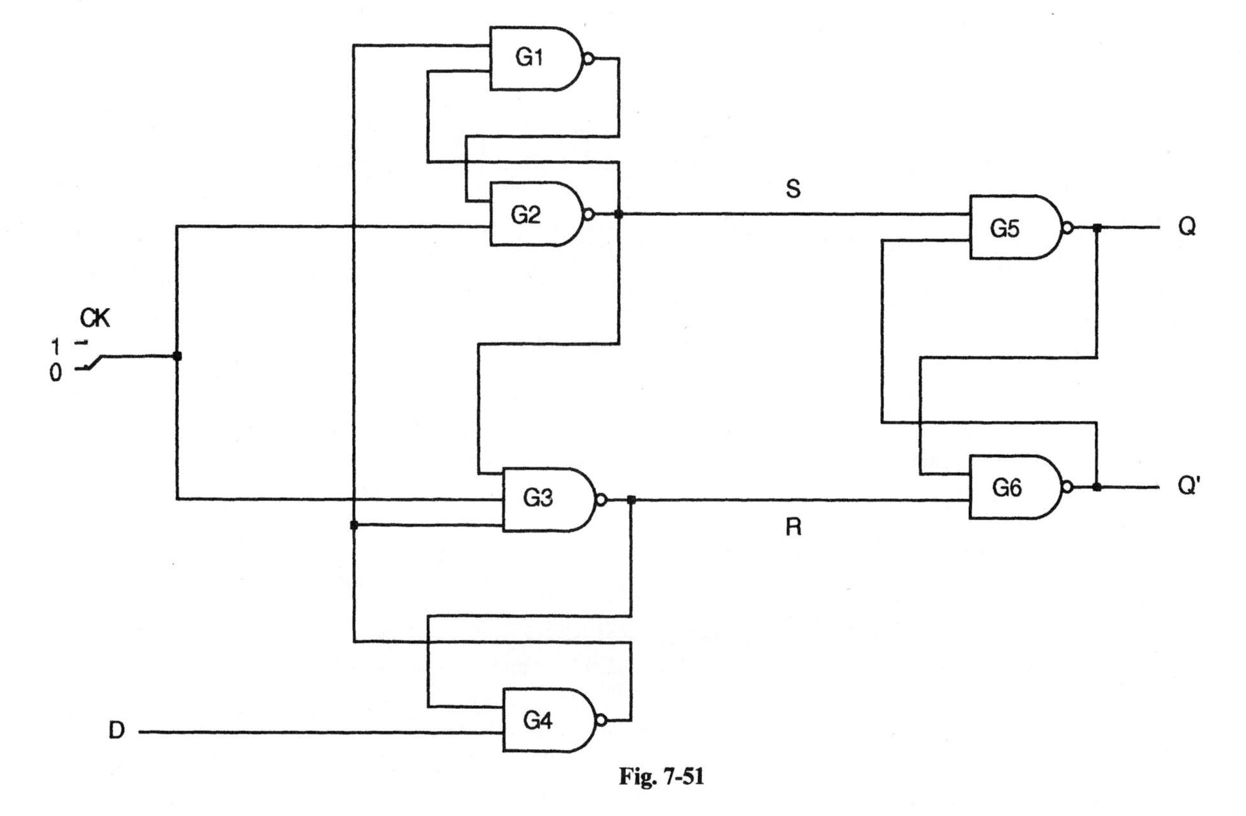

Fig. 7-51

7.22 For the circuit of Fig. 7-51, if D = 0, what happens when the clock goes from 0 to 1?

7.23 Continuing the sequence of Prob. 7.22, what happens to Q when the clock returns to 0?

7.24 Repeat Probs. 7.22 and 7.23 for the case when D = 1.

7.25 The results of Probs. 7.21 to 7.24 indicate that the circuit of Fig. 7-51 is a positive-edge-triggered D flip-flop. Draw a voltage macrotiming diagram for the conditions shown in Fig. 7-52 when the initial value of Q = 1.

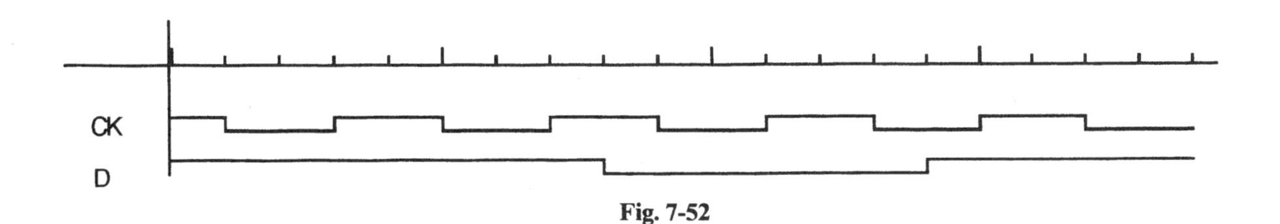

Fig. 7-52

7.26 In order to convert a D to a JK flip-flop, we must add logic to make the device retain its present state when inputs JK = 00 and to toggle when JK = 11. Show the logic.

7.27 The fabrication of a gate or inverter on a custom chip requires the following allocation of resources:

Each gate	2 units
Each input (after the first)	0.5 unit
Each interconnecting wire (including input and output leads)	0.25 unit

If each silicon resource unit costs $0.02, compare the cost of a simple latch (Fig. 7-2) with the JK flip-flop of Fig. 7-11.

7.28 The flip-flop outputs in the circuit of Fig. 7-53 are connected to the select inputs of a multiplexer. This circuit is called a scanner. Assuming negative-edge triggering, create a macrotiming diagram for signals A, B, and C and describe the scanning action.

7.29 In the JK flip-flop circuit shown in Fig. 7-12*a*, illustrate the internal timing for preset action by sketching the voltage microtiming diagram for the individual gates. Assume that, initially, Q = 0 and the clock is disabled (held at 0) when the preset input goes TRUE (LOW).

7.30 For the JK flip-flop circuit shown in Fig. 7-12*a*, assume that J = 0, K = 1, and that the circuit is initially set to Q = 1. Illustrate the internal timing for synchronous clearing by sketching a voltage microtiming diagram for the individual gates when a single positive-going clock pulse is applied.

7.31 In a circuit which exhibits metastability, what is the effect of doubling the clock frequency? Assume a single clock source.

7.32 Use the numerical values of Example 7.6 to obtain a value for the ratio obtained in Prob. 7.31.

7.33 A student has designed the circuit shown in Fig. 7-54 as a JK flip-flop. It is intended that the inverted clock pulse will block feedback from the outputs QF and QF′ until the clock signal returns to 0. Will the circuit toggle as required? Note that the diagram uses positive logic.

7.34 With reference to the previous problem, use a timing diagram of the internal waveforms to identify a simple fix that will enable the student's circuit to meet the design goal.

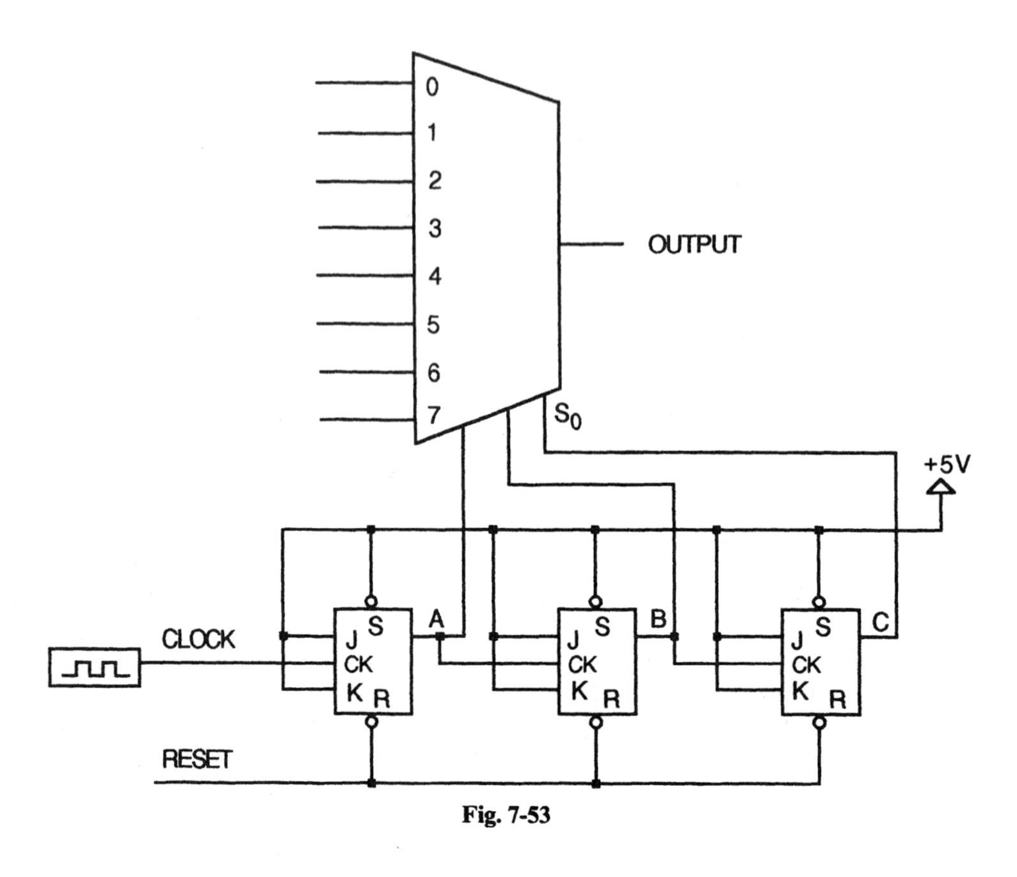

Fig. 7-53

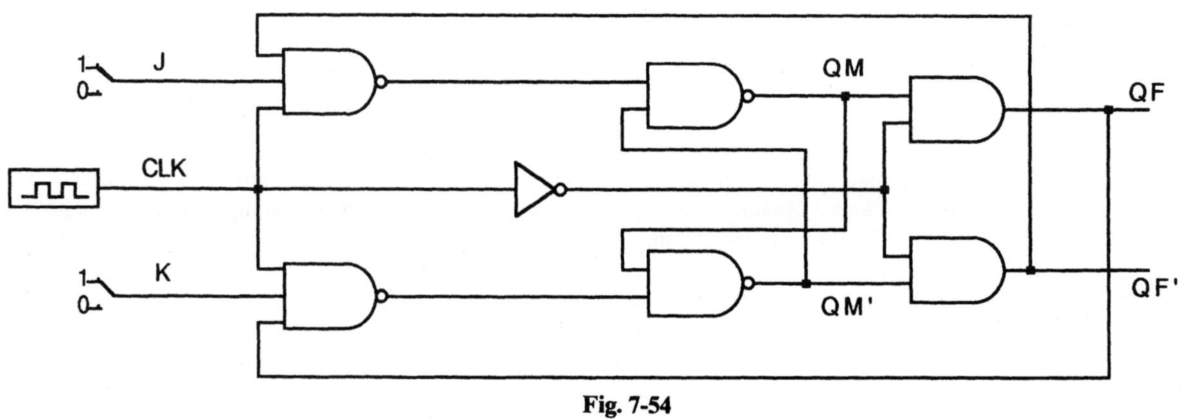

Fig. 7-54

7.35 Can the circuitry shown in Fig. 7-55 convert the D flip-flop into a T flip-flop? Demonstrate your answer with a timing diagram.

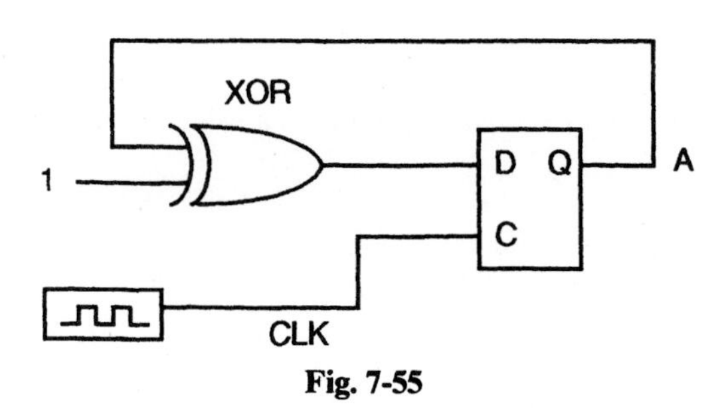

Fig. 7-55

7.36 Design a circuit which will convert a T flip-flop to a D flip-flop.

7.37 Using the techniques of Example 7.5, and with the AND-OR gates used in the combinational logic replaced by multiplexers, find the maximum clock rate for a circuit of the type shown in Fig. 7-18. Assume the following device timing parameters:

Device	Max, ns	Min, ns
Flip-flops:	$t_{pd, HL} = 20$	$t_{pd, HL} = 10$
	$t_{pd, LH} = 25$	$t_{pd, LH} = 15$
	$t_{su} = 10$	$t_{su} = 5$
	$t_h = 5$	$t_h = 0$
Gates:	$t_{pd, HL} = 15$	$t_{pd, HL} = 5$
	$t_{pd, LH} = 25$	$t_{pd, LH} = 15$
Multiplexers:	$t_{pd, HL} = 25$	$t_{pd, HL} = 15$
	$t_{pd, LH} = 15$	$t_{pd, LH} = 10$

Answers to Supplementary Problems

7.16 Use NOR hardware as shown in Fig. 7-56.

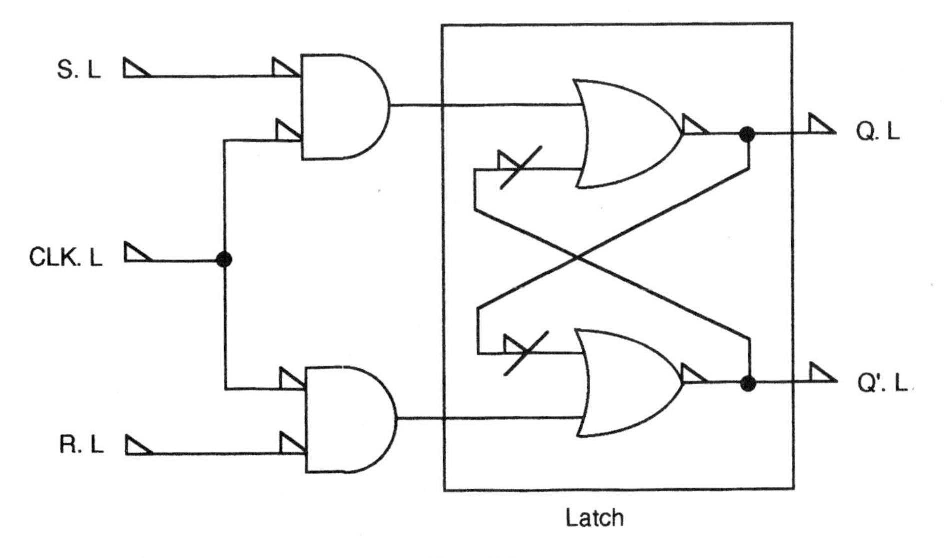

Fig. 7-56

7.17 The logic diagram and state table are shown in Fig. 7-57. Note that this state table is the logical inverse of the one shown in Table 7.1.

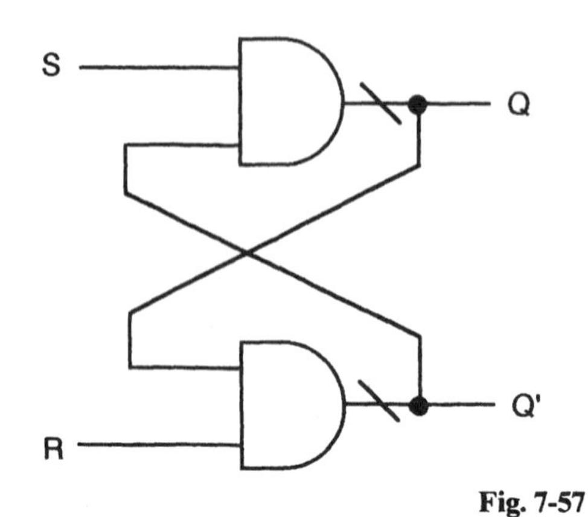

S	R	Qn+1
0	0	*
0	1	1
1	0	0
1	1	Qn

* Both outputs are 1's

Fig. 7-57

7.18 See Table 7.9.

Table 7.9

T	D	Qn+1
0	0	0
0	1	1
1	0	Q'n
1	1	Qn

7.19 See Fig. 7-58.

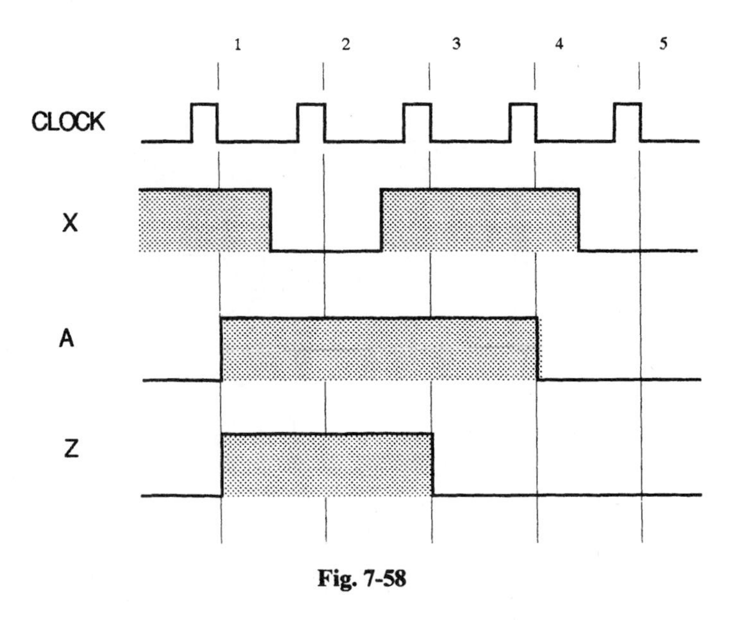

Fig. 7-58

7.20 The output of gate G_7 goes from 1 to 0 as expected, but it only remains there for one delay interval; *the circuit does not latch* (refer to Fig. 7-32).

In actual fact, the situation is far more severe than simple nonfunctionality. The trailing edge of the G_7 pulse affects G_5 and G_8 which, subsequently, feed a signal change back to G_4, and so on. The cross-coupled connections result in instability and the circuit oscillates continuously, as shown in the simulation waveforms of Fig. 7-59.

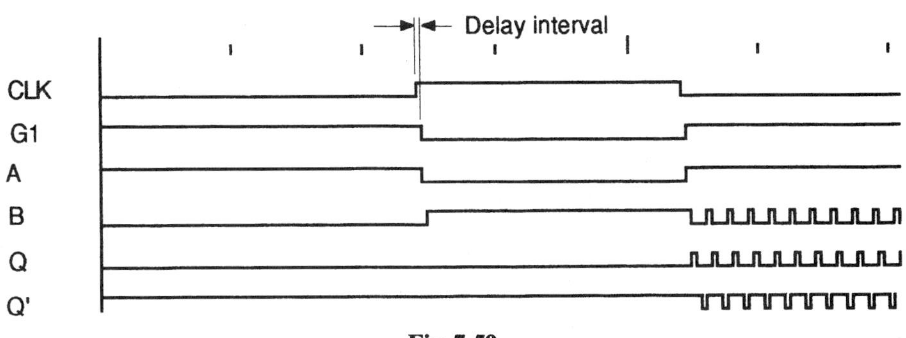

Fig. 7-59

7.21 When the clock is LOW, D has no effect and either output state is allowed. Thus, the value of Q depends upon past history.

7.22 Q goes to 0 or remains there if this was its initial state.

7.23 Q remains at 0.

7.24 Q goes to 1 or remains there if this was its initial state. Its value does not change when the clock returns to 0.

7.25 See Fig. 7-60.

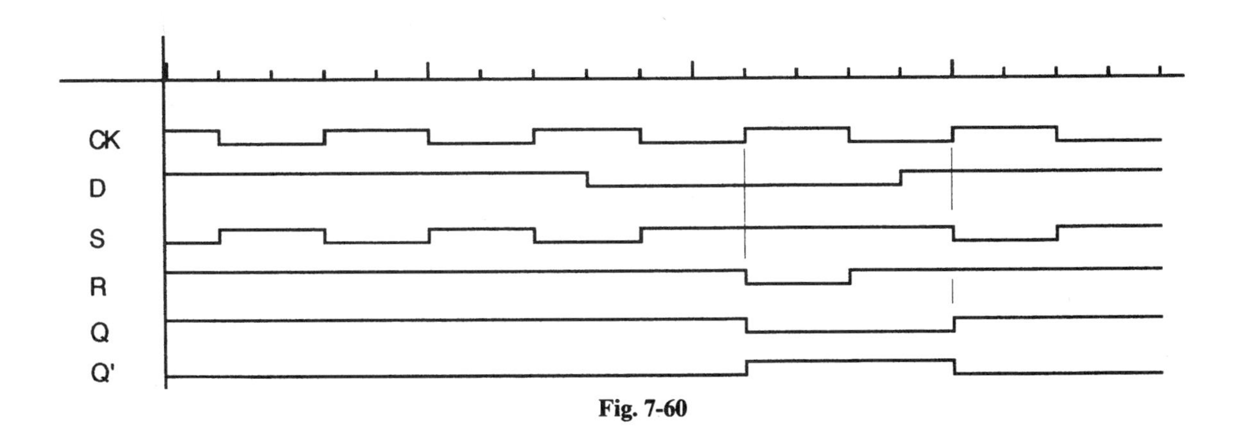

Fig. 7-60

7.26 $D = JQ' + K'Q$. The circuit is shown in Fig. 7-61.

7.27

	Latch	JK Flip-Flop
Resources:	6.5 ($0.13)	36.25 ($0.73)

It is important to note here that the JK flip-flop is capable of more complex operations than the simple latch.

7.28 Multiplexer input lines will be selected in the following order 0-4-2-6-1-5-3-7-0. The timing diagram is shown in Fig. 7-62.

7.29 See Fig. 7-63.

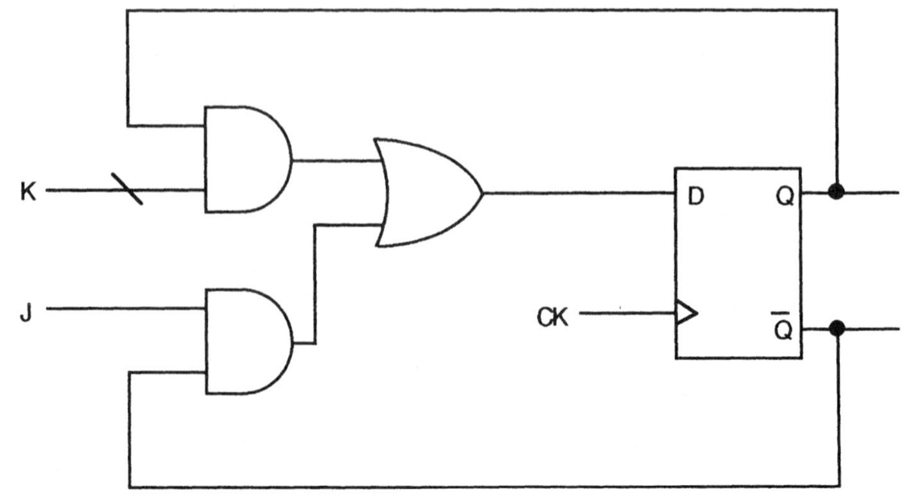

Fig. 7-61

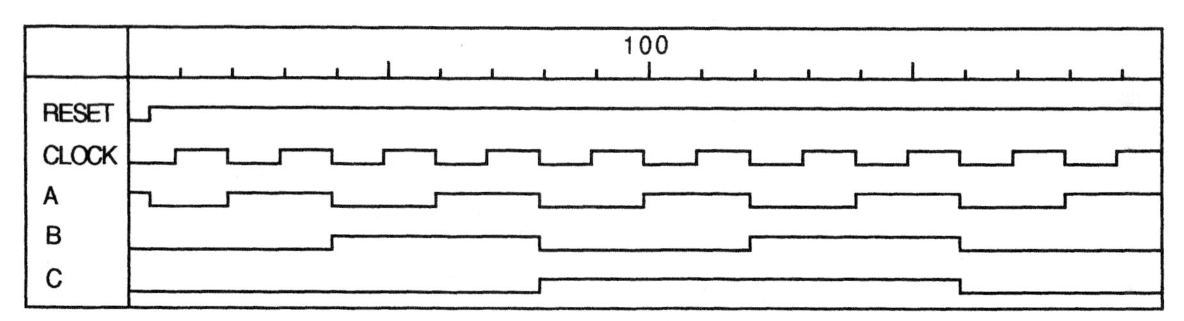

Fig. 7-62 To illustrate reset, flip-flop A has been initialized high in the simulation.

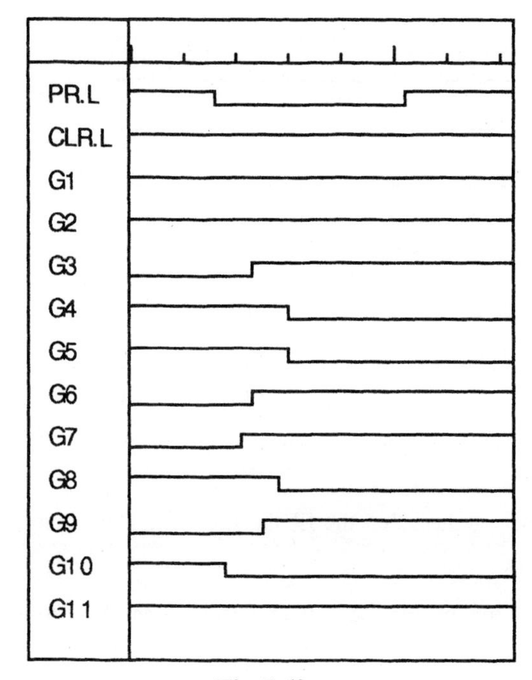

Fig. 7-63

7.30 See Fig. 7-64.

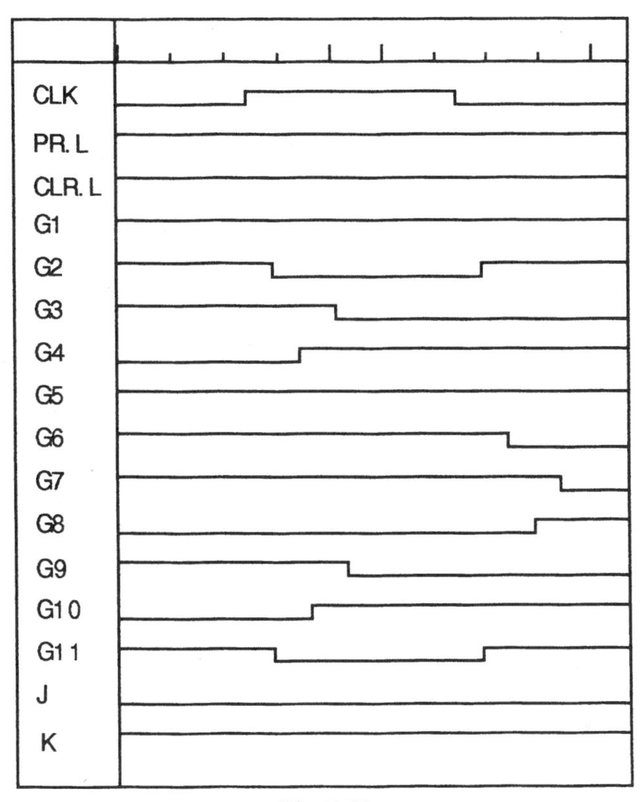

Fig. 7-64

7.31 $$\frac{\text{MTBF(new)}}{\text{MTBF(old)}} = \frac{1}{2}\exp\left(-T_R/2f_{old}\right)$$

7.32 2.59×10^{-17}

7.33 The timing diagram for JK = 11 is shown in Fig. 7-65. The circuit attempts to toggle, but the output change only persists while the inverted clock pulse is present.

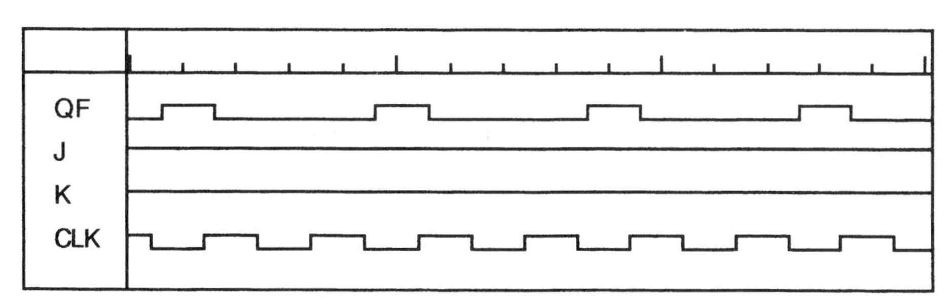

Fig. 7-65

7.34 Use QM and QM′ as outputs (see Fig. 7-66).

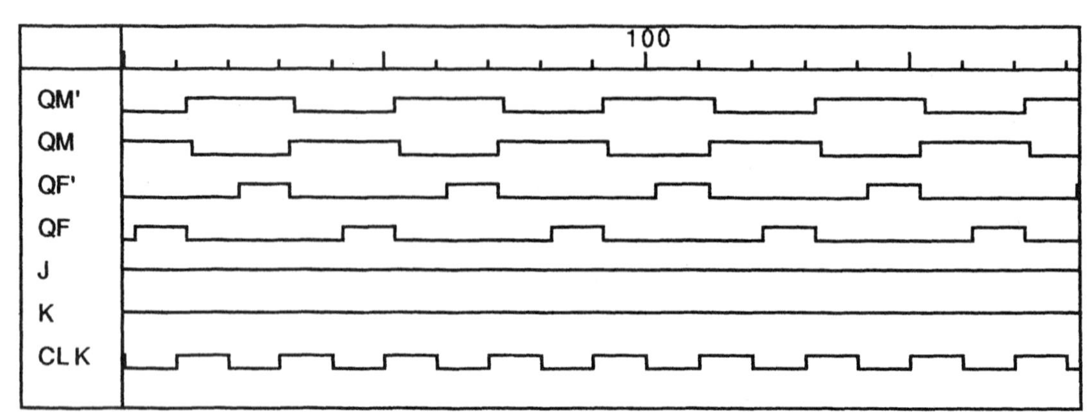

Fig. 7-66

7.35 Yes. See Fig. 7-67.

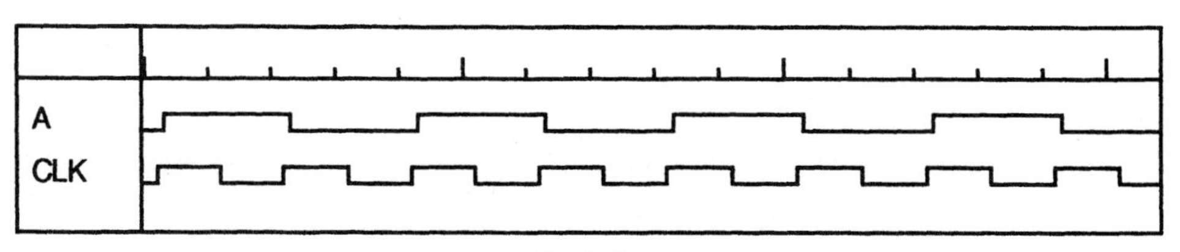

Fig. 7-67

7.36 If the current value of the output differs from D, then the device should toggle. See Fig. 7-68.

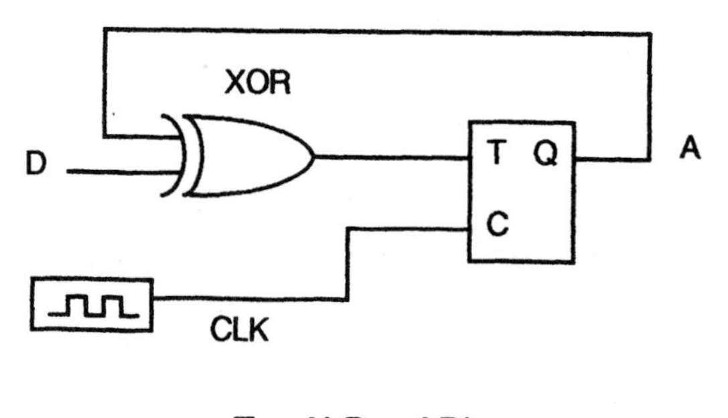

T = A' D + AD'

Fig. 7-68

7.37 $f_{max} = 16.67$ MHz

Chapter 8

Combinations of Flip-Flops

8.1 REGISTERS

A combination of flip-flops, usually served by a common clock pulse and arranged in some sort of systematic order, is referred to as a *register*. Each individual flip-flop in this array is called a *stage*. The register's state is the ordered sequence of flip-flop outputs (1 or 0), customarily expressed as an equivalent binary or hexadecimal number.

EXAMPLE 8.1 Determine the state of the numerically ordered register shown in Fig. 8-1.

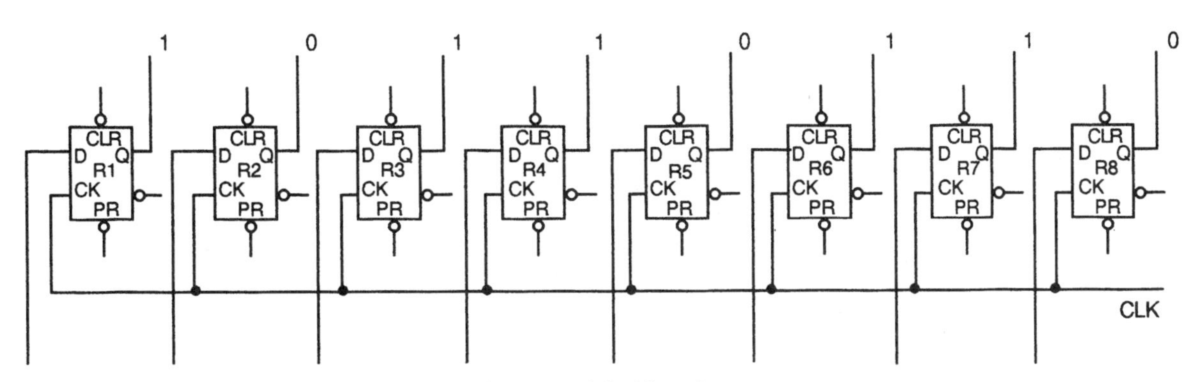

Fig. 8-1 Eight-bit register.

Assuming that stage R_1 contains the most significant bit, the state is 10110110 (binary) or B6 (hex). Note that the register's state does not involve any flip-flop inputs. Connections to the inputs do, however, determine the way the register changes from one state to another.

Latching Register

If the register shown in Fig. 8-1 has its inputs attached to separate data lines, the collective status of these lines can be read into the register and stored (*latched*) by clocking the flip-flops. Stored data will remain unchanged until the register is reclocked at which time new input data, if present, will be latched into the register. An n-stage register can thus be used as a memory device to store or remember a single n-bit binary word.

EXAMPLE 8.2 Four D flip-flops are connected as a latching register as shown in Fig. 8-2 along with a macro-timing diagram of the input signals. Assuming that the flip-flops switch instantaneously on the rising edge of the clock, express the state of the register (as a hex number with R_1 as the most significant bit) after the first, third, and fourth clock pulses.

Just before the first clock pulse, $A = C = 1$ and $B = D = 0$. This data is latched into the flip-flops by the clock's rising edge causing outputs R_1, R_2, R_3, R_4 to equal 1010 which is equal to A_{hex} or \$A. At the rising edge of the third clock pulse, all the inputs are 1s and R_1, R_2, R_3, $R_4 = 1111 =$ \$F. For the fourth clock pulse, $A = 0$ and $B = C = D = 1$ and the register's output is $0111 =$ \$7.

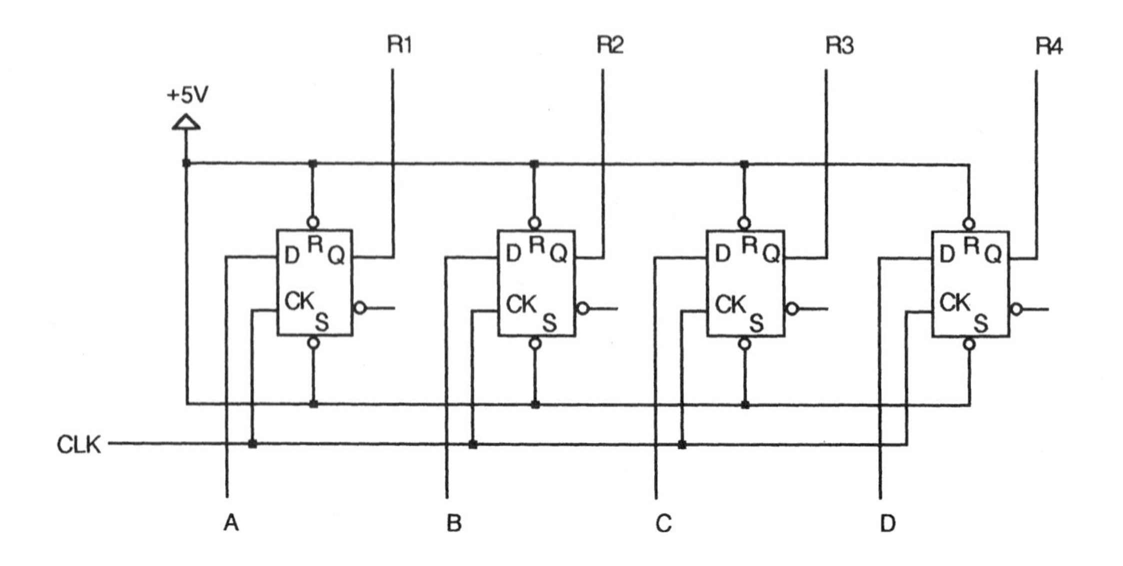

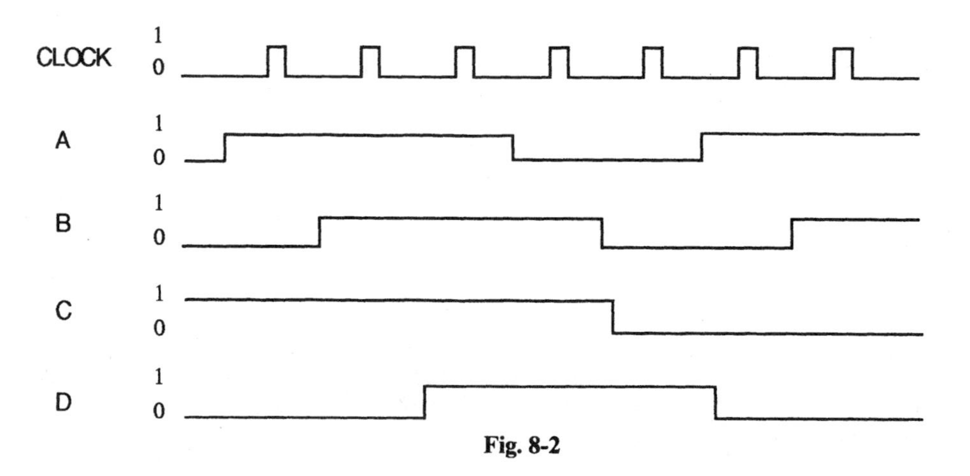

Fig. 8-2

Shift Register

It is possible to interconnect the stages of a register in such a manner as to cause the binary number expressing the register's state to shift right or left upon command. Consider the eight-stage shift register composed of D flip-flops shown in Fig. 8-3*a*. Since the D input of each stage is connected to the output Q of the flip-flop to its left, it follows that the state of each flip-flop after clocking will be equal to the state of its neighbor to the left before clocking. Each clock pulse thus causes a single bit shift to the right.

A shift register can also be made with JK flip-flops if it is recalled that a JK device can be converted to a D flip-flop by arranging for the J and K inputs to be logical inverses (Sec. 7.8). Logical inversion is assured by driving J from the Q output of a flip-flop and K from the Q′ output of the same device. A four-stage shifter using JK stages is shown in Fig. 8-3*b*.

In both cases, data is presented at the left, its bits synchronous with the clock pulses. As shifting progresses, a digital word is pushed 1 bit at a time into the register just as plates are pushed into a spring-loaded stacker in a cafeteria. The plate analogy is so apt that shift registers used for temporary memory in computers are called stacks.

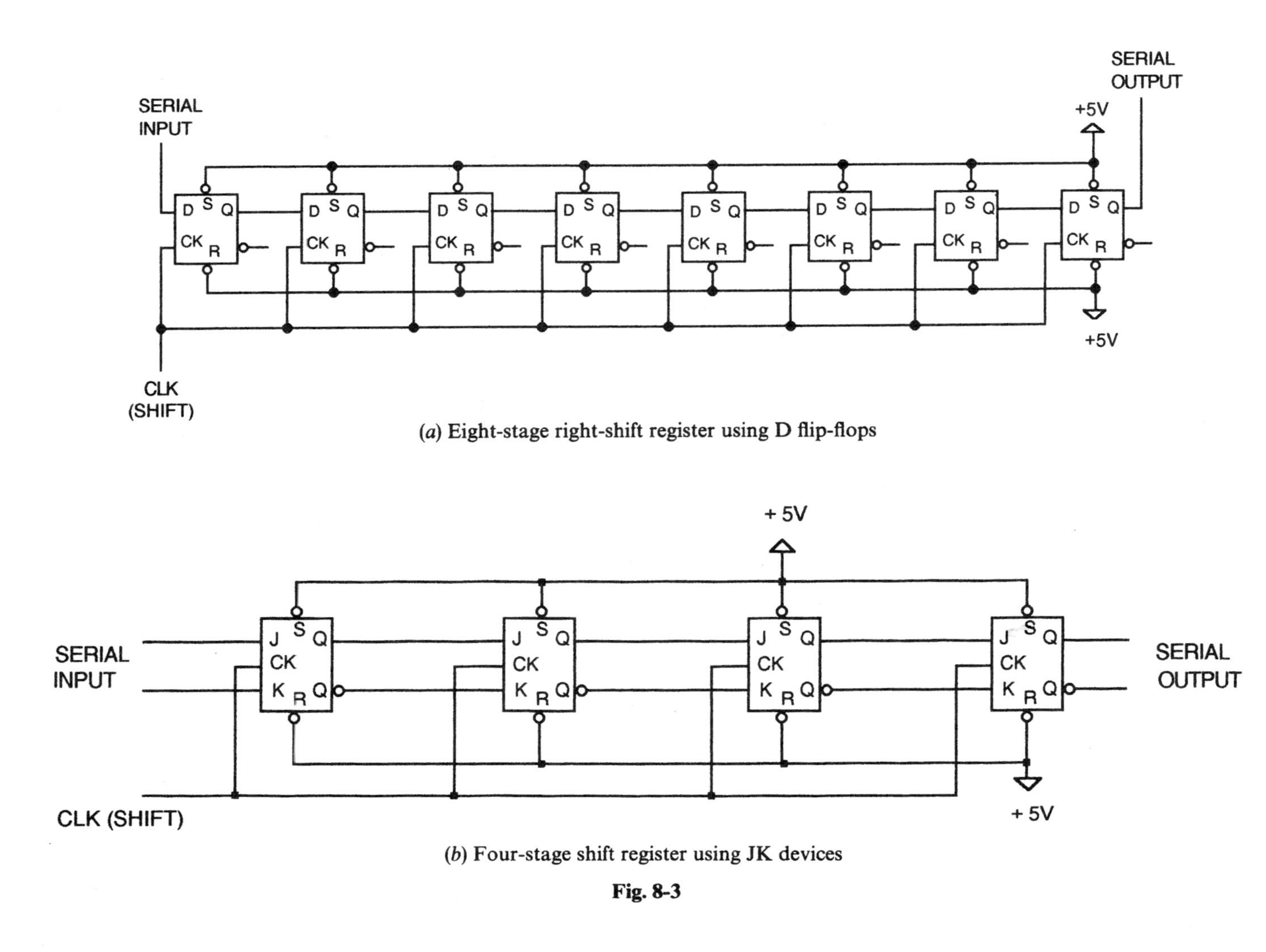

(*a*) Eight-stage right-shift register using D flip-flops

(*b*) Four-stage shift register using JK devices

Fig. 8-3

EXAMPLE 8.3 Design a *reversible shift register* using D flip-flop stages. This example illustrates the possibilities engendered by combining storage elements (flip-flops) and logic elements (gates).

Shift direction is determined by the D input connections. If the output of the first stage is connected to the second, and the second is connected to the third, and so on, data will shift to the right with each clock pulse. On the other hand, if the last output controls the penultimate stage, and this stage's output controls the preceding one, etc., the register will shift left. Clearly, a means is needed by which the input connections can be changed on command.

The multiplexer described in Sec. 5.2 can fulfill this requirement very effectively as shown in Fig. 8-4, which depicts a four-stage register. Each D input is served by a two-input multiplexer. We see that if a control signal, designated R/L, is used as the common select for all multiplexers, then each D input will be connected to the closest *preceding* output when R/L = 1 (shift right) and to the closest *succeeding* output when R/L = 0 (shift left).

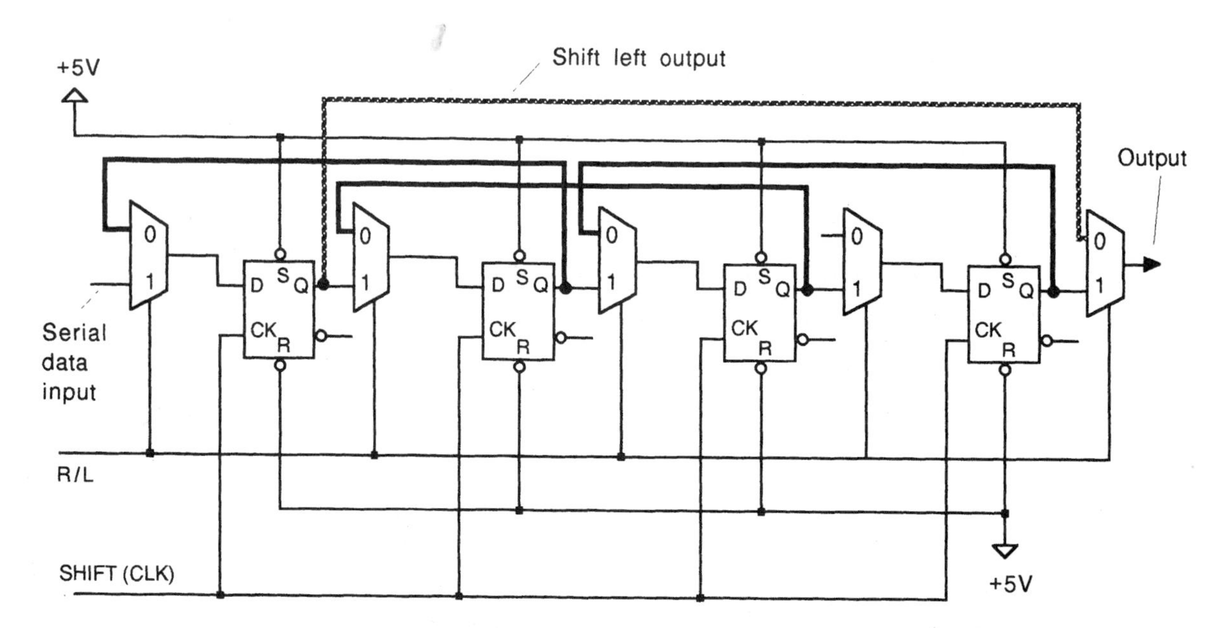

Fig. 8-4 Shift-left connections are shown in bold.

8.2 PARALLEL-SERIAL CONVERSION

If a particular flip-flop type has both synchronous and asynchronous inputs, it is possible to use it in a register arranged to accomplish a form of data manipulation called *parallel-to-serial conversion.* In the system of Fig. 8-5*a*, a 4-bit word is originally in parallel form (i.e., all bits are present simultaneously on four different wires). It is desired to obtain the same data in serial form where the individual bits appear sequentially on a single wire.

The conversion action is best understood by reference to the timing diagram of Fig. 8-5*b*. The asynchronous clear (CLR) signal is first applied, and it sets all the flip-flops in the register to 0. Next, a load pulse is applied. It will pass through those AND gates where the data digit is 1 (D_1 and D_2) and preset the corresponding flip-flops (B and C) to 1. The parallel data has now been read (latched) into the register.

From the standpoint of their synchronous (D) inputs, the flip-flops are connected as a shift register. When clock (CLK) pulses are subsequently applied, the loaded data is shifted to the right, exiting at the output of flip-flop D one bit at a time, synchronous with the clock. The data will be properly converted provided that signals are applied in the order shown. *Note that the synchronous and asynchronous modes of operation are used separately and are not mixed.*

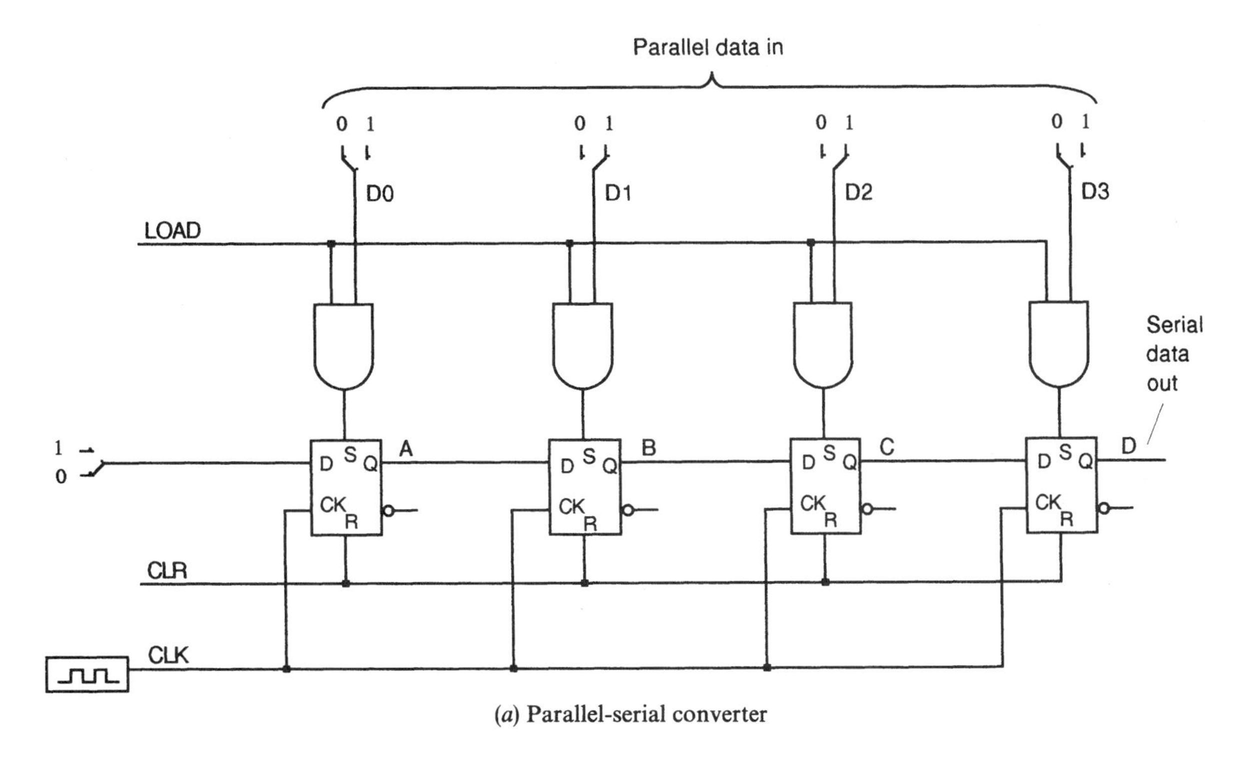

(*a*) Parallel-serial converter

(*b*) Timing diagram (positive-edge triggering)

Fig. 8-5

8.3 RIPPLE COUNTERS

It is possible to connect flip-flops in a manner that causes the circuit to count the number of pulses presented at its input. That is, the flip-flop outputs, taken as a group, will indicate a binary number which is equal to the number of pulses presented at the input from the moment of reset up to the time of observation. Ripple counters are a class of counters which are economical to design, albeit somewhat slow. A typical example is shown in Fig. 8-6.

With their JK inputs tied high, the flip-flops will toggle on each clock pulse (negative edge in this case). Note that unlike the shift register, there is no common clock connection. Instead, with the exception of the first, each flip-flop's clock input is driven from the output of the preceding stage. The first flip-flop (D) will toggle once for each input pulse, and the next flip-flop (C) will toggle once for each negative excursion of D. Similarly, C drives B which, in turn, drives A. Referring to the timing diagram

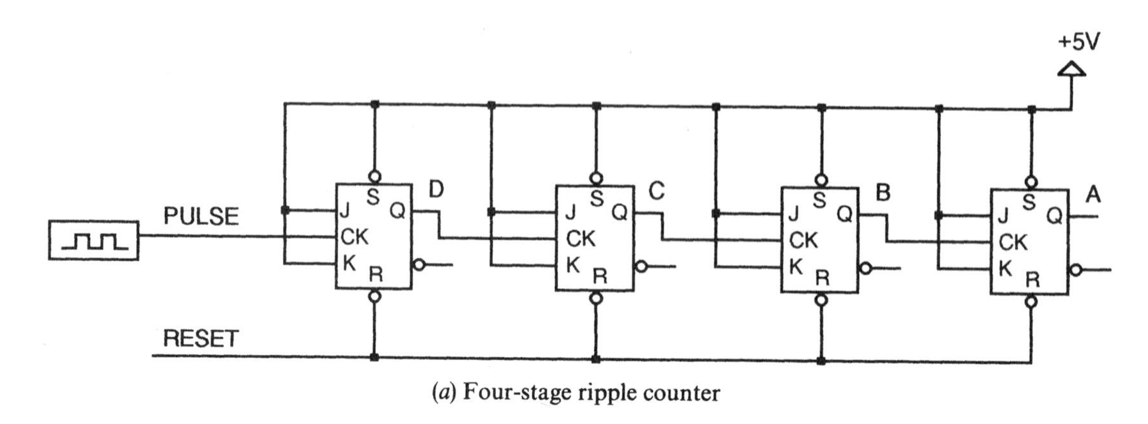

(*a*) Four-stage ripple counter

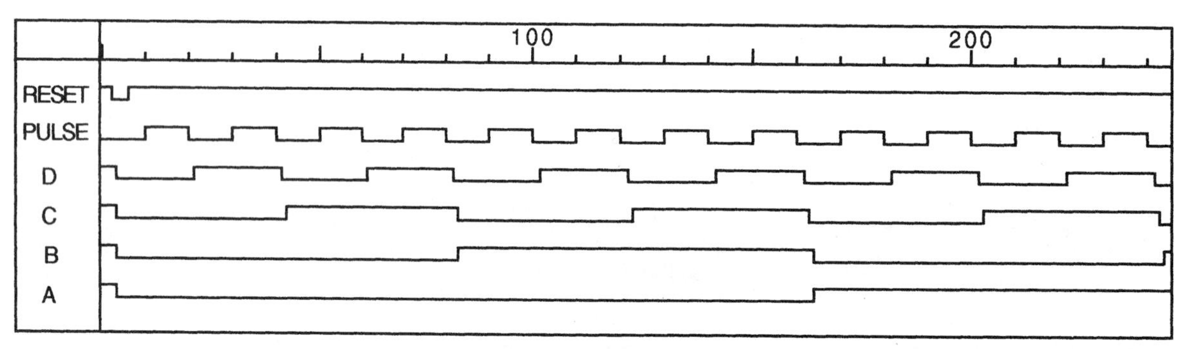

(*b*) Ripple counter timing (negative-edge triggering)

Fig. 8-6

of Fig. 8-6*b* at appropriate intervals, we may determine the flip-flop states following each input pulse (Table 8.1). If output A is considered the most significant bit, we see that the circuit functions as a 4-bit *binary counter*. Though the waveforms stop prematurely, we should expect this circuit to count to 15 and recycle to 0 on the sixteenth pulse. Obviously, one could count higher by adding additional stages.

Since each flip-flop must change before it can affect its successor, the count proceeds like a wave or ripple through the chain of flip-flops; thus the name "ripple counter."

Besides its utility as a counter, the circuit of Fig. 8-6*a* has another very useful characteristic. We see from the waveforms in Fig. 8-6*b* that, progressing left to right, the frequency of each stage is exactly half

Table 8.1

Pulse	A	B	C	D
0	0	0	0	0
1	0	0	0	1
2	0	0	1	0
3	0	0	1	1
4	0	1	0	0
5	0	1	0	1
6	0	1	1	0
7	0	1	1	1
8	1	0	0	0
9	1	0	0	1
10	1	0	1	0
11	1	0	1	1
12	1	1	0	0

that of its predecessor. Thus, the circuit may be used as a precise *frequency divider*. A common application is in digital watches where the output of a very stable high-frequency quartz oscillator is divided down to get accurate 1-s pulses.

EXAMPLE 8.4 A problem inherent to ripple counters is illustrated in Fig. 8-7 where a *recognition gate* has been added to flag the occurrence of the sixth pulse following reset. Its inputs are A'BCD', and it is intended to produce an output only when the counter state is 0110 (decimal 6).

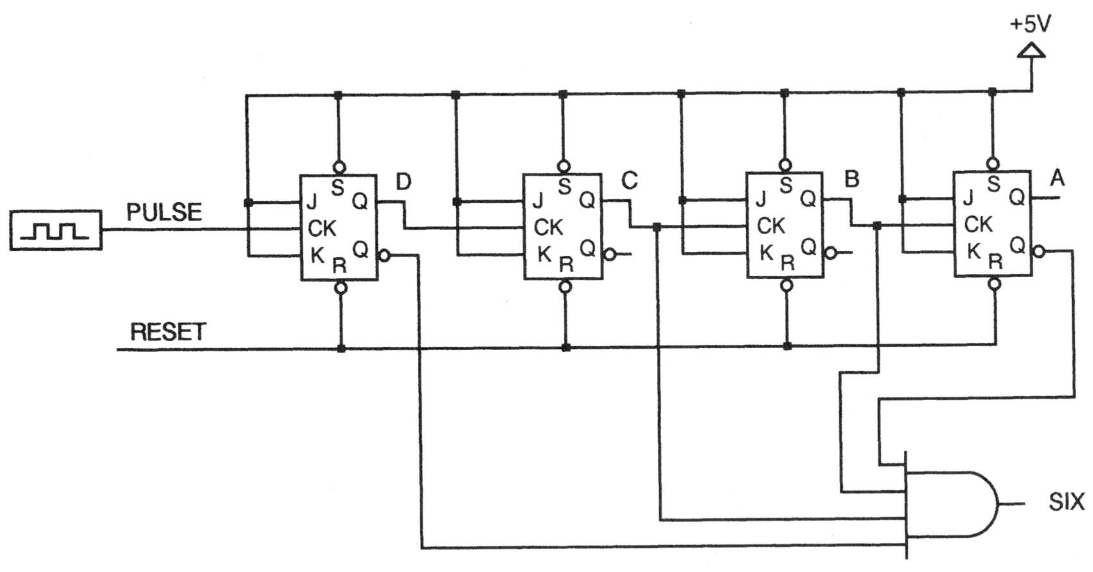

(*a*) Ripple counter with recognition gate

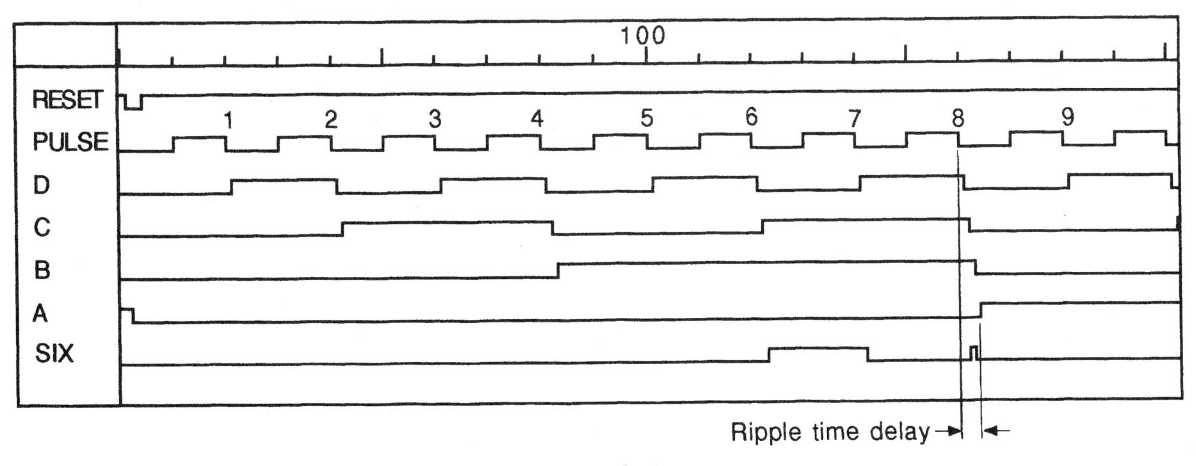

(*b*) Waveforms

Fig. 8-7

Unfortunately, because of the unavoidable propagation-delay time associated with each flip-flop (Sec. 7.9), several transient states can occur during the transition from one count to another as flip-flops trigger sequentially. Table 8.2 illustrates the sequence of states passed through during the transition from count 7 to 8. When D changes from 1 to 0, it triggers C which changes from 1 to 0 triggering B which, in turn, causes A to change from 0 to 1. The stable states are shown in bold type, and the transient states in lighter type. Binary 0110 exists briefly causing the glitch seen in Fig. 8-7*b*.

Table 8.2

Count	A	B	C	D
7	0	1	1	1
6	0	1	1	0
4	0	1	0	0
0	0	0	0	0
8	1	0	0	0

In addition to their susceptibility to glitches, ripple counters have speed limitations since time delays accumulate. For an N-stage counter to work properly, input pulses must be separated by at least the maximum ripple time delay, $(N-1)T_p$, where T_p is the flip-flop propagation time. Thus, a ripple counter is well suited for counting widgets on a conveyor belt but totally inadequate for keeping track of the pulses moving around within a high-speed computer.

8.4 RATE MULTIPLIERS

Another example where use is made of the powerful combination of logic and storage is the rate multiplier. This is a digital device which delivers a train of pulses whose average number, per unit time interval, may be accurately chosen by the setting of control switches to be a fixed fraction of an input pulse rate. Such devices are often used in power controllers for stepper motors whose shafts rotate a fixed number of degrees per pulse. By selecting the number of pulses delivered, highly precise mechanical positioning can be achieved for use in machine tools, computer printer paper drives, robots, etc.

A logic diagram for a typical rate multiplier is shown in Fig. 8-8*a*. Here, a 4-bit binary counter is used to generate eight signals which are related to each other in time (the Q and Q′ outputs of each stage). These signals are combined with clock pulses and control inputs at four recognition AND gates, each of which will produce an output only if the correct logic conditions are satisfied and an input pulse is present (TRUE).

Referring to the circuit diagram in Fig. 8-8*a*, we see that the gate G_8 will pass a pulse only if C_8 and A are simultaneously TRUE; a condition which holds for every other input pulse. Thus, when C_8 is TRUE, half of the input pulses occur at the output of G_8. Similarly, G_4 will only pass an input pulse if A′, B, and C_4 are simultaneously TRUE. This condition will hold for every fourth input pulse, and none of these pulses from G_4 will occur at the same time as any of the pulses from G_8 since A and A′ cannot be TRUE simultaneously. In like manner, G_2 and G_1 pass every eighth and sixteenth pulse, respectively, without any overlapping. The recognition gate outputs are ORed together to produce system output F which can be fed to a power amplifier and thence to a motor.

We see that it is possible, by an appropriate choice of control values, to create output pulse rates equal to any of the sixteen fractions from 0 to 15/16 of the input pulse rate in steps of one-sixteenth. In the example shown in Fig. 8-8*a*, control variables C_8 and C_2 are TRUE, while C_4 and C_1 are FALSE, and the resulting binary 1010 causes the output to produce 10 pulses for every 16 input pulses. Figure 8-8*b* shows that the output pulses are not uniformly spaced but occur, instead, at an irregular rate which is investigated quantitatively in Prob. 8.12.

An equation expressing rate multiplication as a functon of the logical control variables may be written as

$$M = \tfrac{1}{2} \times C_8 + \tfrac{1}{4} \times C_4 + \tfrac{1}{8} \times C_2 + \tfrac{1}{16} \times C_1$$

If *finer fractional increments* are needed, additional counter stages can be added, along with an equal number of additional recognition gates and control inputs. To increase the resolution to one thirty-second, for example, a flip-flop E is added. Its J and K inputs are connected to logic 1, and its clock input is driven from output D. The additional recognition gate is connected to A′, B′, C′, D′, and E

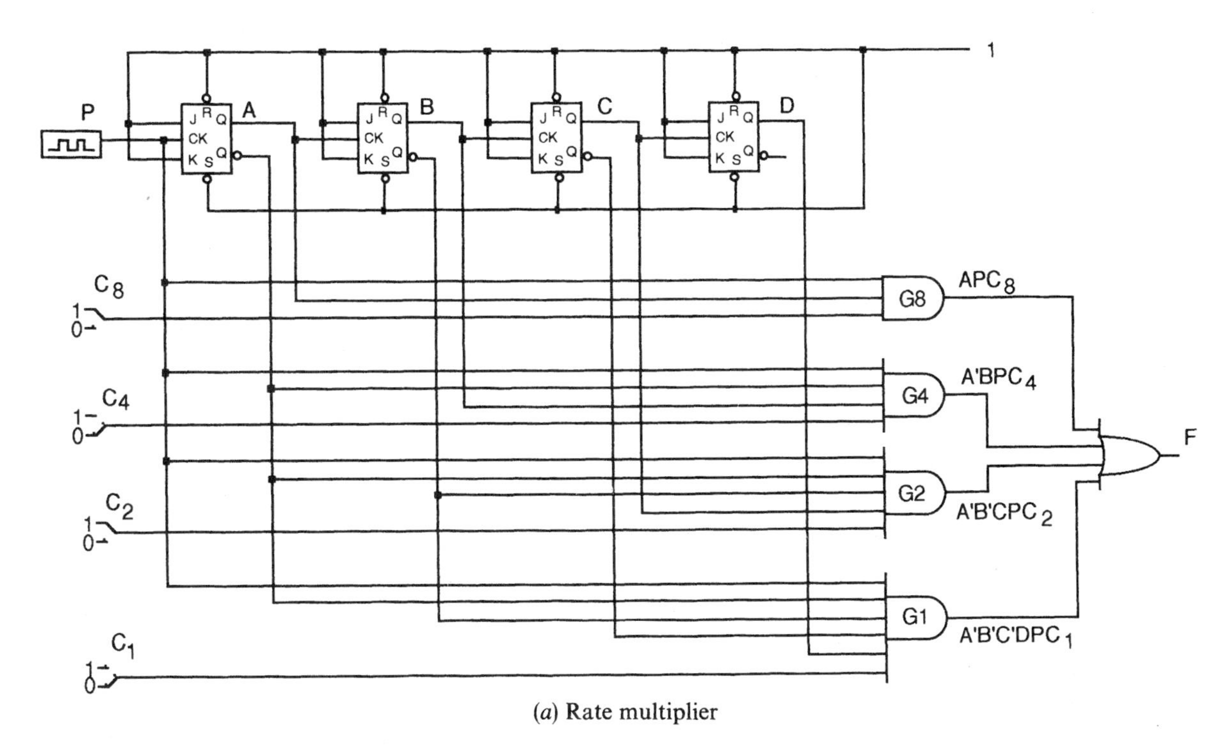

(a) Rate multiplier

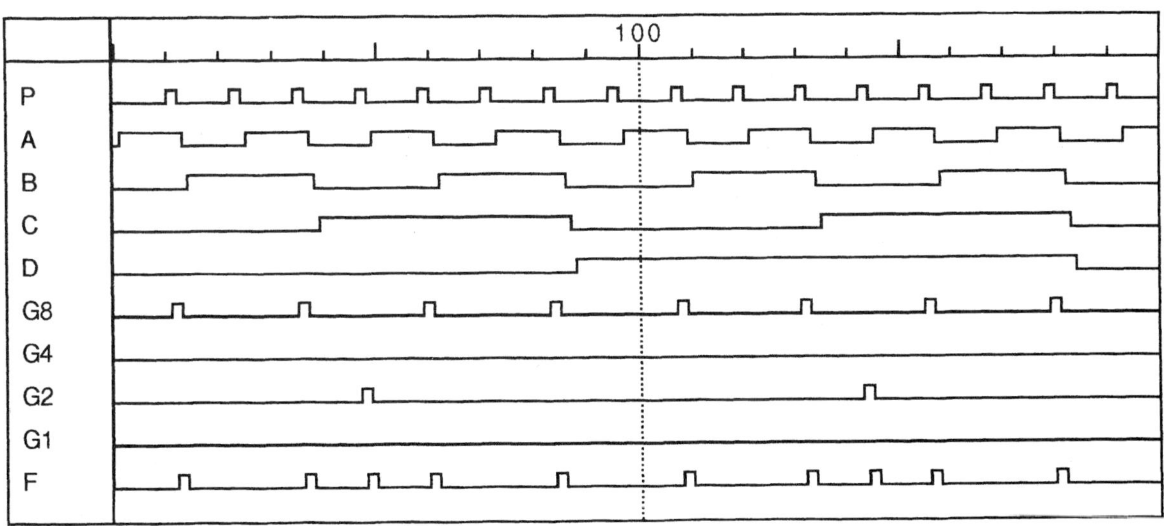

(b) Waveforms

Fig. 8-8

along with the input pulse and a new control line. Control lines are relabeled C_1 to C_{16}, and the rate multiplication equation becomes

$$M = \tfrac{1}{2} \times C_{16} + \tfrac{1}{4} \times C_8 + \tfrac{1}{8} \times C_4 + \tfrac{1}{16} \times C_2 + \tfrac{1}{32} \times C_1$$

8.5 RANDOM-ACCESS MEMORY (RAM)

RAM is the name given to a class of memory elements whose contents can be altered (written to) as well as read at electronic speeds. It is, perhaps, best visualized as an array of addressable registers, each one of which holds an n-bit binary word. Figure 8-9 shows the logical equivalent of a 4-bit RAM

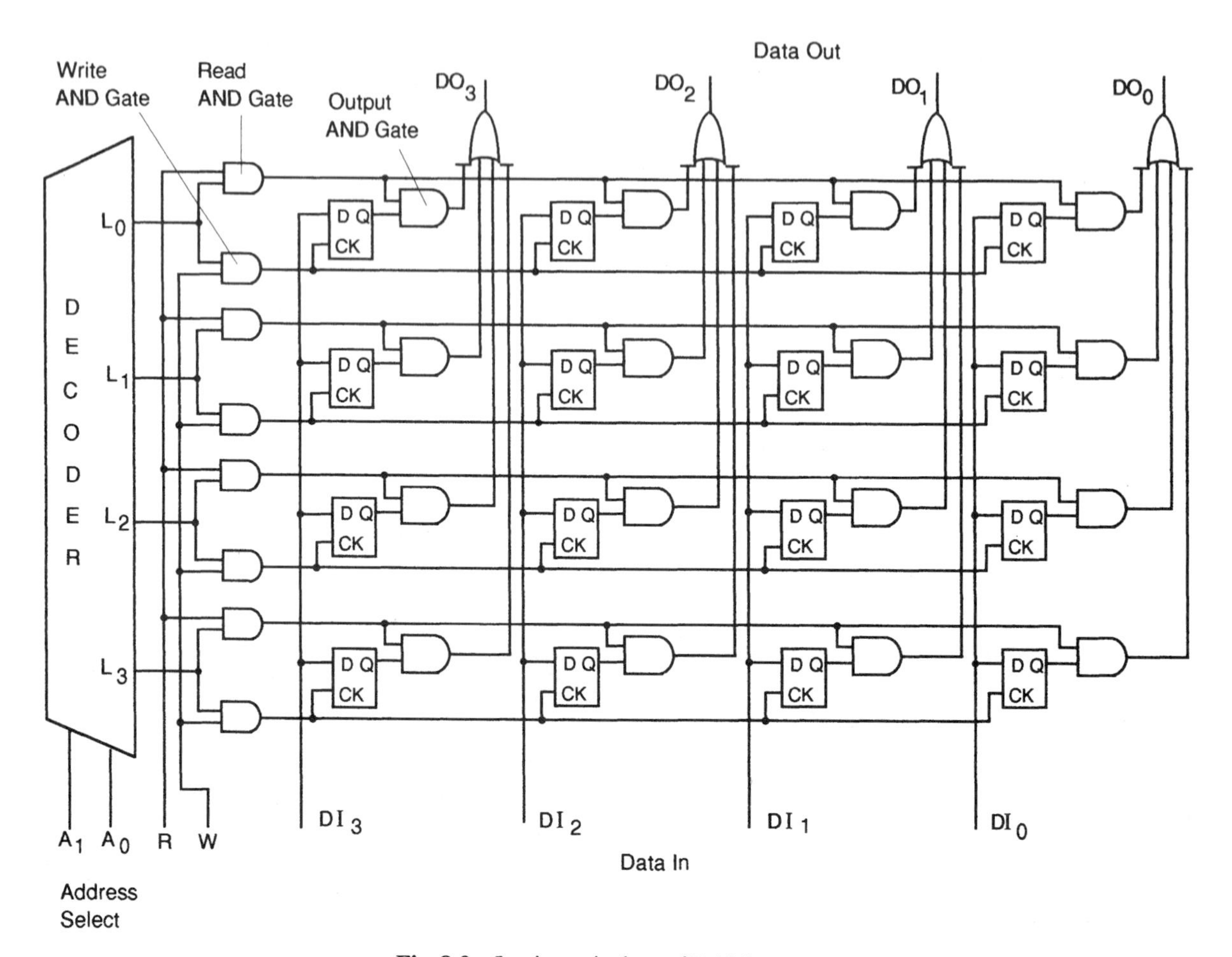

Fig. 8-9 Logic equivalent of RAM memory.

structure, one of whose prominent features is an address decoder, functioning exactly as in a ROM, which selects the appropriate register (horizontal row of flip-flops).

A brief definition is in order at this point. An AND gate is said to be *primed* if all but one of its inputs are connected to logic 1 (TRUE). In this case, its *output will assume the logic value of the remaining input* and we may think of the gate as a sentinel which passes signals when it is primed and blocks them otherwise. We see that the address decoder output primes two AND gates in the selected row so that if a read (R) signal is present, the read gate will produce a TRUE output as will the write gate if a write (W) signal is present.

Each column of flip-flops corresponds to the bit rank (most significant at the left and least significant at the right). The D inputs of all flip-flops in the same column are connected together and to an input data line, as shown in Fig. 8-9. If a write pulse is applied, it will pass through the appropriate primed AND gate to clock data into all the flip-flops in the selected (addressed) row, thus storing an n-bit binary word at the specified location.

In the read mode, an AND gate connected to the output of each flip-flop in a selected row is primed by the address decoder so that, for each column, the value of the bit stored is passed to a large OR gate having as many inputs as the RAM's addressable storage capacity. The outputs of similarly connected AND gates from the *same order bit* (column) in all of the stored words constitute the other inputs to this OR gate. Since the selected word (row) is the only one with primed output AND gates, all other inputs to the output OR gate in each column must be 0 and the data outputs will be determined solely by the bits in the selected word.

In practice, all data entering or leaving the memory passes through an Input/Output (I/O) register as shown in Fig. 8-10. This register must communicate with the off-chip data inputs as well as send data to the outside world. A typical I/O cell is shown in Fig. 8-11. Observe that data enters and leaves the memory chip on the *same* path (called a data bus). Data is latched into the I/O register via its D inputs which are connected to the data bus when a write signal is present and to the memory chip's output OR gates when a read signal is applied. Connection of each I/O register output bit to the data bus is made via a circuit called a TRI-STATE®* buffer which, when enabled by the read signal, provides a conductive path and when not enabled, acts like an open switch. By this means, it is possible to prevent the I/O register's output circuitry from interfering with input data during a write operation. In general, by acting as controlled switches, TRI-STATE® buffers permit various circuits to share a common data bus without interacting. Typical TRI-STATE® symbols are shown in Fig. 8-12.

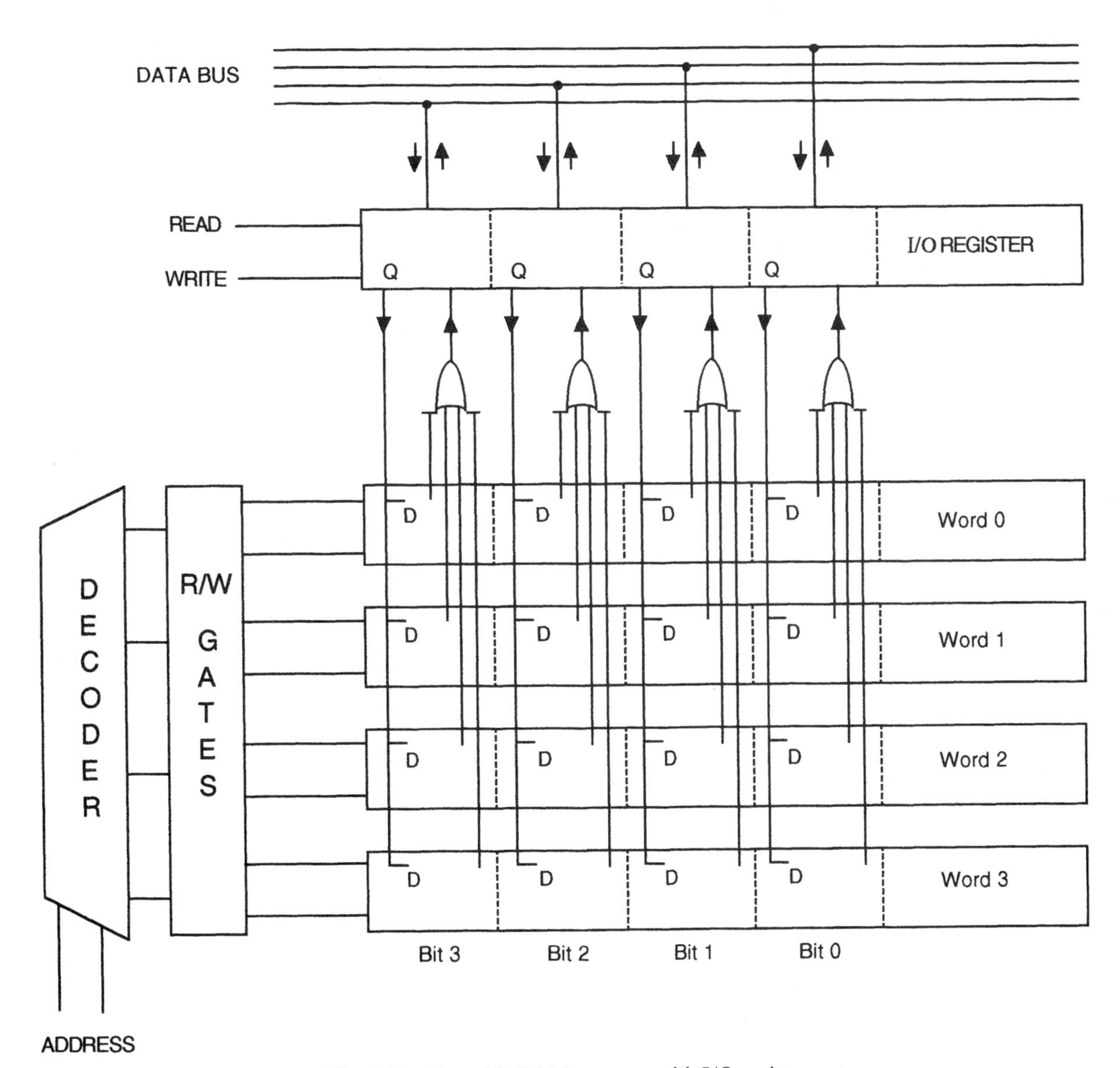

Fig. 8-10 Four-bit RAM memory with I/O register.

---

* TRI-STATE® is a registered trademark of National Semiconductor Corporation.

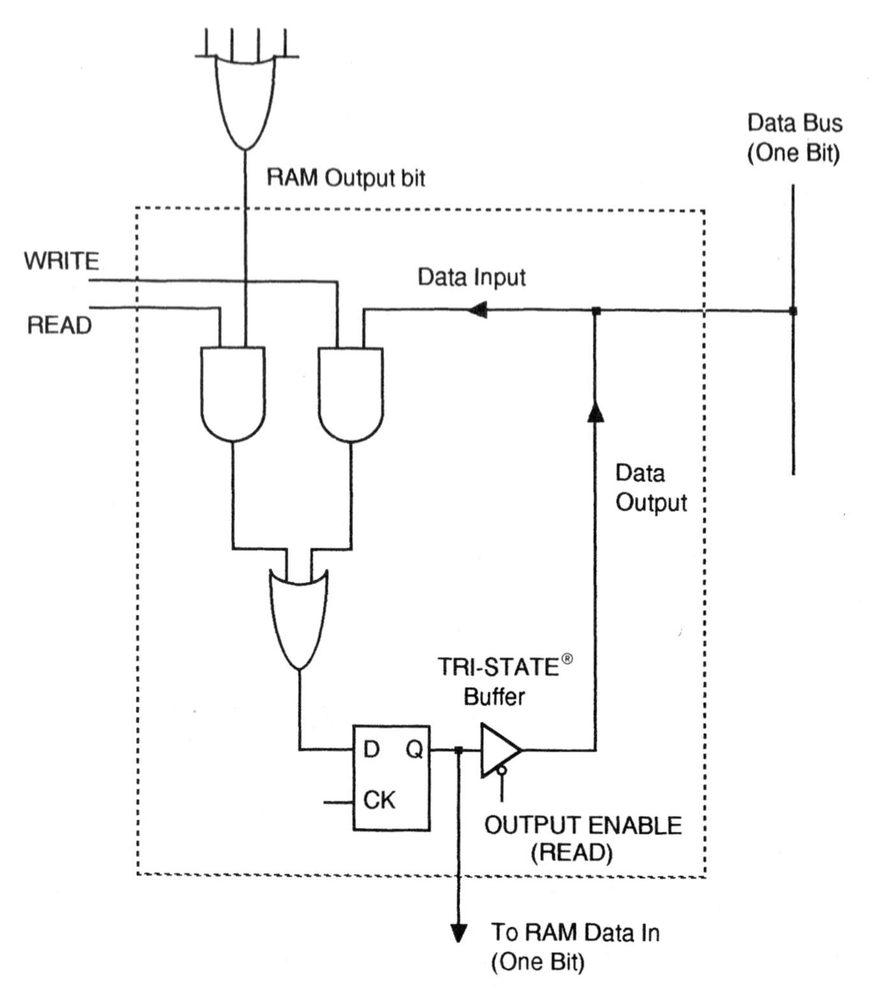

Fig. 8-11 I/O register cell.

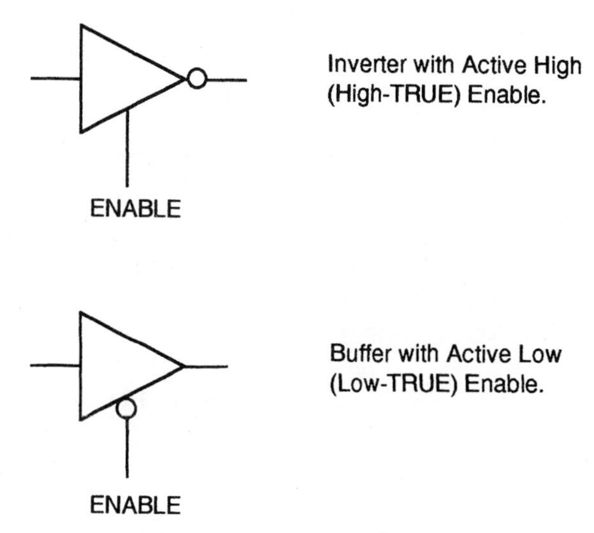

Fig. 8-12 TRI-STATE® devices.

RAM is used in computers and other digital systems for the temporary storage of data which must be recalled for later use, its role being much like that of a scratchpad in hand calculations. It is also used for storing computer and/or device programs which can be modified as required by particular operating conditions. This is contrasted with ROM devices which are used for the permanent storage of computer instructions and lists of data and are installed at manufacture. The stored information can only be changed during shutdown by removing chips and replacing or reprogramming them (refer to Chap. 9).

Solved Problems

Note that several of the problems which follow involve sequential circuits that require careful analysis of states prior to the receipt of a clock pulse, and the tabulation of resulting outputs. The methods discussed in this chapter are adequate for relatively simple circuits. A more powerful systematic method for analyzing sequential circuits in general is described in connection with the study of state machines in Chap. 10.

8.1 A shift register which is connected in a loop having a logical inversion at one point is called a Mobius or Johnson counter. A four-stage version using D flip-flops is shown in Fig. 8-13. Assuming that it is initially cleared and transitions occur on rising clock edges, show the macro-timing waveforms at the outputs of each stage.

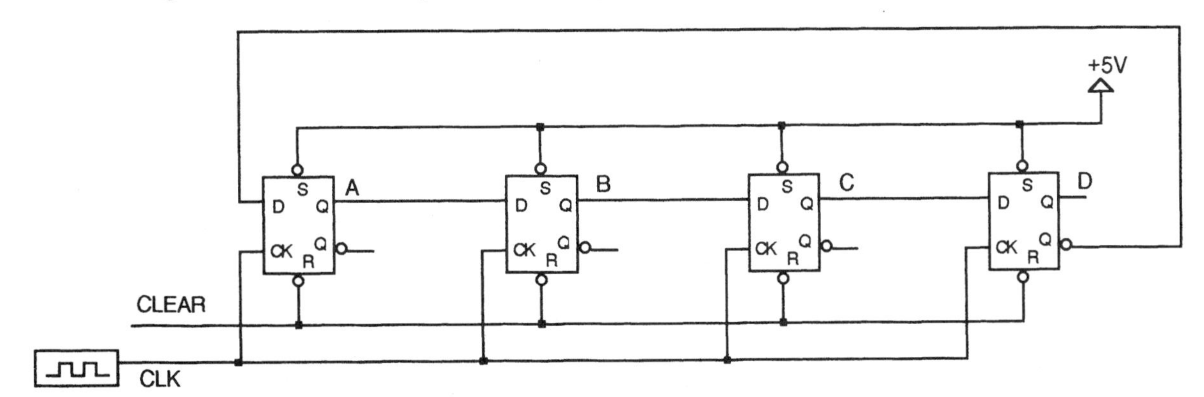

Fig. 8-13 Mobius counter.

When reset, all Q outputs will be at logic 0. D′ will be at logic 1 which will cause A to go to 1 upon the first rising clock signal. Output A, being the data input for B, will cause this flip-flop to go to logic 1 on the second clock pulse. This change will be transferred to C which, in turn, will cause D to change at pulse 4. At this point, D′ goes to 0 causing A to go to logic 0 at pulse 5. A HIGH-to-LOW transition now ripples through the counter until it reaches D, at which time, the process repeats. The waveforms are shown in the simulation output of Fig. 8-14.

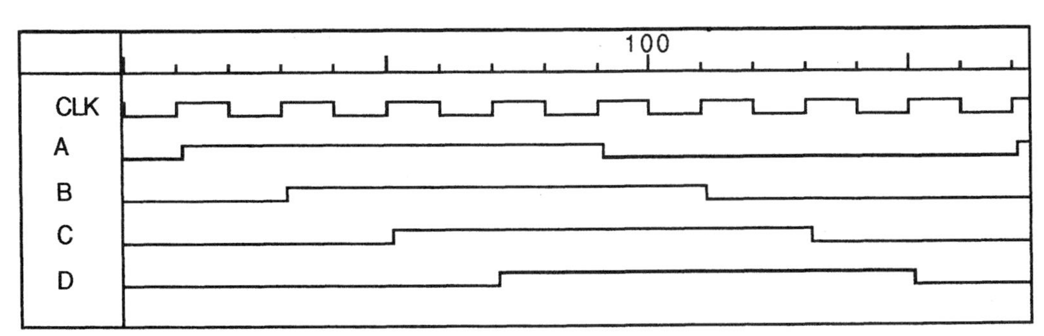

Fig. 8-14 Mobius counter waveforms.

8.2 For the counter of Fig. 8-15(*a*), assume that the flip-flops are master-slave devices of the type discussed in Chap. 7. If all flip-flops trigger on the positive-going edge of a clock pulse and are assumed to be initially cleared, sketch the waveshapes at A and C for 10 clock pulses (ignoring propagation delays).

As in Prob. 7.9, tabulate the values of flip-flop JK inputs after each pulse (see Table 8.3). The waveforms shown in Fig. 8-15(*b*) can be drawn by reading logic values from the table.

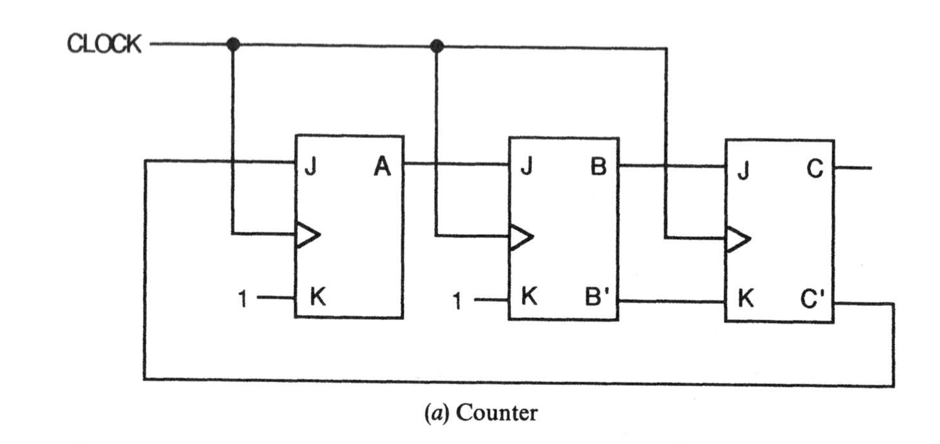

(*a*) Counter

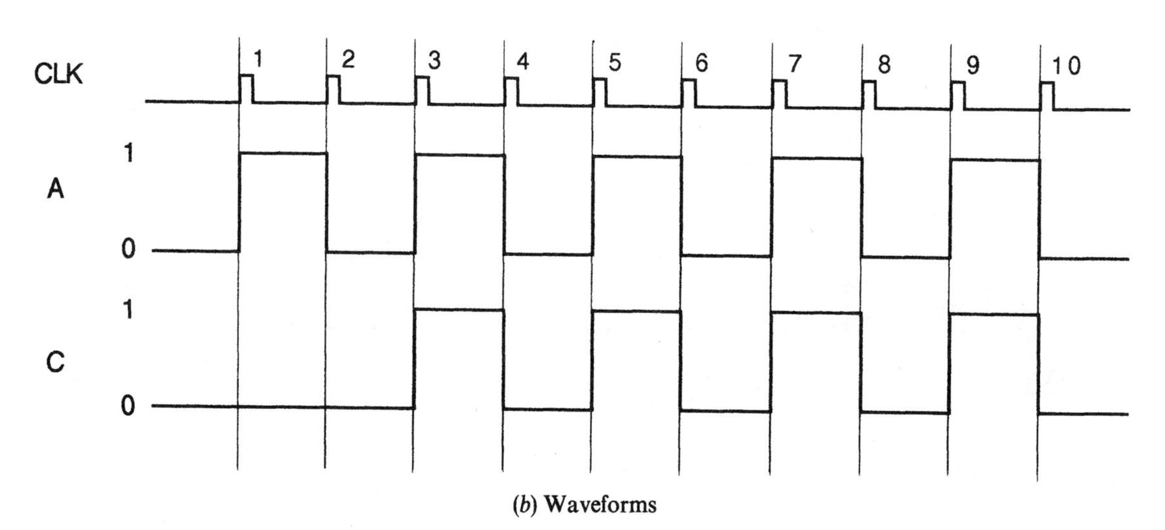

(*b*) Waveforms

Fig. 8-15

Table 8.3

Pulse	A	B	C	JA(= C')	KA	JB(=A)	KB	JC(=B)	KC(=B')
0	0	0	0	1	1	0	1	0	1
1	1	0	0	1	1	1	1	0	1
2	0	1	0	1	1	0	1	1	0
3	1	0	1	0	1	1	1	0	1
4	0	1	0	1	1	0	1	1	0

Repeats steps 3 and 4 endlessly

8.3 If the flip-flops in the circuit of Fig. 8-16*a* are initially cleared and trigger on 1-to-0 clock transitions, sketch the macrotiming diagram at F for the D waveform given in Fig. 8-16*b*. Assume that the A multiplexer select line is most significant.

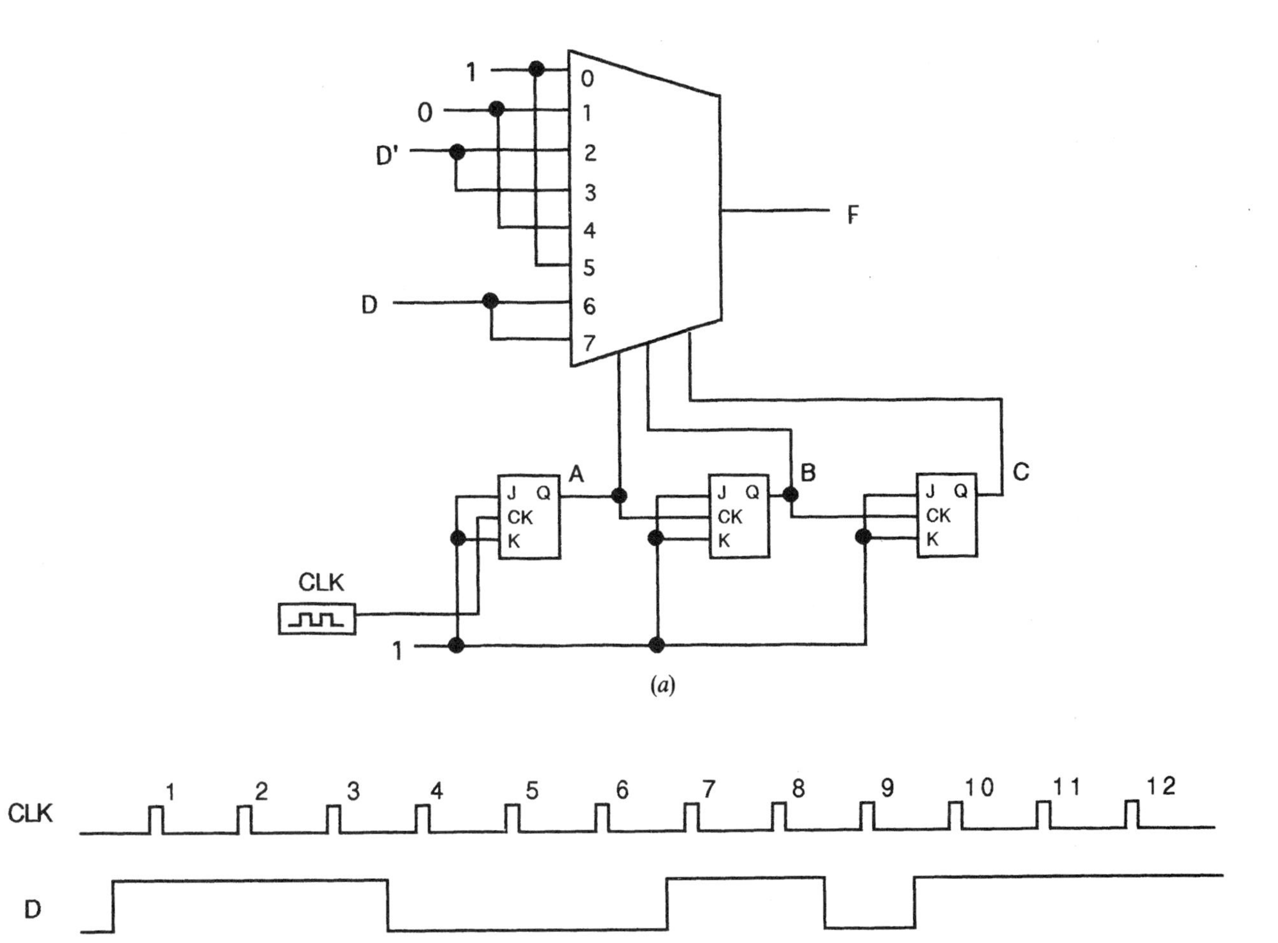

(*a*)

(*b*)

Fig. 8-16

Notice that a 1-to-0 transition of a flip-flop causes the stage to its right to toggle, thus, creating a ripple counter as described in Sec. 8.3. The sequence of states is shown in Table 8.4 from which the desired waveform of Fig. 8-17 is derived.

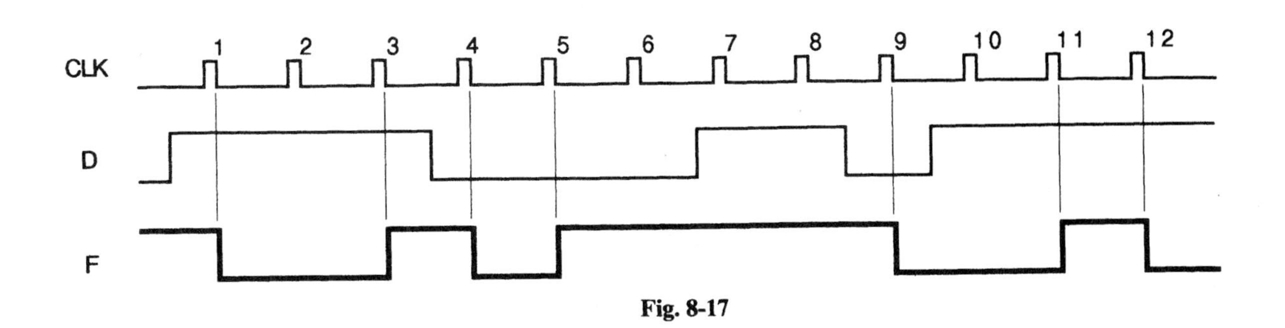

Fig. 8-17

Table 8.4

PULSE	D	SELECTS A B C	MUX INPUT	MUX OUTPUT (F)
1	1	1 0 0	4 (0)	0
2	1	0 1 0	2 (D')	0
3	1	1 1 0	6 (D)	1
4	0	0 0 1	1 (0)	0
5	0	1 0 1	5 (1)	1
6	0	0 1 1	3 (D')	1
7	1	1 1 1	7 (D)	1
8	1	0 0 0	0 (1)	1
9	0	1 0 0	4 (0)	0
10	1	0 1 0	2 (D')	0
11	1	1 1 0	6 (D)	1
12	1	0 0 1	1 (0)	0

8.4 The circuitry shown in Fig. 8-18 is a counter. Assuming that it is initially cleared, find the counting sequence. (Note that material on state machines in Chap. 10 is relevant.)

From the logic diagram, the equations for the flip-flop D inputs are found to be

$$D_W = W'$$
$$D_X = X \oplus W + Y \oplus Z = XW' + X'W + YZ' + Y'Z$$
$$D_Y = WXY' + X'Z + W'Z$$
$$D_Z = W'Z + Y'Z' + X'Y$$

Mapping these equations yields Fig. 8-19.

A given WXYZ coordinate for all the maps may be interpreted as a count state, while individual map entries represent D inputs to the corresponding flip-flops. Initially, the counter is at 0000. This means that the four D inputs are read from the upper-left corner of the maps yielding, in this case, 1001. This tells us that, after clocking, the flip-flop outputs will assume the combined state WXYZ = 1001. The map coordinates now move to 1001 (bottom row, second column) where the entries are WXYZ = 0110. The process continues, yielding Table 8.5.

Table 8.5

Pulse	W	X	Y	Z
0	0	0	0	0
1	1	0	0	1
2	0	1	1	0
3	1	1	0	0
4	0	0	1	1
5	1	0	1	1
6	0	1	1	1
7	1	1	1	1

Repeats

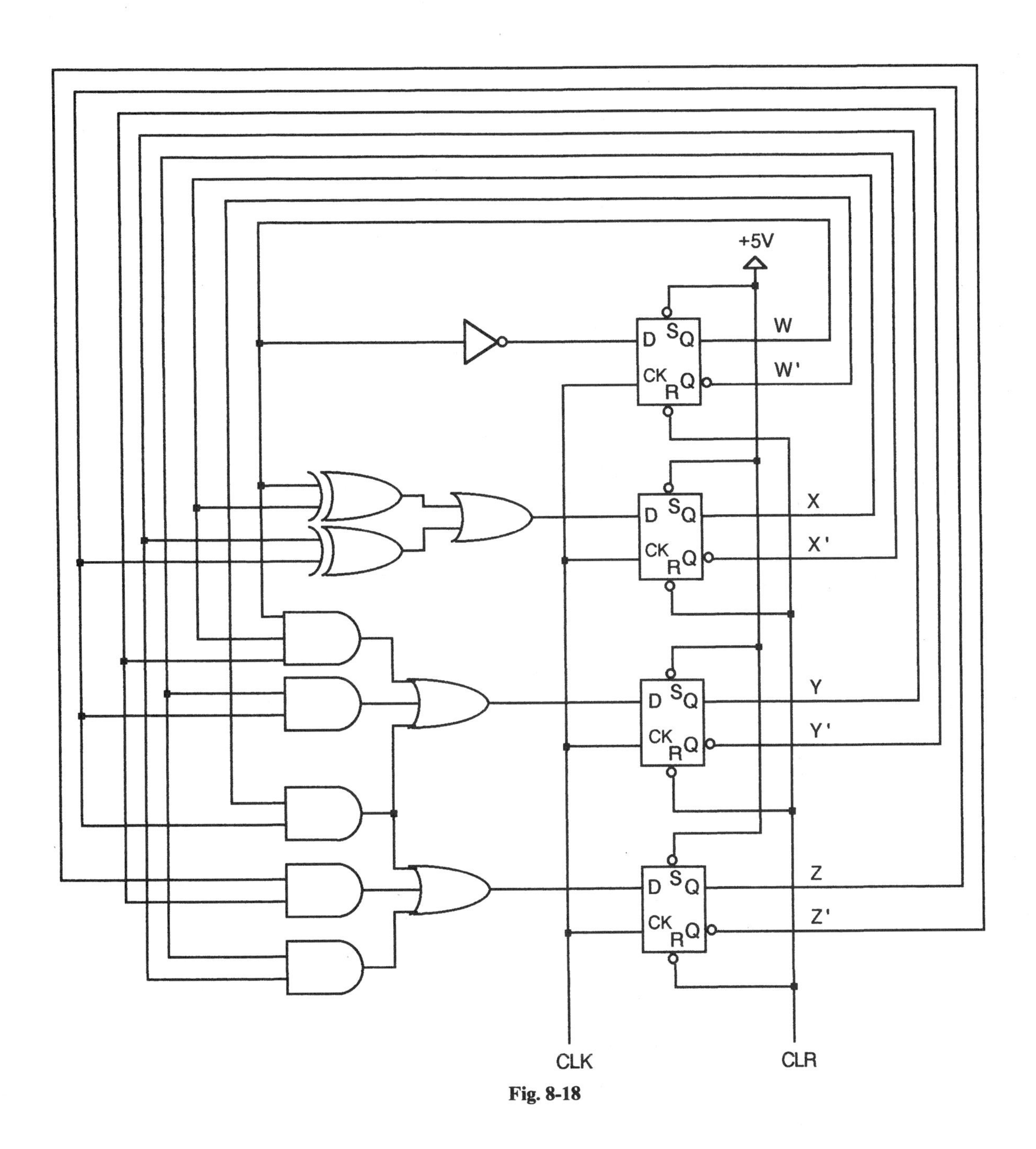

Fig. 8-18

8.5 We have previously demonstrated the glitch problem that can arise in a ripple counter with a recognition gate (Fig. 8-7). The circuit is reproduced in Fig. 8-20 for convenience.

Show that for the case where the flip-flops are toggled by 1-to-0 clock transitions, a recognition gate wired to respond to an odd number cannot suffer from a glitch.

An odd number is obtained each time the least significant bit (D) changes to 1. Since this is a 0-to-1 transition, it will not initiate a transition in flip-flop C or, consequently, in any other flip-flops. Thus, no ripple will occur and, hence, no glitches.

8.6 Referring to Prob. 8.5, under what circumstances will there be the possibility of a glitch in the output of a recognition gate set to respond to count state 0000?

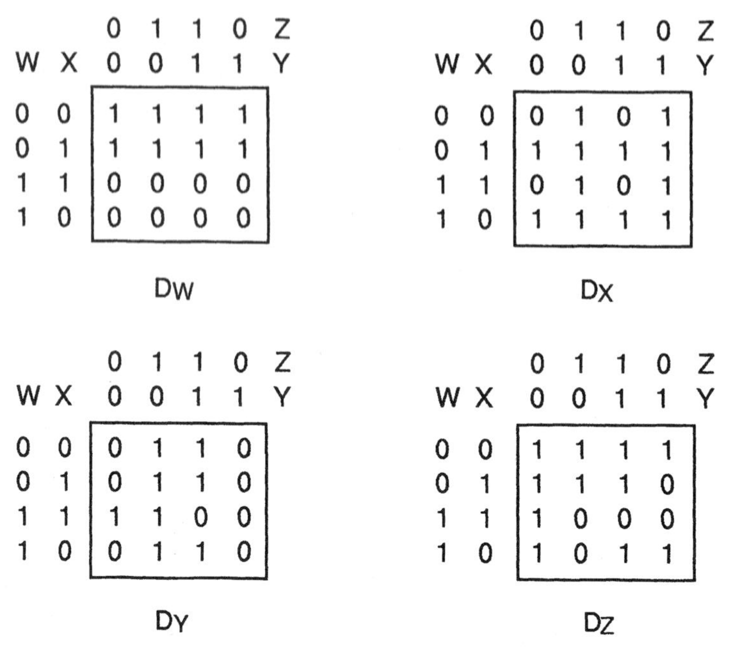

Dw

WX \ ZY	00	10	11	01
00	1	1	1	1
01	1	1	1	1
11	0	0	0	0
10	0	0	0	0

Dx

WX \ ZY	00	10	11	01
00	0	1	0	1
01	1	1	1	1
11	0	1	0	1
10	1	1	1	1

Dy

WX \ ZY	00	10	11	01
00	0	1	1	0
01	0	1	1	0
11	1	1	0	0
10	0	1	1	0

Dz

WX \ ZY	00	10	11	01
00	1	1	1	1
01	1	1	1	0
11	1	0	0	0
10	1	0	1	1

Fig. 8-19

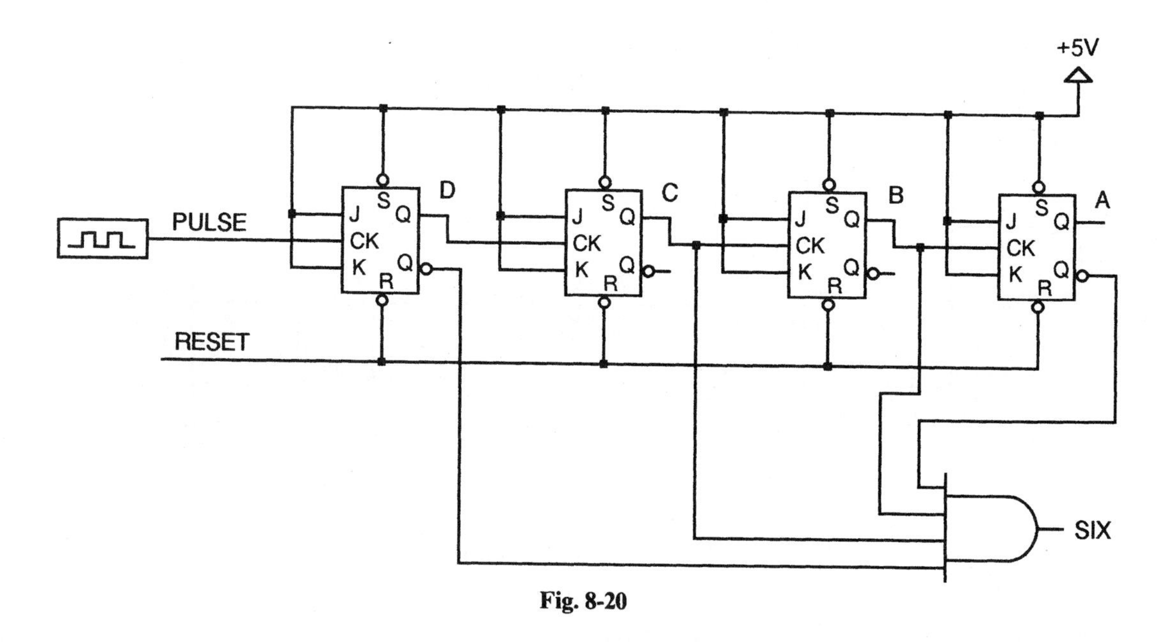

Fig. 8-20

Consider when rippling can occur. As demonstrated previously, the creation of an odd number causes no ripple. We therefore investigate the creation of an even number, which must, of course, immediately follow an odd number. All these odd-to-even transitions are shown in detail, step by step, in Table 8.6. Glitch states of 0000 are indicated with an (X).

8.7 The system shown in Fig. 8-21 is intended to be a waveshape generator. Assuming that the clear and clock pulses occur in the order shown, that triggering occurs on the leading edge, and that A is the most significant select bit, sketch the waveshapes at the flip-flop outputs (macrotiming).

Table 8.6

1 to 2	3 to 4	5 to 6	7 to 8
ABCD	ABCD	ABCD	ABCD
0001	0011	0101	0111
0000 (X)	0010	0100	0110
0010	0000 (X)	0110	0100
	0100		0000 (X)
			1000

9 to 10	11 to 12	13 to 14	15 to 0
ABCD	ABCD	ABCD	ABCD
1001	1011	1101	1111
1000	1010	1100	1110
1010	1000	1110	1100
	1100		1000
			0000 Desired gate output

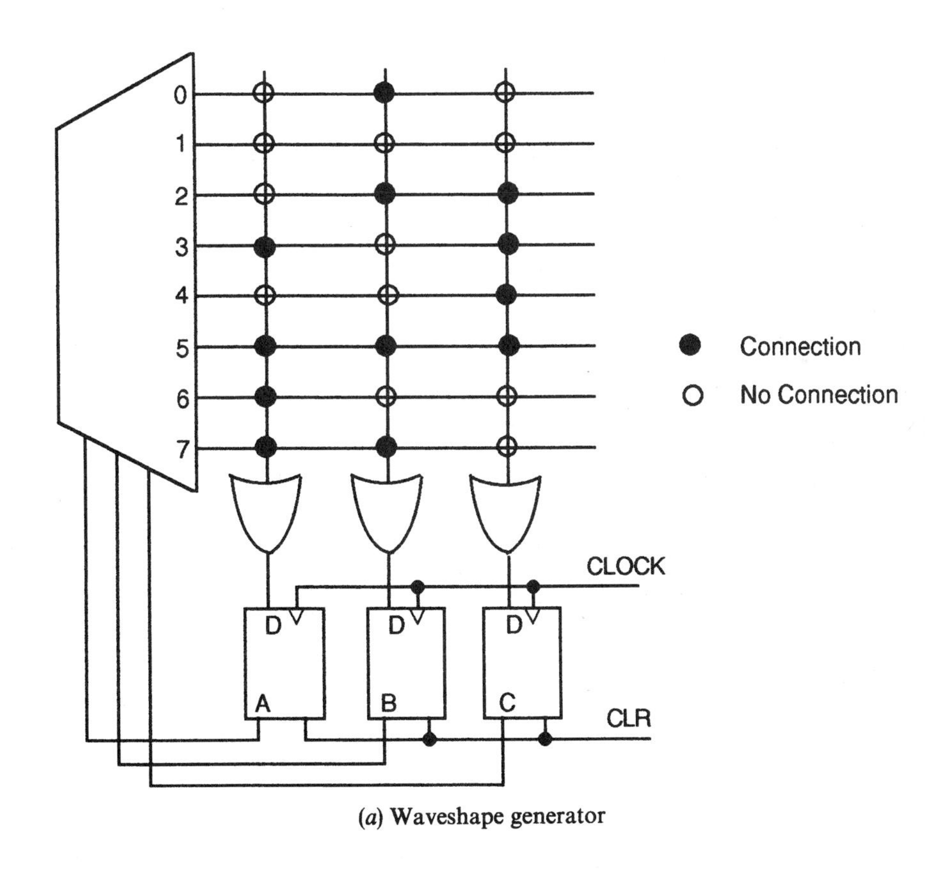

(*a*) Waveshape generator

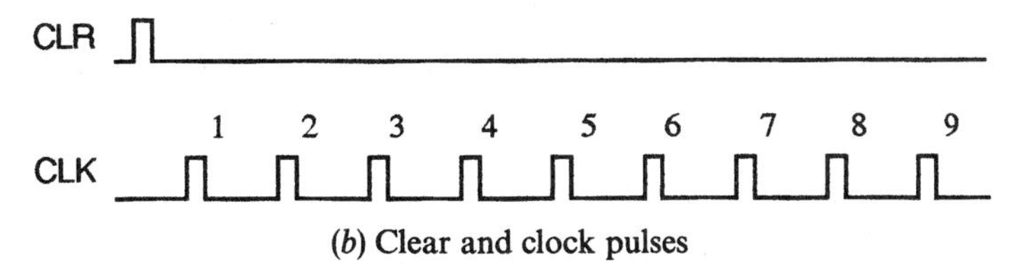

(*b*) Clear and clock pulses

Fig. 8-21

Following receipt of the CLR pulse, the ROM address will be 000 which selects line 0. Data stored at the selected address is connected to the D inputs of the flip-flops and specifies what the next state will be after clocking. In the present case, the ROM has 010 stored at address 000 indicating that line 2 will be selected when the next clock pulse arrives. The process continues, yielding the following decimal equivalent of the binary sequence:

$$0 \to 2 \to 3 \to 5 \to 7 \to 6 \to 4 \to 1 \to 0$$

The required waveforms are easily constructed in vertical strips, assigning 0 and 1 levels corresponding to the binary numbers representing each state in the sequence as shown in Fig. 8-22.

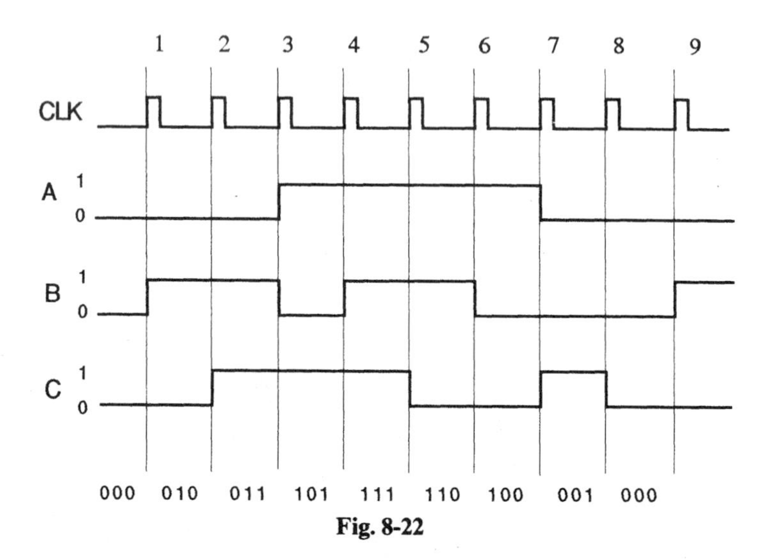

Fig. 8-22

8.8 Ignoring the recognition gate, consider a four-stage binary ripple counter of the type shown in Fig. 8-20. Suppose there are N stages and each flip-flop has a HIGH-to-LOW propagation delay of 20 ns and a LOW-to-HIGH propagation delay of 10 ns (refer to Prob. 7.1). If each distinct count must be held for 50 ns, find an expression for the maximum rate at which pulses may be counted. Assume that the counter is allowed to roll over; that is, it can go from its maximum count back to 0.

The longest time required for the reaction to a clock pulse occurs when the response must ripple through every stage. This occurs when the count goes from its maximum value ($2^N - 1$) to 0 as seen in Table 8.6. Since, in this case, all flip-flops experience a 1-to-0 transition, it will take 20N ns for the 0 count to appear in final form. The next count will be 1, and only the first stage (LSB) will change. Since this will be a 0-to-1 transition, it will require 10 ns to complete. A step-by-step analysis follows:

1. Counter initially reads $2^N - 1$.
2. Next pulse enters counter.
3. Pulse ripples through the counter, requiring 20N ns for completion.
4. Since the count of 0 must be displayed for 50 ns, as per specification, 40 ns (50 less the 10 ns delay of the first stage) must be allowed to elapse before the next pulse can be applied.

The minimum time between pulses T_{min} is (20N + 40) ns, so the maximum allowable frequency for N stages is given by

$$F_{max} = \frac{1}{T_{min}} = \frac{1}{20N + 40} \times 10^9 \text{ pulses per second}$$

For N = 4

$$F_{max} = \frac{10^9}{120} = 8.33 \text{ MHz}$$

The relationships may be more easily understood by referring to Fig. 8-23.

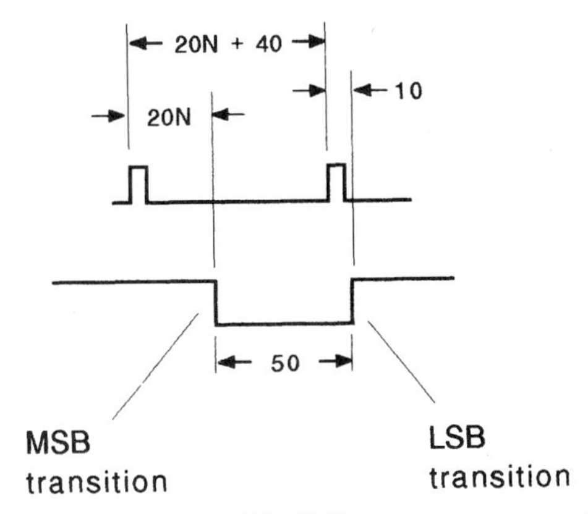

Fig. 8-23

8.9 Three interconnected master-slave JK flip-flops are shown in Fig. 8-24. Assume that they are designed to trigger on the trailing edge of the clock pulse and that they are initially cleared.

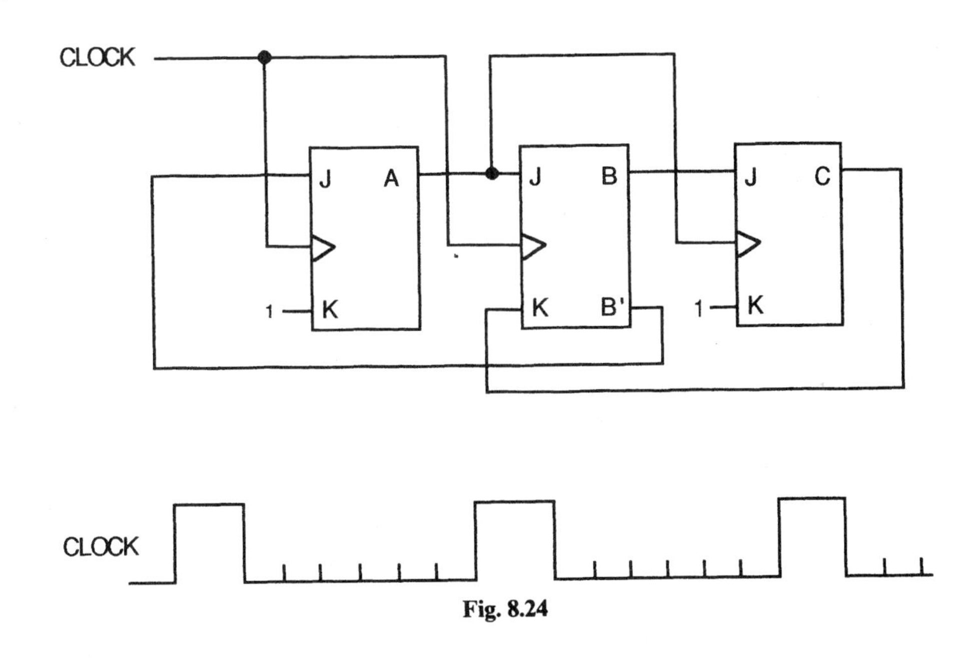

Fig. 8.24

Ignoring propagation delays, determine if the circuit will count and, if it does, determine the counting sequence. If the circuit does not count, determine the state at which a hang-up occurs.

The solution is obtained by determining the status of the flip-flop control signals (J and K) before each pulse as in Prob. 7.9. Note that output A serves as the clock for flip-flop C. The results are shown in Table 8.7.

Table 8.7

Pulse	A	B	C	J_A (=B')	K_A	J_B (=A)	K_B (=C)	J_C(=B)	K_C
0	0	0	0	1	1	0	0	0	1
1	1	0	0	1	1	1	0	0	1
2	0	1	0	0	1	0	0	1	1
3	0	1	0	0	1	0	0	1	1

The circuit hangs up after the second clock pulse since the JK inputs to flip-flop A call for A = 0 (where it already is), the 00 JK inputs to flip-flop B indicate status quo, and there is no clock transition at flip-flop C since A has not changed.

8.10 Again assuming initial clearing, sketch a microtiming diagram for the outputs of flip-flops B and C in Prob. 8.9. Assume that each flip-flop has 1 unit of delay corresponding to the tick marks in the waveform of Fig. 8-24.

The waveshapes can be inferred from Table 8.7 where we see that C remains at logic 0 and a change occurs in output B at the second clock pulse. This transition occurs one delay interval following the clock pulse trailing edge as shown in Fig. 8-25.

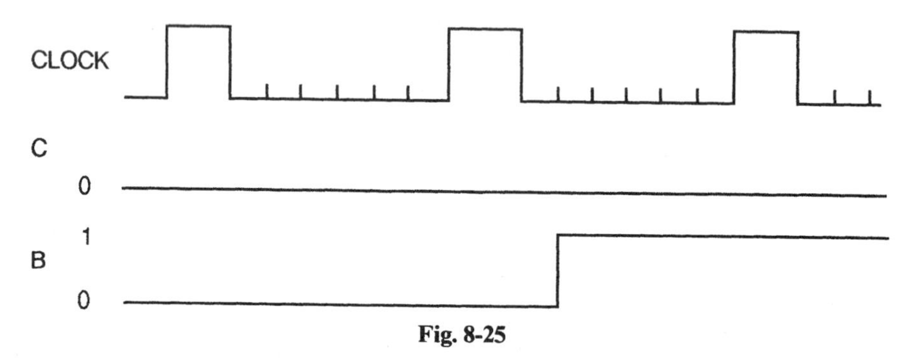

Fig. 8-25

8.11 Assuming that the flip-flops in Fig. 8-26 are initially cleared, determine the counting sequence (sequential states of outputs ABC, as clock pulses are applied).

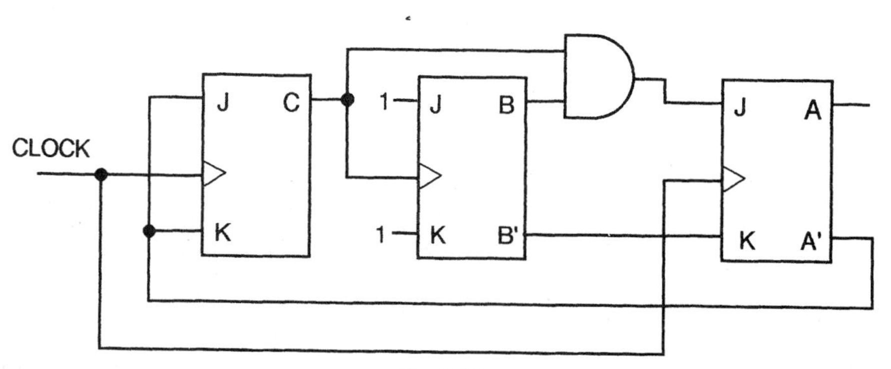

Fig. 8-26

Assume that the circuit responds to clock pulse trailing edges. As in Prob. 8.2, tabulate the JK inputs after each pulse to determine the following state of each flip-flop (Table 8.8). Since flip-flop B is triggered from C, B will only change state following a 1-to-0 transition at output C, regardless of its JK inputs. Therefore, the state of C must be analyzed first. The circuit counts from 0 to 4 and repeats.

Table 8.8

Pulse	A	B	C	J_A(= BC)	K_A(=B')	J_B	K_B	J_C(=A')	K_C(=A')
0	0	0	0	0	1	1	1	1	1
1	0	0	1	0	1	1	1	1	1
2	0	1 ←	0	0	0	1	1	1	1
3	0	1	1	1	0	1	1	1	1
4	1	0 ←	0	0	1	1	1	0	0
5	0	0	0						

It is interesting to note that if we attempt to start the counter without resetting to 000, there are *two initial states where the circuit is essentially locked and will not respond to clock pulses.* The first such state is ABC = 110. In this case, JK = 00 for flip-flops A and C (no change will occur), and, since B cannot toggle unless there is an output from C, the circuit will remain static. The second case is ABC = 111. Here, flip-flop A's inputs are JK = 10 which programs for a state the flip-flop is already in. Flip-flop C has JK = 00, indicating that no change will occur in C, and consequently, in B also so that the circuit will not start to count.

The only remaining state outside of the normal count progression shown in Table 8.8 is ABC = 101. It is easily shown that starting from this state, the circuit will enter its normal sequence at 001 after one clock pulse. The circuit's sequential counting behavior may be diagrammed as shown in Fig. 8-27.

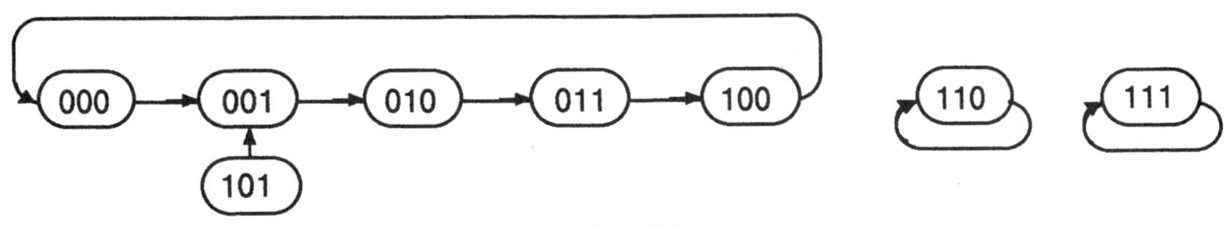

Fig. 8-27

8.12 A term used in conjunction with the output of rate multipliers is *asynchronism* which refers to the ratio of the maximum to minimum separation between pulses. Find the asynchronism of the rate multiplier of Fig. 8-8 when it is set for multiplications of 5/16 and 11/16, respectively.

When the ratio is 5/16, G_1 and G_4 will be active and a G_1 pulse will be inserted among the G_4 pulses. Referring to Fig. 8-28, it is seen that the G_1 pulse will occur exactly halfway between a pair of G_4 pulses. If the pulse period is T, the spacing between G_4 pulses is 4T and the spacing between a G_1 pulse and an adjacent G_4 pulse is half of that, or 2T. Thus, the asynchronism is 4T/2T = 2.

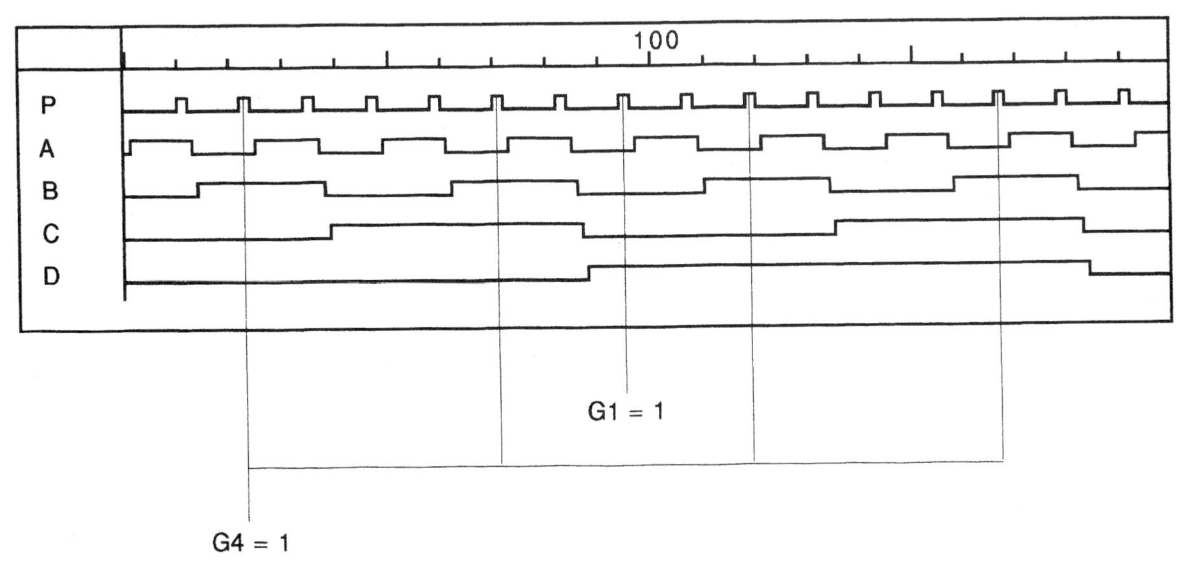

Fig. 8-28

For a multiplication of 11/16, the pulses come from G_8 with added pulses interspersed from G_2 and G_1. The greatest spacing is between two consecutive G_8 pulses, or 2T; the lesser spacing is the distance between two consecutive input pulses, or T. Again, the asynchronism is 2.

8.13 Consider the latching register of Fig. 8-2. Suppose that the data on input lines A to D are changing at a rate of 2 MHz, while the register is clocked at a 10-MHz rate. Compute the mean time between metastable failures using $T_W = 0.5$ s and $T_R = 0.75 \times 10^9\ s^{-1}$. Repeat for the case where T_R is decreased by an order of magnitude.

Using the equation for MTBF from Sec. 7.10,

$$\text{MTBF} = \frac{1}{F_{clk} \times F_{data} \times T_W \times \exp(-T_R\, \Delta t)}$$

where $F_{clk} = 10^7\ s^{-1}$

$F_{data} = 2 \times 10^6\ s^{-1}$

$\Delta t = 10^{-7}$ s (the clock period)

MTBF $= 3.73 \times 10^{19}$ s and, for all practical purposes, there is no metastable behavior.

In the second case, $T_R = 0.75 \times 10^8\ s^{-1}$, MTBF $= 1.8 \times 10^{-10} = 0.18$ ns and, metastable behavior is now a serious problem. Note the sensitivity of the failure rate to the technology-dependent parameter T_R.

8.14 The circuit shown in Fig. 8-29 is called a pseudo-random binary sequence (PRBS) generator. Prior to any clock pulses, a single negative-going initializing pulse is applied. After clocking begins, show that the state of the register changes through all possible states (except 0000) before repeating. The generation of such a sequence determines that the circuit functions as a *maximal-length* PRBS generator.

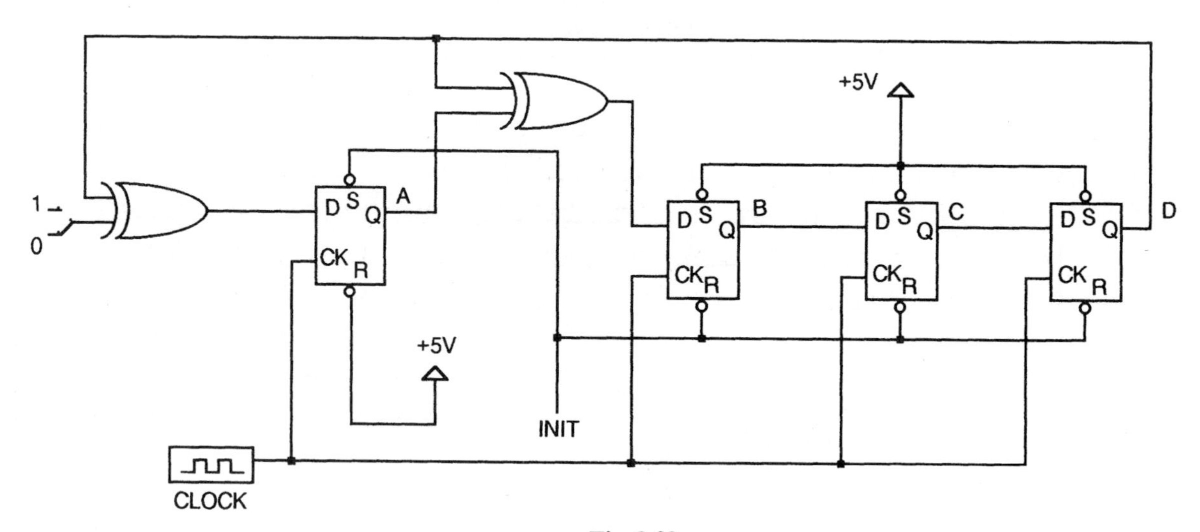

Fig. 8-29

The initializing pulse sets flip-flop A and resets the rest so that the initial state is ABCD = 1000. Before each clock pulse, we examine the XOR gate outputs and determine the values at the D inputs as follows:

$$D_A = Q_D$$
$$D_B = Q_A \oplus Q_D$$
$$D_C = Q_B$$
$$D_D = Q_C$$

These become the flip-flop values after each clock pulse, resulting in the sequence listed in Table 8.9. Note, for example, that following the third clock pulse, the logic 1 at the output of flip-flop D causes inputs D_A and D_B to equal 1 and that these values are transferred to outputs A and B at pulse 4. Following pulse 7, where A and D are both equal to logic 1, gate G_1 has a 0 output which is reflected in the 0 value of B at pulse 8.

Table 8.9

Pulse	A	B	C	D	Pulse	A	B	C	D
0	1	0	0	0	8	1	0	1	0
1	0	1	0	0	9	0	1	0	1
2	0	0	1	0	10	1	1	1	0
3	0	0	0	1	11	0	1	1	1
4	1	1	0	0	12	1	1	1	1
5	0	1	1	0	13	1	0	1	1
6	0	0	1	1	14	1	0	0	1
7	1	1	0	1	15	1	0	0	0

Since fifteen pulses elapse before a repeat occurs, the circuit is a maximal-length PRBS generator. A simulation timing diagram is shown in Fig. 8-30 where an arbitrary state ABCD = 0100 has been chosen to illustrate reset action.

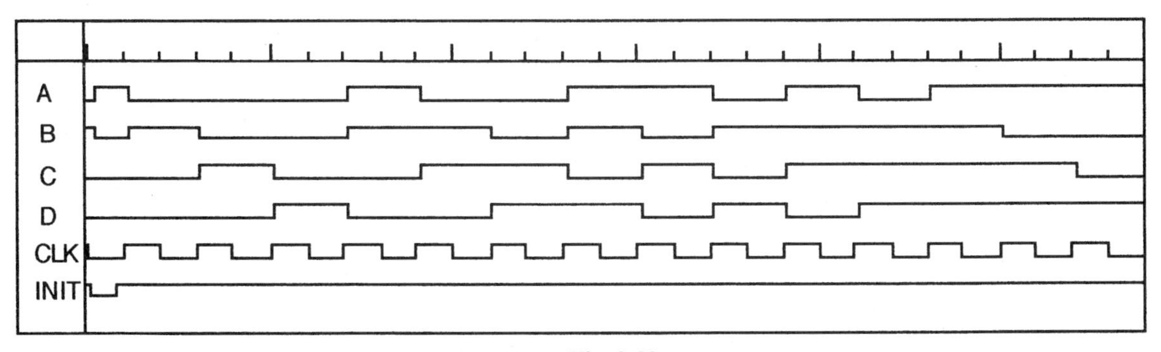

Fig. 8-30

8.15 The circuit shown in Fig. 8-31 represents an alternate form of a PRBS generator. The mod-2 adder may be considered a circuit which checks the parity of its inputs and produces a 1 at its output when the parity is odd. The shift register is not shown in detail; instead, its five stages are represented by blocks from which the signals connected to the AND gate and fed back to the adder are assumed to be taken from the Q outputs of the flip-flops in the stages involved.

If the register is initially loaded with 10000, show that the circuit functions as a maximal-length PRBS generator, as defined in Prob. 8.14, and that the output of the AND gate produces 1s with a probability equal to 8/31.

Denote the shift register stages from left to right by ABCDE. The status after each clock pulse is shown in Table 8.10

Since it takes 31 clock cycles to repeat a sequence, the circuit is seen to be a maximal-length PRBS generator, and, since there are eight 1s produced by the 31 clock pulses, the probability of getting a 1 at the output is 8/31.

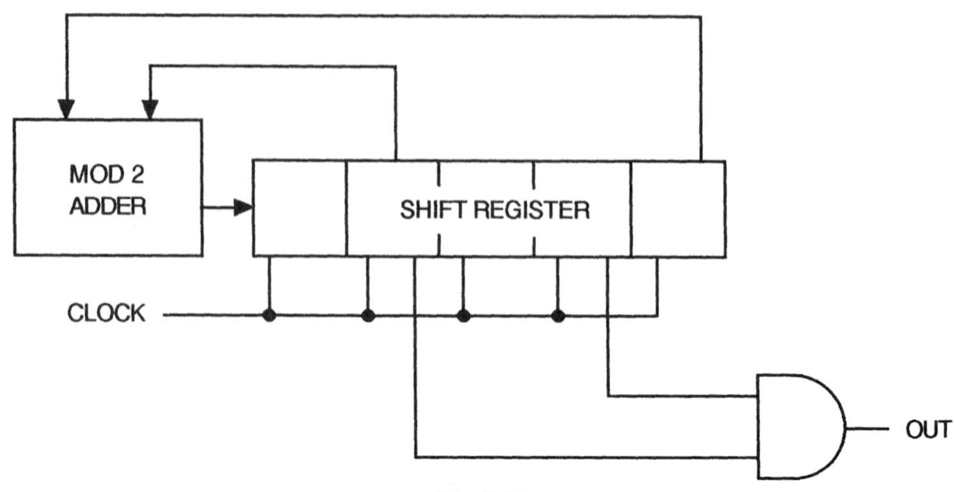

Fig. 8-31

Table 8.10

Clock	ABCDE	Output	Clock	ABCDE	Output
0	10000	0	16	11110	1
1	01000	0	17	11111	1
2	10100	0	18	01111	1
3	01010	1	19	00111	0
4	10101	0	20	10011	0
5	11010	1	21	11001	0
6	11101	0	22	01100	0
7	01110	1	23	10110	0
8	10111	0	24	01011	1
9	11011	1	25	00101	0
10	01101	0	26	10010	0
11	00110	0	27	01001	0
12	00011	0	28	00100	0
13	10001	0	29	00010	0
14	11000	0	30	00001	0
15	11100	0	31	10000	0

Supplementary Problems

8.16 For the binary rate multiplier of Fig. 8-8, determine the output waveforms for $C_1 C_2 C_4 C_8 = 1001$.

8.17 Repeat Prob. 8.16 using $C_1 C_2 C_4 C_8 = 1011$.

8.18 A student decides to use a recognition gate output to reset a ripple counter to 0000 after it reaches a count of six, as shown in Fig. 8-32. The flip-flops chosen trigger on a 1-to-0 clock transition and are cleared by a low reset input. When it is built, the circuit sometimes operates as intended and other times begins, on power-up, at a random count. What is the design flaw?

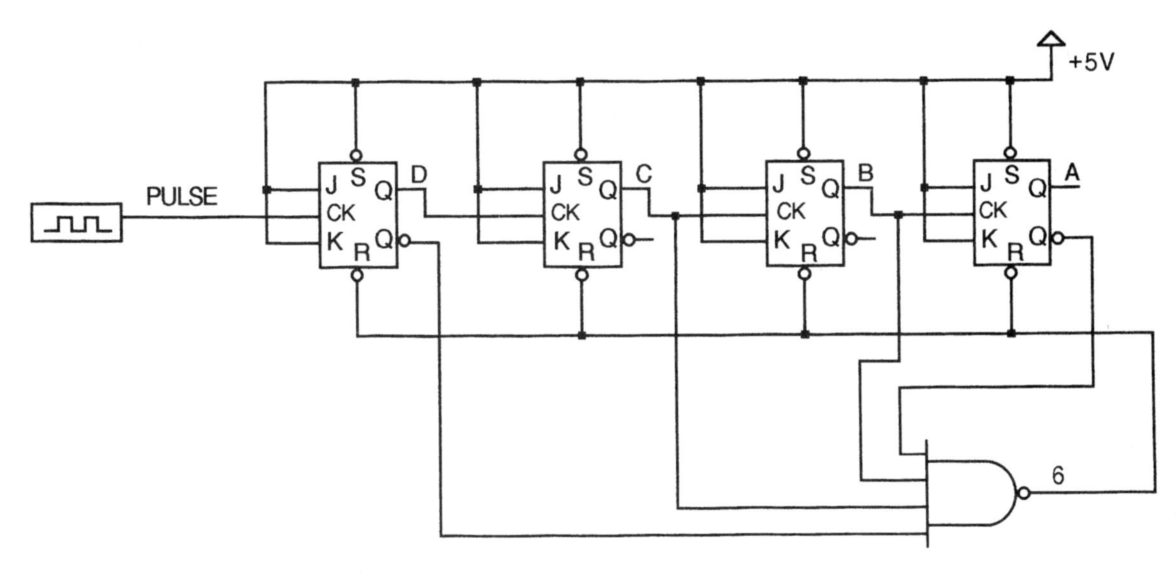

Fig. 8-32

8.19 Redesign the circuit of Prob. 8.18 using an additional reset signal to initialize the counter to 0. Evaluate your design with a microtiming diagram.

8.20 The circuit in Fig. 8-33 is called a cellular automaton. Show that it functions as a maximal-length PRBS generator (see Prob. 8.14).

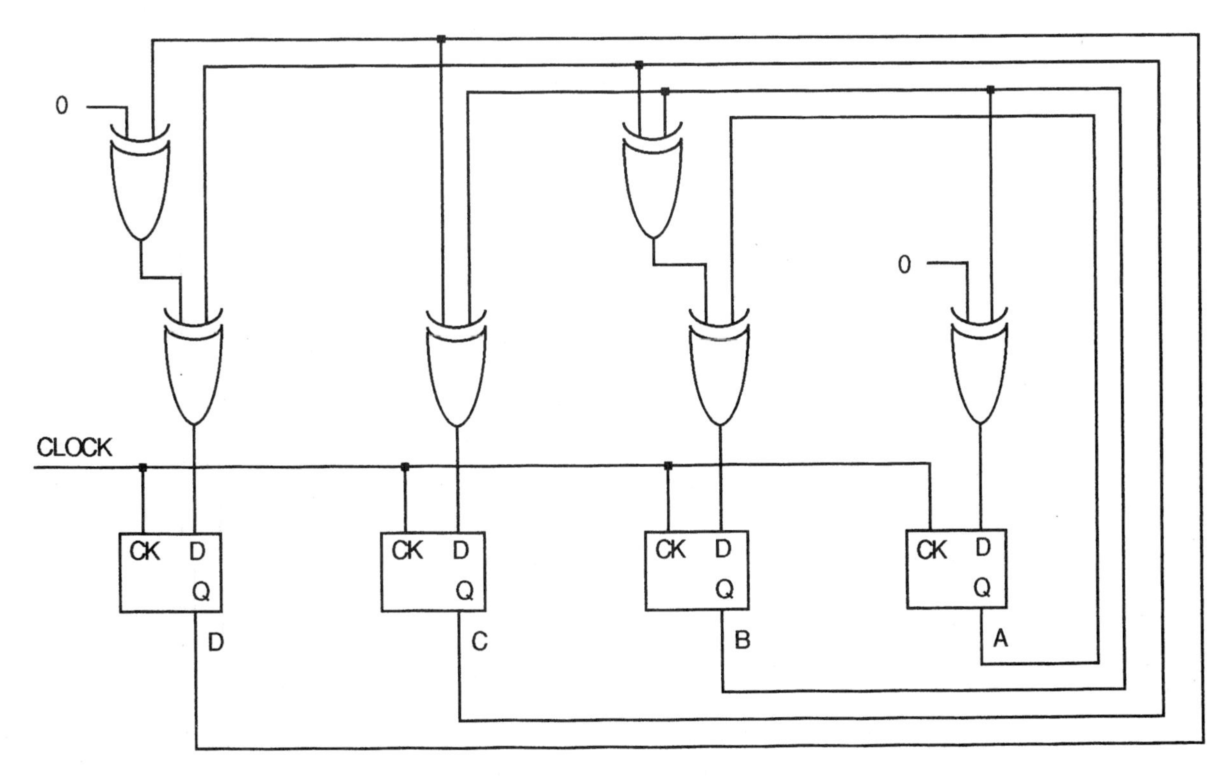

Fig. 8-33

8.21 The arrangement shown in Fig. 8-34 has been previously studied in Prob. 7.8 where we were told that the T flip-flops are negative-edge-triggered and the D flip-flop is positive-edge-triggered. It is required that the circuit be modified so that when ABC = 011, operation will freeze in this state until a manual clear occurs. Implement the toggle flip-flops with JK's and use a two-input recognition gate to achieve the desired performance. Demonstrate operation with a timing diagram.

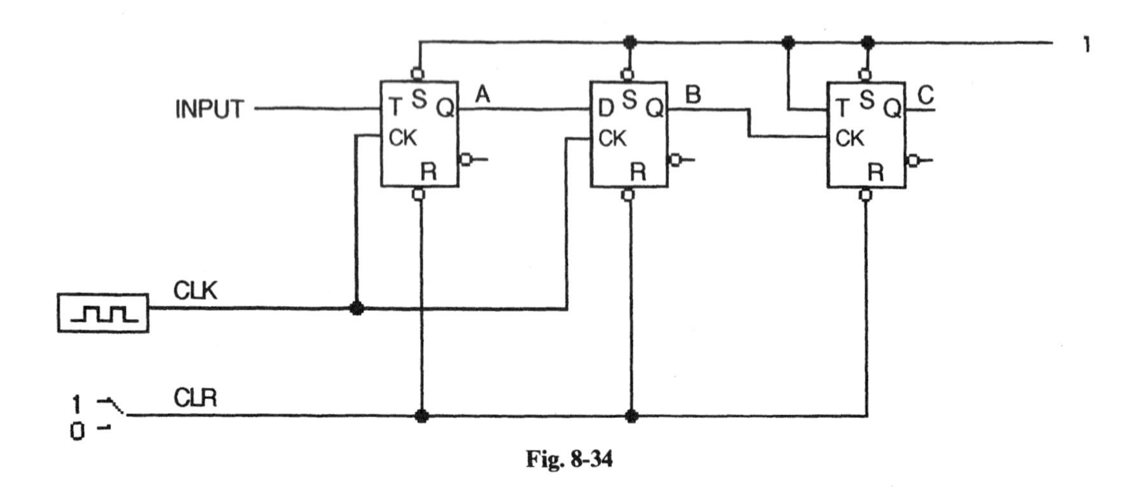

Fig. 8-34

8.22 A student wants to design a circuit which will count from 0 (0000) to 9 (1001) and then reset itself to 0000 at the tenth clock pulse. A ripple counter is combined with a recognition gate and an inverter to achieve this function. Referring to Fig. 8-35, we see that when the count reaches 9, flip-flops A and D are simultaneously 1s for the first time in the sequence and these signals are used to prime an AND gate to steer the next clock pulse to the reset line. Create the microtiming diagram and comment on the circuit's operation.

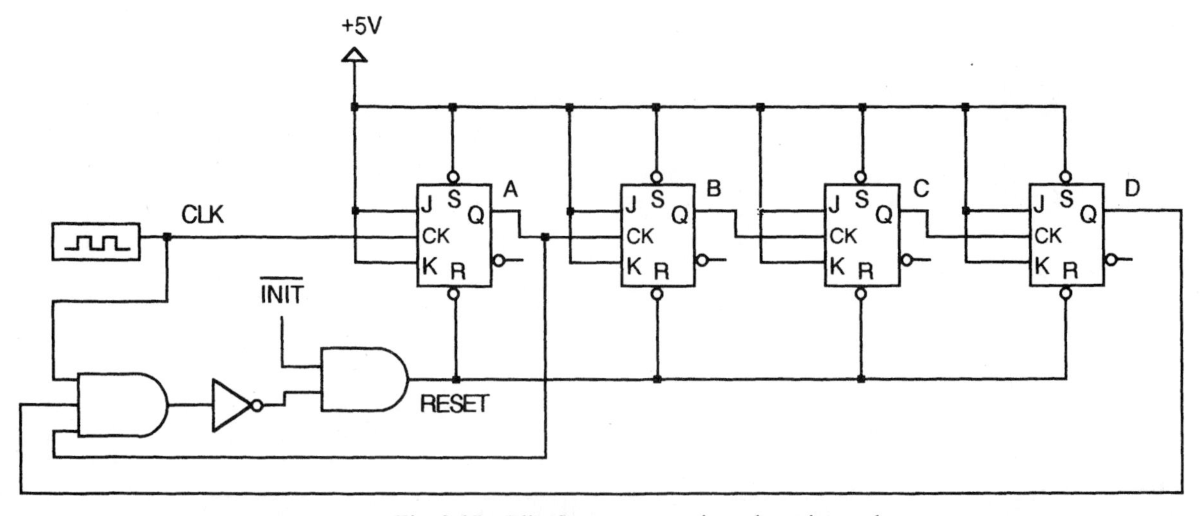

Fig. 8-35 Flip-flops are negative-edge-triggered.

8.23 Modify the diagram of Prob. 8.22 so that it functions as a divide-by-10 circuit in which the reset line serves as an output providing 1 pulse for every 10 clock pulses. Validate your design with a timing chart.

8.24 Modify the circuit of Fig. 8-45 on page 265 to accommodate flip-flops which have set and reset inputs that are active-HIGH rather than active-LOW.

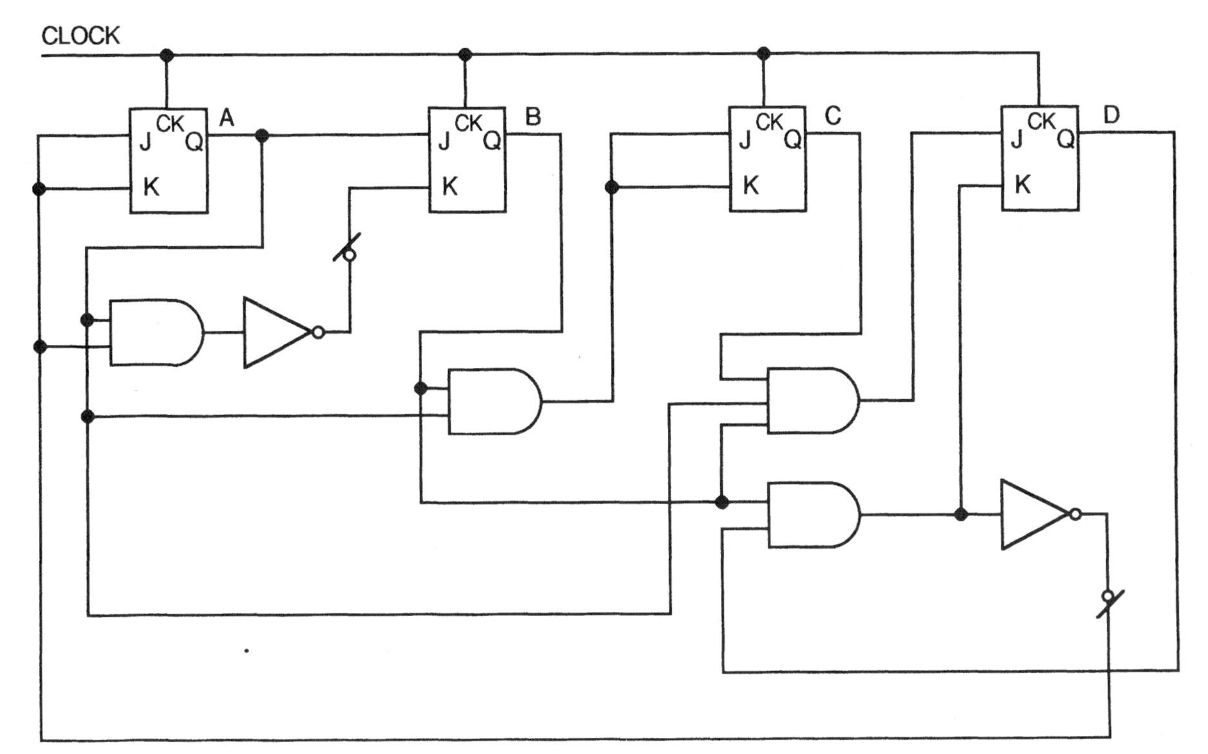

Fig. 8-51

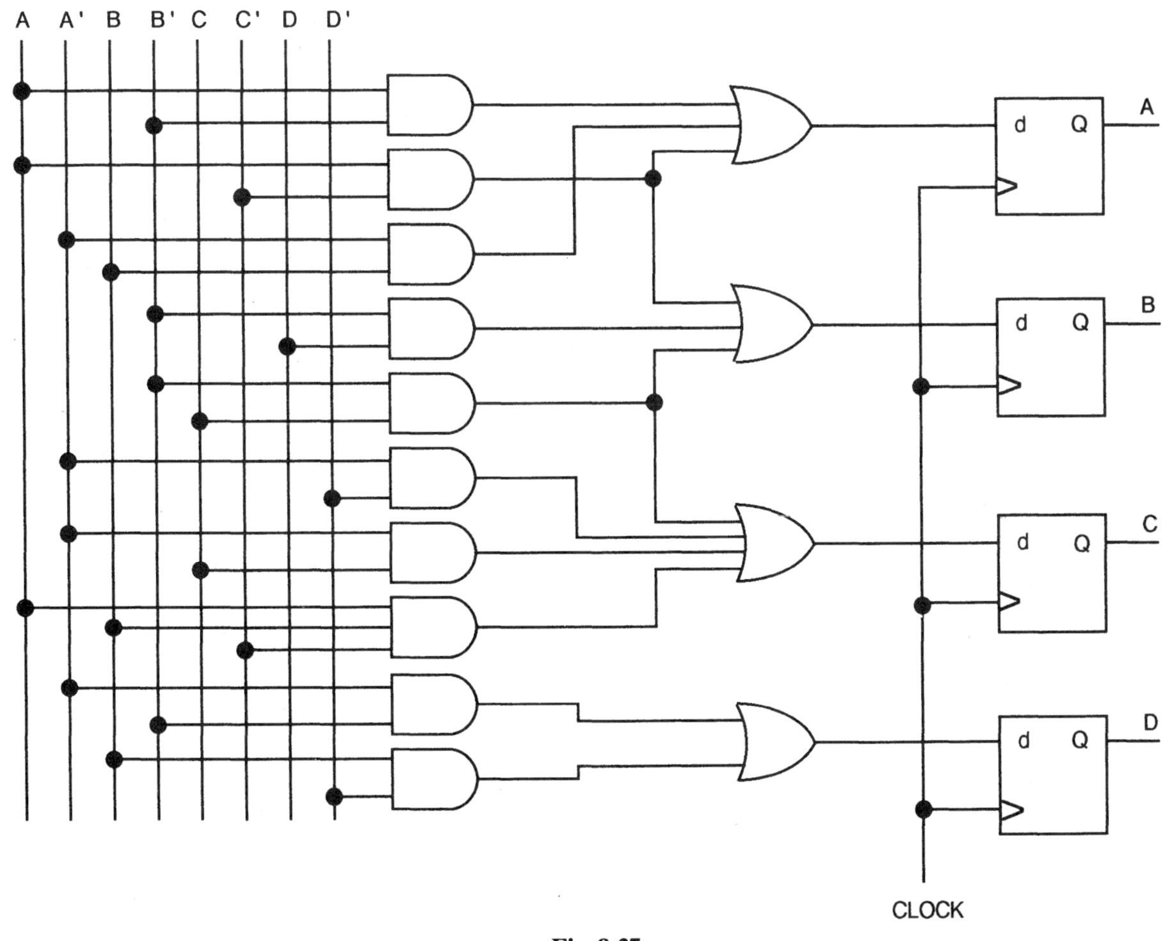

Fig. 8-37

8.25 Does the hardware change in Prob. 8.24 have any significant effect on system timing?

8.26 Design a 10-state Mobius counter and add recognition gates to identify states 3 and 7.

8.27 Validate the operation of the circuit in Fig. 8-49 with a timing diagram drawn for positive-edge-triggered flip-flops. Assume that the counter is initially cleared.

8.28 Design a 10-state binary ripple counter with appropriate recognition gates to identify states 3 and 7.

8.29 State how a chain of JK flip-flops should be interconnected to act as an n-stage shift register that will shift right when control variable R = 1 and left when R = 0.

8.30 Assuming that the circuit shown in Fig. 8-36 is initially cleared to 0000, determine the counting sequence if D is the most significant bit.

8.31 Repeat Prob. 8.30 when the count starts from binary 1111.

8.32 The inputs to a D flip-flop register are driven from logic as shown in Fig. 8-37. Assuming that the flip-flops are cleared at t = 0, determine the counting sequence.

Answers to Supplementary Problems

8.16 See Fig. 8-38 (microtiming diagram).

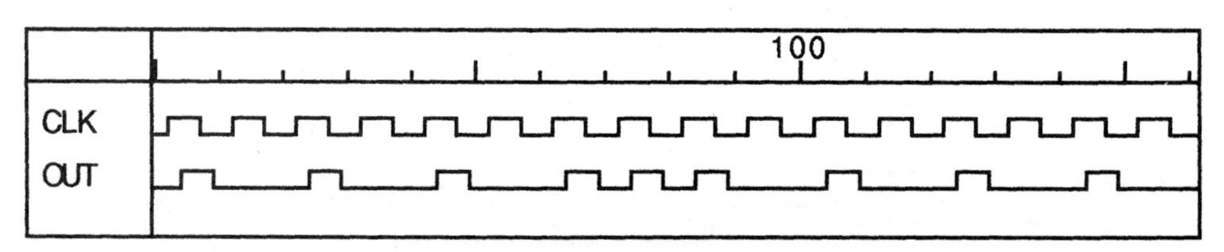

Fig. 8-38

8.17 See Fig. 8-39.

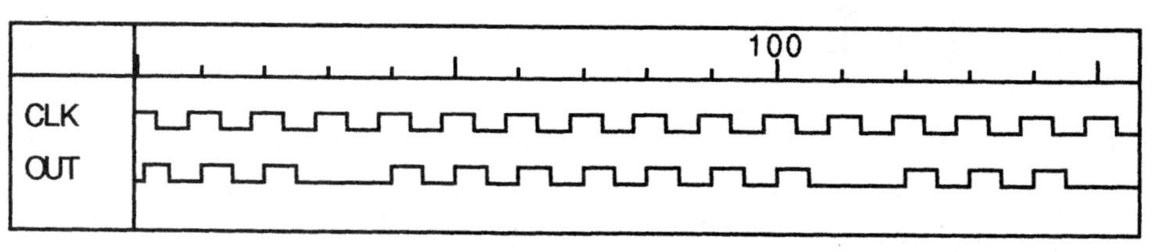

Fig. 8-39

8.18 There is no provision to initialize the circuit and, at power-up, it can assume any count from 0 (0000) to 15 (1111).

8.19 A modified version of the circuit is shown in Fig. 8-40 in which the additions are highlighted. Reset can be forced by either the recognition gate or an initialize signal. The microtiming diagram is obtained by simulation (see Fig. 8-41). Note that the state of 6 exists only long enough for the value to be recognized, propagate through the AND and NOR gates, and reset the ripple counter. This may be a serious problem, particularly at high clock frequencies, and the designer must account for it.

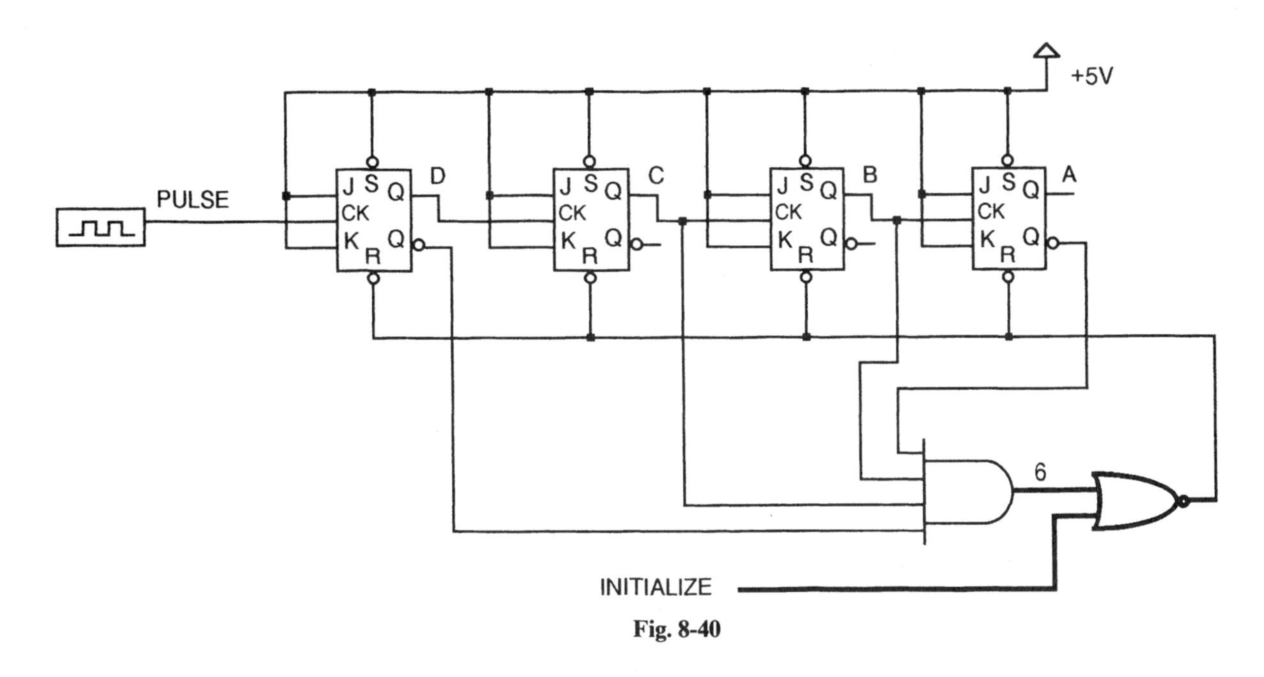

Fig. 8-40

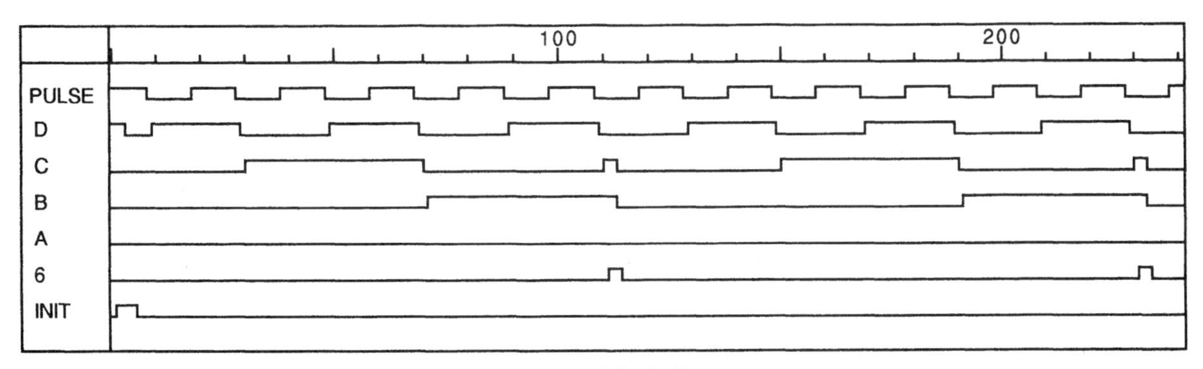

Fig. 8-41

8.20 As a nonzero starting point, assume that ABCD = 1000. The clocked sequence, shown in Table 8.11, takes 15 steps before a repeat occurs, and it is therefore maximal length for four flip-flops.

Table 8.11

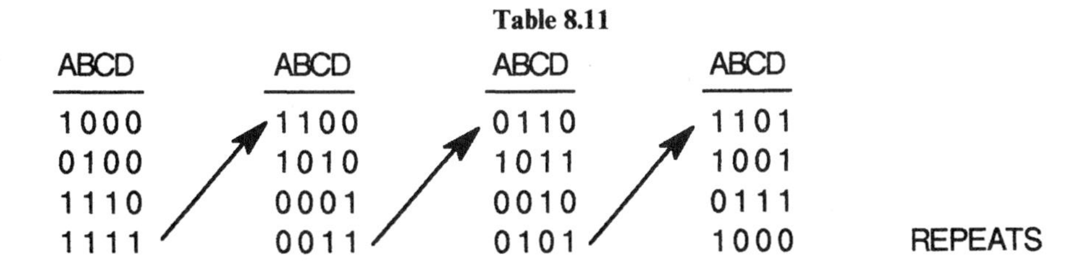

ABCD	ABCD	ABCD	ABCD	
1000	1100	0110	1101	
0100	1010	1011	1001	
1110	0001	0010	0111	
1111	0011	0101	1000	REPEATS

8.21 See Figs. 8-42 and 8.43. Comments on the simulation of this problem can be found in App. C.

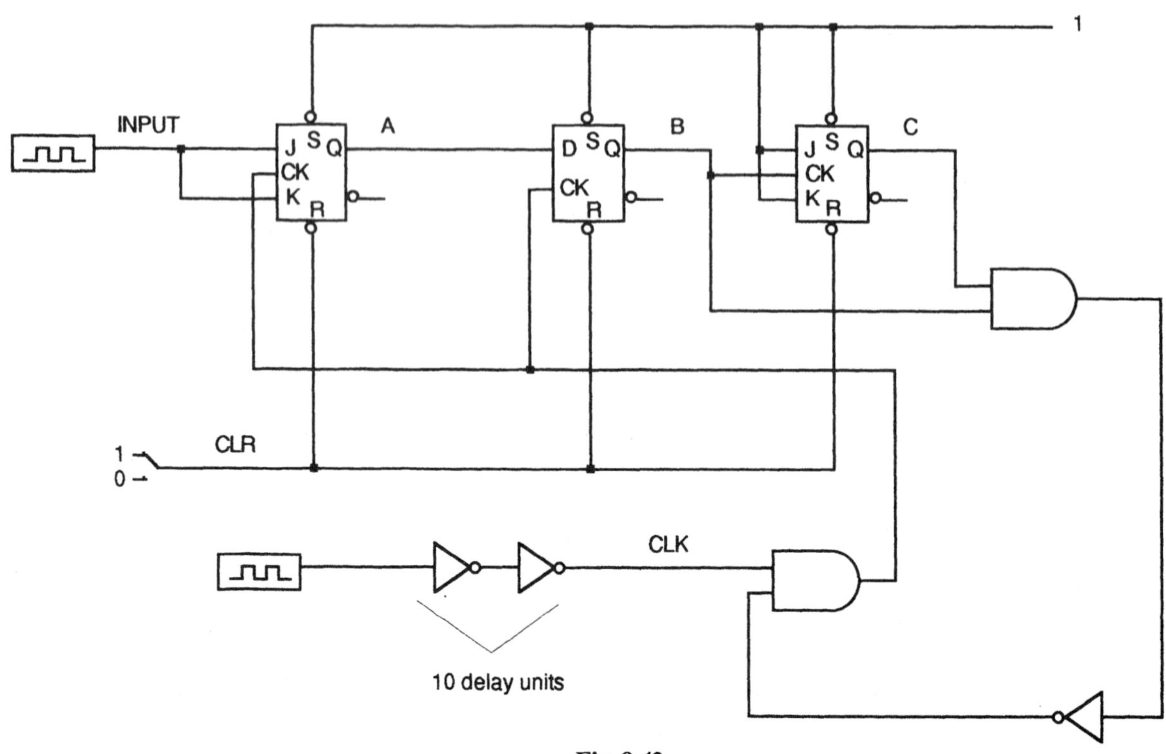

Fig. 8-42

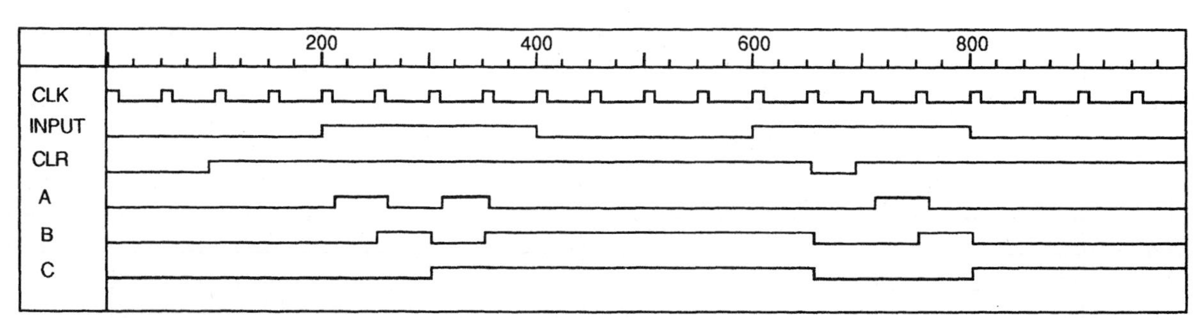

Fig. 8-43

8.22 Refer to Fig. 8-44. The same clock pulse which activates reset also advances the count to 1, skipping the 0 count. Since a reset pulse is generated for every nine clock pulses, the counter makes an excellent divide-by-9 circuit.

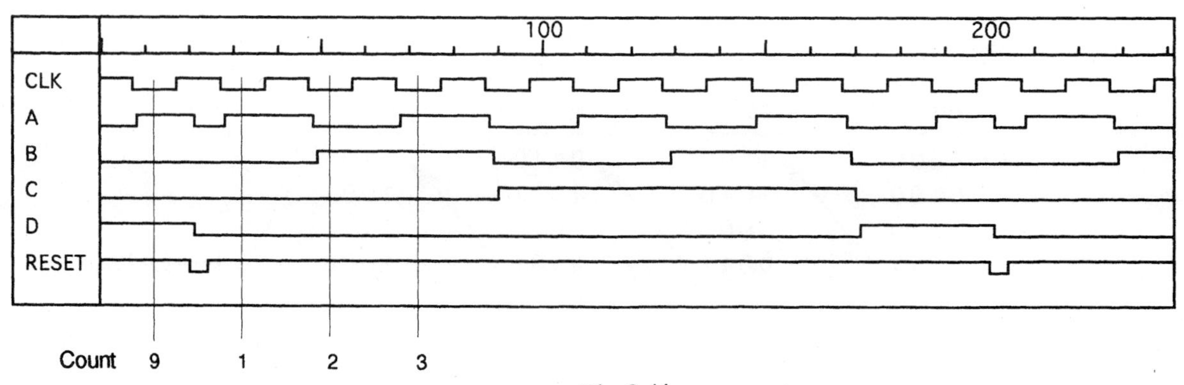

Fig. 8-44

8.23 See Figs. 8-45 and 8-46.

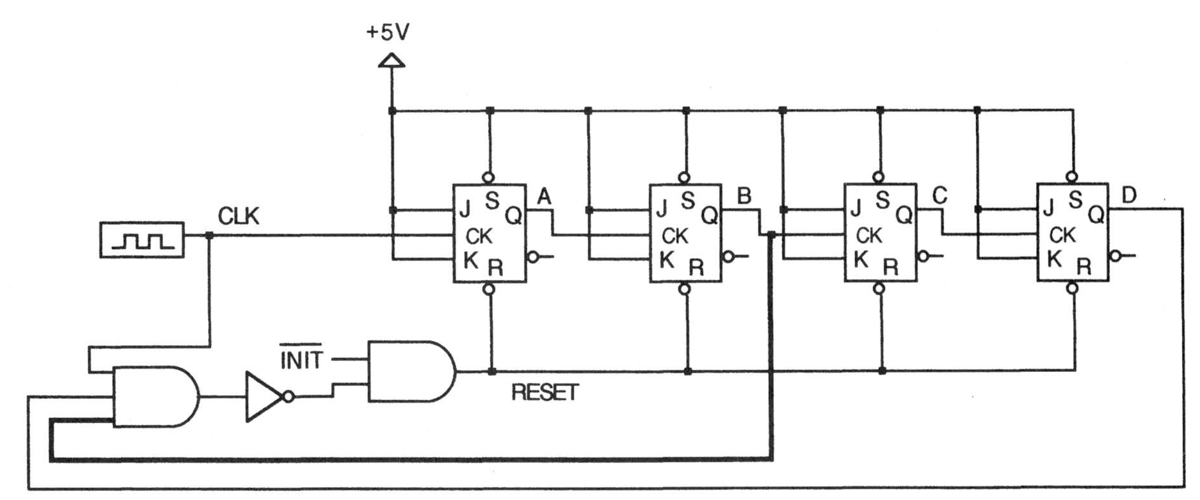

Fig. 8-45

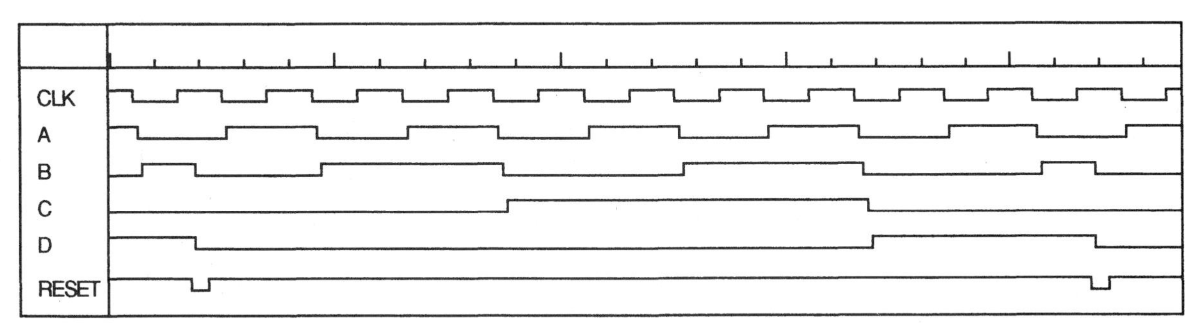

Fig. 8-46

8.24 See Fig. 8-47 and refer to Prob. 4.39.

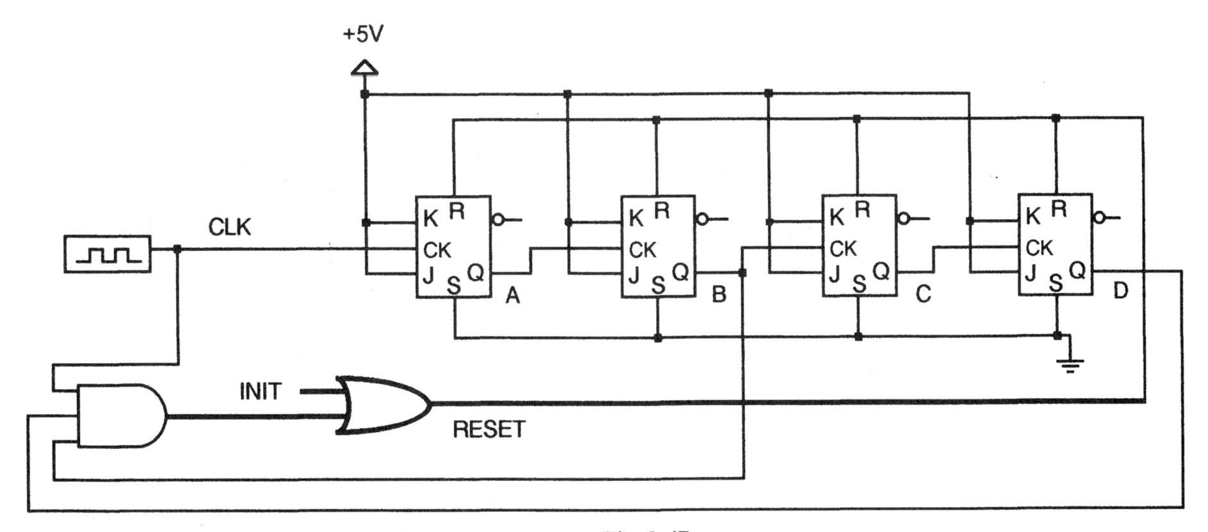

Fig. 8-47

8.25 Comparing the microtiming diagrams of Figs. 8-46 and 8-48, we see that in both cases, the circuits remain in the count of 10 for somewhat longer than half a clock cycle before resetting to 0. The reset pulse is narrower by one propagation-delay interval in the second case, and if clock frequency is increased sufficiently, a metastable problem may arise sooner.

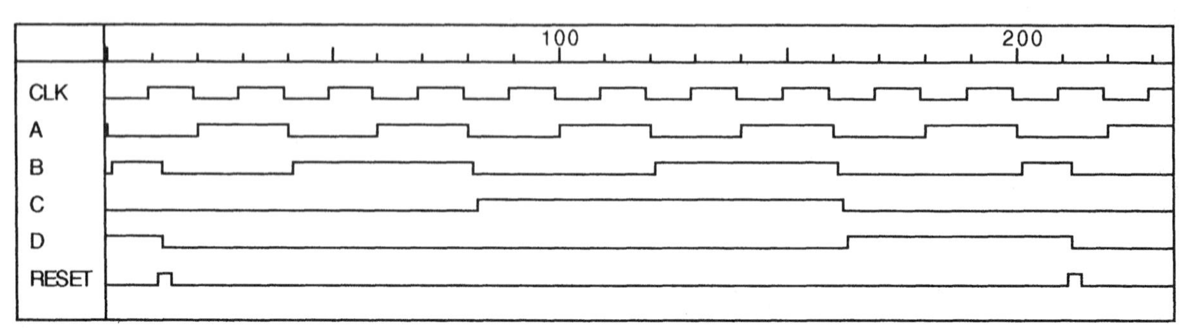

Fig. 8-48

8.26 See Fig. 8-49. Note that two-input gates are sufficient.

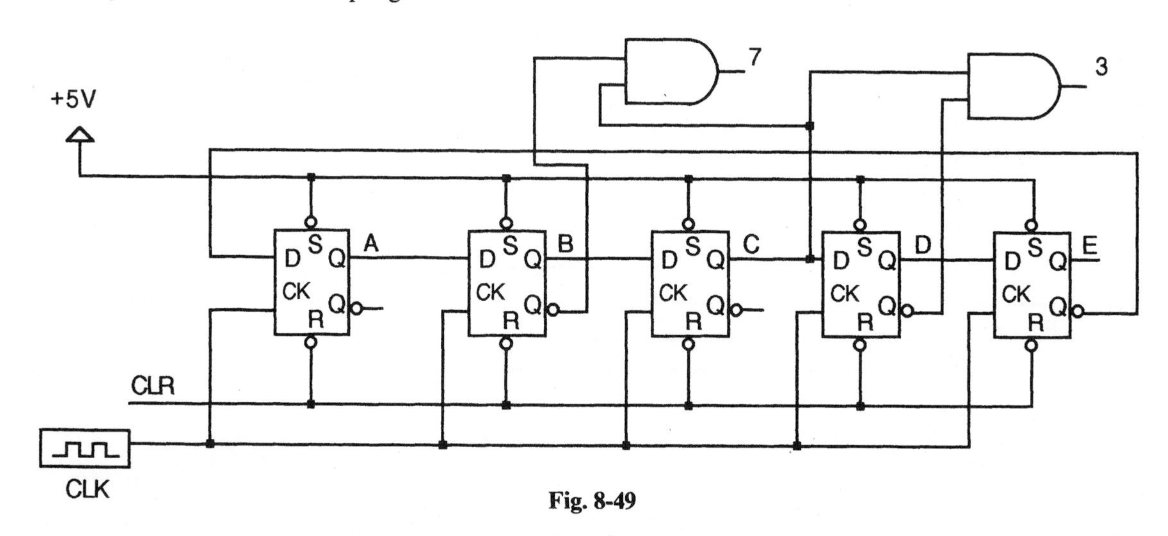

Fig. 8-49

8.27 See Fig. 8.50.

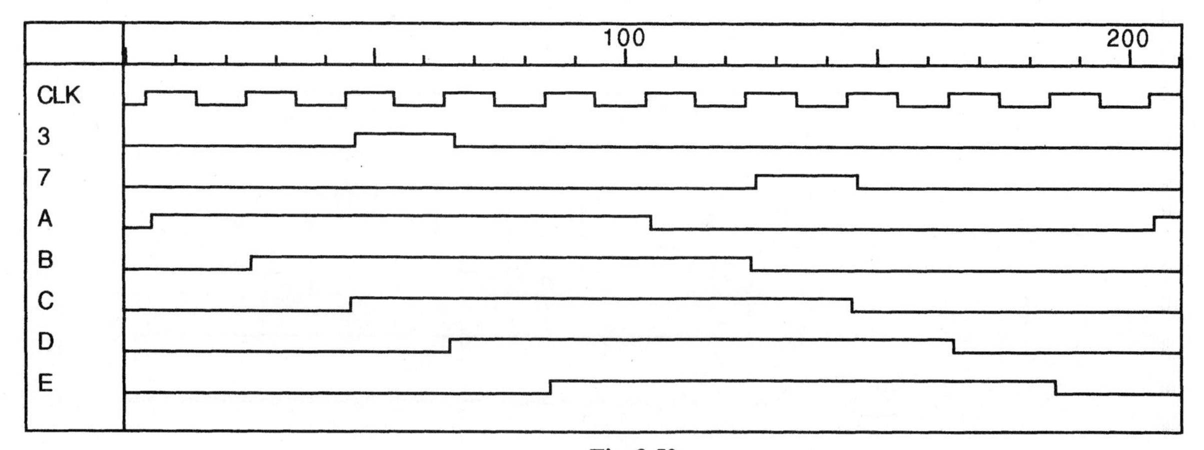

Fig. 8-50

8.28 See Fig. 8-51.

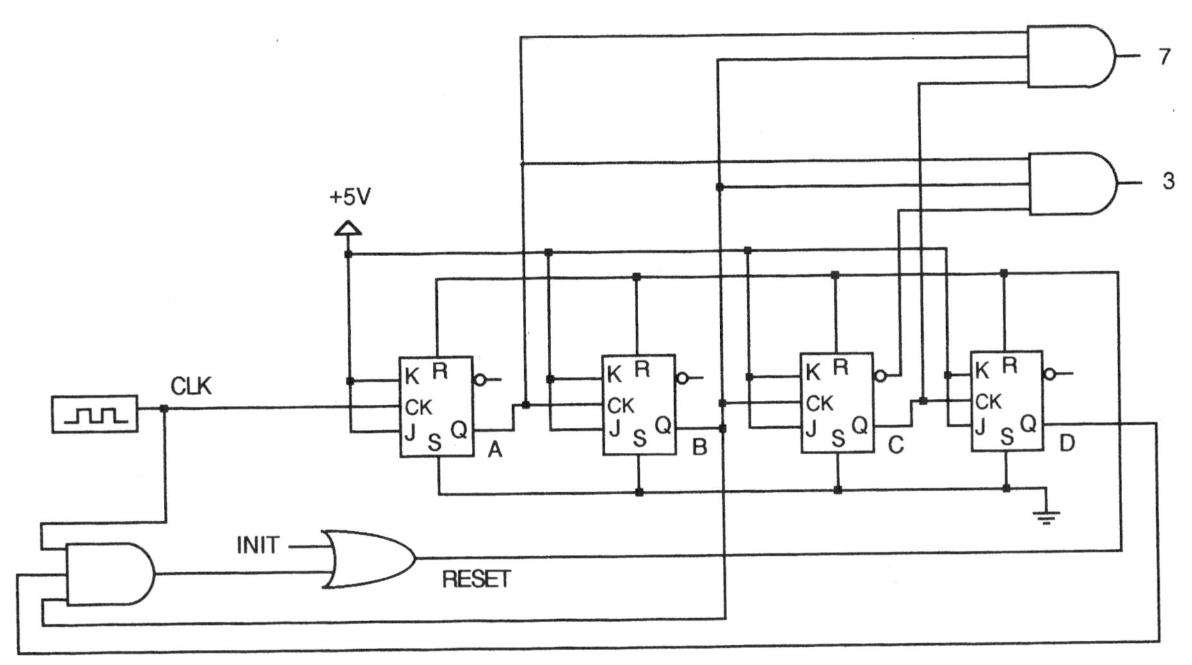

Fig. 8-51

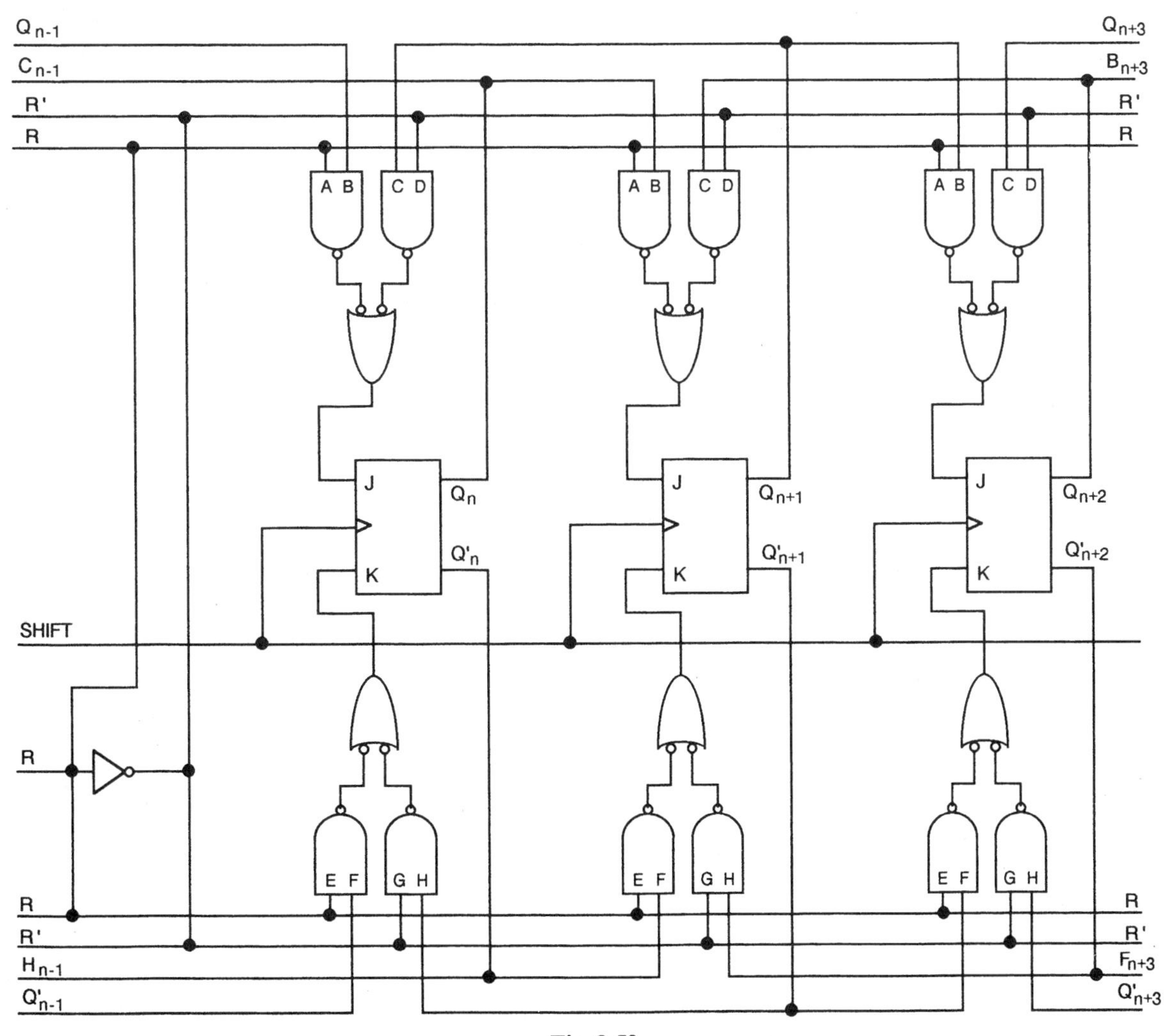

Fig. 8-52

8.29 The coupling equations are $J_n = RQ_{n-1} + R'Q_{n+1}$ and $K_n = RQ'_{n-1} + R'Q_{n+1}$. An implementation of this arrangement is shown in Fig. 8-52 which should be compared to the circuit of Fig. 8-4.

8.30 Starting from zero, the decimal equivalent count alternates continuously 1, 2, 1, 2, ...

8.31 The decimal equivalent count is 15, 1, 2, 1, 2, ...

8.32 The decimal equivalent count is 0, 3, 7, 10, 14, 1, 5, 8, 12, 15, 0 (repeats).

Chapter 9

Application-Specific Devices

9.1 INTRODUCTION

There are many applications for digital logic where the market is not great enough to develop a special-purpose MSI or LSI chip. The *total* market for custom devices of all kinds is, however, quite large. This situation has led to the development of *programmable logic* devices (PLDs) which can be produced in high volume and can be easily configured by the individual user for specialized applications.

Programmable logic devices (*PLDs*) can be divided into two rather broad categories. The first, circuits in which programming for a specific application causes a physical change within the device, is discussed in this chapter. The second type, where logic functions are controlled by electronic signals, is discussed in Chap. 11.

Another approach to the custom IC market is to create a standard library of components which can be readily fabricated on silicon. This library, in computer-readable form, contains photolithography information needed to fabricate the components and design rules for interconnecting them. Powerful computer hardware and software are combined to create a customized silicon chip for a user. This approach yields circuits which are termed custom ICs or, more generally, *application-specific integrated circuits* (*ASICs*).

Note that the PLDs mentioned above are also application-specific devices. Distinction between various types is usually made on the basis of the type of structure incorporated by the vendor, the nature of programming technique required, the ability to erase programming, and the time required to do so.

9.2 PROGRAMMING TECHNOLOGIES

Various techniques are used to implement programmable logic:

Fusible Links

In this case, the logic is constructed with all possible user-determined internal connections preestablished at manufacture. These connections are created with a fusible material, and the device is made application-specific by blowing the fuses on the connections which are to be removed. The chip is designed to allow relatively high currents to be passed through specific fusible links selected by the user during the programming operation. This method of programming is obviously not reversible.

Ultraviolet (UV) Erasable Programming

Devices are fabricated with all possible user-selectable connection sites left unconnected. These sites can be selectively activated during programming by the application of appropriate voltages to pairs of addressable conductors which intersect at each desired connection point. The voltages create stable "packets of charge" in the silicon to establish a connection. In UV erasable devices, the IC packages have a small window located above the chip which permits the chip to be exposed to UV light which causes the charge packets forming programmed connections to bleed off, essentially "dissolving" them. Approximately 10 min of exposure to UV light of suitable intensity will usually produce complete erasure, leaving a device free for reprogramming.

Electronically Erasable Programming

There are many applications where quicker erasure times than 10 min are necessary for an application. In this case, erasure can be accomplished electronically in what is essentially a reversal of the programming process. In some devices of this type, it is possible to selectively erase only a portion of the device, whereas, in others, the entire chip must be cleared. High-speed electronic erasure is termed "flash" erasure and can be accomplished in seconds.

As far as the user is concerned, the programming processes for erasable and fusible link devices are essentially the same.

9.3 PROMS AND EPROMS

The programmable read-only memory (PROM) is a programmable device with a specific structure. It is a ROM (see Sec. 5.4) in which connections to OR gate inputs are user-controllable. These devices are available in fusible link versions (PROMs), UV erasable versions (EPROMs), and electronically erasable versions (EEPROMs). All PROMs contain a full decoder whose outputs are constructed so that a connection between each one and any OR gate input is possible. The structure is shown in Fig. 9-1.

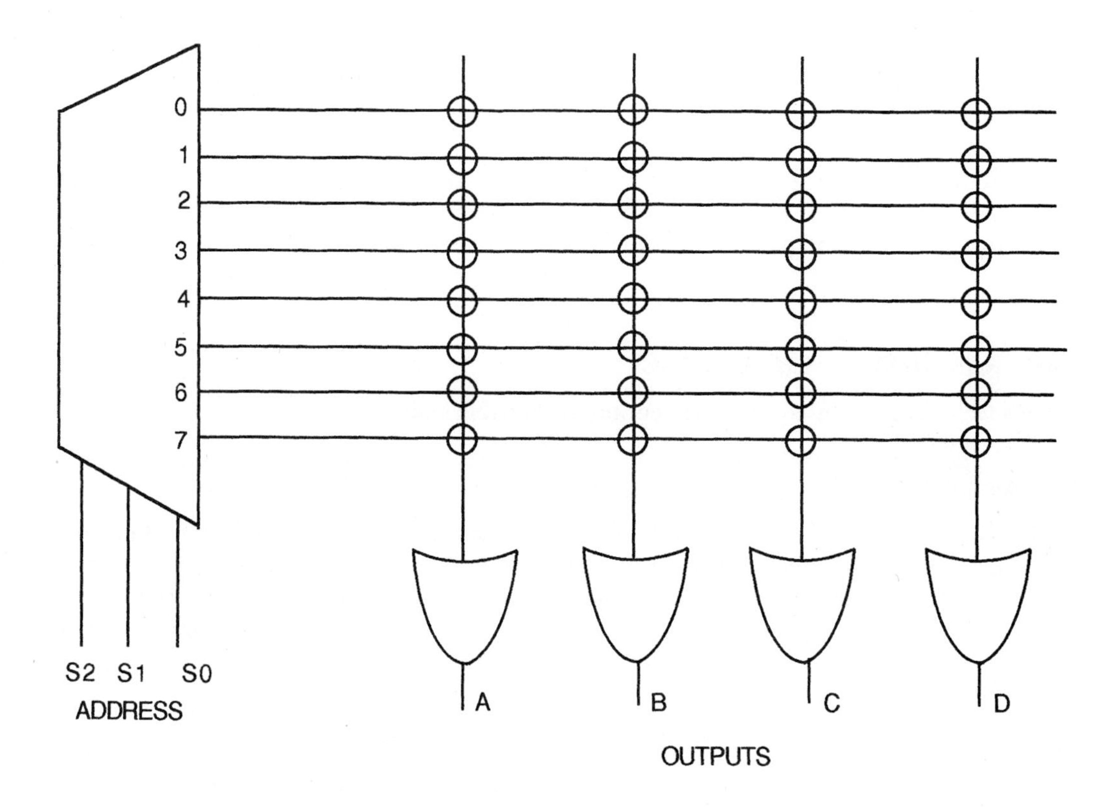

○ Represents a user selectable connection point.

Fig. 9-1 PROM architecture (N = 3).

The full decoder with N address bits can produce a signal on any one of its 2^N output lines, thus accounting for every row in the truth table. Since inputs to the logic serve as address bits and the decoder has 2^N output lines for N address bits, each time a new logic input is introduced, the size of the decoder doubles. PROMs are not particularly useful for implementing logic which requires a large number of inputs because of the excessive silicon area needed for the decoder. Furthermore, the structure does not allow the designer to take advantage of any "don't care" conditions. The major application of ROMs, PROMs, EPROMs, etc., is in the storage of instructions for computer applications rather than in the implementation of logic. PROMs are generally programmed by determining the desired interconnection pattern and treating it as a group of hexadecimal numbers to be stored in a memory device.

Programming a PROM or EPROM requires some computer software and a "PROM burner" which attaches to a port of the programming computer or to a terminal. It is usually accomplished by treating the OR gate connections as binary numbers (a 1 representing connection and a 0 no connection). The PROM or EPROM program is simply a list of hexadecimal numbers representing connection patterns and addresses into which they are to be loaded. The program data is entered via a keyboard, and software converts it into a set of electrical signals which are routed by the computer to the PROM burner which causes the desired connection pattern to be created.

EXAMPLE 9.1 Program an EPROM to convert a 4-bit binary number into an equivalent Gray code number. The inverse of this problem and the appropriate truth table is shown in Prob. 5.8.

Strategy: Treat the device as a lookup table in which the binary digits are used to specify addresses whose contents are the corresponding Gray code numbers. The connections are shown in Fig. 9-2. The actual program starts in address 0000 and is listed in hexadecimal as 0, 1, 3, 2, 6, 7, 5, 4, C, D, F, E, A, B, 9, 8.

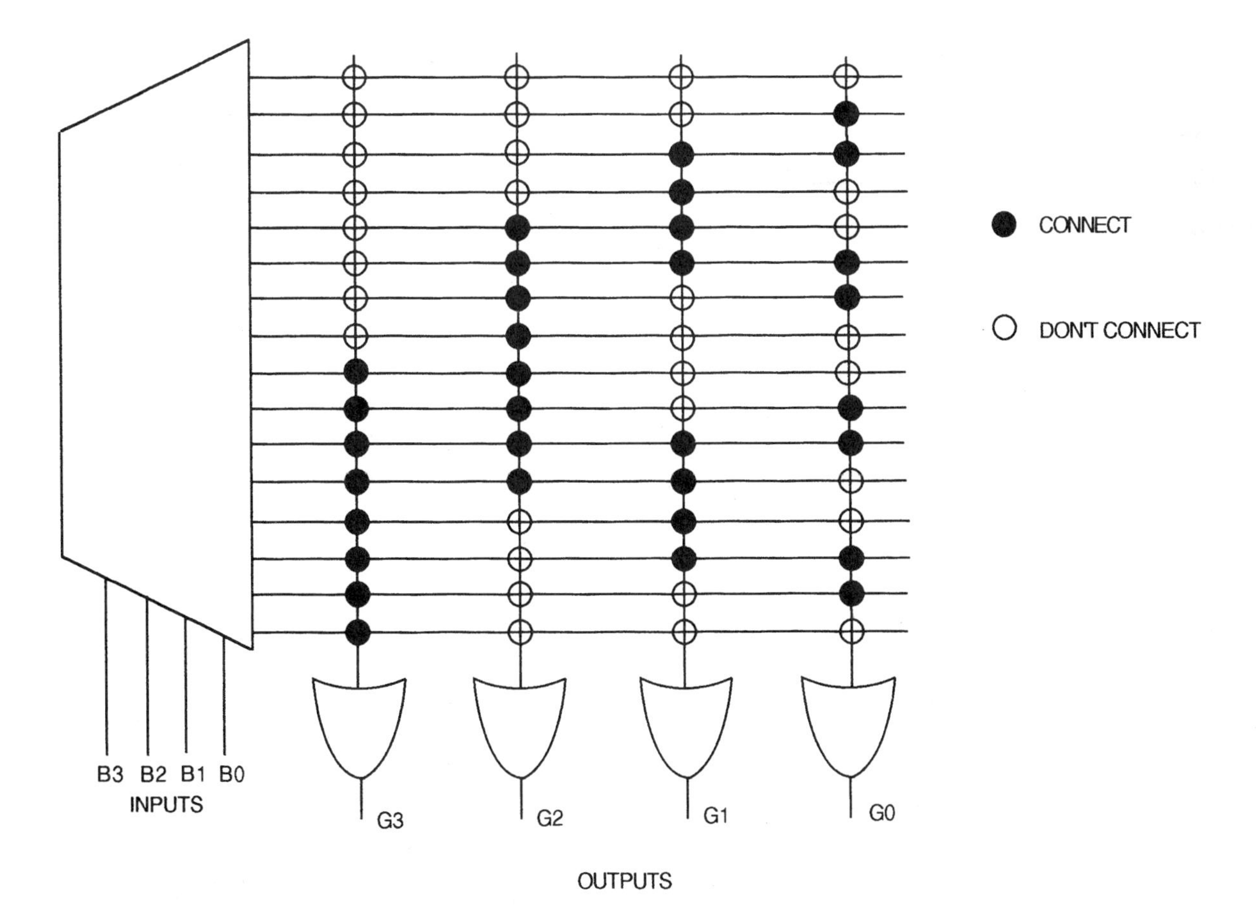

Fig. 9-2 Programmed PROM for binary to Gray code conversion.

9.4 PROGRAMMABLE ARRAY LOGIC (PAL®*)

The PAL is a programmable device which is especially useful in applications where there are a relatively large number of inputs and, at the same time, a significant number of constraints on them. These constraints function to create "don't cares," meaning that a number of truth table rows need not be implemented.

As in the case of ROMs, the basic structure is that of AND gates driving OR gates. In a PAL IC, the OR gate inputs are preconnected to a limited number of AND gates, leaving the designer freedom to select AND gate inputs only. This restriction is often acceptable since only a limited number of OR gate connections are required because of the presence of "don't cares." Since inputs to the logic can occur in either direct or logically inverted form, provision is made internally to produce the logical inversions.

Some PAL devices are made with only combinational elements (ANDs, ORs, and inverters), while others contain a few flip-flops so that an entire small-state machine (Chap. 10) may be placed on a single chip. A generic type of PAL architecture, containing flip-flops, is shown in Fig. 9-3. Note that some AND gate inputs come directly from external pins on the chip, while others are fed back from internal flip-flops. The OR gate outputs are either passed directly to an output pin (unregistered) or passed through flip-flop latches (registered). The designer selects the desired configuration by choosing to use or exclude appropriate AND gates. In cases where signals are fed back from flip-flops, provison is usually made for both direct and inverted forms.

The internal structure of a commercial device (the PAL16R6, manufactured by Advanced Micro Devices, Inc.) is shown in Fig. 9-4. It contains eight OR gates, each of which is driven by a dedicated set of 8 of the 64 AND gates.

Each AND gate has 32 possible user-selectable input connections. Some of the OR gate outputs are registered, while others are not; all of them, however, are capable of being fed back to AND gate inputs. There are six D flip-flops, as indicated by the last digit in the part number.

The 16R6 chip makes use of the TRI-STATE® buffer element (see Fig. 8-12) which serves as a switch, controlled by an enable signal, that selectively connects or disconnects (isolates) its associated output pin from any circuitry to which it is attached.

Figure 9-5 shows another symbol, used on the 16R6 diagram, which is a shorthand means of indicating an amplifier having a single input and two outputs. One output is inverted (as indicated by the bubble) and the other is not.

PAL Device Programming

The PAL device is programmed using sophisticated computer-aided engineering software and PAL burner hardware attached to a computer or terminal. The internal structure of a PALIC, with AND gates driving OR gates, lends itself to the expression of programming information in basic Boolean equations derived from a truth table. There are many specialized software languages available for programming PAL devices. One of the earliest is called PALASM®† which is based on positive logic notation and uses Boolean entry very effectively. In this language, the logic operation symbols are modified so as to be useful on a general keyboard. A short list (not complete) of operational symbols, in order of precedence, is given below:

;	Comment follows
/	Logical inversion or active-LOW
*	Logical AND
+	Logical OR

* PAL is a registered trademark of Advanced Micro Devices, Inc.

† PALASM® is a registered trademark of Advanced Micro Devices, Inc.

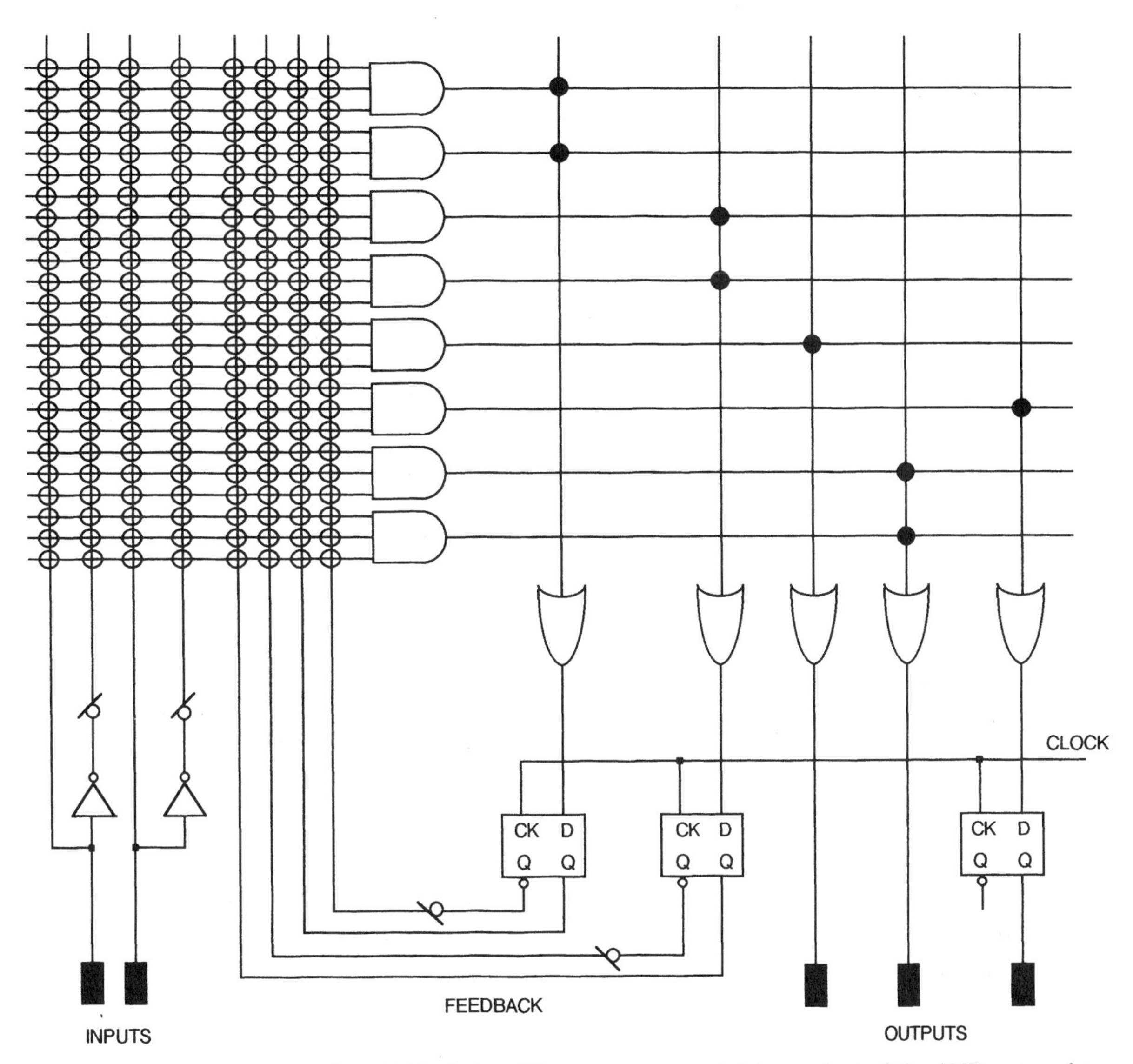

Fig. 9-3 Generic architecture for a PAL device. OR gates are connected to a subset of the AND gates whose inputs are user-selectable.

:+:	Exclusive OR
:*:	Coincidence (inverted Exclusive OR)
=	Equality (combinational output equation operator)
:=	Registered output equation operator

If it is desired to enter the Boolean equation, $F = AB' + CD + (B + C')'$, De Morgan's theorem is first used to convert the equation to the AND-OR form (sum of products) required by PALASM software. Syntax, as defined above, is then applied to obtain:

$F = AB' + CD + B'C$	AND-OR
F = A*/B + C*D + /B*C	PALASM software syntax

If the output F is to be registered, the equals sign would be preceded by a colon.

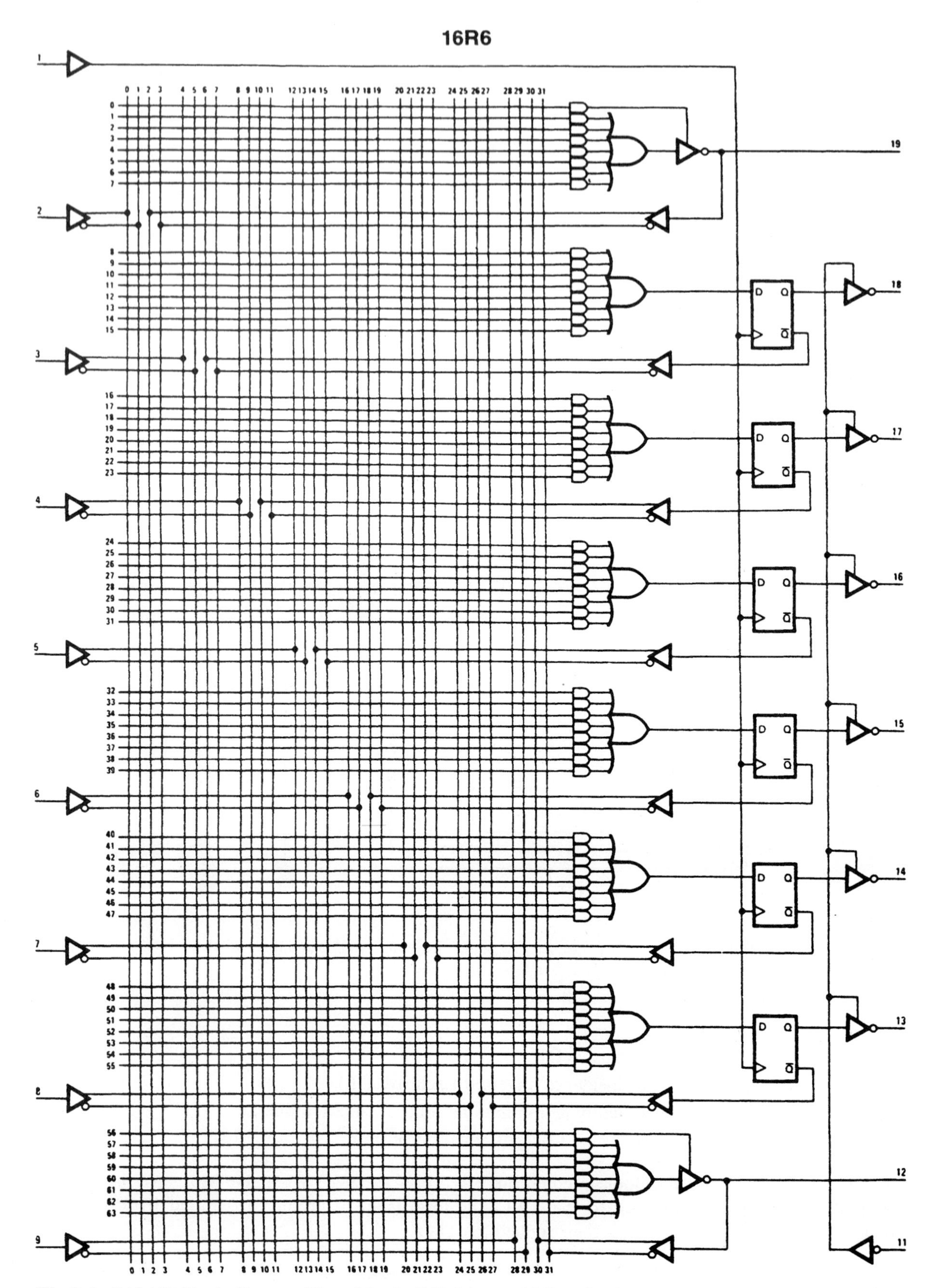

Fig. 9-4 PAL16R6 logic diagram. (Copyright © 1991 Advanced Micro Devices, Inc. Reprinted with persmision of copyright owner. All rights reserved.)

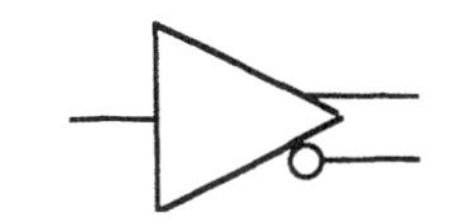

Fig. 9-5 Amplifier pair.

Output Pin Truth Value Considerations

Inspection of the PAL16R6 circuit diagram in Fig. 9-4, shows that the logic, in sum of products form, is produced by networks of AND gates connected to OR gates which drive the D inputs of corresponding flip-flops. The arrival of a clock pulse transfers these values to the Q outputs which are, in turn, connected to package pins via inverting amplifiers. Thus, the truth value at each output pin is opposite to that associated with the affiliated flip-flop's Q output. By designating a pin low-TRUE, we are establishing that Q be defined high-TRUE and the flip-flop D input must be driven by a 1's covering sum of products. If, on the other hand, as in Example 9.2 which follows, the output pins are designated high-TRUE, then the bubbles can only be balanced by the addition of a logical inversion meaning that Q (and hence D) must be driven from the inverted logical equation or, equivalently, a 0's covering.

Often, logic design procedures produce Boolean equations which correspond to a 1's covering for the sum of products (see Chapter 10 for examples). If the designer is willing to accept the outputs in low-TRUE form (i.e., pins designated by /X), then the Boolean design equations can be directly converted to PALASM form using the relationships tabulated earlier in this section. If, however, high-TRUE outputs are required, it is necessary to use logically inverted Boolean expressions. In many versions of PALASM software, this can be done by simply placing parentheses around the equations, preceded by a slash mark (/). With the exception of Example 9.2, all problems in this book involving the PAL16R6 or PAL16R8 will assume that low-TRUE output pins are specified so that the Boolean design equations can be converted directly to PALASM syntax.

It is important to note that not all PAL devices couple the output pins via inverting amplifiers. Furthermore, there are many other ways in which design equations, in various forms, can be implemented in PAL devices by the PALASM software. The reader is advised to consult an appropriate PALASM manual for further details.

EXAMPLE 9.2 Programming a simple binary bit adder with registered outputs on a PAL16R8. The truth table, initially described in Sec. 2.4, is shown in Table 9.1, and equations for the sum and carry-out variables, in PALASM form, are

$$F = /X*/Y*C_i + /X*Y*/C_i + X*/Y*/C_i + X*Y*C_i$$
$$C_o = /X*Y*C_i + X*/Y*C_i + X*Y*/C_i + X*Y*C_i$$

Here, the SUM output is given by F and the carry-out by C_o.

Table 9.1

INPUTS			OUTPUTS	
X	Y	Ci	F	Co
0	0	0	0	0
0	0	1	1	0
0	1	0	1	0
0	1	1	0	1
1	0	0	1	0
1	0	1	0	1
1	1	0	0	1
1	1	1	1	1

Following basic syntax rules, the designer types a source file for the PALASM program using any convenient text editor. This file, which defines variables, specifies the Boolean relations and assigns pin numbers as shown in Fig. 9-6. Any line beginning with a semicolon is ignored by software; its function is for annotation and clarifcation. The first few lines are largely documentation entries. The lines following each line beginning with PINS designate chip pin number assignments which must be entered in sequence. All pins must be accounted for and, consequently, NC is used to indicate the absence of a connection. The Boolean equations are entered next using PALASM syntax.

The simulation portion of the input file provides instructions for the software to simulate circuit operation and check it against the Boolean specification. The command SETF indicates to the program that the input variables listed following this key word are to be applied. The first simulation entry SETF/OE indicates that an LT input is

```
               EXAMPLE PDS FILE FOR PALASM USING BINARY BIT ADDER


TITLE           ADDER
PATTERN         A
REVISION        1
AUTHOR          JIM SWATCH
COMPANY         RIT
DATE            MAY 1992

CHIP            ADDER    PAL16R8

;PINS           1        2        3        4        5        6        7
                CLK      X        Y        CI       NC       NC       NC

;PINS           8        9        10       11       12       13       14
                NC       NC       GND      OE       NC       NC       NC

;PINS           15       16       17       18       19       20
                NC       NC       NC       F        CO       VCC

EQUATIONS

     /F:=/X*/Y*/CI+/X*Y*CI+X*/Y*CI+X*Y*/CI        ;SINCE THE OUTPUTS FROM
                                                   ;THE PAL16R8 ARE INVERED,
     /CO:=/X*/Y*/CI+/X*/Y*CI+/X*Y*/CI+X*/Y*/CI   ;WRITE THE EQUATIONS FOR
                                                   ;THE ZERO TERMS NOT THE ONES.
SIMULATION

        SETF /OE                                 ;THIS ENABLES THE OUTPUTS
        CLOCKF CLK                               ;GENERATES A CLOCK PULSE
        SETF /X /Y /CI                           ;THIS FIRST LINE OF THE
        CLOCKF CLK                               ;TRUTH TABLE
        SETF /X /Y  CI
        CLOCKF CLK
        SETF /X  Y /CI
        CLOCKF CLK
        SETF /X  Y  CI
        CLOCKF CLK
        SETF X  /Y /CI
        CLOCKF CLK
        SETF X  /Y  CI
        CLOCKF CLK
        SETF X   Y /CI
        CLOCKF CLK
        SETF X   Y  CI
        CLOCKF CLK
```

Fig. 9-6

to be applied to the TRI-STATE® buffers, thereby enabling the outputs (output enable). The command CLOCKF indicates that a clock pulse is to be applied to the following specified pin(s) which, in the current example, is the single pin labeled CLK. The simulation proceeds one step at a time, a clock pulse being applied following the application of each set of inputs corresponding to the truth table input permutations.

When the program is run, a successful simulation will be indicated by the statements "No errors," "No warnings," "File processed successfully."

The results of the simulation are stored in a file named ADDER.HST shown in Fig. 9-7. Each vertical column represents a time interval (as in a microtiming diagram). The letter "c" indicates the CLKF command and "g" indicates a SETF command. Clocking action proceeds in three steps: First, the clock pin voltage is raised, following which new output pin values are recorded. Last, the clock pin voltage is lowered. Each column contains the values high (H), low (L), or undefined (X) at each pin that result from simulation commands.

```
PALASM89  PLDSIM    - DEVELOPEMENT VERSION (26-SEP-1989)
 (C) - COPYRIGHT ADVANCED MICRO DEVICES INC., 1989

PALASM SIMULATION HISTORY LISTING

Title      : ADDER                              Author    : JIM SWATCH
Pattern    : A                                  Company   : RIT
Revision   : 1                                  Date      : MAY 1992

PAL16R8
Page : 1
      g  cg   cg   cg   cg   cg   cg   cg   cg  c
 CLK XHHLLHHLLHHLLHHLLHHLLHHLLHHLLHHLLHHL
 X   XXXXLLLLLLLLLLLLLLLLHHHHHHHHHHHHHHHH
 Y   XXXXLLLLLLLLHHHHHHHHLLLLLLLLHHHHHHHH
 CI  XXXXLLLLHHHHLLLLHHHHLLLLHHHHLLLLHHHH
 GND LLLLLLLLLLLLLLLLLLLLLLLLLLLLLLLLLLLL
 OE  LLLLLLLLLLLLLLLLLLLLLLLLLLLLLLLLLLLL
 F   XXXXXXLLLLHHHHHHHHLLLLHHHHLLLLLLLLHH
 CO  XXXXXXLLLLLLLLLLLLHHHHLLLLHHHHHHHHHH
 VCC HHHHHHHHHHHHHHHHHHHHHHHHHHHHHHHHHHHH
```

Fig. 9-7

The fuse pattern is stored in a file called ADDER_ex.XPT which is shown in Fig. 9-8. Here, the horizontal index runs from 0 to 31 and is associated with the numbered data columns shown in the circuit diagram of Fig. 9-4. The vertical index runs from 0 through 63, each line corresponding to an AND gate in the circuit diagram. Blown fuses (no connection) are indicated by (-) and connections are indicated by (X). Unused gates show all fuses unblown. In the current example, AND gates 0 to 3 and 8 to 11 are used to do logic, while the other AND gates are held at 0. Actual circuit connections may be traced by reference to the indices and the circuit diagram of Fig. 9-4.

9.5 THE PROGRAMMED LOGIC ARRAY (PLA)

The PLA provides more degrees of freedom to designers because both the AND gate array and the OR gate array connections are available. It combines the AND gate freedom of the PAL IC with the OR gate freedom of the PROM. Programmed logic arrays often contain on-board flip-flops for registering outputs and/or feedback variables. A simplified generic logic diagram is shown in Fig. 9-9.

As in the case of the PAL device, the PLA contains input data in direct and complemented form. The ability to control the connections to OR gate inputs gives the designer an extra degree of freedom. For example, the same AND gate can drive several OR gates, while in a PAL device, a duplicate AND gate must be used for each of the ORs.

The PLA is programmed on a computer or workstation in a manner similar to the PAL IC, and a significant amount of relatively user-friendly software is available. The program is entered in Boolean form, as demonstrated in Example 9.2, or with state table entry as described in Chap. 10. The PLA is often used as a basic building block in the realization of very large scale (VLSI) ICs.

```
PALASM89  PAL ASSEMBLER   - DEVELOPEMENT VERSION (28-AUG-1989)
 (C) - COPYRIGHT ADVANCED MICRO DEVICES INC., 1989

TITLE   :ADDER                            AUTHOR :JIM SWATCH
PATTERN :A                                COMPANY:RIT
REVISION:1                                DATE   :MAY 1992

PAL16R8
ADDER

                    11   1111  1111  2222  2222  2233
      0123  4567  8901  2345  6789  0123  4567  8901

0     -X--  -X--  -X--  ----  ----  ----  ----  ----
1     -X--  -X--  X---  ----  ----  ----  ----  ----
2     -X--  X---  -X--  ----  ----  ----  ----  ----
3     X---  -X--  -X--  ----  ----  ----  ----  ----
4     XXXX  XXXX  XXXX  XXXX  XXXX  XXXX  XXXX  XXXX
5     XXXX  XXXX  XXXX  XXXX  XXXX  XXXX  XXXX  XXXX
6     XXXX  XXXX  XXXX  XXXX  XXXX  XXXX  XXXX  XXXX
7     XXXX  XXXX  XXXX  XXXX  XXXX  XXXX  XXXX  XXXX

8     -X--  -X--  -X--  ----  ----  ----  ----  ----
9     -X--  X---  X---  ----  ----  ----  ----  ----
10    X---  -X--  X---  ----  ----  ----  ----  ----
11    X---  X---  -X--  ----  ----  ----  ----  ----
12    XXXX  XXXX  XXXX  XXXX  XXXX  XXXX  XXXX  XXXX
13    XXXX  XXXX  XXXX  XXXX  XXXX  XXXX  XXXX  XXXX
14    XXXX  XXXX  XXXX  XXXX  XXXX  XXXX  XXXX  XXXX
15    XXXX  XXXX  XXXX  XXXX  XXXX  XXXX  XXXX  XXXX

16    XXXX  XXXX  XXXX  XXXX  XXXX  XXXX  XXXX  XXXX
17    XXXX  XXXX  XXXX  XXXX  XXXX  XXXX  XXXX  XXXX
18    XXXX  XXXX  XXXX  XXXX  XXXX  XXXX  XXXX  XXXX
19    XXXX  XXXX  XXXX  XXXX  XXXX  XXXX  XXXX  XXXX
20    XXXX  XXXX  XXXX  XXXX  XXXX  XXXX  XXXX  XXXX
21    XXXX  XXXX  XXXX  XXXX  XXXX  XXXX  XXXX  XXXX
22    XXXX  XXXX  XXXX  XXXX  XXXX  XXXX  XXXX  XXXX
23    XXXX  XXXX  XXXX  XXXX  XXXX  XXXX  XXXX  XXXX

24    XXXX  XXXX  XXXX  XXXX  XXXX  XXXX  XXXX  XXXX
25    XXXX  XXXX  XXXX  XXXX  XXXX  XXXX  XXXX  XXXX
26    XXXX  XXXX  XXXX  XXXX  XXXX  XXXX  XXXX  XXXX
27    XXXX  XXXX  XXXX  XXXX  XXXX  XXXX  XXXX  XXXX
28    XXXX  XXXX  XXXX  XXXX  XXXX  XXXX  XXXX  XXXX
29    XXXX  XXXX  XXXX  XXXX  XXXX  XXXX  XXXX  XXXX
30    XXXX  XXXX  XXXX  XXXX  XXXX  XXXX  XXXX  XXXX
31    XXXX  XXXX  XXXX  XXXX  XXXX  XXXX  XXXX  XXXX

32    XXXX  XXXX  XXXX  XXXX  XXXX  XXXX  XXXX  XXXX
33    XXXX  XXXX  XXXX  XXXX  XXXX  XXXX  XXXX  XXXX
34    XXXX  XXXX  XXXX  XXXX  XXXX  XXXX  XXXX  XXXX
35    XXXX  XXXX  XXXX  XXXX  XXXX  XXXX  XXXX  XXXX
36    XXXX  XXXX  XXXX  XXXX  XXXX  XXXX  XXXX  XXXX
37    XXXX  XXXX  XXXX  XXXX  XXXX  XXXX  XXXX  XXXX
38    XXXX  XXXX  XXXX  XXXX  XXXX  XXXX  XXXX  XXXX
39    XXXX  XXXX  XXXX  XXXX  XXXX  XXXX  XXXX  XXXX

40    XXXX  XXXX  XXXX  XXXX  XXXX  XXXX  XXXX  XXXX
41    XXXX  XXXX  XXXX  XXXX  XXXX  XXXX  XXXX  XXXX
42    XXXX  XXXX  XXXX  XXXX  XXXX  XXXX  XXXX  XXXX
```

```
43   XXXX XXXX XXXX XXXX XXXX XXXX XXXX XXXX
44   XXXX XXXX XXXX XXXX XXXX XXXX XXXX XXXX
45   XXXX XXXX XXXX XXXX XXXX XXXX XXXX XXXX
46   XXXX XXXX XXXX XXXX XXXX XXXX XXXX XXXX
47   XXXX XXXX XXXX XXXX XXXX XXXX XXXX XXXX

48   XXXX XXXX XXXX XXXX XXXX XXXX XXXX XXXX
49   XXXX XXXX XXXX XXXX XXXX XXXX XXXX XXXX
50   XXXX XXXX XXXX XXXX XXXX XXXX XXXX XXXX
51   XXXX XXXX XXXX XXXX XXXX XXXX XXXX XXXX
52   XXXX XXXX XXXX XXXX XXXX XXXX XXXX XXXX
53   XXXX XXXX XXXX XXXX XXXX XXXX XXXX XXXX
54   XXXX XXXX XXXX XXXX XXXX XXXX XXXX XXXX
55   XXXX XXXX XXXX XXXX XXXX XXXX XXXX XXXX

56   XXXX XXXX XXXX XXXX XXXX XXXX XXXX XXXX
57   XXXX XXXX XXXX XXXX XXXX XXXX XXXX XXXX
58   XXXX XXXX XXXX XXXX XXXX XXXX XXXX XXXX
59   XXXX XXXX XXXX XXXX XXXX XXXX XXXX XXXX
60   XXXX XXXX XXXX XXXX XXXX XXXX XXXX XXXX
61   XXXX XXXX XXXX XXXX XXXX XXXX XXXX XXXX
62   XXXX XXXX XXXX XXXX XXXX XXXX XXXX XXXX
63   XXXX XXXX XXXX XXXX XXXX XXXX XXXX XXXX

     SUMMARY
     -------

       TOTAL FUSES BLOWN     = 232
```

Fig. 9-8 (Copyright © 1991 Advanced Micro Devices, Inc. Reprinted with permission of copyright owner. All rights reserved.)

9.6 GATE ARRAYS

Chips called *gate arrays* (or full gate arrays) contain very large numbers of uncommitted (unconnected) gates. Often, they contain an array of individual cells in which an added interconnection creates a specific NAND or NOR gate. Programming the gate array causes these gates to be connected so as to create combinational logic, flip-flops, registers, counters, memory, etc. The original chip, which can be made in very large quantities (with consequent economies of scale), can later be customized by specifying the interconnections between gates. Programming these devices usually requires the overhead of rather complex software and a high degree of computer power to accomplish circuit placement and connection routing. There are design rules, for example, which limit the number of crossovers in the computer-generated interconnection pattern, and adherence to these rules often requires many iterations in the design process. Additionally, interconnections, once made, cannot be modified. The result is that the use of gate arrays is limited to higher-volume products (where the overhead can be more easily amortized) and requires a relatively longer development cycle.

In one version, the gate array is fabricated up to, but not including, the metal interconnections. Programming, which customizes the design, results in artwork for the photolithographic masks needed for creation of the final interconnection step in the fabrication process. The device is then completed in a few weeks and delivered to the end user.

In a second version, two interconnection layers are fabricated in a grid design with vias built to connect the layers. The user's program is converted into a pattern of top-layer metal which is to be etched away to form interconnections for both layers. This second case requires fewer added fabrication steps and less time for design turnaround.

Gate arrays usually make use of only a small variety of standard transistor cells and consequently, the design cannot be fully optimized with regard to power dissipation or speed for all applications. The gate array generally does not make as efficient use of the silicon resources as would a fully customized design. It can, however, be fabricated less expensively and more quickly than a full custom design.

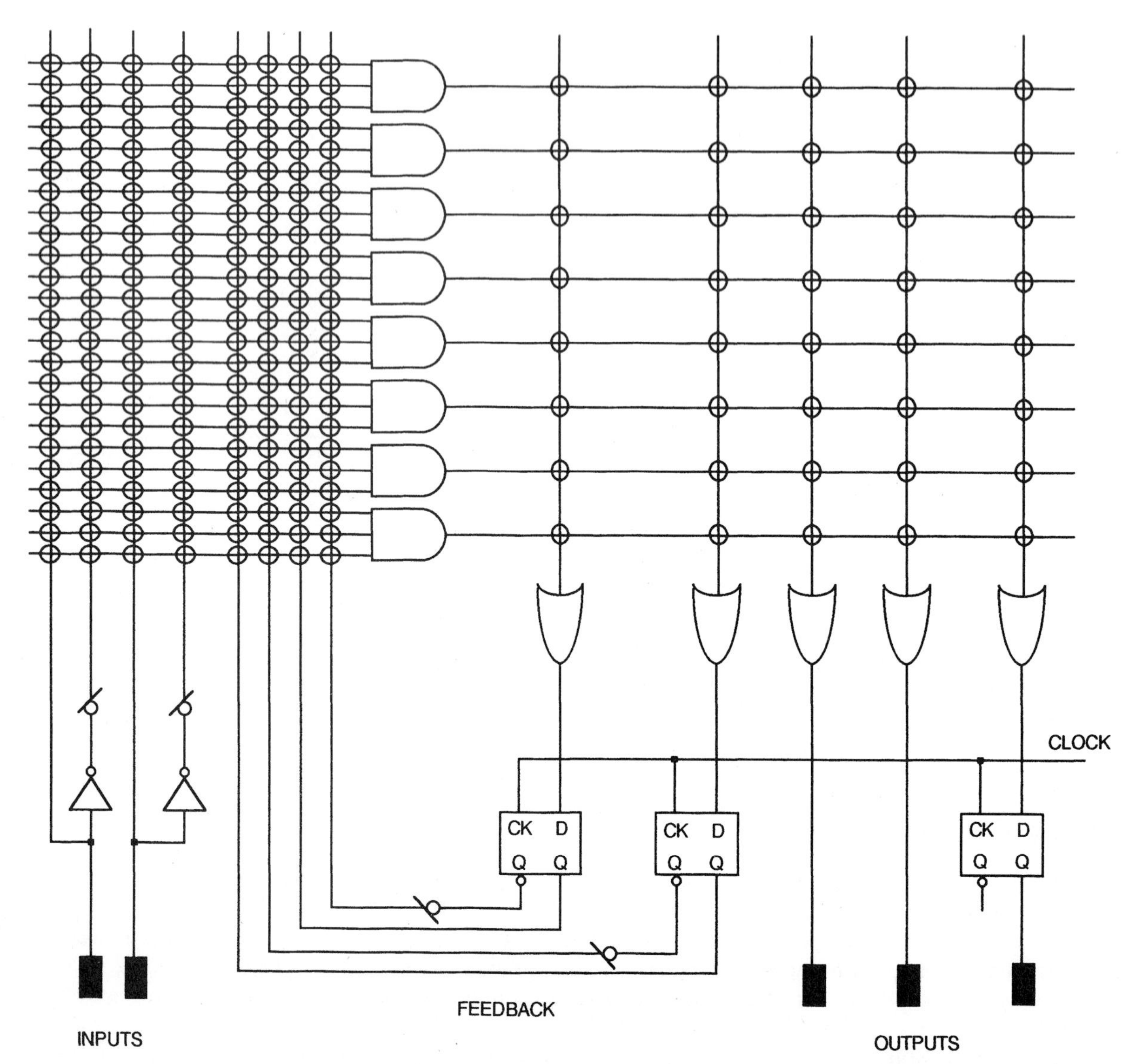

Fig. 9-9 Generic PLA architecture.

The gate array is programmed on a computer or workstation in a manner similar to the PAL or PLA devices, but it requires a significantly greater amount of software and computer power. In addition to handling connections and routing, consideration of thermal problems such as "hot spots" is included. The designer's program can be entered in Boolean form, as described previously, or with state table entry. Another common programming method is that of *schematic capture.* Here, the designer creates the logic diagram on the monitor of a workstation by calling up components from a software library and connecting them as desired using a mouse or equivalent pointing device. There is a considerable amount of software assistance in the form of wire routing and placement algorithms and in *macros* which treat frequently used combinations of gates (such as registers) as fixed named groups. It is usually possible for the user to create his or her own macros in software.

Gate array programming results in a software-generated set of computer instructions for the creation of photolithographic artwork necessary to form the required connections between all gates in the circuit. The gate array can often be used to implement entire systems on VLSI chips.

It is very important that a design be evaluated for performance before it is actually constructed.

Since a trial build or breadboard is often too expensive to construct, and does not accurately predict timing delays because of its size difference, the major tool for performance evaluation is simulation software of the type described in Chap. 6. Almost all design software for the creation of large ICs includes a sophisticated simulation package which can be used to evaluate performance and to prepare test signals for use during production.

9.7 PROGRAMMABLE GATE ARRAYS

A product which is somewhat intermediate between PAL ICs and PLAs, on the one hand, and the full gate array on the other, has been developed. Called the programmable gate array (PGA) or field programmable gate array (FPGA), it represents considerably less fixed internal structure than a PLA but has more structure than the full gate array. The FPGA, sometimes referred to as a logic cell array (LCA®),* is a PLD whose structure is not changed by programming. Instead, electronic control signals are used to modify the function of fixed circuitry. A discussion of the FPGA and its control circuitry may be found in Chap. 11.

The discussion of PGAs is complicated by the fact that some vendors market products called PGAs which make use of connection techniques which permanently modify the device and cannot be deprogrammed as can the LCA. It is perhaps best to consider a generic PGA as a device which is intermediate between PLA devices and full gate arrays. The user should investigate the products of specific vendors for structure and programming techniques.

9.8 FULL CUSTOM DESIGN

It is often the case, in high-volume or high-speed applications, that an entire chip is custom-designed from scratch. This is a rather complex process which involves a comparatively long design cycle and sophisticated designer training. The technology to be used is studied extensively and a library of detailed electronic component models is created. These models include parameters describing power dissipation, speed properties, silicon utilization, and the photolithography masks used in fabrication. The models may be those of individual transistors, resistors, etc., or they can consist of gate-level integration. They may also contain various subsystem macros as mentioned in Sec. 9.6. If the circuit is designed using high-level macros and a restricted set of gates, it is described as a *standard cell* design.

Solved Problems

9.1 A generic PLA which has no on-board flip-flops is shown in Fig. 9-10. Write Boolean logic equations for the outputs.

The inputs are A, B, C and the outputs are W, X as shown. Each OR gate is driven by a subset of the AND gates whose input connections are shown in the left-hand portion of Fig. 9-10. The combined inputs to each AND gate can be represented by a Boolean expression, e.g., A′B′C′ for the top AND gate. Combining all the AND terms connected to each OR yields

$$W = A'B'C' + AB'C + ABC$$
$$X = A'BC + ABC' + ABC$$

* LCA is a registered trademark of XILINX, Inc.

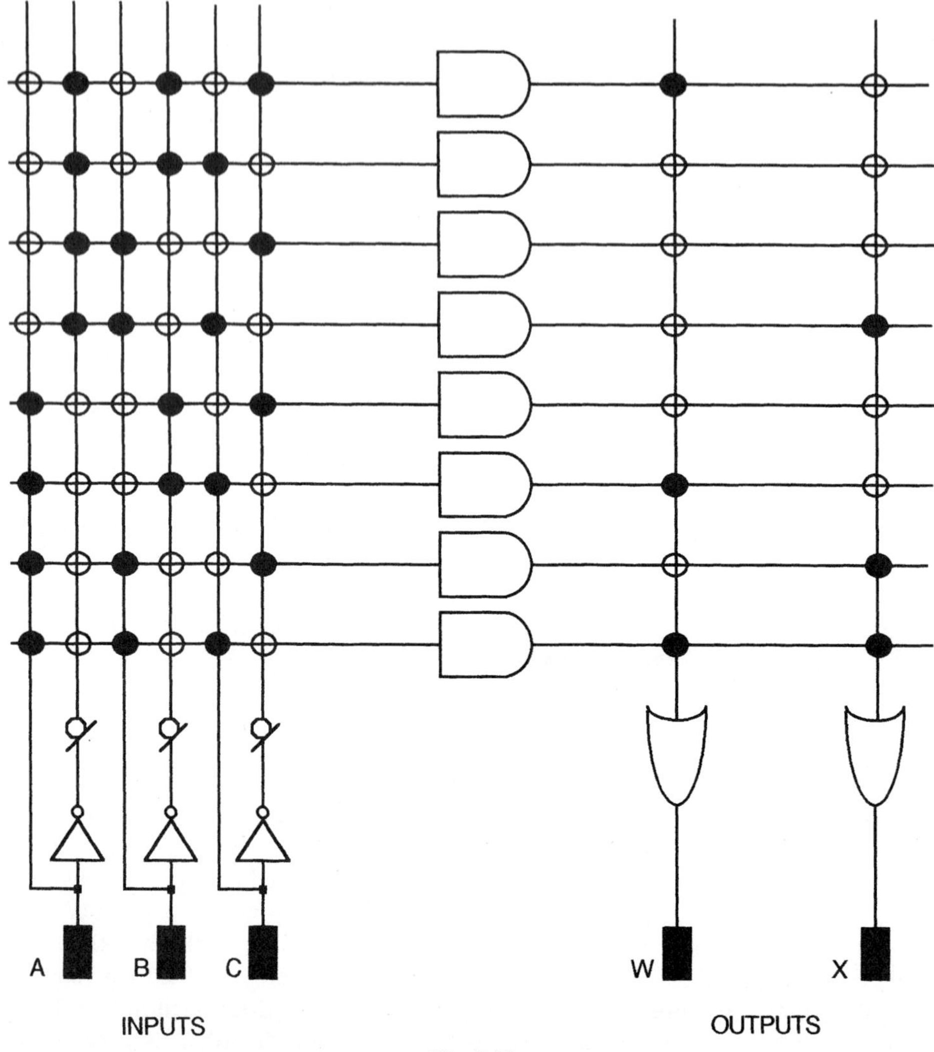

Fig. 9-10

9.2 Design a reversible shift register whose stages are connected in a ring, making use of the following specifications:

- The hardware is to be a 16R6 PAL, as shown in Fig. 9-4.
- Each clock pulse will cause the register to shift left when a control signal LR is 1 and to shift right when LR is 0.
- The shift register is to be loaded in parallel from data appearing on four input pins under the control of a signal LS which will cause loading when TRUE and shifting when FALSE.

Show the PALASM equations for Boolean entry, and sketch the connection pattern.

The shift sequence is shown in Fig. 9-11 and a block diagram of the specified system in Fig. 9-12.

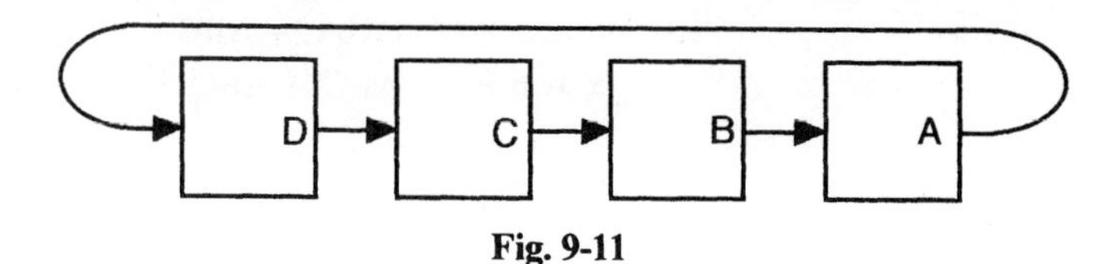

Fig. 9-11

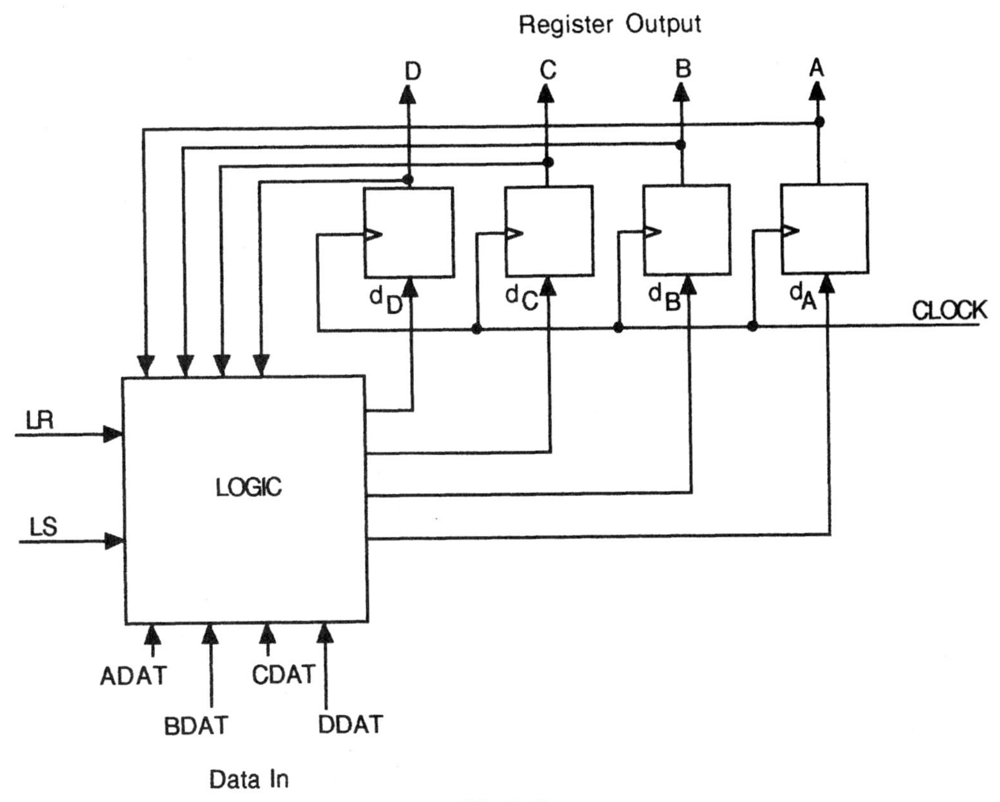

Fig. 9-12

1. When LS = 0, each flip-flop is to receive its d input from an adjacent right-hand neighbor when LR = 1 and from an adjacent left-hand neighbor when LR = 0. In Boolean form, for a given flip-flop k:

$$d_{in(k)} = \text{(FF RIGHT) (LR)} + \text{(FF LEFT) (LR)}'$$

2. During the loading mode (LS = 1), the register's d inputs are connected to input data. When LS = 0, the connections should be as in step 1, above. Again, in Boolean form:

$$d_{in(k)} = [\text{(FF RIGHT) (LR)} + \text{(FF LEFT) (LR)}'] \text{(LS)}' + (k_{dat}) \text{(LS)}$$

3. The PALASM equations are

$$A := D*LR*/LS + B*/LR*/LS + A_{dat}*LS$$
$$B := A*LR*/LS + C*/LR*/LS + B_{dat}*LS$$
$$C := B*LR*/LS + D*/LR*/LS + C_{dat}*LS$$
$$D := C*LR*/LS + A*/LR*/LS + D_{dat}*LS$$

Refer to the 16R6 diagram in Fig. 9-13 for one of the many possible interconnection schemes.

Sequential logic circuits of the type considered in this, and the following two problems, are called *state machines*, and they are considered in more detail in Chap. 10.

9.3 A simple stepping motor requires two drive voltages (A and B) which must be sequenced in a particular pattern. Voltages remain constant until a clock pulse occurs, at which time, the signals A and B change as shown in Table 9.2, and the motor shaft advances one step. Program a PAL16R6 (Boolean entry) so that the motor will move, under command of external signal F/R, in either a forward or reverse direction. The PAL IC should have three inputs: F/R, MOVE, and a clock. Motion is to take place only when MOVE is TRUE and the rotation rate will be determined by the clock frequency.

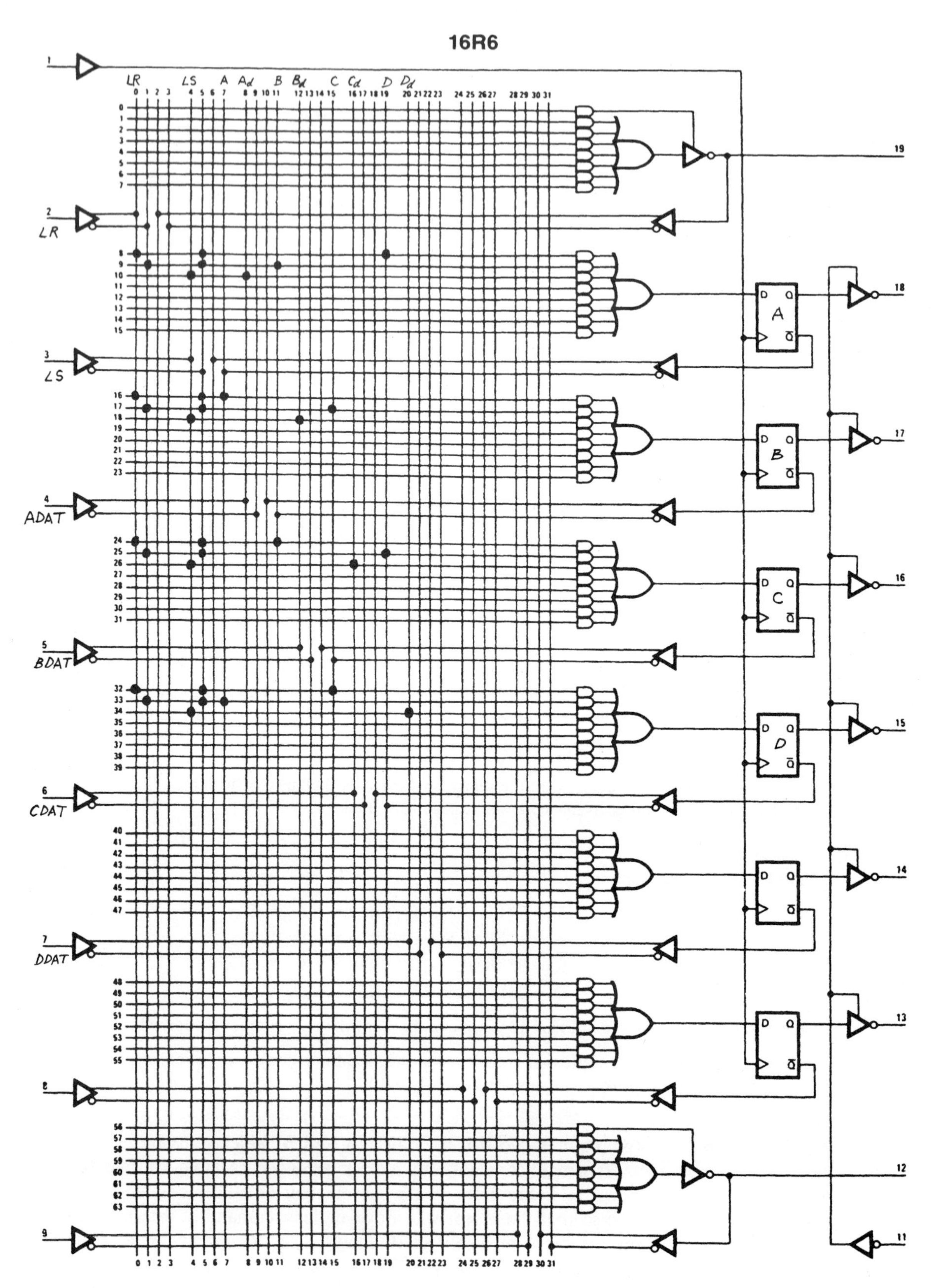

Fig. 9-13 Connection pattern for Problem 9.2 drawn on a PAL16R6 circuit diagram (Copyright © 1991 Advanced Micro Devices, Inc. Reprinted with permission of copyright owner. All rights reserved.)

Table 9.2 Drive Voltages for a Stepping Motor*

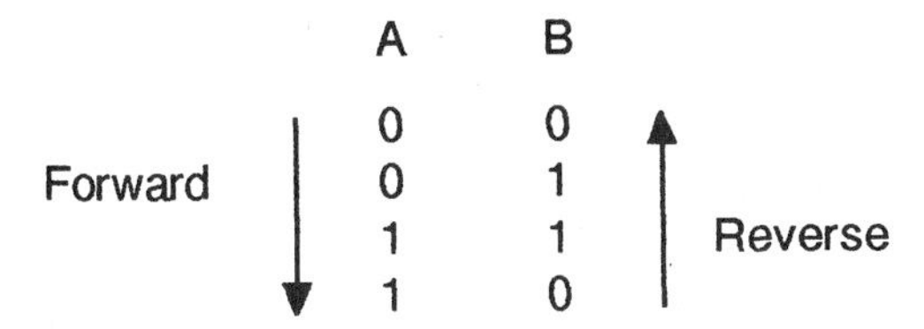

	A	B	
Forward ↓	0	0	↑ Reverse
	0	1	
	1	1	
	1	0	

* The sequence folds so that state 00 follows state 10 in the forward direction. Logic 0 represents 0 V and logic 1 represents a suitable dc voltage level.

We will choose a design combining D flip-flops and logic, whose Boolean description will constitute PAL device input programming. Assume F/R = 1 means forward motion and that flip-flops A and B will be used to store stepping motor signals between clock pulses.

A block diagram of the system is shown in Fig. 9-14. Each pair of d inputs constitute the next state of the flip-flops and is on deck awaiting a clock pulse before being applied to the motor and producing a shaft step. Since, at any given time, the next motor drive voltages A and B to be applied depend on the motion and direction signals as well as upon the status of present drive signals, logic must be designed to produce D flip-flop inputs which are functions of the move command, the direction signal F/R, and the present flip-flop states. A truth table (Table 9.3) may be constructed directly from the given specifications.

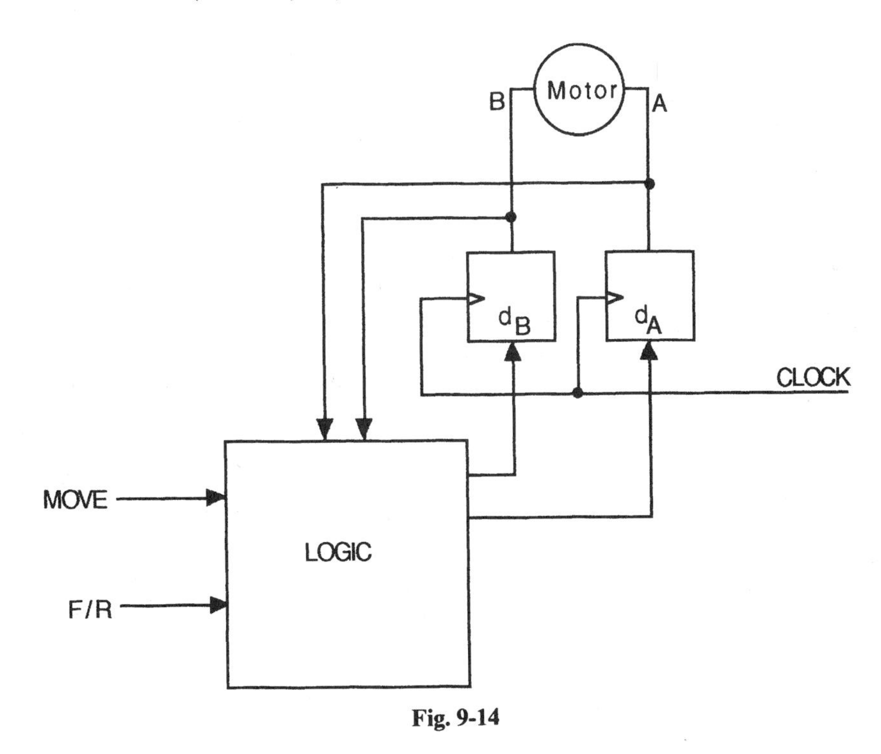

Fig. 9-14

Note how motor outputs A, B are used as logic inputs to determine the next state in accordance with the sequencing information from Table 9.2. When M = 0, next-state outputs do not change, indicating that the motor shaft remains stationary.

Boolean equations, in PALASM form, may be written directly from the truth table:

$$A := /F*/M*A*/B + /F*/M*A*B + /F*M*/A*/B + /F*M*A*/B + F*/M*A*/B + F*/M*A*B + F*M*/A*B + F*M*A*B$$

$$B := /F*/M*/A*B + /F*/M*A*B + /F*M*A*/B + /F*M*A*B + F*/M*/A*B + F*/M*A*B + F*M*/A*/B + F*M*/A*B$$

Table 9.3

Inputs to Logic				Outputs (Next State)	
F/R	M	A	B	d_A	d_B
0	0	0	0	0	0
0	0	0	1	0	1
0	0	1	0	1	0
0	0	1	1	1	1
0	1	0	0	1	0
0	1	0	1	0	0
0	1	1	0	1	1
0	1	1	1	0	1
1	0	0	0	0	0
1	0	0	1	0	1
1	0	1	0	1	0
1	0	1	1	1	1
1	1	0	0	0	1
1	1	0	1	1	1
1	1	1	0	0	0
1	1	1	1	1	0

The PALASM symbol (:=) is used to indicate that outputs A and B are registered. The fact that these variables appear on both sides of their respective equations indicates that next-state values are functions of their present values, as described above.

There are eight AND terms for each OR, and only two D flip-flops are required. Thus, the design will fit on a 16R6, and no simplification need be attempted.

9.4 It is desired to implement the sequence shown in Table 9.4 with the flip-flops of a PAL 16R6. When a control input U/D = 0, the sequence is to move up and, when U/D = 1, it is to reverse direction and move down. Determine whether the circuit can fit on the specified PAL IC, and, if it can, develop the control logic equations. Indicate, also, how one would apply a signal to initialize the flip-flops to 0000.

Table 9.4

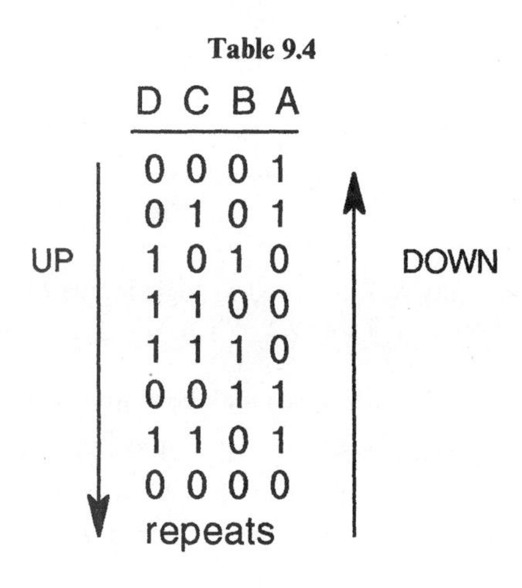

D	C	B	A
0	0	0	1
0	1	0	1
1	0	1	0
1	1	0	0
1	1	1	0
0	0	1	1
1	1	0	1
0	0	0	0
repeats			

As in the previous problem, the flip-flop outputs only change when a clock pulse is received. Each next state is determined by the logic, using present state information and the U/D variable as inputs. An appropriate truth table may be written from the given sequence. For example, we see from Table 9.4, that when the present state is 1110 (fifth row), then the next state will be 0011 when U/D = 0 and 1100 when U/D = 1. The result is shown in Table 9.5 in which nonoccurring states have been omitted.

Table 9.5

INPUTS TO LOGIC (Present State)					LOGIC OUTPUTS (Next State)			
UD	D	C	B	A	dD	dC	dB	dA
0	0	0	0	0	0	0	0	1
0	0	0	0	1	0	1	0	1
0	0	0	1	1	1	1	0	1
0	0	1	0	1	1	0	1	0
0	1	0	1	0	1	1	0	0
0	1	1	0	0	1	1	1	0
0	1	1	0	1	0	0	0	0
0	1	1	1	0	0	0	1	1
1	0	0	0	0	1	1	0	1
1	0	0	0	1	0	0	0	0
1	0	0	1	1	1	1	1	0
1	0	1	0	1	0	0	0	1
1	1	0	1	0	0	1	0	1
1	1	1	0	0	1	0	1	0
1	1	1	0	1	0	0	1	1
1	1	1	1	0	1	1	0	0

Each output variable has no more than eight 1s; therefore, eight AND gates are sufficient for each output, which happens to be the number available on the 16R6. There are four output variables and four flip-flops are needed. Since these are available also, the design will fit.

Taking the equations directly from the truth table:

$$\begin{aligned}
D &:= (U/D)'D'C'BA + (U/D'D'CB'A + (U/D)'DC'BA' + (U/D)'DCB'A' + (U/D)D'C'B'A' \\
&\quad + (U/D)D'C'BA + (U/D)DCB'A' + (U/D)DCBA' \\
C &:= (U/D)'D'C'B'A + (U/D)'D'C'BA + (U/D)'DC'BA' + (U/D)'DCB'A' + (U/D)D'C'B'A' \\
&\quad + (U/D)D'C'BA + (U/D)DC'BA' + (U/D)DCBA' \\
B &:= (U/D)'D'CB'A + (U/D)'DCB'A' + (U/D)'DCBA' + (U/D)D'C'BA + (U/D)DCB'A' \\
&\quad + (U/D)DCB'A \\
A &:= (U/D)'D'C'B'A' + (U/D)'D'C'B'A + (U/D)'D'C'BA + (U/D)'DCBA' + (U/D)D'CB'A \\
&\quad + (U/D)DC'BA' + (U/D)DCB'A
\end{aligned}$$

As far as initialization is concerned, a signal called IN may be ANDed with all product terms. When IN is 0, the D inputs to all flip-flops are 0 and they all assume the 0 state when the next clock pulse arrives. When IN goes to 1, the AND gates are free to function in the manner defined by the logic equations. The connection diagram for a PAL16R6 implementation is shown in Fig. 9-15.

9.5 The connection diagram for a generic PAL device is shown in Fig. 9-16, and signals which are applied to the various inputs appear in Fig. 9-17. Sketch the macrotiming waveshapes at outputs A, B, C, D, and E. Neglect any internal propagation delays and assume that all flip-flops are

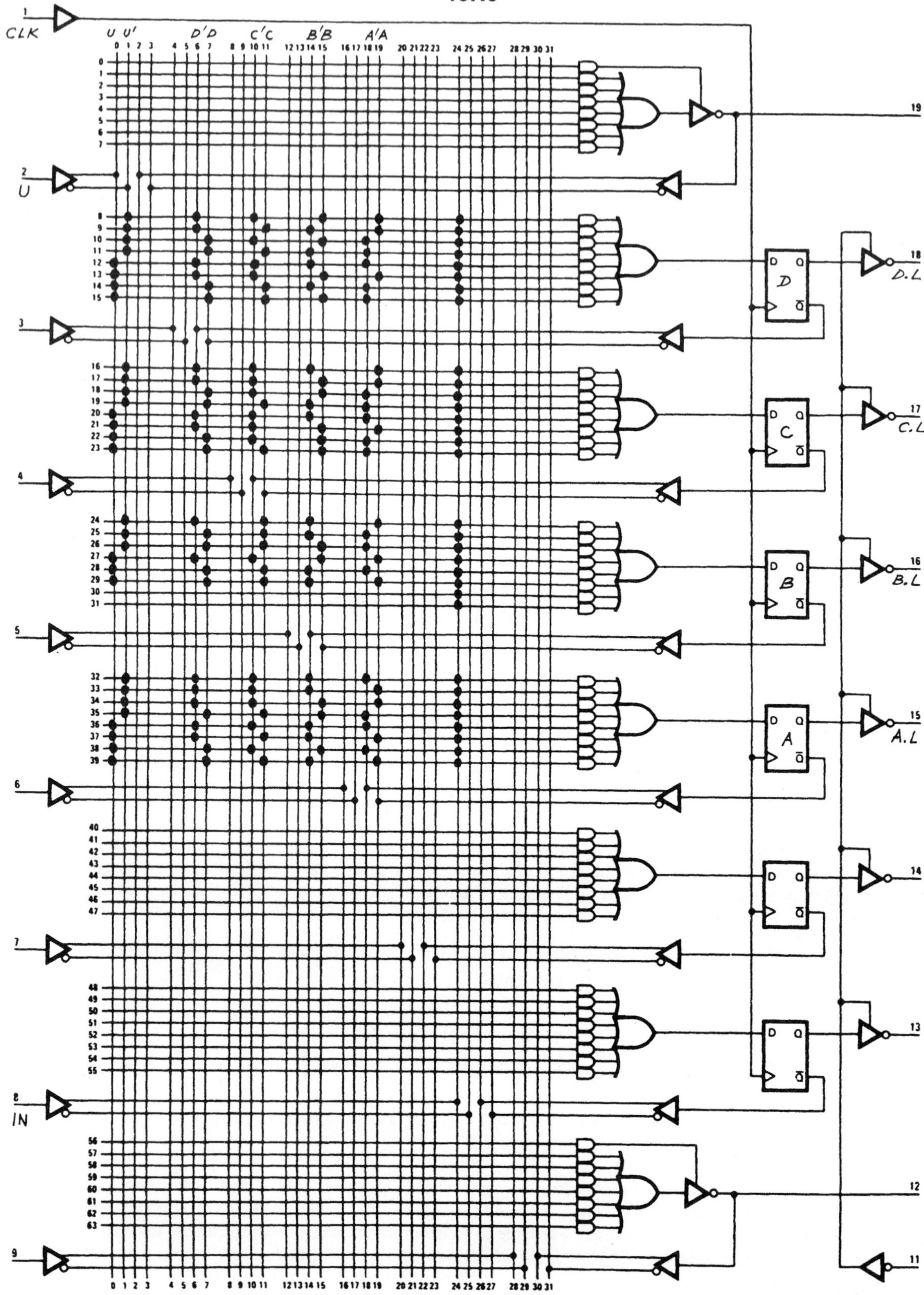

Fig. 9-15 Connection pattern for Problem 9.4 drawn on a PAL16R6 circuit diagram (Copyright © 1991 Advanced Micro Devices, Inc. Reprinted with permission of copyright owner. All rights reserved.)

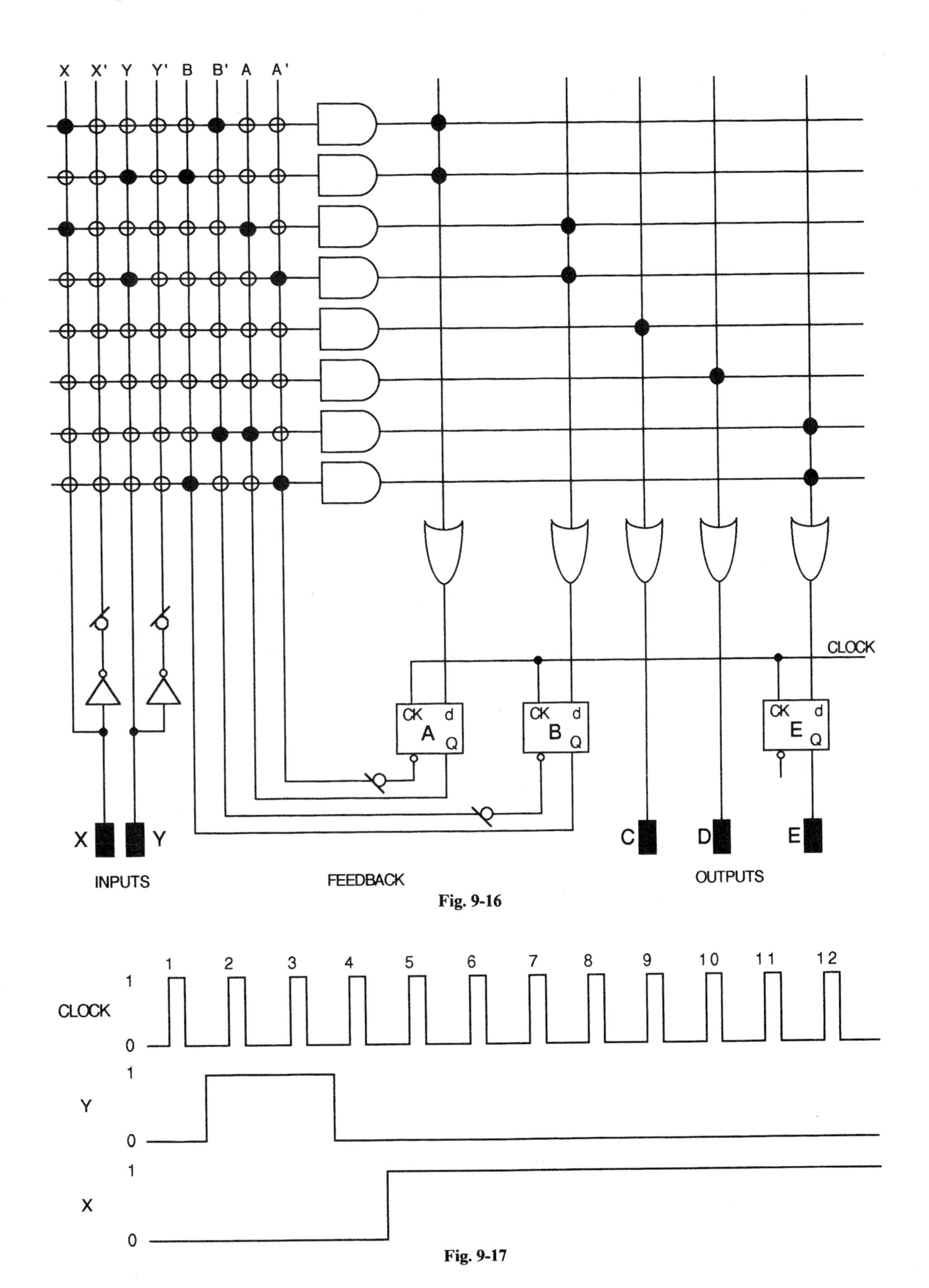

Fig. 9-16

Fig. 9-17

cleared prior to the first clock pulse. The flip-flops are positive-edge-triggered D type (refer to Chap. 7).

1. From Fig. 9-16, we see that the flip-flop input equations are

$$d_A = B'X + BY \qquad d_B = AX + A'Y \qquad d_E = AB' + A'B$$

2. C and D are unconnected internally and do not affect flip-flop states.
3. Initially A, B, and E are 0 (cleared) and X, Y are 0 (given waveforms).

Tabulating the results as clock pulses occur yields Table 9.6.

Table 9.6

			Present		Next			
Pulse	X	Y	A	B	d_A	d_B	d_E	E
1	0	0	0	0	0	0	0	0
2	0	1	0	0	0	1	0	0
3	0	1	0	1	1	1	1	0
4	0	0	1	1	0	0	0	1
5	1	0	0	0	1	0	0	0
6	1	0	1	0	1	1	1	0
7	1	0	1	1	0	1	0	1
8	1	0	0	1	0	0	1	0
9	1	0	0	0	1	0	0	1
10	1	0	1	0	1	1	1	0
11	1	0	1	1	0	1	0	1
12	1	0	0	1	0	0	1	0

The required waveforms may be obtained by reading down columns A, B, and E in Table 9.6. The result is shown in Fig. 9-18.

9.6 The PAL device circuit diagram of Fig. 9-19 shows the connections for a system containing both flip-flops and combinational logic. Assuming that the signal IN is initially FALSE (low) and becomes TRUE immediately following the first clock pulse at $t = 0$, find the flip-flop sequence and express it as a series of decimal numbers assuming that A is the most significant digit, followed by B, C, D. Determine when output E becomes true.

Since there are no AND gate connections to E, outputs A, B, C, and D do not depend on it, and, consequently, they may be treated as a separate system. The logic yields

$$d_A = AB' + AC' + A'B \qquad d_C = B'C + A'D' + A'C' + ABC'$$
$$d_B = AC' + B'D + B'C \qquad d_D = A'B' + BD'$$

which may be mapped as shown in Fig. 9-20.

Since IN is initially FALSE and since it is connected to all the AND gates which drive flip-flops, the first clock pulse reads 0000 into ABCD. This is the situation denoted by the upper-left corner of the maps. Congruent map entries collectively represent the next state for ABCD which, from Fig. 9-20, is seen to be ABCD = 0011. Similarly, following the second clock pulse, the map coordinates change to first row, third column. Entries at this point show a next state of ABCD = 0111 which will occur at clock pulse 3. This procedure can be continued until a previously obtained value is encountered, indicating that the sequence repeats (refer to Table 9.7).

The connections to output flip-flop E are A'B'C'D' (ABCD = 0000) and A'CD (ABCD = 0111 or 0011). Thus, E is TRUE during the first three clock cycles of the counting sequence.

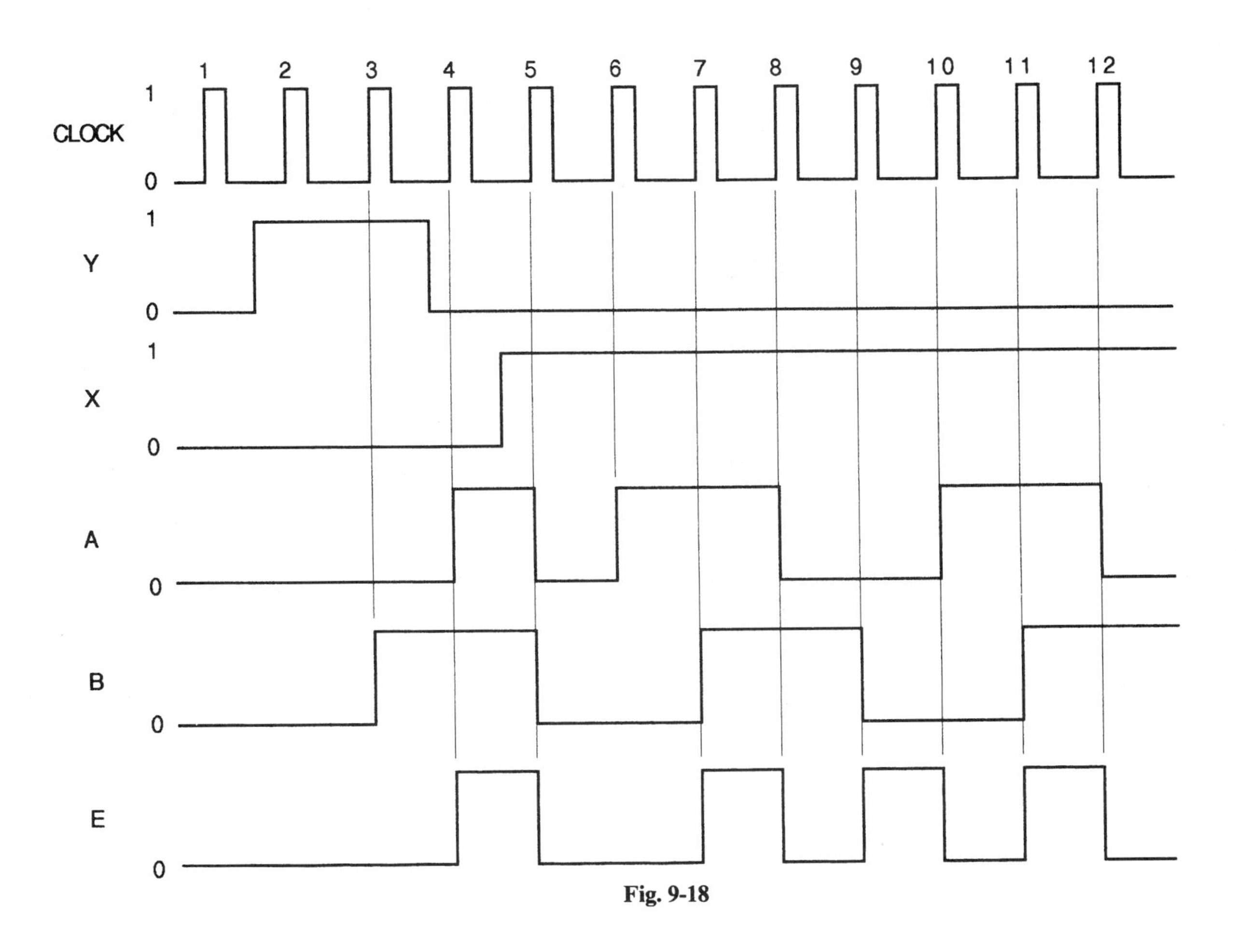

Fig. 9-18

Table 9.7 Representation of the Sequence 0-3-7-8-12-15-0

A	B	C	D
0	0	0	0
0	0	1	1
0	1	1	1
1	0	0	0
1	1	0	0
1	1	1	1
0	0	0	0
repeats			

9.7 Use the 1-bit binary adder described in Sec. 2.4 to demonstrate how the generic PLA shown in Fig. 9-21 can be programmed directly from a truth table. Recall that each AND gate corresponds to a line in the truth table while each OR gate is associated with an output column.

Table 9.8 shows the logic. Program the AND gates in the order of the truth table rows (as in a full decoder) and then connect each term (row) to the appropriate output OR gate. The result is shown in Fig. 9-22. Note the correspondence between truth table 1s and connections made in the diagram.

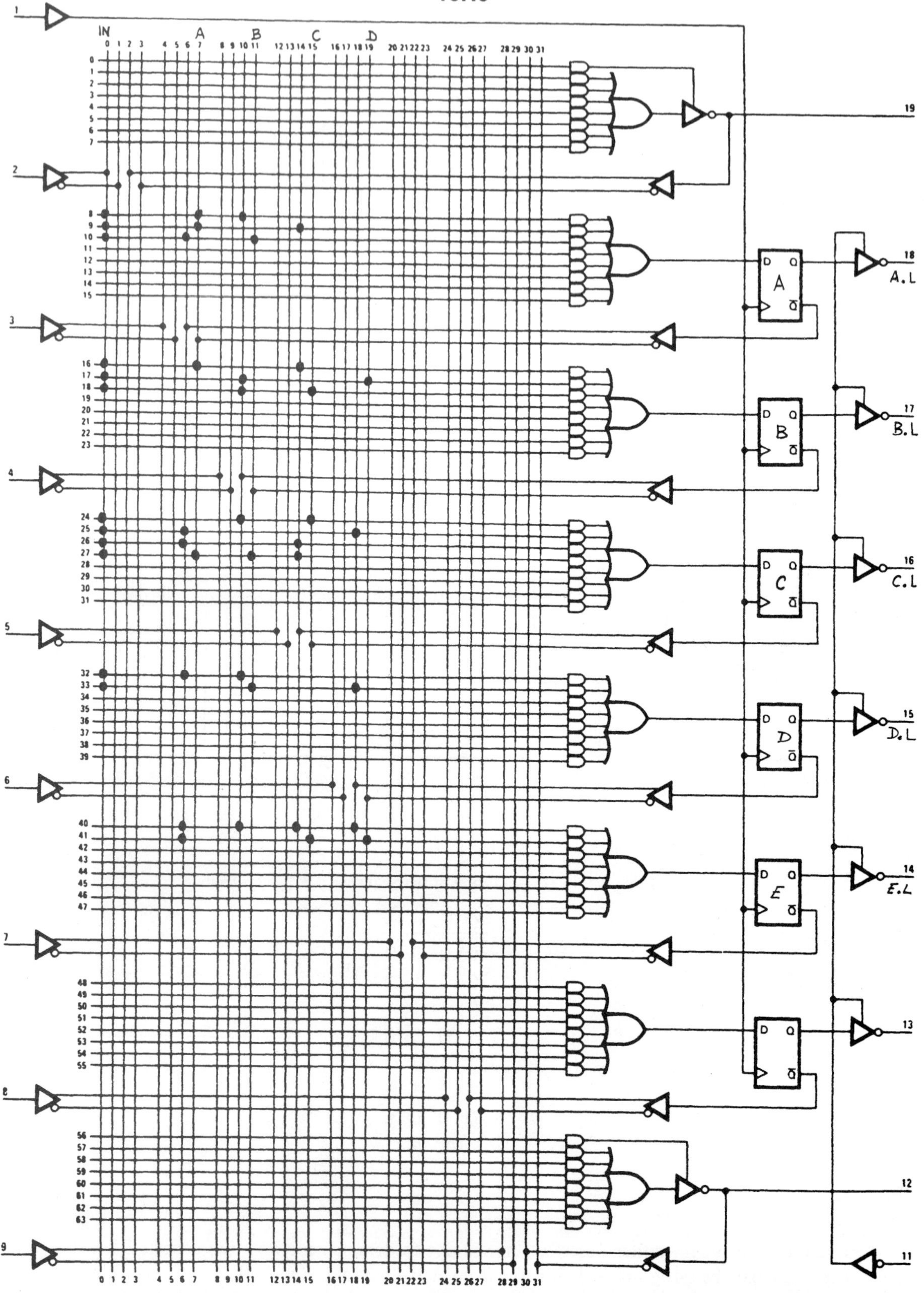

Fig. 9-19 Connection pattern for Problem 9.6 drawn on a PAL16R6 circuit diagram (Copyright © 1991 Advanced Micro Devices, Inc. Reprinted with permission of copyright owner. All rights reserved.)

A B \ D C	0 0	1 0	1 1	0 1
0 0	0	0	0	0
0 1	1	1	1	1
1 1	1	1	0	0
1 0	1	1	1	1

d_A

A B \ D C	0 0	1 0	1 1	0 1
0 0	0	1	1	1
0 1	0	0	0	0
1 1	1	1	0	0
1 0	1	1	1	1

d_B

A B \ D C	0 0	1 0	1 1	0 1
0 0	1	1	1	1
0 1	1	1	0	1
1 1	1	1	0	0
1 0	0	0	1	1

d_C

A B \ D C	0 0	1 0	1 1	0 1
0 0	1	1	1	1
0 1	1	0	0	1
1 1	1	0	0	1
1 0	0	0	0	0

d_D

Fig. 9-20

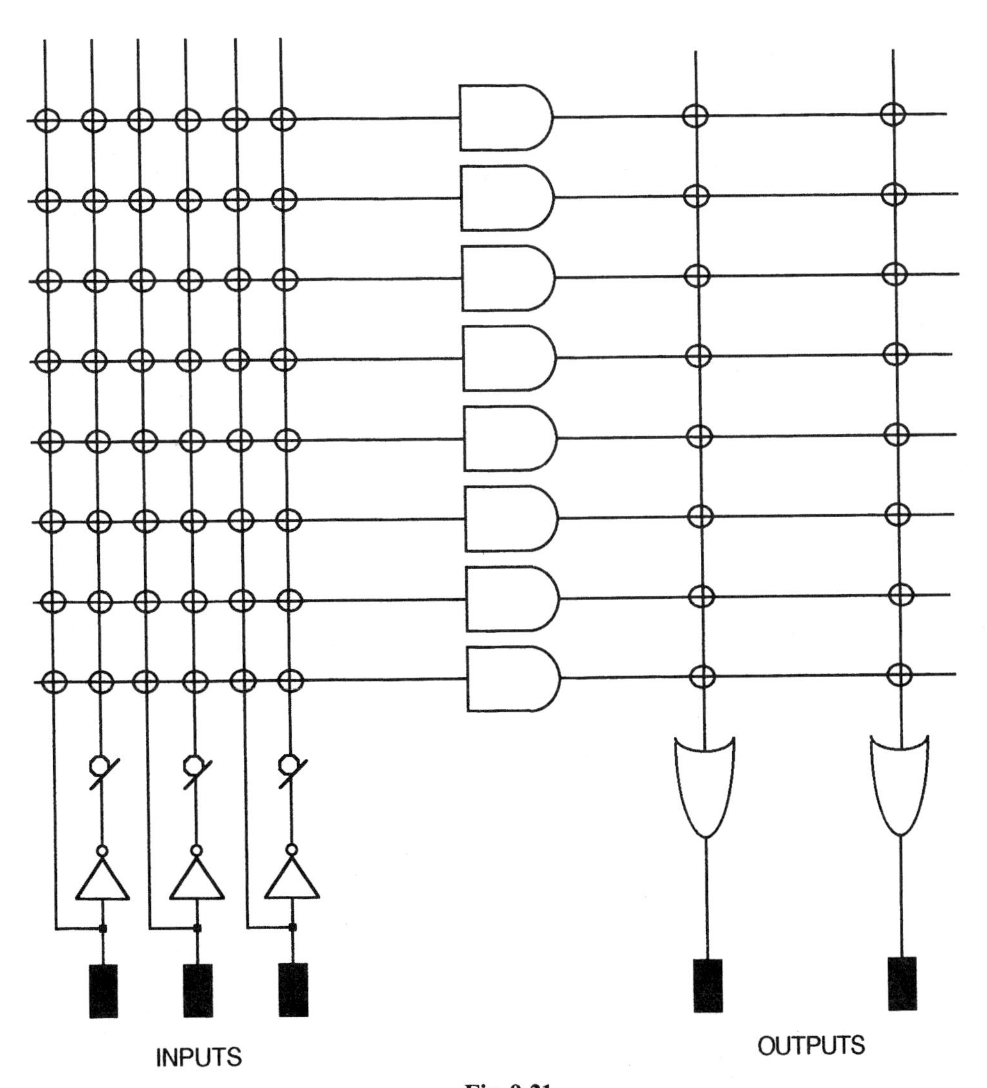

Fig. 9-21

Table 9.8

X	Y	Ci	S	Co
0	0	0	0	0
0	0	1	1	0
0	1	0	1	0
0	1	1	0	1
1	0	0	1	0
1	0	1	0	1
1	1	0	0	1
1	1	1	1	1

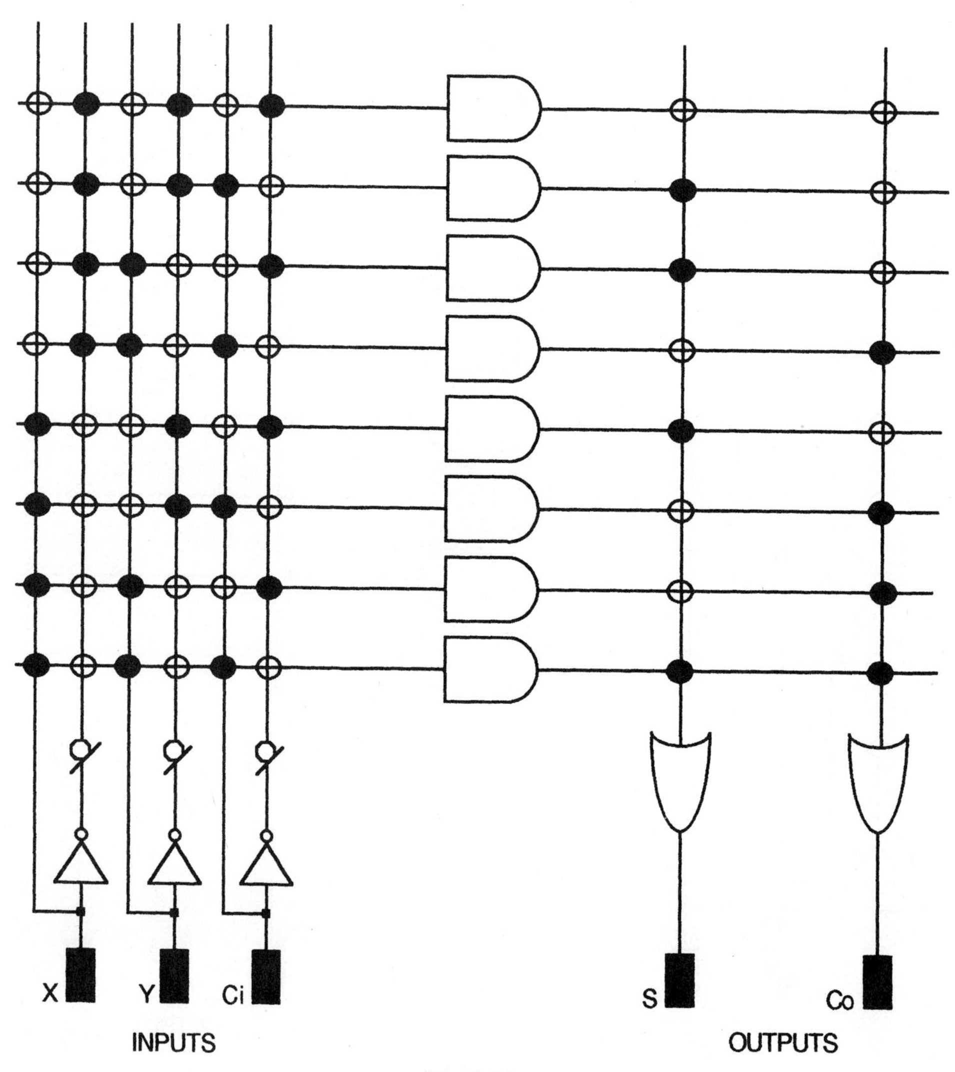

Fig. 9-22

9.8 Implement the adder of Prob. 9.7 on the generic PAL shown in Fig. 9-23. Demonstrate that while it is possible to implement the function, it is not possible to program it directly from a standard truth table.

Note the PAL IC's structure. The AND gates are grouped in sets, and each group is hard wired to the inputs of a single OR gate. Only if the entire group of AND gates is used as consecutive entries in a truth table is there a correspondence. In the current problem solution, the AND gates must be rearranged relative to the truth table.

Truth table rows having decimal equivalents 1, 2, 4, and 7 are grouped for output S, while decoder outputs corresponding to rows 3, 5, 6, and 7 are grouped for output C_o (the second OR gate). Since connections represent 1s, we see that the truth table has had its rows rearranged, and one of them, row 7, is used twice (see Fig. 9-24).

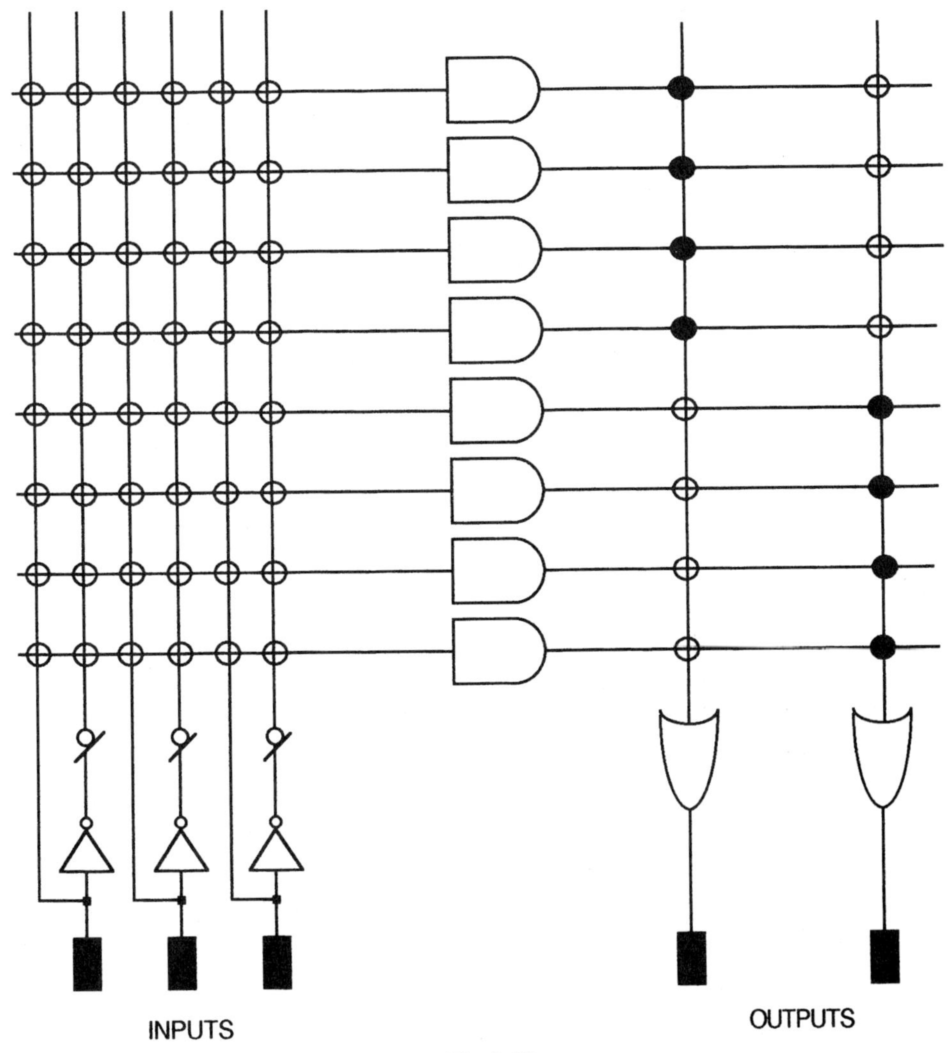

Fig. 9-23

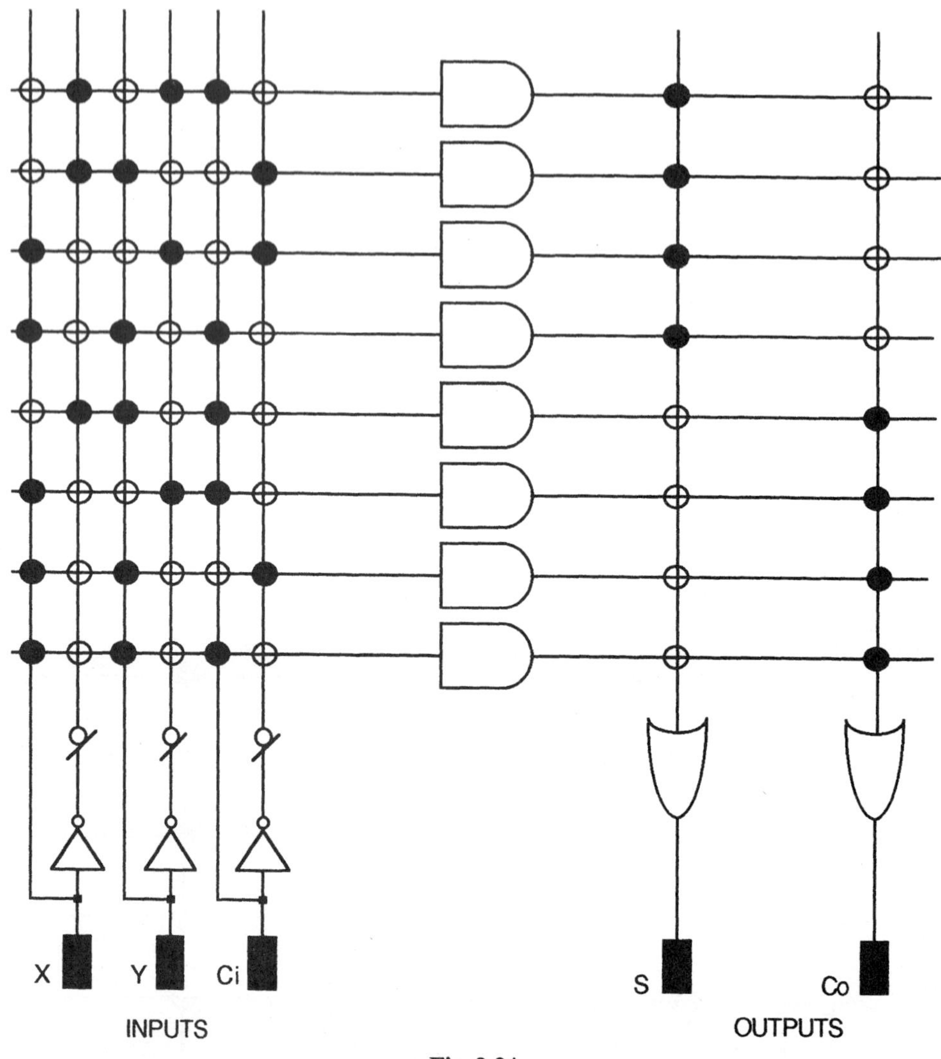

Fig. 9-24

Supplementary Problems

9.9 Given the programmed generic PAL device shown in Fig. 9-25, determine the logic equation for outputs C, D, and E.

9.10 Using the system of Fig. 9-25, determine the output sequence for clocked operation given that X and Y are both maintained at 0. The initial values of A and B are both 0s.

9.11 What is the effect on the sequence of Prob. 9.10 if, after the third clock pulse but before the fourth, Y goes to 1 and remains there?

9.12 The PAL IC connection diagram for a counting circuit is shown in Fig. 9-26. Assuming that pin 11 is tied low and that the flip-flops are initially cleared, write in sequential order the decimal equivalent output values of ABCF for as many clock pulses as required for a repeat to occur.

9.13 This problem demonstrates how the connection pattern for a PLA device may be compactly defined by means of a simple hexadecimal code. Referring to the generic PLA shown in Fig. 9-27, each row of possible

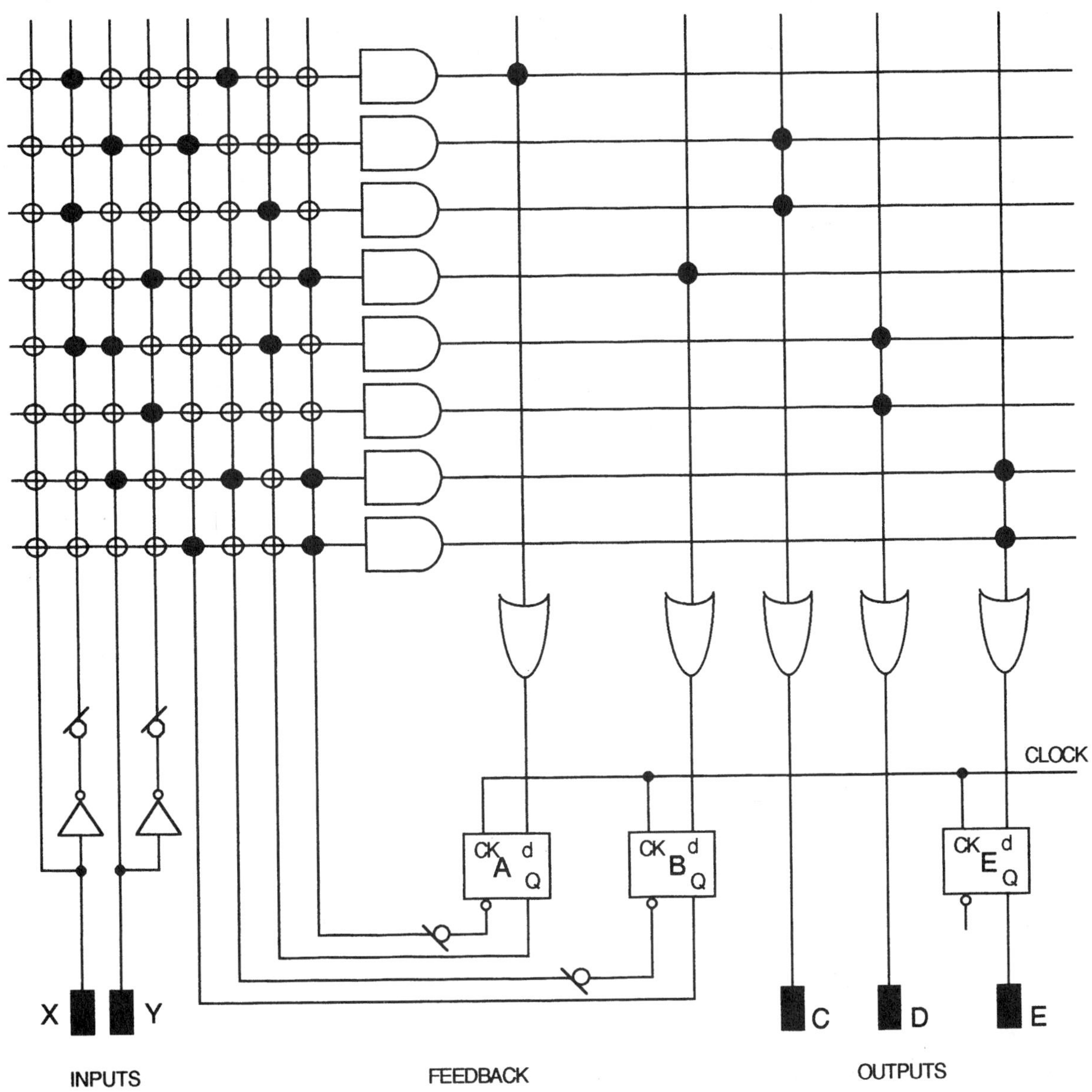

Fig. 9-25

interconnects can be described by a three-digit hex word, the first two digits being assigned to the eight input lines and the last digit to the four output lines. Using this scheme, write an eight-line code sequence that describes the connections in Fig. 9-27.

9.14 With reference to Prob. 9.13, are there any disallowed hexadecimal codes?

9.15 For the PLA of Prob. 9.13, specify the logic defined by the following hexadecimal code words:

49A
4AA
86A
85A
454
494
864
8A4

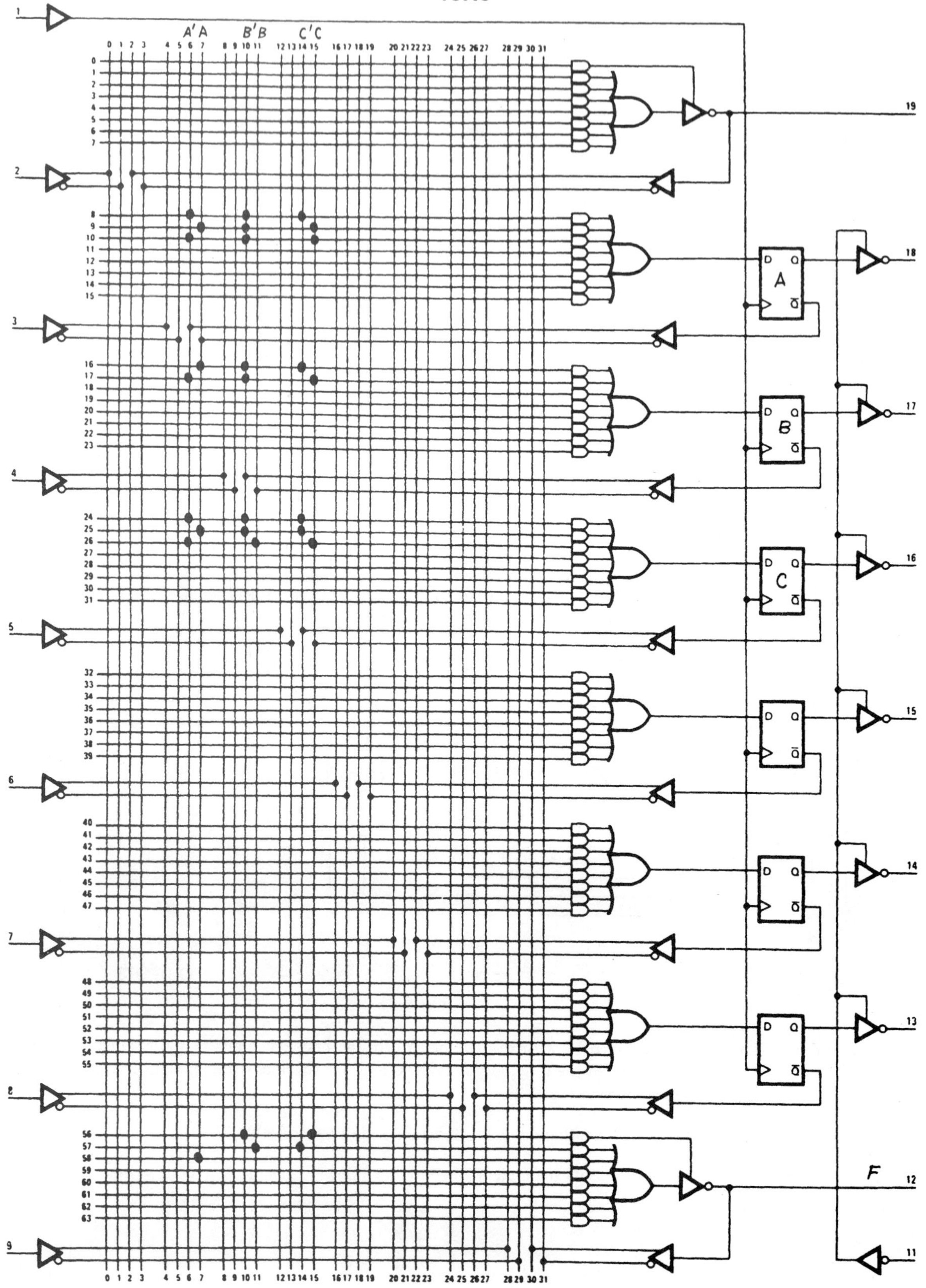

Fig. 9-26 Connection pattern for Problem 9.12 drawn on a PAL16R6 circuit diagram (Copyright © 1991 Advanced Micro Devices, Inc. Reprinted with permission of copyright owner. All rights reserved.)

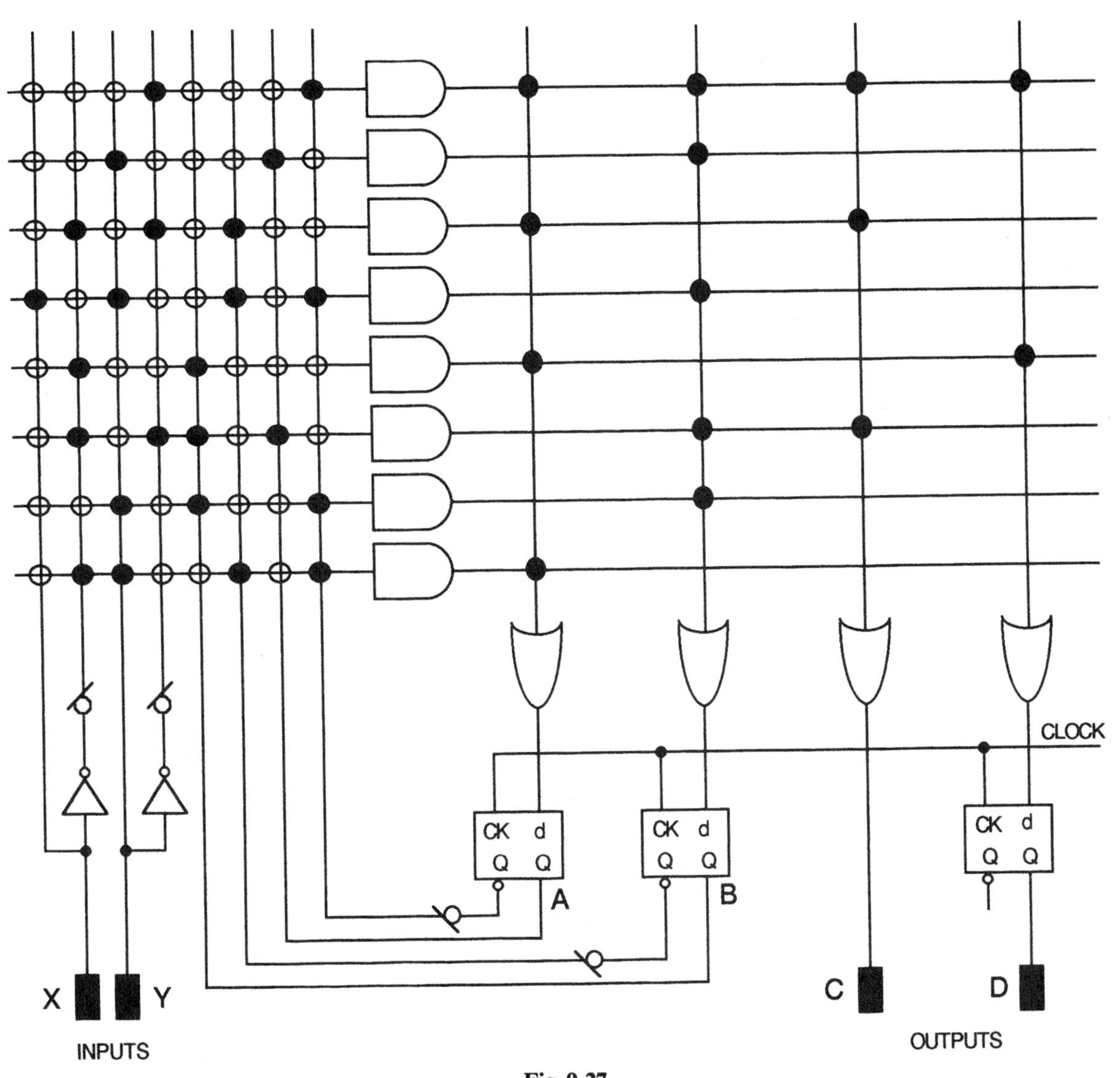

Fig. 9-27

9.16 Using the hex program of Prob. 9.15, tabulate the clocked performance of the PLA assuming that A and B are initially both 0 and that (*a*) X is always 0 and (*b*) X is always 1.

9.17 Draw, on a PAL16R6 connection diagram, the implementation of a synchronous up/down counter which produces the sequence shown in Table 9.9. There is to be a control input F which, when 1, causes the circuit to count down and, when 0, causes the circuit to count up.

9.18 Using a PAL16R6 device as the hardware, show all connections required to implement the sequence given in Table 9.10, assuming that the register is initially at 0011.

9.19 If the counter of Prob. 9.18 is started with ABCD = 0001, what will the register outputs be following the next clock pulse?

9.20 The counting sequence shown in Table 9.11 is to be implemented on a PAL16R6 device. Write the PALASM equations required for programming.

Table 9.9

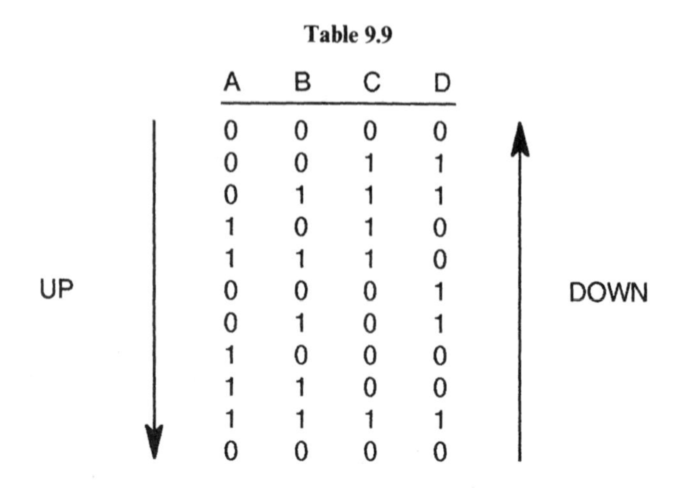

A	B	C	D
0	0	0	0
0	0	1	1
0	1	1	1
1	0	1	0
1	1	1	0
0	0	0	1
0	1	0	1
1	0	0	0
1	1	0	0
1	1	1	1
0	0	0	0

Table 9.10

A	B	C	D
0	0	1	1
0	1	0	0
0	1	0	1
0	1	1	0
0	1	1	1
1	0	0	0
1	0	0	1
1	0	1	0
1	0	1	1
1	1	0	0

repeats

Table 9.11

D	C	B	A
0	0	0	0
0	0	0	1
0	0	1	0
0	1	0	0
0	1	0	1
1	0	1	0
1	0	1	1
1	1	0	1
1	1	1	0
1	1	1	1

repeats

9.21 Refer to Prob. 9.20. Using output pins 18, 17, 16, and 15 for register elements D, C, B, and A, respectively, and assuming that AND gates are assigned in numerical order within an OR-gated group, indicate on a PAL16R6 diagram which of the possible connections should be left connected.

Answers to Supplementary Problems

9.9 $C = YB + X'A$; $D = X'YA + Y'$; $d_E = A'B + A'B'Y$

9.10 See Table 9.12.

After the first clock pulse, C will synchronously oscillate between 1 and 0 while D and E will remain at 1 and 0, respectively.

9.11 See Table 9.13. No change occurs after the sixth clock pulse.

Table 9.12

Clock #	C	D	E
0	0	1	- ← unspecified
1	1	1	0
2	0	1	0
3	1	1	0

Table 9.13

Clock #	C	D	E
4	0	0	0
5	1	1	1
6	1	1	0
7	1	1	0

9.12 0, 11, 9, 6, 3, 13, 0.

9.13 11F
224
54A
A54
489
5A6
294
658

Note how physical information about connections has been converted to a set of hexadecimal numbers which can be easily processed by a digital computer.

9.14 Since a digital signal and its logical inverse should not be simultaneously connected to the inputs of the same AND gate, the first two digits of the hex word may not contain 3, 7, B, C, D, E, or F. There is no such restriction on the third hex digit.

9.15 $d_A = C = X'A'B + X'AB + XAB' + XA'B'$
$d_B = X'A'B' + X'A'B + XAB' + XAB$
$d_D = 0$

9.16 (*a*)

A	B
0	0
0	1
1	1
1	0

(*b*)

A	B
0	0
1	0
1	1
0	1

9.17 Only eight AND terms are possible with the specified PAL device, and the truth table requires ten. Thus, some sort of simplification is required prior to transferring the design to a fuse map. A solution is

$$A_N = F'AB' + F'AC' + F'A'B + FAB + FA'B'C'$$
$$B_N = F'AB' + F'AC' + F'B'D + FB'C' + FAD + FB'D'$$
$$C_N = F'A'D' + F'B'C + F'A'C + F'ABC' + FCD' + FA'BC + FA'B'C'$$
$$D_N = F'A'B' + F'BD' + FB'D' + FA'B$$

The PAL16R6 connection diagram is shown in Fig. 9-28.

9.18 If K-map simplification is used

$$A_N = AB' + BCD$$
$$B_N = A'D' + A'C' + B'CD$$
$$C_N = AB + C'D + CD'$$
$$D_N = D'$$

The connection diagram is shown in Fig. 9-29.

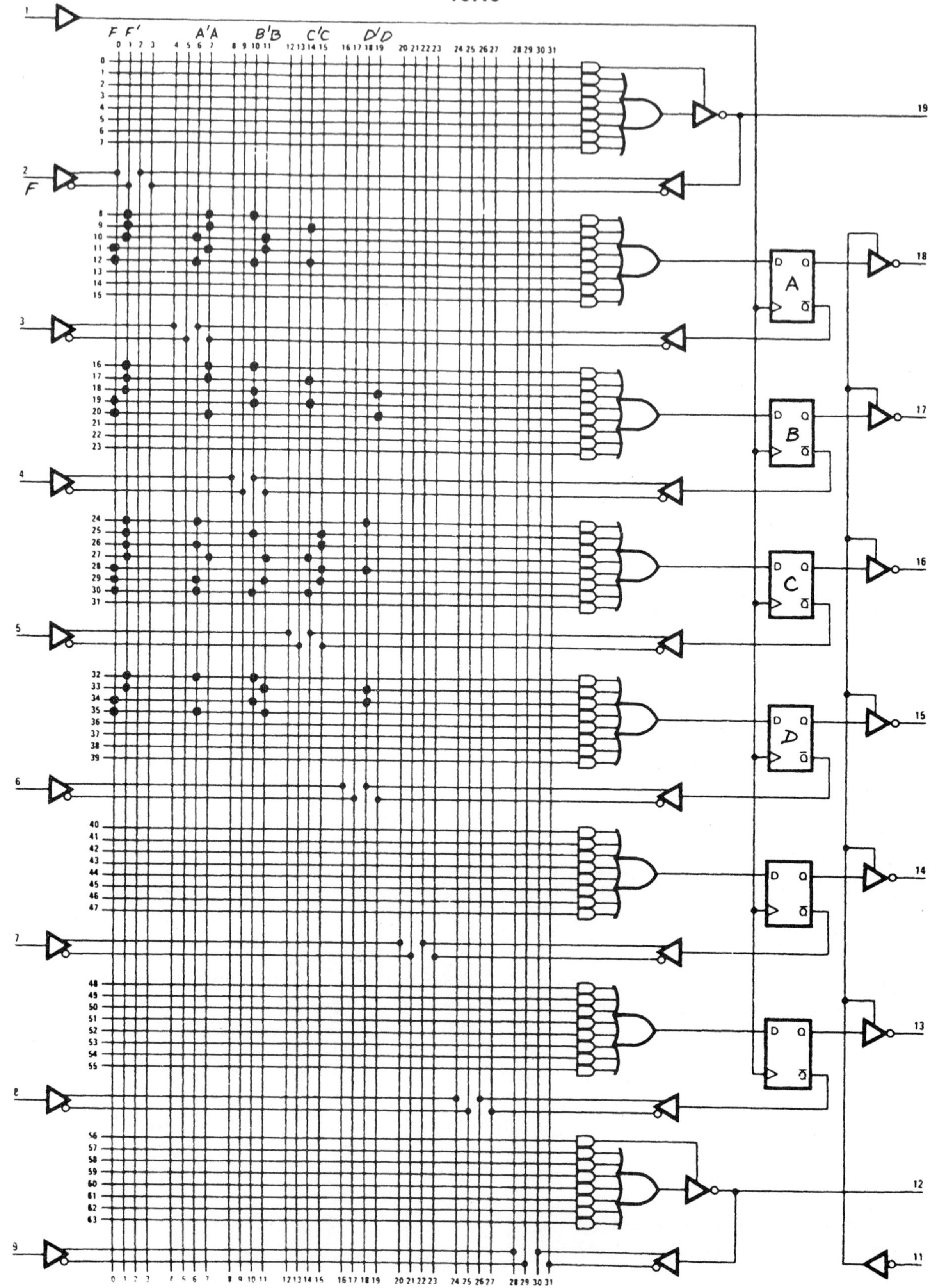

Fig. 9-28 Connection pattern for Problem 9.17 drawn on a PAL16R6 circuit diagram

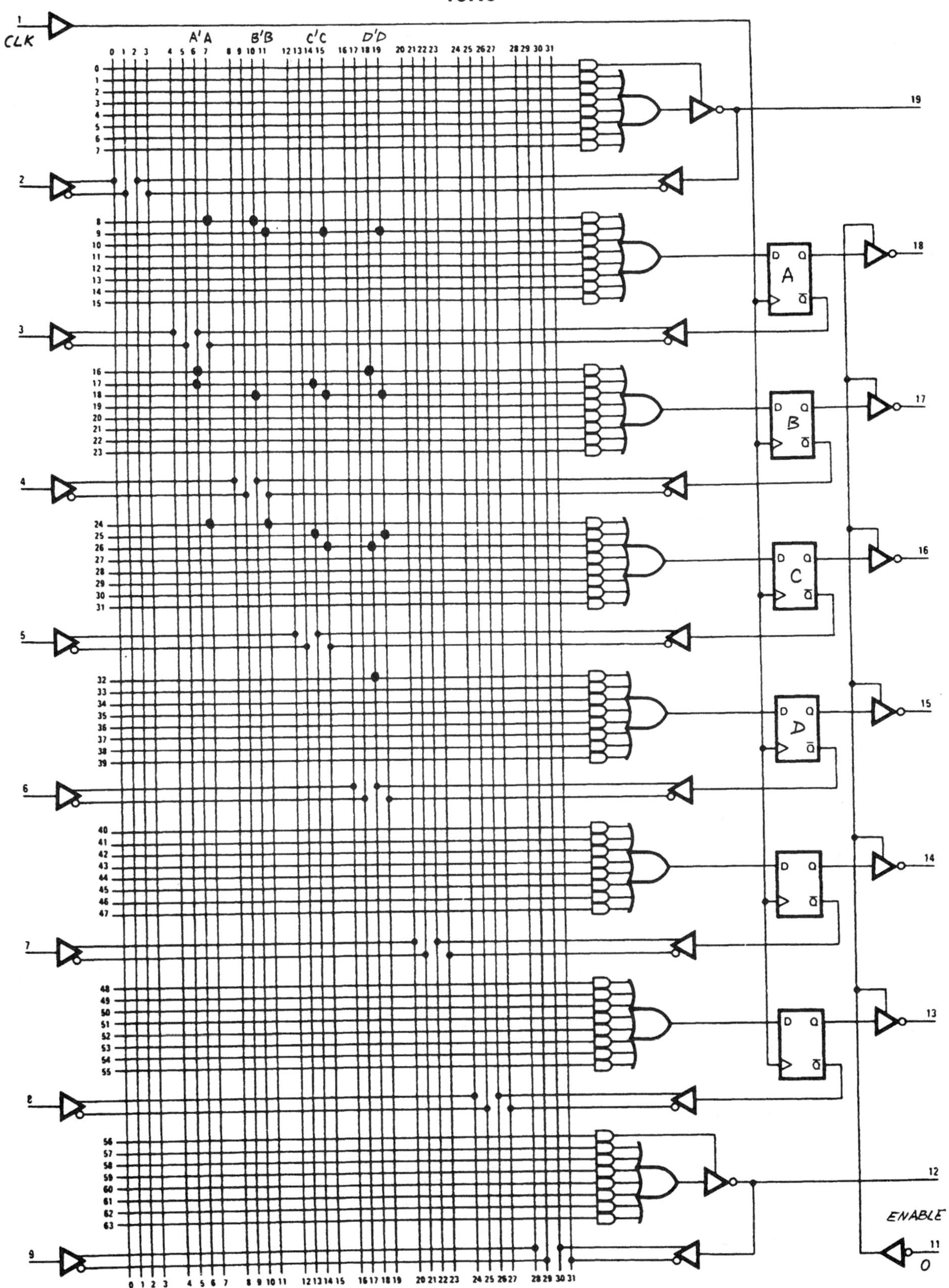

Fig. 9-29 Connection pattern for Problem 9.18 drawn on a PAL16R6 circuit diagram (Copyright © 1991 Advanced Micro Devices, Inc. Reprinted with permission of copyright owner. All rights reserved.)

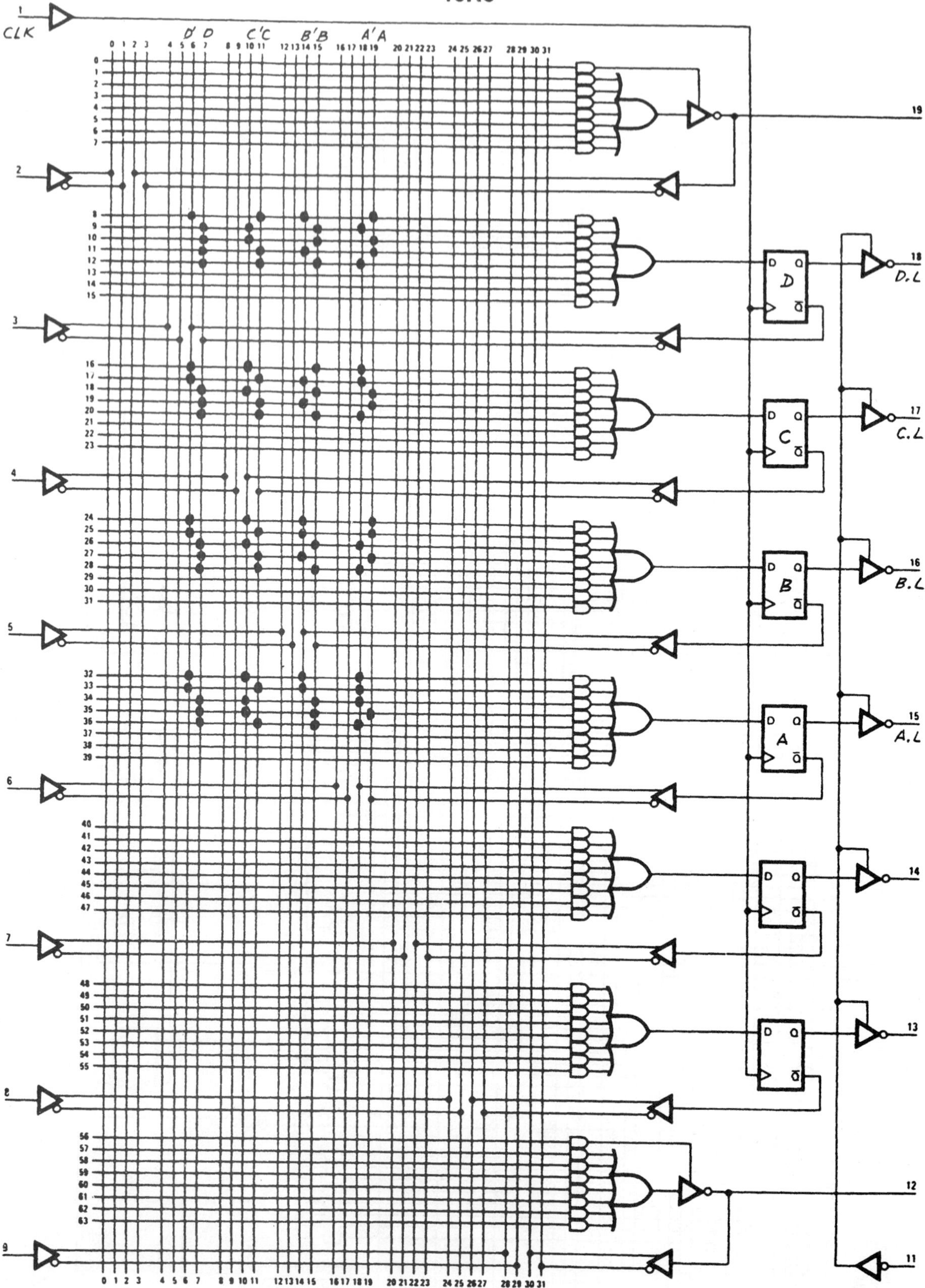

Fig. 9-30 Connection pattern for Problem 9.21 drawn on a PAL16R6 circuit diagram (Copyright © 1991 Advanced Micro Devices, Inc. Reprinted with permission of copyright owner. All rights reserved.)

9.19 If all unspecified values are treated as 0 (no "don't cares"), the next state will be 0000. If "don't cares" are used in a K-map 1s covering, as in the given solution to Prob. 9.18, the next state will be 0110.

9.20 D := /D*C*/B*A + D*/C*B*/A + D*/C*B*A + D*C*/B*A + D*C*B*/A
C := /D*/C*B*/A + /D*C*/B*/A + D*/C*B*A + D*C*/B*A + D*C*B*/A
B := /D*/C*/B*A + /D*C*/B*A + D*/C*B*/A + D*C*/B*A + D*C*B*/A
A := /D*/C*/B*/A + /D*C*/B*/A + D*/C*B*/A + D*/C*B*A + D*C*B*/A

9.21 See Fig. 9-30.

Chapter 10

Design of Simple State Machines

10.1 INTRODUCTION

The *state machine* constitutes an important class of sequential digital logic circuitry that is characterized by the following properties:

1. It has a well-defined portion containing only combinational logic.
2. It has a synchronously clocked latching register whose logic outputs can be used directly or can be fed back to the combinational logic inputs.
3. The combinational logic is capable of receiving, as input, either the fed-back signals or external inputs (or both). It produces outputs which may be accessed via external pins and/or used as latching register inputs.

The basic components of a state machine are illustrated in Fig. 10-1.

The purpose of the combinational logic is to provide, prior to the receipt of each clock pulse, those commands which are required to make the state machine perform its intended function.

The strict separation of a state machines logic and register portions allows the reduction of its design to a straightforward combinational logic problem. An important characteristic of the state machine is the synchronous clocking of logic into the register which ensures that timing differences in the combinational logic circuitry cannot affect performance.

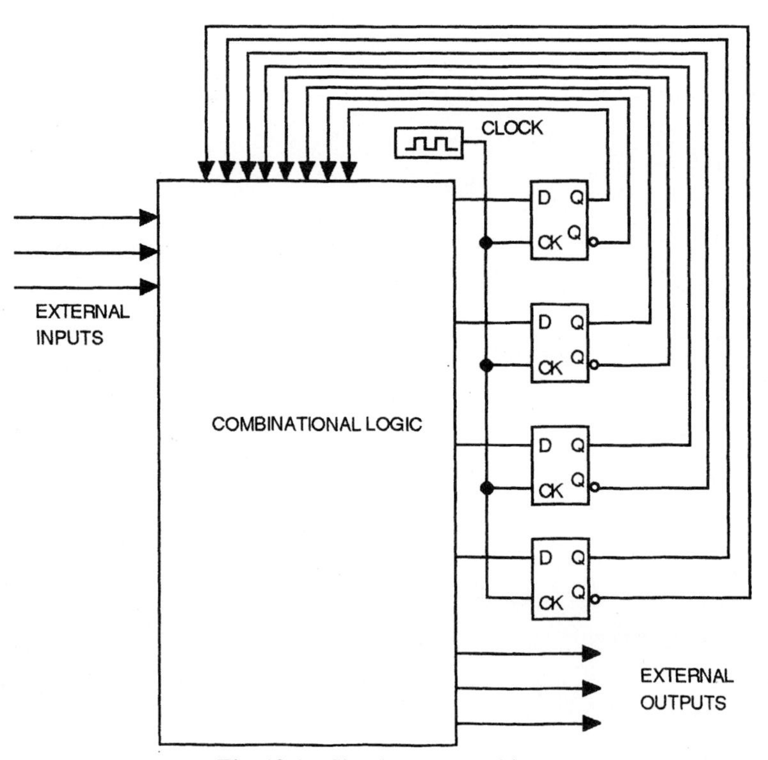

Fig. 10-1 Simple state machine.

10.2 TRADITIONAL STATE MACHINE DESIGN WITH D FLIP-FLOPS

The traditional method of state machine design uses a truth table for basic system specification. The current, or *present state* of the system is stored in the register flip-flops. The combinational logic accepts present-state data as inputs (possibly along with external inputs) and produces the *next state* which is presented to the register's D inputs. Upon receipt of a clock pulse, this data is latched into the register to become the new present state and the process repeats. Because next-state data is static between clock pulses, sequential state machine design can be reduced to a combinational logic design problem in which a truth table may be used to specify the future state of a given flip-flop as a function of the values of present-state variables.

EXAMPLE 10.1 **Synchronous counter specification.** In synchronous counter design, the count and the state are one and the same. Consider the 4-bit Gray code counter which counts in the sequence shown in Table 10.1. The individual flip-flops of the register are identified by D, C, B, and A.

Table 10.1 Counting Sequence

D	C	B	A
0	0	0	0
0	0	0	1
0	0	1	1
0	0	1	0
0	1	1	0
0	1	1	1
0	1	0	1
0	1	0	0
1	1	0	0
1	1	0	1
1	1	1	1
1	1	1	0
1	0	1	0
1	0	1	1
1	0	0	1
1	0	0	0
repeats			

This sequence is easily converted to a present-state/next-state truth table (Table 10.2*a* and *b*).

The *state table* (Table 10.2*a*) is created directly from the counting sequence and is converted to a conventional logic design truth table (Table 10.2*b*) by arranging the input (present state) side in ascending binary order. Notice that the vehicle for state-to-state transitions is the combinational logic specified by the present-state/next-state truth table and that any of the design techniques described in Chaps. 3 through 5 can be used to create it, along with the use of programmable devices discussed in Chap. 9. Several methods will be pursued in the problems of this chapter.

If an external input is required, an additional variable must be added to the input side of the truth table, which effectively doubles its length. If an output other than that from a register flip-flop is needed, then an additional output variable must be added to the truth table.

EXAMPLE 10.2 Suppose that an external forward/reverse command (F/R) is introduced to the system of Example 10.1. The function of variable F/R is to modify the sequence of transitions from state to state. If F/R is 0, the circuit counts in the sequence shown in Table 10.1 and, if F/R is 1, the count progresses in reverse order (i.e., from bottom to top). Additionally, suppose that a supplemental output is specified to provide an odd-parity bit for the next state. Create the new logic design truth table.

Table 10.2a State Table Derived from Counting Sequence

Present D	Present C	Present B	Present A	Next D	Next C	Next B	Next A
0	0	0	0	0	0	0	1
0	0	0	1	0	0	1	1
0	0	1	1	0	0	1	0
0	0	1	0	0	1	1	0
0	1	1	0	0	1	1	1
0	1	1	1	0	1	0	1
0	1	0	1	0	1	0	0
0	1	0	0	1	1	0	0
1	1	0	0	1	1	0	1
1	1	0	1	1	1	1	1
1	1	1	1	1	1	1	0
1	1	1	0	1	0	1	0
1	0	1	0	1	0	1	1
1	0	1	1	1	0	0	1
1	0	0	1	1	0	0	0
1	0	0	0	0	0	0	0

Table 10.2b Corresponding Truth Table (Inputs Arranged in Binary Order)

Inputs (Present) D	C	B	A	Outputs (Next) D	C	B	A
0	0	0	0	0	0	0	1
0	0	0	1	0	0	1	1
0	0	1	0	0	1	1	0
0	0	1	1	0	0	1	0
0	1	0	0	1	1	0	0
0	1	0	1	0	1	0	0
0	1	1	0	0	1	1	1
0	1	1	1	0	1	0	1
1	0	0	0	0	0	0	0
1	0	0	1	1	0	0	0
1	0	1	0	1	0	1	1
1	0	1	1	1	0	0	1
1	1	0	0	1	1	0	1
1	1	0	1	1	1	1	1
1	1	1	0	1	0	1	0
1	1	1	1	1	1	1	0

Table 10.3a State Table for Reverse Counting

Present D	Present C	Present B	Present A	Next D	Next C	Next B	Next A
0	0	0	0	1	0	0	0
0	0	0	1	0	0	0	0
0	0	1	1	0	0	0	1
0	0	1	0	0	0	1	1
0	1	1	0	0	0	1	0
0	1	1	1	0	1	1	0
0	1	0	1	0	1	1	1
0	1	0	0	0	1	0	1
1	1	0	0	0	1	0	0
1	1	0	1	1	1	0	0
1	1	1	1	1	1	0	1
1	1	1	0	1	1	1	1
1	0	1	0	1	1	1	0
1	0	1	1	1	0	1	0
1	0	0	1	1	0	1	1
1	0	0	0	1	0	0	1

Table 10.3b Truth Table for Reverse Counting

Inputs (Present) D	C	B	A	Outputs (Next) D	C	B	A
0	0	0	0	1	0	0	0
0	0	0	1	0	0	0	0
0	0	1	0	0	0	1	1
0	0	1	1	0	0	0	1
0	1	0	0	0	1	0	1
0	1	0	1	0	1	1	1
0	1	1	0	0	0	1	0
0	1	1	1	0	1	1	0
1	0	0	0	1	0	0	1
1	0	0	1	1	0	1	1
1	0	1	0	1	1	1	0
1	0	1	1	1	0	1	0
1	1	0	0	0	1	0	0
1	1	0	1	1	1	0	0
1	1	1	0	1	1	1	1
1	1	1	1	1	1	0	1

For reverse counting, the sequence must be converted to a state table and then rearranged into a logic design truth table (as per Example 10.1). Tables 10.2*b* and 10.3*b* are then combined in series, the variable F/R is added to the input side in the most significant column, and a parity output column is appended as shown in Table 10.4.

Table 10.4 Truth Table for the Up/Down Counter with Parity

Inputs (Present)					Outputs (Next)					Inputs (Present)					Outputs (Next)				
F/R	D	C	B	A	D	C	B	A	P	F/R	D	C	B	A	D	C	B	A	P
0	0	0	0	0	0	0	0	1	0	1	0	0	0	0	1	0	0	0	0
0	0	0	0	1	0	0	1	1	1	1	0	0	0	1	0	0	0	0	1
0	0	0	1	0	0	1	1	0	1	1	0	0	1	0	0	0	1	1	1
0	0	0	1	1	0	0	1	0	0	1	0	0	1	1	0	0	1	0	0
0	0	1	0	0	1	1	0	0	1	1	0	1	0	0	0	1	0	1	1
0	0	1	0	1	0	1	0	0	0	1	0	1	0	1	0	1	1	1	0
0	0	1	1	0	0	1	1	1	0	1	0	1	1	0	0	0	1	0	0
0	0	1	1	1	0	1	0	1	1	1	0	1	1	1	0	1	1	0	1
0	1	0	0	0	0	0	0	0	1	1	1	0	0	0	1	0	0	1	1
0	1	0	0	1	1	0	0	0	0	1	1	0	0	1	1	0	1	1	0
0	1	0	1	0	1	0	1	0	1	1	1	0	1	0	1	1	1	0	0
0	1	0	1	1	1	0	0	1	1	1	1	0	1	1	1	0	1	0	1
0	1	1	0	0	1	1	0	1	1	1	1	1	0	0	0	1	0	0	0
0	1	1	0	1	1	1	1	1	1	1	1	1	0	1	1	1	0	0	1
0	1	1	1	0	1	0	1	0	1	1	1	1	1	0	1	1	1	1	1
0	1	1	1	1	1	1	1	0	0	1	1	1	1	1	1	1	0	1	0

The present-state/next-state truth table forms the basis for design of the combinational logic associated with the synchronous state machine. This logic can be implemented by any of the techniques discussed in earlier chapters. If it is to be realized with individual gates in small-scale integration, then five-variable K maps might be created for each of the output variables, simplified, and implemented in the available hardware. If preferred, a multiplexer could be programmed for each output variable.

Another choice would be the use of a programmed logic device such as a PALIC or PLA. In this case, one could create the logic by entering the Boolean equations for each output variable. A small PAL device is the simpler of the two and should be tried first. The major design decision here is estimating the proper size of the programmable device. If the parity output is to be registered, then at least five on-board flip-flops must be available. The maximum number of AND terms for a single OR is estimated from the logic equations, perhaps using maps, and compared with the number contained on the chip. Alternately, the equations could be entered directly from the truth table without simplification and the software will attempt to simplify the logic and will inform the user if the design is too large for the chip. In this case, a larger PALIC might be considered or a PLA implementation attempted.

10.3 DESIGN WITH JK FLIP-FLOPS

In state machine design, the designer often has the option of selecting between D and JK flip-flops; the choice often being dependent on the specific application. The following example illustrates the differences in design approach.

EXAMPLE 10.3 Consider the counter specified in Table 10.5. The state table is developed (Table 10.6), from which next-state maps for each output variable are created. The appropriate Boolen equations are derived making use of "don't care" states as shown in Fig. 10-2. Map coordinates represent each possible present state, and entries at a given location indicate the corresponding next-state value of the variable represented by the particular map being

Table 10.5

D	C	B	A
0	0	0	0
0	0	0	1
0	0	1	1
0	1	1	0
0	1	1	1
1	0	0	0
1	0	0	1
1	1	0	0
1	1	1	0
1	1	1	1

repeats

(Sequence proceeds downward, as indicated).

Table 10.6

Present State				Next State			
D	C	B	A	D	C	B	A
0	0	0	0	0	0	0	1
0	0	0	1	0	0	1	1
0	0	1	1	0	1	1	0
0	1	1	0	0	1	1	1
0	1	1	1	1	0	0	0
1	0	0	0	1	0	0	1
1	0	0	1	1	1	0	0
1	1	0	0	1	1	1	0
1	1	1	0	1	1	1	1
1	1	1	1	0	0	0	0

viewed. Thus, for example, if the present state is DCBA = 1100, we see from the D map that the value of D following the next clock pulse will be 1 and, from the A map, that the next value of A will be 0. Since all possible counts are not included in the given sequence, there are, of course, "don't care" states (represented by X's).

D Implementation

The Boolean equations of Fig. 10-2 may be used to implement the combinational logic whose outputs drive the D inputs of the flip-flops in Fig. 10-1.

JK Implementation

Consider the JK synchronous truth table from Chap. 7 (Table 10.7*a*). If output Q is 0, it will remain at 0 following a clock pulse if JK = 00 or JK = 01; i.e., it is only required that J = 0; the value of K is irrelevant. If output Q is 0, it will change to 1 following a clock pulse if JK = 10 or JK = 11 (again, the value of K is irrelevant). If Q = 1, it will change to 0 if JK = 01 or JK = 11 (K must be 1 and J is irrelevant). If Q = 1, it will remain at 1 if JK = 10 or 00 (K must be 0 regardless of J).

		0	1	1	0	A
D	C	0	0	1	1	B
0	0	0	0	0	X	
0	1	X	X	1	0	
1	1	1	X	0	1	
1	0	1	1	X	X	

Dnext = D B' + D A' + D' C A

		0	1	1	0	A
D	C	0	0	1	1	B
0	0	0	0	1	X	
0	1	X	X	0	1	
1	1	1	X	0	1	
1	0	0	1	X	X	

Cnext = C' B + C A' + D C' A

		0	1	1	0	A
D	C	0	0	1	1	B
0	0	0	1	1	X	
0	1	X	X	0	1	
1	1	1	X	0	1	
1	0	0	0	X	X	

Bnext = C A' + D' C' A

		0	1	1	0	A
D	C	0	0	1	1	B
0	0	1	1	0	X	
0	1	X	X	0	1	
1	1	0	X	0	1	
1	0	1	0	X	X	

Anext = D' B' + B A' + D C' A'

Fig. 10-2

Table 10.7*a* JK Truth Table

J	K	Qn+1
0	0	Qn
0	1	0
1	0	1
1	1	Qn'

Table 10.7*b* JK Sequential-State-Change Table

Transition	J	K
0 to 0	0	X
0 to 1	1	X
1 to 0	X	1
1 to 1	X	0

This behavior may be summarized in a *sequential-state-change* (transition) table as shown in Table 10.7*b*.

The JK sequential-state-change table can be used with the next-state maps of Fig. 10-2 to form a J map and a K map for each variable.

The map for flip-flop D, for example, is decomposed into a J_D and K_D map. In each of these, the present states are the map coordinates and the entries are control variable values. The nature of each transition can be easily read from the next-state maps (Fig. 10-2). For map D_{next}, the present value of flip-flop D is 0 for the top two rows and 1 for the bottom two rows. With this in mind, we see that the second row, third column represents a 0-to-1 transition. Thus, in Fig. 10-3, the entry at 0111 is 1 in the J_D map and X in the K_D map as per Table 10.7*b*.

		0	1	1	0	A
D	C	0	0	1	1	B
0	0	0	0	0	X	
0	1	X	X	1	0	
1	1	X	X	X	X	
1	0	X	X	X	X	

$J_D = C A$

		0	1	1	0	A
D	C	0	0	1	1	B
0	0	X	X	X	X	
0	1	X	X	X	X	
1	1	0	X	1	0	
1	0	0	0	X	X	

$K_D = C A$

Fig. 10-3

Note that for stage D, a single two-input gate is needed which replaces two two-input gates and two three-input gates. The remaining J and K maps and corresponding logic are developed in Prob. 10.4.

Though, in the present example, JK implementation leads to simpler combinational logic than the D flip-flop design, this is not always the case, and a designer must investigate both if gate count is an important design factor.

10.4 DESIGN FOR PROGRAMMABLE LOGIC DEVICES

The architecture shown in Fig. 10-1 is exactly that of a PAL device or PLA as described in Chap. 9. The present-state/next-state table defines the combinational-logic outputs which are to be registered (stored) in the flip-flops and which constitute the circuit's state. As we have seen, the relation between present state and next state may be expressed as Boolean equations derived from the truth table, and, quite often, chip-burning software accepts these equations directly. As an example, consider the case where the counter of Example 10.1 is to be placed on a PAL16R6.

EXAMPLE 10.4 PALIC design of a Gray code counter. The count sequence and related present-state/next-state truth table (Tables 10.1 and 10.2*b* respectively) are reproduced below for convenience.

D C B A
0 0 0 0
0 0 0 1
0 0 1 1
0 0 1 0
0 1 1 0
0 1 1 1
0 1 0 1
0 1 0 0
1 1 0 0
1 1 0 1
1 1 1 1
1 1 1 0
1 0 1 0
1 0 1 1
1 0 0 1
1 0 0 0
repeats

Counting Sequence

Inputs (Present) D C B A	Outputs (Next) D C B A
0 0 0 0	0 0 0 1
0 0 0 1	0 0 1 1
0 0 1 0	0 1 1 0
0 0 1 1	0 0 1 0
0 1 0 0	1 1 0 0
0 1 0 1	0 1 0 0
0 1 1 0	0 1 1 1
0 1 1 1	0 1 0 1
1 0 0 0	0 0 0 0
1 0 0 1	1 0 0 0
1 0 1 0	1 0 1 1
1 0 1 1	1 0 0 1
1 1 0 0	1 1 0 1
1 1 0 1	1 1 1 1
1 1 1 0	1 0 1 0
1 1 1 1	1 1 1 0

Corresponding Truth Table

The Boolean equations in PALASM syntax are created directly from the truth table as follows:

```
D := /D*C*/B*/A + D*/C*/B*A + D*/C*B*/A + D*/C*B*A + D*C*/B*/A + D*C*/B*A
     + D*C*B*/A + D*C*B*A
C := /D*/C*B*/A + /D*C*/B*/A + /D*C*/B*A + /D*C*B*/A + /D*C*B*A + D*C*/B*/A
     + D*C*/B*A + D*C*B*A
B := /D*/C*/B*A + /D*/C*B*/A + /D*/C*B*A + /D*C*B*/A + D*/C*B*/A + D*C*/B*A
     + D*C*B*/A + D*C*B*A
A := /D*/C*/B*/A + /D*/C*/B*A + /D*C*B*/A + /D*C*B*A + D*/C*B*/A + D*/C*B*A
     + D*C*/B*/A + D*C*/B*A
```

These equations, when entered, will cause the chip burner to create the desired connections within a PAL16R6 connected to it, thereby creating the desired counter.

10.5 THE ASM CHART

In Example 10.2, where an external forward/reverse variable is applied to the state machine, we see that the added input causes a doubling in the length of the design state table. If there are a large number of external inputs, the state table can grow quite long and often becomes unwieldly. In this situation, designers usually make use of an *algorithmic state machine* (*ASM*) *chart*. This chart, which is intentionally constructed to resemble a computer programming flow diagram, uses a relatively small number of symbols which are logically interconnected to indicate the state machine's progress from one state to the next. It serves as a step-by-step detailed description of the system's desired performance.

ASM Chart Symbols

The *rectangle* is used to represent a specific system state. There is often a mnemonic label adjacent to the rectangle to name the state, and inside the rectangle are listed any outputs which occur while the machine is in that particular state. These outputs are termed *unconditional outputs*. The machine exists in the specified state for exactly one clock cycle.

The rectangle shown in Fig. 10-4 represents state AA, and, while the system is in this state, it will produce outputs OUT1 and XX. If the machine proceeds from state to state in a predetermined manner, independent of any inputs, the states are joined by directed line segments as shown in Fig. 10-5.

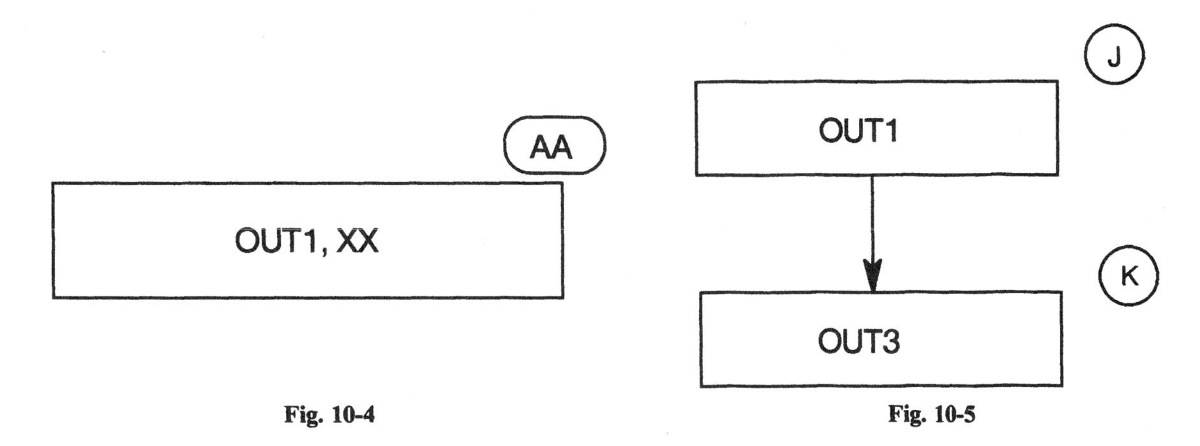

Fig. 10-4   Fig. 10-5

In this case, after existing in state J for one clock cycle, the machine moves to state K when the next clock pulse arrives. Note the change of outputs in this case.

The *diamond* or *extended diamond* is used to indicate *conditional branching* where the next state assumed depends on the value of one or more input variables. The Boolean function(s) involving these variables are shown inside the extended diamond, and branch paths are labeled with the corresponding values that the function(s) may have. The diamond, itself, is called a *decision block* and is not separately labeled.

In the ASM chart segment shown in Fig. 10-6, a machine in state J will proceed, on the next clock pulse, to either state K or state L depending on whether the Boolean function A + B is TRUE or

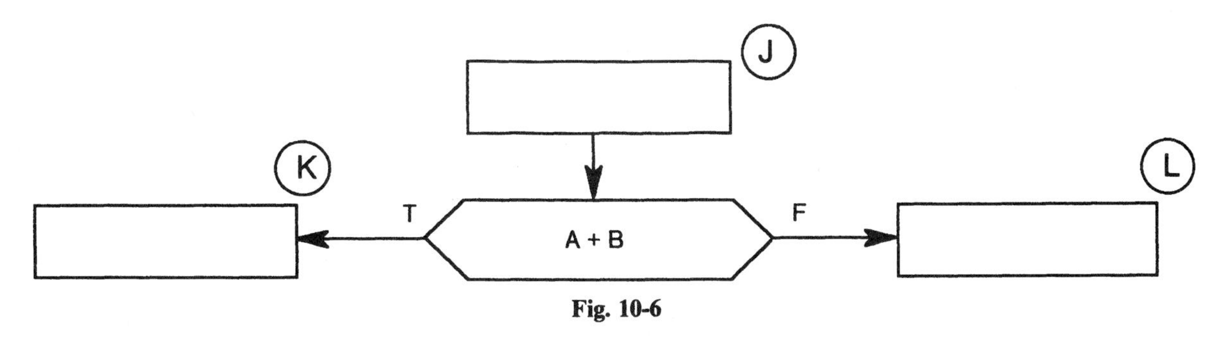

Fig. 10-6

FALSE. It is important to note that the decision is made during the clock cycle in which the machine is in state J. In this sense, the decision block belongs to the state from which it is entered.

The *oval* is used to indicate a *conditional output* which depends not only on the state but also on the status of one or more inputs. It is always associated with an exit from a decision block and is called a *conditonal output block.*

For the chart shown in Fig. 10-7, the machine begins in state J. During this clock cycle, the Boolean function A(B + C) is evaluated. Output OUT1 occurs unconditionally during this clock cycle, at which time *either* OUT2 or OUT3 will also occur depending on the evaluation of the Boolean function. In Fig. 10-7, OUT2 occurs conditionally in state J and unconditionally in state K. Note how the decision block determines which outputs occur in a given state while also determining the next state. Note also that *only those inputs which actually influence these decisions occur in the diagram*; it is not necessary to evaluate all the inputs before a change in state is determined.

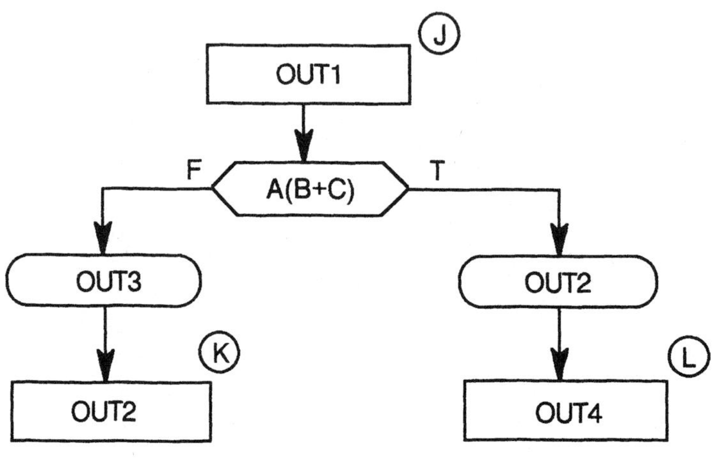

Fig. 10-7

ASM Chart Construction

How the step-by-step development of an ASM chart progresses from a system specification is, perhaps, best described by means of an example.

EXAMPLE 10.5 Incremental encoder logic. An incremental encoder is widely used in control system applications. It can be thought of as a two-track Gray code shaft encoder of the type described in Example 1.11. The output signals produced have the form of two square waves, one of which is delayed by a quarter cycle relative to the other, as shown in Fig. 10-8. The up and down directions indicated refer to the direction of apparent Gray code counting as the shaft encoder rotates either clockwise or counterclockwise.

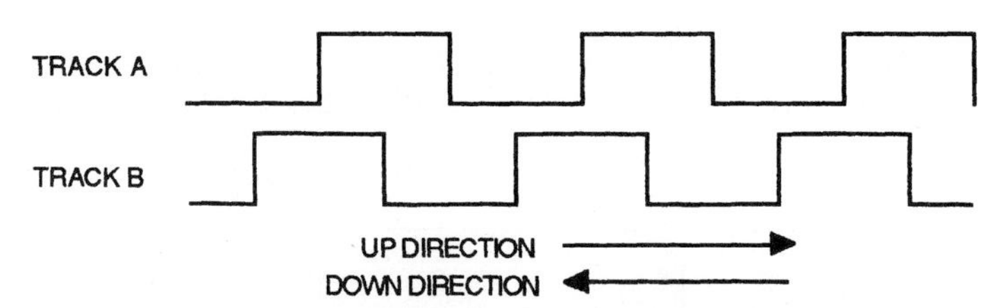

Fig. 10-8 Encoder output waveshapes.

It is desired to produce a pulse which occurs at any transition of either track output. This pulse is to appear at output PU if the encoder is moving in the up direction or at PD if the encoder rotates in the down direction. Thus, there should be four pulses for every output cycle (see Fig. 10-9).

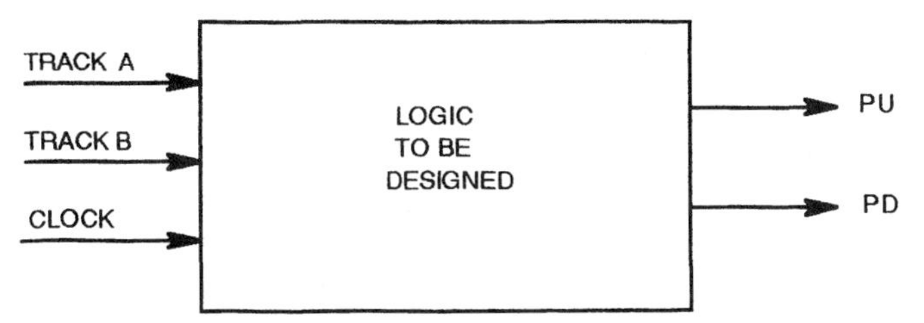

Fig. 10-9 Example block diagram.

The design ASM chart is most easily constructed by equating every proposed state with a known "physical" situation. The resultant chart is shown in Fig. 10-10.

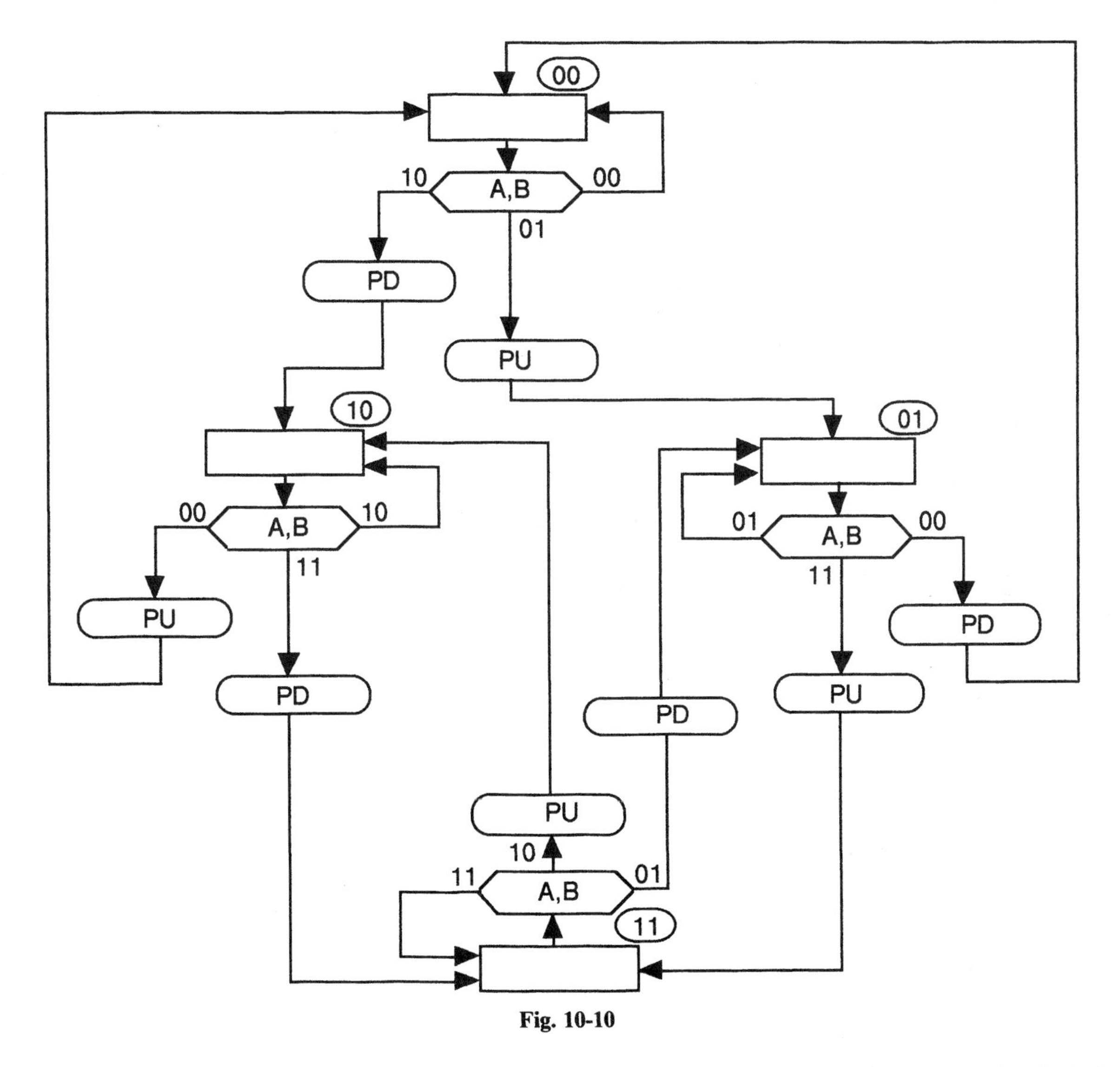

Fig. 10-10

Consider its construction. The initial state (labeled 00) was arbitrarily chosen and represents the physical situation where both encoder tracks produce 0s. As long as no changes occur on either track, no output is desired and the system stays in the same state. It will only move out of that state when an input changes. Suppose, now, that track B changes to 1. Reference to the waveforms in Fig. 10-8 shows that the motion must be in the up direction and a PU pulse is to be generated. Thus, the next state corresponds to the physical situation where track

A is 0 and track B is 1 (state 1). If track A had changed to 1, a PD pulse would have been required and the next state would have corresponded to the physical situation where track A produces 1 and track B produces a 0 (state two). Note that the construction of the encoder does not permit tracks A and B to simultaneously change to 1s. This is indicated on the design ASM chart by the absence of a 11 output from the uppermost decision block which examines the track outputs and directs the machine to the next state.

The fact that outputs appear only in conditional output symbols (the ovals) indicates that the system only produces outputs as the consequence of state-to-state transitions. These conditional outputs exist for only that portion of the clock cycle between the determination of the states of A and B and the next clock pulse. For this reason, the inputs are often *synchronized* (i.e., passed through a D flip-flop which is clocked by the machine clock). Unsynchronized inputs can result in very narrow output pulses which may not contain sufficient energy to trigger any downstream circuitry.

The transitions from states 1 and 2 are controlled by inputs A and B in a manner similar to that outlined for the transitions from state 0. Since the physical situations are finite in extent, it follows that this machine ASM chart will close on itself.

The timing of the clock pulses is critical. Output pulses are specified to occur at encoder transitions; in actuality, they occur at the first clock pulse following a transition. If delay is to be minimized, the clock pulse period should be very small compared to the encoder pulse widths. On the other hand, the width of the PU and PD pulses is equal to a single clock period, and there is usually a minimum width requirement for these pulses.

10.6 DESIGN FROM AN ASM CHART: BOOLEAN IMPLEMENTATION FOR MINIMAL NUMBER OF FLIP-FLOPS

If a suitable ASM chart can be constructed, it is possible to determine the logic in a straightforward manner. The ASM chart of Fig. 10-10 will be used to illustrate the procedures.

Step 1. Create a state-transition table

(a) *Determine the number of states by counting the rectangles.* In this case, there are four.

(b) *Determine the number of flip-flops needed to designate all possible states and assign their outputs to states identified on the ASM chart.* Two flip-flops can represent four states, three flip-flops up to eight states, etc. In the example of Fig. 10-10, two flip-flops are required, and we will designate their Q outputs as X and Y, respectively. For state 0, XY = 00, state 1 is represented by XY = 01, and so on. The first column of the state-transition table is a list of the states both by name and by flip-flop representation (see Table 10.8).

(c) *Tabulate destination states.* In the second column, list for each state the possible destination states which can be reached from the given state.

(d) *Tabulate conditions.* List any conditions, in column 3, which are associated with each state-to-state transition.

In a sense, the completed state-transition table is merely a logically organized list of directions which serve as the equivalent of a road map.

Step 2. Obtain controller design equations.

(a) Read down the second column of the state-transition table and *note when a flip-flop is expected to take on the value 1 in the next state* (i.e., after the next clock pulse).

(b) *Determine the conditions* from the third column. For example, the third line tells us that the next X should be 1 if the present state is 0 and if AB′ is TRUE.

(c) *Express the conditions, which call for the next X or next Y to be 1, in terms of present states and inputs.* We can obtain a Boolean equation set, in the same way that logic equations are written from a truth table, to specify the conditions required for all state-to-state transitions. This logic is called the *controller*. The controller equations obtained from Table 10.8 are

$$X_{next} = \text{“zero” } AB' + \text{“one” } AB + \text{“two” } A + \text{“three” } A$$
$$Y_{next} = \text{“zero” } A'B + \text{“one” } B + \text{“two” } AB + \text{“three” } B$$

Table 10.8 State-Transition Table

PRESENT STATE X Y		NEXT STATE X Y		CONDITIONS
zero	0 0	zero	0 0	A' B'
		one	0 1	A' B
		two	1 0	A B'
one	0 1	zero	0 0	A' B'
		one	0 1	A' B
		three	1 1	A B
two	1 0	zero	0 0	A' B'
		two	1 0	A B'
		three	1 1	A B
three	1 1	one	0 1	A' B
		two	1 0	A B'
		three	1 1	A B

(*d*) *Encode the state names* with flip-flop output values. The resulting controller design equations represent the logic to be presented to the state flip-flops:

$$X_{next} = X'Y'AB' + X'YAB + XY'A + XYA \tag{10.1}$$

$$Y_{next} = X'Y'A'B + X'YB + XY'AB + XYB \tag{10.2}$$

If D flip-flops are used, connections are made to the D inputs; if JK flip-flops are used, the equations can be divided into a J and a K portion as illustrated in Sec. 10.3.

Step 3. Obtain output design equations.

(*a*) *Locate every reference to an output* in the ASM chart and determine the conditions, if any, associated with that reference.

(*b*) *Join all references with a Boolean OR connection.*

In Fig. 10-10, variable PU occurs in four places, all of which happen to be conditional outputs. The topmost appearance of PU occurs when the system is in state 0 AND when A'B is TRUE. The other instances are similarly defined and lead to the following Boolean equation:

$$PU = \text{“zero” } A'B + \text{“one” } AB + \text{“two” } AB' + \text{“three” } A'B'$$

After coding the named states into flip-flop values, we have

$$PU = X'Y'A'B + X'YAB + XY'AB' + XYA'B' \tag{10.3}$$

Similarly, the equation for PD is determined to be

$$PD = \text{“zero” } AB' + \text{“one” } A'B' + \text{“two” } AB + \text{“three” } A'B$$
$$= X'Y'AB' + X'YA'B + XY'AB + XYA'B \tag{10.4}$$

Equations (10.1) through (10.4) constitute a complete set of design equations for the state machine, and the designer can implement them in any form desired. The equations can be simplified with maps if small-scale integration is to be used. More likely, a programmed logic device will be used, and it is only a question of translating the Boolean equations to the syntax needed for the design software. Problem 10.9 illustrates a method using multiplexers, which is preferred by some designers.

10.7 DESIGN FROM AN ASM CHART: ONE-HOT CONTROLLER IMPLEMENTATION

For small controllers, some designers prefer an arrangement where there are as many state flip-flops as there are states. Each flip-flop is identified with a state and, when in that state, the corresponding flip-flop is the only one which is "hot," i.e., that has its Q output set to TRUE. This results in a controller with simpler decoding logic than the minimal flip-flop implementation. Design of a one-hot controller begins, as before, with a state-transition table drawn from the ASM chart. This time, however, the information is tabulated to indicate all previous states which lead to a given present state.

Controller Design Equations

With reference to the top block of Table 10.9 and confirmation from Fig. 10-10, we see that the D input of flip-flop "zero" should be a 1 if inputs A and B are both FALSE, regardless of the prior state. Continuing, we observe from the table that all input conditions for a given present state are the same, independent of any previous state. Thus the four design equations are "zero" = A′B′, "one" = A′B, "two" = AB′, and "three" = AB.

Table 10.9

PREVIOUS STATE	PRESENT STATE	CONDITIONS
zero	zero	A' B'
one		A' B'
two		A' B'
three		don't care
zero	one	A' B
one		A' B
three		A' B
two		don't care
zero	two	A B'
two		A B'
three		A B'
one		don't care
three	three	A B
one		A B
two		A B
zero		don't care

Output Equations

The output equations are similar to those obtained in Sec. 10.6 except that it is not necessary to encode the states:

PU = "zero" A′B + "one" AB + "two" AB′ + "three" A′B′
PD = "zero" AB′ + "one" A′B′ + "two" AB + "three" A′B

Logic for the complete one-hot solution is shown in Fig. 10-11.

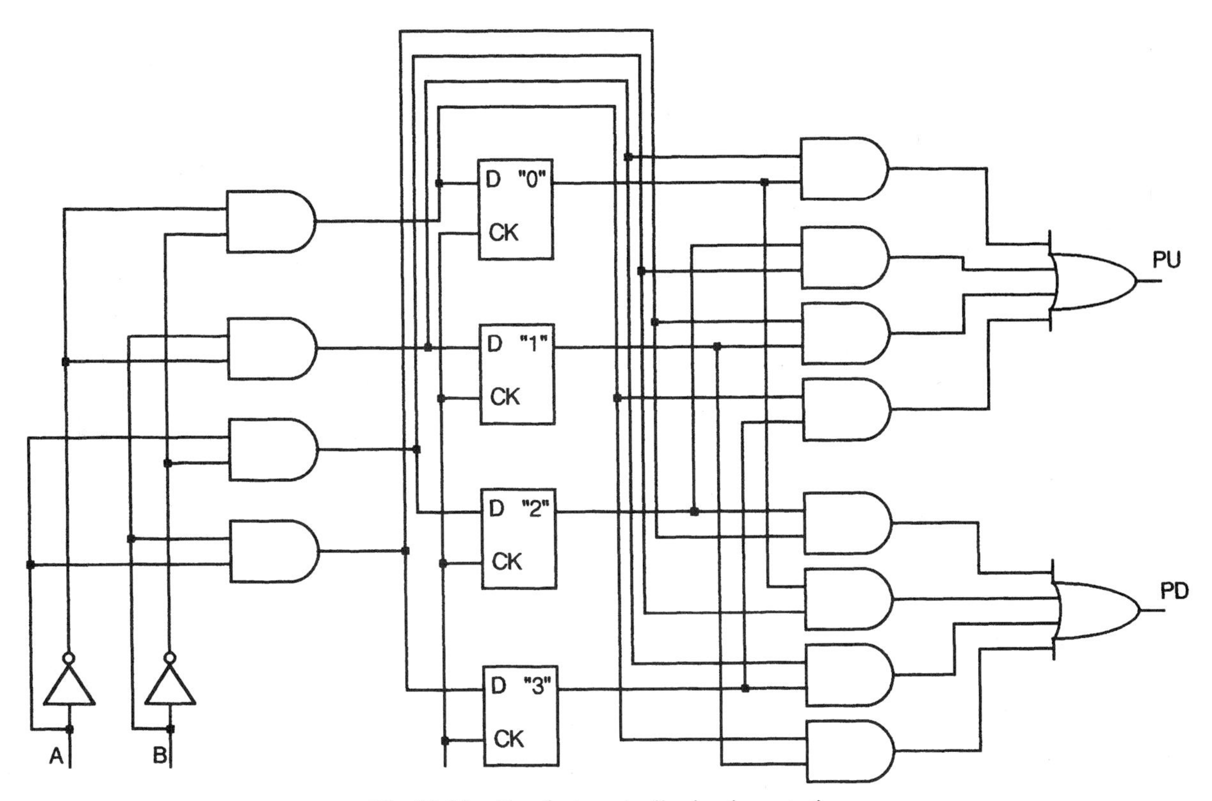

Fig. 10-11 One-hot controller implementation.

10.8 DESIGN FROM AN ASM CHART: STATE TABLE ENTRY TO A PROGRAMMABLE LOGIC DEVICE

When a state machine is to be implemented on a programmable device, as indicated in Sec. 10.2, Boolean entry can be used. Many designers, however, prefer to use a software technique called *state table entry* in which definitions of all states and ASM chart information about them are written in structured programming form.

EXAMPLE 10.6 Consider the incremental encoder design specification of Example 10.5 arranged for state table entry using INTEL iPLSII software.*

ENTRY	COMMENT
HEADER	Not included in this example
MACHINE: Counter Example	System name
CLOCK: CLK	Clock name
STATES: [FF1 FF2]	
ZERO [0 0]	
ONE [0 1]	State definitions
TWO [1 0]	
THREE [1 1]	

* Permission to use iPLSII examples granted by Intel Corporation.

ENTRY	COMMENT
ZERO:	
IF A'*B' THEN ZERO	Conditional exits to next states
IF A'*B THEN ONE	
TWO	Default exit. Since two encoder track outputs never change simultaneously, a transition from 00 to 11 can't occur. Thus, state TWO is the only remaining possibility
ASSERT:	
IF A'*B THEN PU	Conditional outputs
IF A*B' THEN PD	
ONE:	
CASE	Alternative to IF_THEN form
A'*B': ZERO	Conditional exits to next states
A*B: THREE	
ENDCASE	
ONE	Default exit (AB = 10 cannot occur)
ASSERT:	
IF A'*B' THEN PD	Conditional outputs
IF A*B THEN PU	
TWO:	
CASE	
A'*B': ZERO	Conditional exits to next states
A*B: THREE	
ENDCASE	
TWO	Default exit (state 01 cannot occur)
ASSERT:	
IF A'*B' THEN PU	Conditional outputs
IF A*B THEN PD	
THREE:	
CASE	
A'*B: ONE	Conditional exits to next states
A*B': TWO	
ENDCASE	
THREE	Default exit (state 00 cannot occur)
ASSERT:	
IF A'*B THEN PD	Conditional outputs
IF A*B' THEN PU	
END$	

States are defined in terms of the state flip-flops by means of a matrix. Then, for each state in sequence, exits and outputs are indicted. IF_THEN or CASE statements are used for conditional exits; unconditional exits, when present, are listed alone.

Outputs are listed next, within the state declaration, following the word ASSERT. Conditional outputs can be listed either by IF (condition), THEN (output), or CASE statements exactly as were the exits.

When the state table entry program is completed, it is entered into a host programming computer which, is often a personal computer. The iPLSII software translates state table entries into an appropriate set of Boolean equations which can usually be inspected by the designer, if desired. Device programming then takes place, just as if Boolean entry had been used. The programming operation has been made more user friendly since it allows the designer to enter programming information directly from his or her design tool, the ASM chart.

10.9 CLOCK SKEW IN STATE MACHINES

The heartbeat of a synchronous state machine is the clock pulse which is assumed to appear simultaneously at all points in the circuit. In many practical situations, it is not possible to maintain precise synchronism because of differing propagation delays in the various clock pathways. This difference in time between the appearance of two supposedly coincident clock signals is referred to as *clock skew* which, in high-speed circuitry, can be a significant source of error.

EXAMPLE 10.7 To illustrate the effect of clock skew, the circuit of Fig. 7-18 is redrawn in Fig. 10-12.

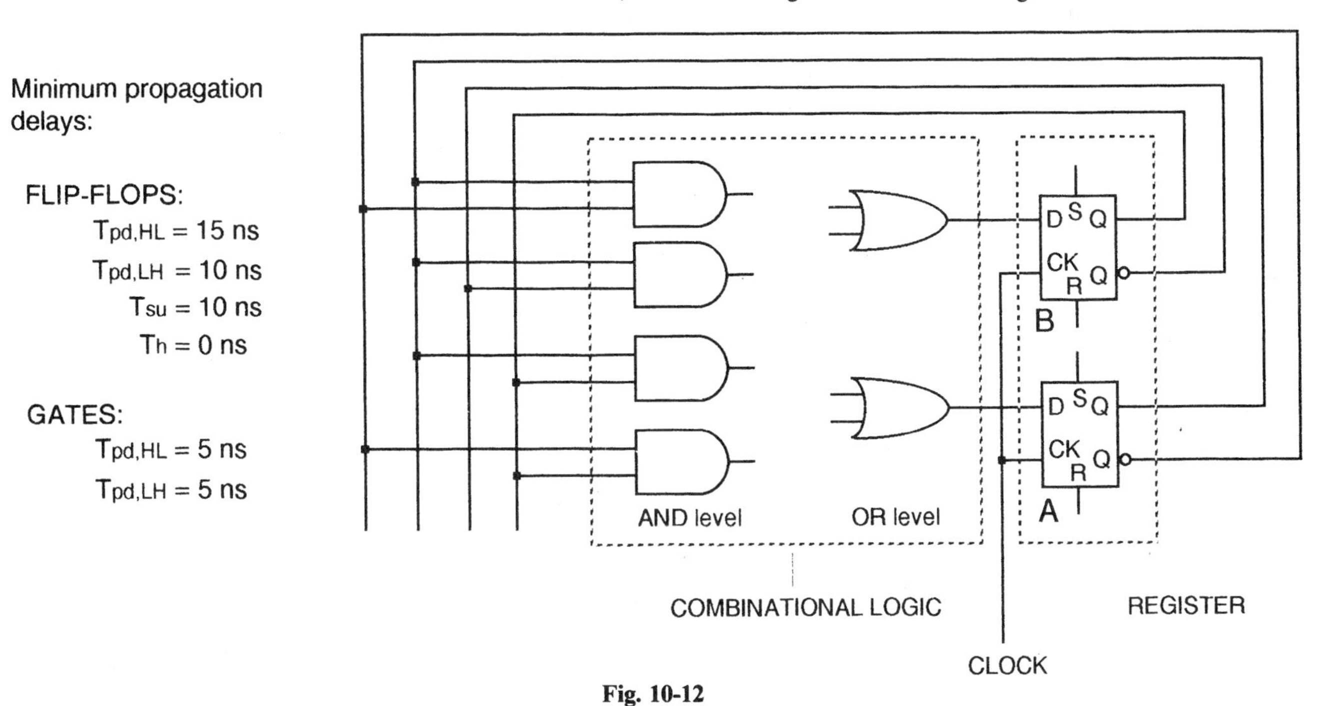

Fig. 10-12

The first and last flip-flops have been labeled A and B, respectively, for convenience. Let us assume that the clock pulse arrives at flip-flop B sometime after the "same" clock pulse arrives at flip-flop A. It is important that the earlier clock pulse does not cause any change to appear at the D input of flip-flop B before or coincident with the arrival of the clock pulse at this flip-flop. Consider the sequence of events:

1. Clock pulse arrives at flip-flop A.
2. After a propagation delay, flip-flop A changes causing a logic change at the AND level input (s).
3. After another propagation delay, the change in one or more AND gates causes a change at the OR level input(s).
4. After a further propagation delay, the change in an OR gate causes a change at the D input of another flip-flop (B in this case).
5. After the minimum setup time, the B flip-flop acts on erroneous information.

The minimum time taken to complete the five steps represents the maximum delay (clock skew) that the system can tolerate. Using the numbers listed in the figure, the maximum clock skew which can be tolerated by this system is $10 + 5 + 5 + 10 = 30$ ns.

Notice that computation of the maximum allowable clock frequency in Example 7.5 required knowledge of the maximum propagation delays, while the computation of maximum clock skew involves use of minimum delays.

It is also worth noting that the change at the D input of flip-flop B (step 4) causes a set-up time violation for that flip-flop. If metastability failures are a problem for the design in question, the maximum acceptable skew is reduced to the cumulative time delay of the first four steps which, in this example, is 20 ns.

Designers of high-speed digital systems have to be concerned with clock skew and often go to great lengths to equalize clock path delays and/or reduce minimum propagation delays. While manufacturers have always specified maximum values of timing parameters, many are now beginning to specify minimum values as well.

10.10 INITIALIZATION AND LOCKOUT IN STATE MACHINES

It is usually very important to have a means for initializing the state machine to a known state after equipment turn on, and it is often convenient to let this state be designated by setting all state flip-flops to 0. A common clear or reset line can be used for initialization, as shown in Fig. 10-13.

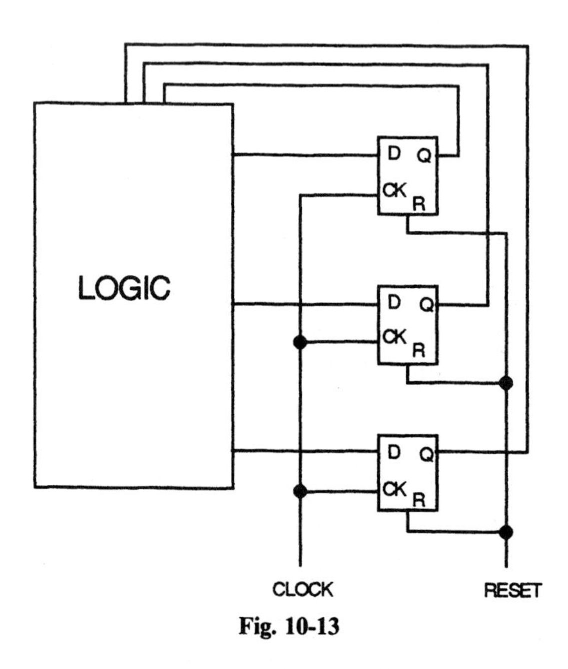

Fig. 10-13

In the one-hot implementation (Sec. 10.7) the all-zero situation cannot occur. In this case, the designer often resorts to interchanging the roles of the Q and Q′ sides of the flip-flop which is to be hot in the initialized state. The Q′ output is connected to all points in the logic where Q would normally have been connected. The common clear line, when pulsed, then puts the system in the proper initial one-hot state. Note that it is also necessary to logically invert the signal applied to the D input of the flip-flop so that it produces a 0 when a 1 is desired. The modification is shown in Fig. 10-14.

Fig. 10-14

If the state machine is not initialized, then, when the device is powered up, it may randomly assume a "don't care" or unspecified state from which it may be impossible to reach an allowed state regardless of the number of clock pulses applied. Such a situation is termed *lockout* and is illustrated in Prob. 8.11. The most effective solution for lockout problems is to avoid them by providing a self-applied or externally applied initialization signal.

Solved Problems

10.1 Implement the Gray code counter of Example 10.1 using small-scale integration packages for the state machine logic. Estimate the number of chips required.

The truth table from which the logic is designed has been derived previously (Table 10.2) and is reproduced below.

Present D C B A	Next D C B A
0 0 0 0	0 0 0 1
0 0 0 1	0 0 1 1
0 0 1 0	0 1 1 0
0 0 1 1	0 0 1 0
0 1 0 0	1 1 0 0
0 1 0 1	0 1 0 0
0 1 1 0	0 1 1 1
0 1 1 1	0 1 0 1
1 0 0 0	0 0 0 0
1 0 0 1	1 0 0 0
1 0 1 0	1 0 1 1
1 0 1 1	1 0 0 1
1 1 0 0	1 1 0 1
1 1 0 1	1 1 1 1
1 1 1 0	1 0 1 0
1 1 1 1	1 1 1 0

Using the set of present states as independent variable coordinates, K maps describing the next state for each of the four variables are created as described in Example 10.3.

The simplified Boolean expression for d_A shown in Fig. 10-15 is obtained by making use of the fact $X \oplus Y = XY' + X'Y$ and $(X \oplus Y)' = XY + X'Y'$. The exclusive-OR form is predictable from the map's checkerboard pattern (refer to Sec. 3.5).

A	0	1	1	0
DC \ B	**0**	**0**	**1**	**1**
0 0	1	1	0	0
0 1	0	0	1	1
1 1	1	1	0	0
1 0	0	0	1	1

$d_A = (B \oplus C \oplus D)'$

A	0	1	1	0
DC \ B	**0**	**0**	**1**	**1**
0 0	0	1	1	1
0 1	0	0	0	1
1 1	0	1	1	1
1 0	0	0	0	1

$d_B = BA' + DCA + D'C'A$

A	0	1	1	0
DC \ B	**0**	**0**	**1**	**1**
0 0	0	0	0	1
0 1	1	1	1	1
1 1	1	1	1	0
1 0	0	0	0	0

$d_C = CB' + CA + D'BA'$

A	0	1	1	0
DC \ B	**0**	**0**	**1**	**1**
0 0	0	0	0	0
0 1	1	0	0	0
1 1	1	1	1	1
1 0	0	1	1	1

$d_D = CB'A' + DA + DB$

Fig. 10-15

SSI chip count: 2 exclusive ORs (1 chip)
5 two-input ANDs (2 chips)
4 three-input ANDs (2 chips)
3 three-input ORs (1 chip)

All variables appear direct and complemented (four inverters), and there is an inverter required for the XORs. The five inverters together require one more chip, for a total of seven. Note that it is possible to save a chip by wiring the two spare three-input ANDs as two-input ANDs. The complete counter logic circuit is shown in Fig. 10-16.

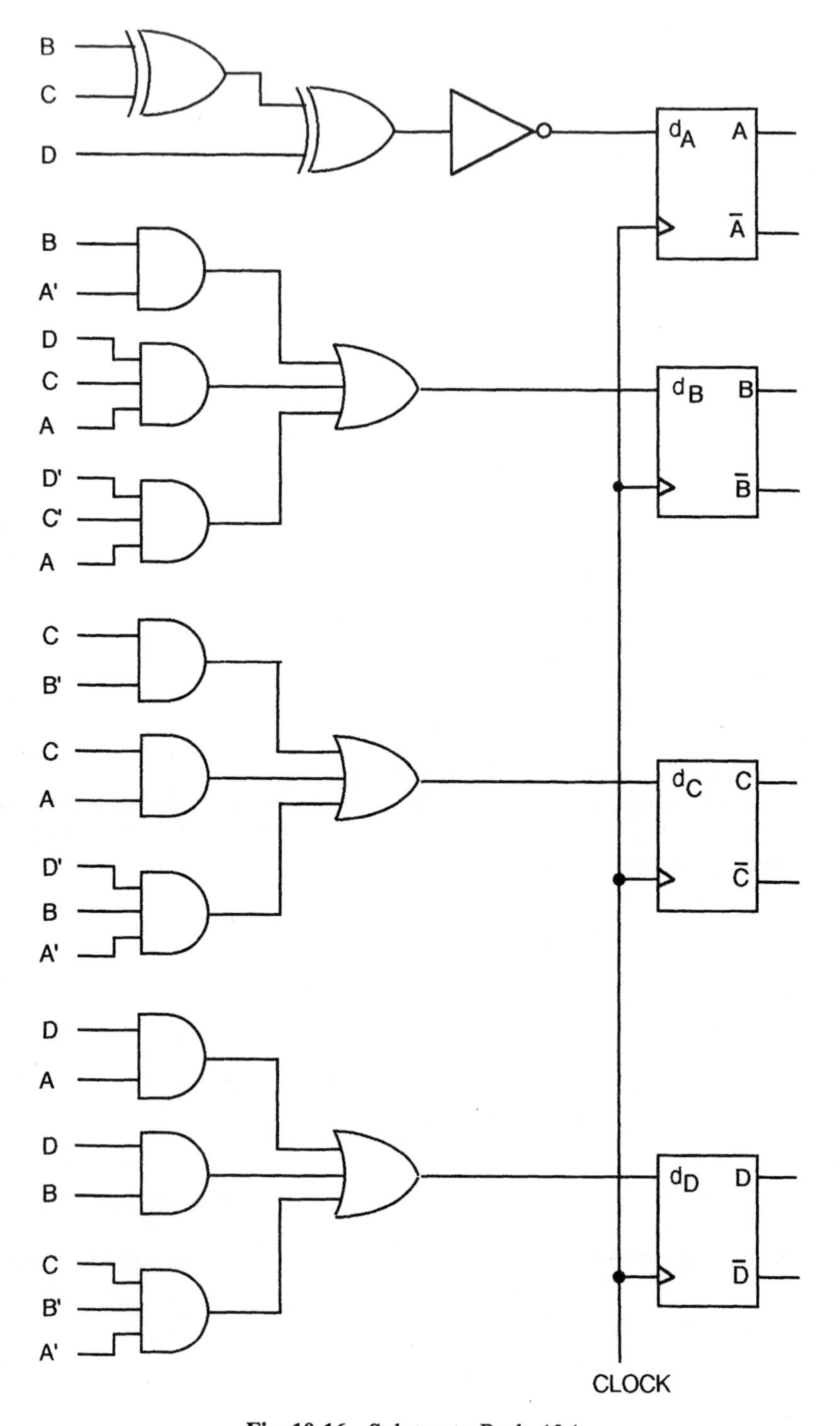

Fig. 10-16 Soluton to Prob. 10.1.

10.2 Table 10.10 shows the counting sequence for an excess-three counter. This device counts in binary from the decimal equivalent of 3 to the decimal equivalent of 12 and then repeats. Design the counter as a state machine using a ROM and D flip-flops to register present states.

Table 10.10

A	B	C	D
0	0	1	1
0	1	0	0
0	1	0	1
0	1	1	0
0	1	1	1
1	0	0	0
1	0	0	1
1	0	1	0
1	0	1	1
1	1	0	0
repeats			

Table 10.11

Decimal ROM Address	Present State ABCD	Next State ABCD
3	0 0 1 1	0 1 0 0
4	0 1 0 0	0 1 0 1
5	0 1 0 1	0 1 1 0
6	0 1 1 0	0 1 1 1
7	0 1 1 1	1 0 0 0
8	1 0 0 0	1 0 0 1
9	1 0 0 1	1 0 1 0
10	1 0 1 0	1 0 1 1
11	1 0 1 1	1 1 0 0
12	1 1 0 0	0 0 1 1

Use the state table as a basis for ROM programming (see Table 10.11). The resulting state machine is shown in Fig. 10-17.

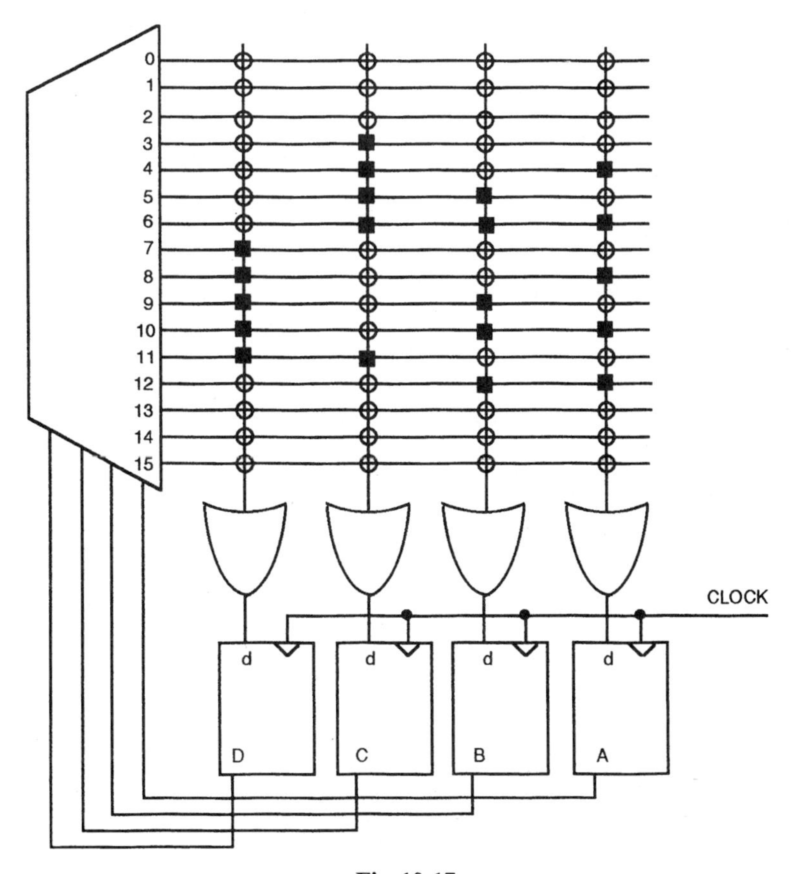

Fig. 10-17

10.3 Referring to Prob. 10.2, determine how a 4-bit excess-three up/down counter can be built using a PAL16R6 IC as the hardware.

A good design strategy is the use of the flip-flop outputs to represent states and to design using a present-state/next-state table. Such a table was developed in Chap. 2 (Prob. 2.33).

There are six flip-flops on the PAL16R6, and only four are needed for counter states. However, each output of the combinational logic contains ten AND terms and will not fit onto the PAL device since only eight AND gates are available for driving each flip-flop. It is therefore necessary to attempt simplification. A five-variable K-map simplification has been done for two of the variables in Prob. 3.28 where the functions A_n and C_n were reduced to six AND terms. The remaining two variables can be easily simplified to eight terms or less.

Following simplification, the counter can be implemented with a PAL16R6 with room to spare. One set of design equations that works is

$$A_n = (U/D)'BCD + AD + AC + (U/D)'AB' + (U/D)AB + (U/D)A'B'$$
$$B_n = (U/D)'A'B'CD + (U/D)'A'BC' + (U/D)'A'BCD' + (U/D)'AB'CD$$
$$+ (U/D)A'B'CD + (U/D)A'BC'D + (U/D)A'BC + (U/D)'AB'C'D'$$
$$C_n = (U/D)'CD' + (U/D)'C'D + (U/D)C'D' + (U/D)'AB + (U/D)ACD + (U/D)BCD$$
$$D_n = D'$$

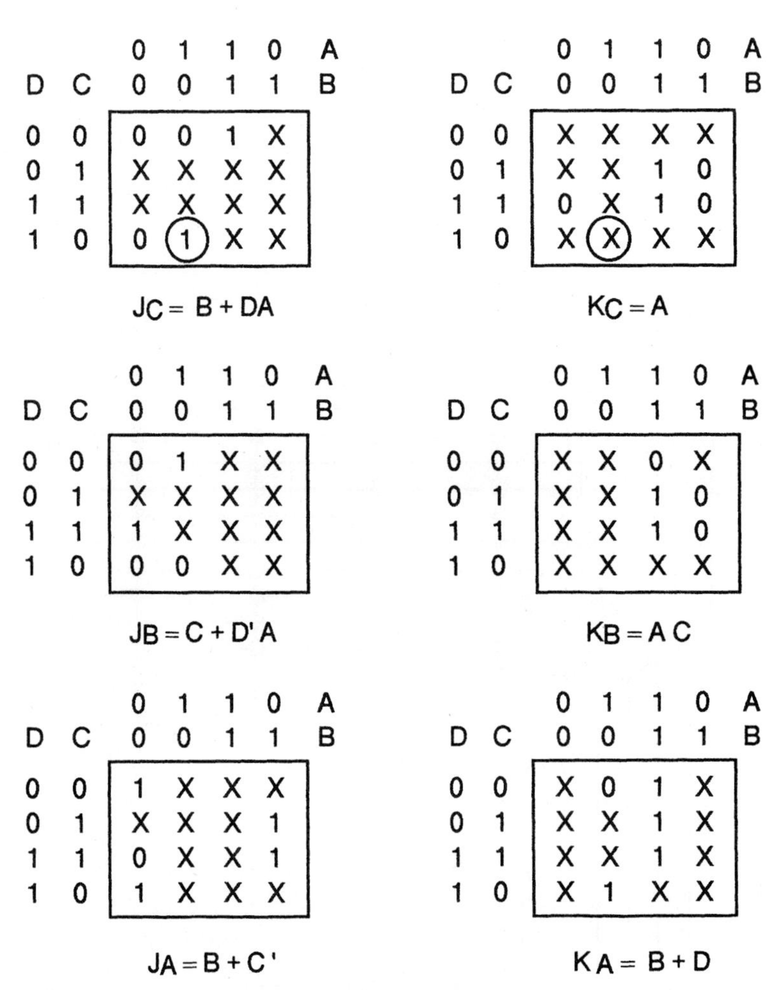

Fig. 10-18

10.4 Finish the J and K tables of Example 10.3 and draw the circuit configuration.

The required maps are created with reference to Fig. 10-2 and Table 10.7. As a first step, all "don't care" entries in the D maps are transferred to the JK map pairs. A typical JK entry is now generated as follows. Consider row 4, column 2 of the C_{next} map in Fig. 10-2. It tells us that if the present state is 1001, the next value of C will be 1. Thus, C will change from 0 to 1 which Table 10.7 indicates will occur if $J_C = 1$ and $K_C = X$. These entries are circled in the first pair of maps in Fig. 10-18. The circuit configuration is shown in Fig. 10-19.

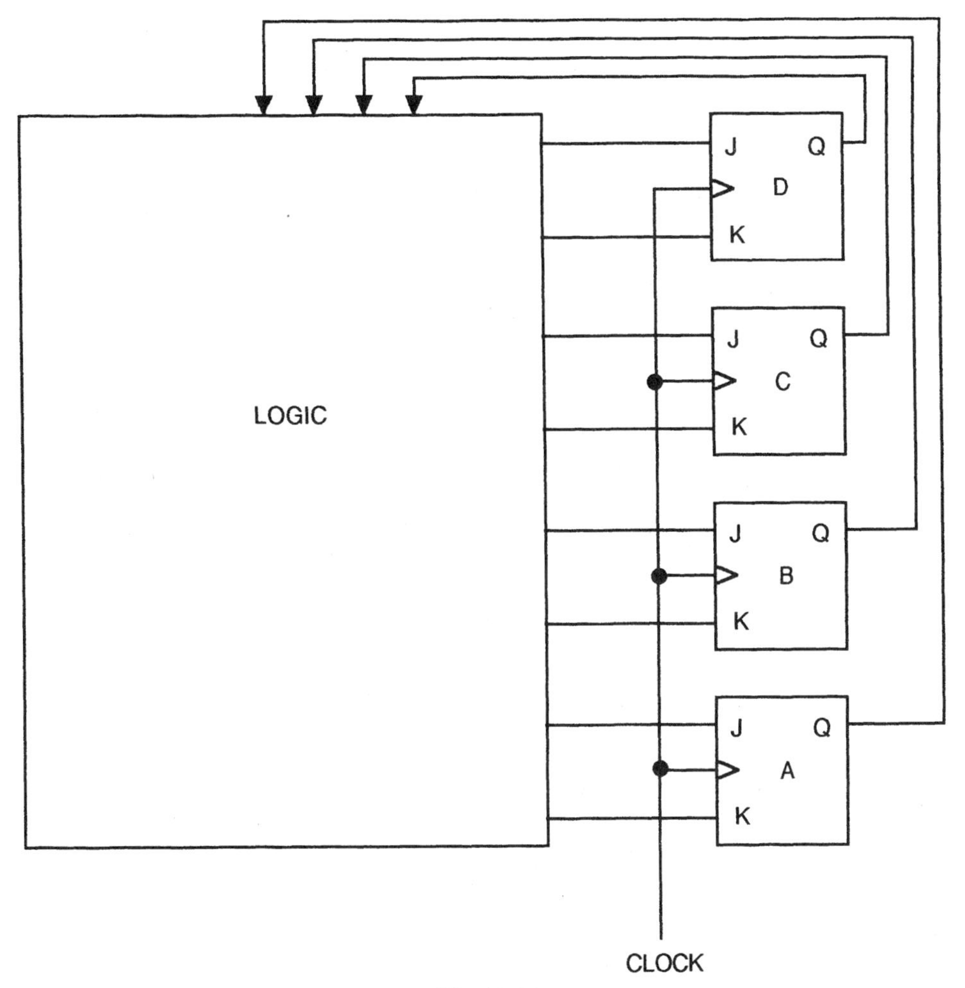

Fig. 10-19

10.5 For the counter specified in Table 10.12, design the logic required for a state machine using JK flip-flops.

The state table is created (Table 10.13), noting that there are six "don't care" states: 0010, 0100, 0101, 1010, 1011, and 1101. D maps are drawn for each variable and then expanded to corresponding JK maps using the transition relationships of Table 10.7 which are summarized as follows:

If the next state is to be 1:
When Q = 1, then J = X and K = 0
When Q = 0, then J = 1 and K = X

If the next state is to be 0:
When Q = 1, then J = X and K = 1
When Q = 0, then J = 0 and K = X

Table 10.12

DCBA
0 0 0 0
0 0 0 1
0 0 1 1
0 1 1 0
0 1 1 1
1 0 0 0
1 0 0 1
1 1 0 0
1 1 1 0
1 1 1 1
repeats

Table 10.13

Present State DCBA	Next State DCBA
0 0 0 0	0 0 0 1
0 0 0 1	0 0 1 1
0 0 1 1	0 1 1 0
0 1 1 0	0 1 1 1
0 1 1 1	1 0 0 0
1 0 0 0	1 0 0 1
1 0 0 1	1 1 0 0
1 1 0 0	1 1 1 0
1 1 1 0	1 1 1 1
1 1 1 1	0 0 0 0

In the resulting set of maps (Fig. 10-20), specified "don't care" states are shown in bold type to distinguish them from "don't cares" associated with the JK transition process.

10.6 Consider the unidirectional counter specified in Table 10.14. Assuming that both direct and complemented flip-flop outputs are available, develop the combinational logic system equations for use with D flip-flops. Repeat for JK flip-flops and compare the two designs from the standpoint of silicon used and maximum clocking rate.

Table 10.14

A	B	C
0	0	0
0	1	0
0	1	1
1	0	1
1	1	0
0	0	1
1	0	0
1	1	1
repeats		

Use the following specifications:

1. The D and JK flip-flops each occupy 10 units of silicon and both have a setup time of 20 time units, a hold time of 5 time units, and propagation delays of 25 time units.
2. Gates have a propagation delay of 10 time units and use 4 silicon units plus 0.5 units per input.

Maps relevant to the D and JK designs are shown in Figs. 10-21 and 10-22, respectively.

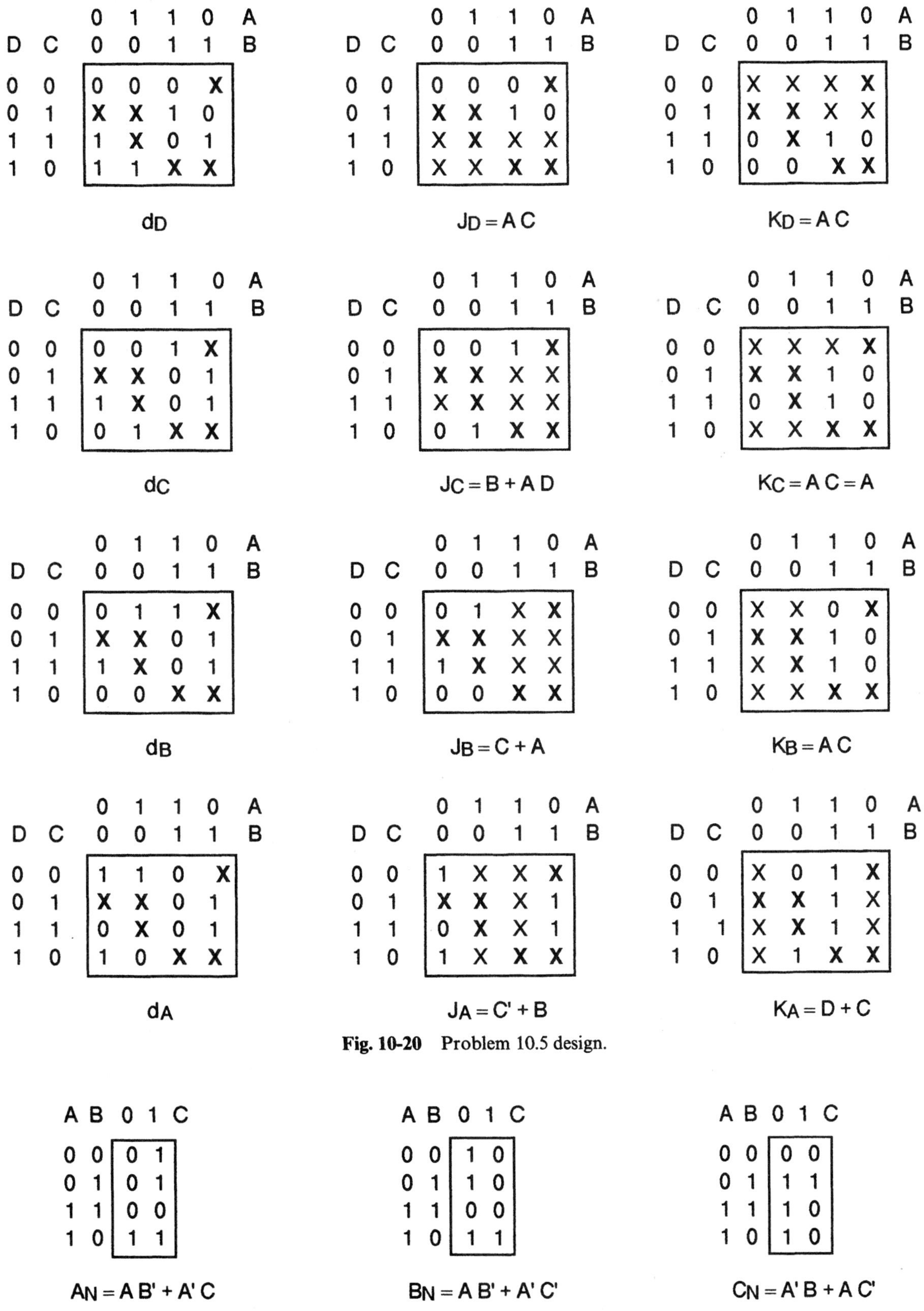

Fig. 10-20 Problem 10.5 design.

Fig. 10-21 Problem 10.6 D flip-flop design.

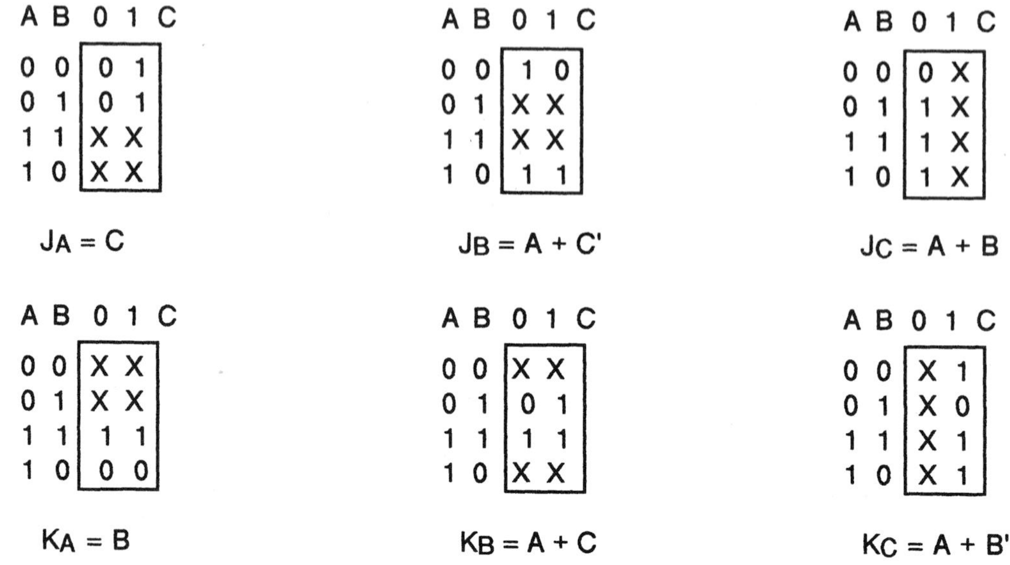

Fig. 10-22 Problem 10.6 JK flip-flop design.

D Flip-Flop Design

Silicon used:

A_N	15 silicon units	$(4 \times 3 + 6 \times 0.5)$
B_N	10	(for A′C′ only since AB′ is already available)
C_N	15	
D flip-flops	30	(3×10)
	70 silicon units	

Timing:

Two levels of logic	20 time units
Flip-flop setup time	20
Propagation time	25
	65 time units

JK Flip-Flop Design

Silicon used:

J_A, K_A	0 silicon units	(direct connection)
J_B, K_B	10	(1 two-input OR gate for each)
J_C, K_C	10	(1 two-input OR gate for each)
JK flip-flops	30	
	50 silicon units	(29% less than the D design)

Timing:

Gate delay	10 times units
Flip-flop setup time	20
Propagation time	25
	55 time units (15% faster than the D design)

10.7 Implement the system of Fig. 10-10 using multiplexers for the logic and a minimal number of D flip-flops.

An effective strategy is to use the present-state outputs as select inputs for the multiplexers. If each multiplexer output is connected to the D input of a state flip-flop, then the problem becomes one of ensuring that the data input of the multiplexer associated with a present state provides the proper next-state value for the flip-flop.

We make use of the transition table developed in Sec. 10.6:

PRESENT STATE X Y	NEXT STATE X Y	CONDITIONS
zero 0 0	zero 0 0	A' B'
	one 0 1	A' B
	two 1 0	A B'
one 0 1	zero 0 0	A' B'
	one 0 1	A' B
	three 1 1	A B
two 1 0	zero 0 0	A' B'
	two 1 0	A B'
	three 1 1	A B
three 1 1	one 0 1	A' B
	two 1 0	A B'
	three 1 1	A B

From the table, it is clear that if the system is in state 0, the next state of flip-flop X will be 1 only if AB′ is TRUE. Thus, logic implementing AB′ is attached to input 0 of the multiplexer serving this flip-flop. Similarly, logic for AB is connected to data input 1 (which is selected by state 1 outputs). In the case of data input 2, the logic is AB′ + AB which can be simplified to A. The final implementation for both X and Y multiplexers is shown in Fig. 10-23. The output logic is formed from Eqs. (10.3) and (10.4) in Sec. 10.6.

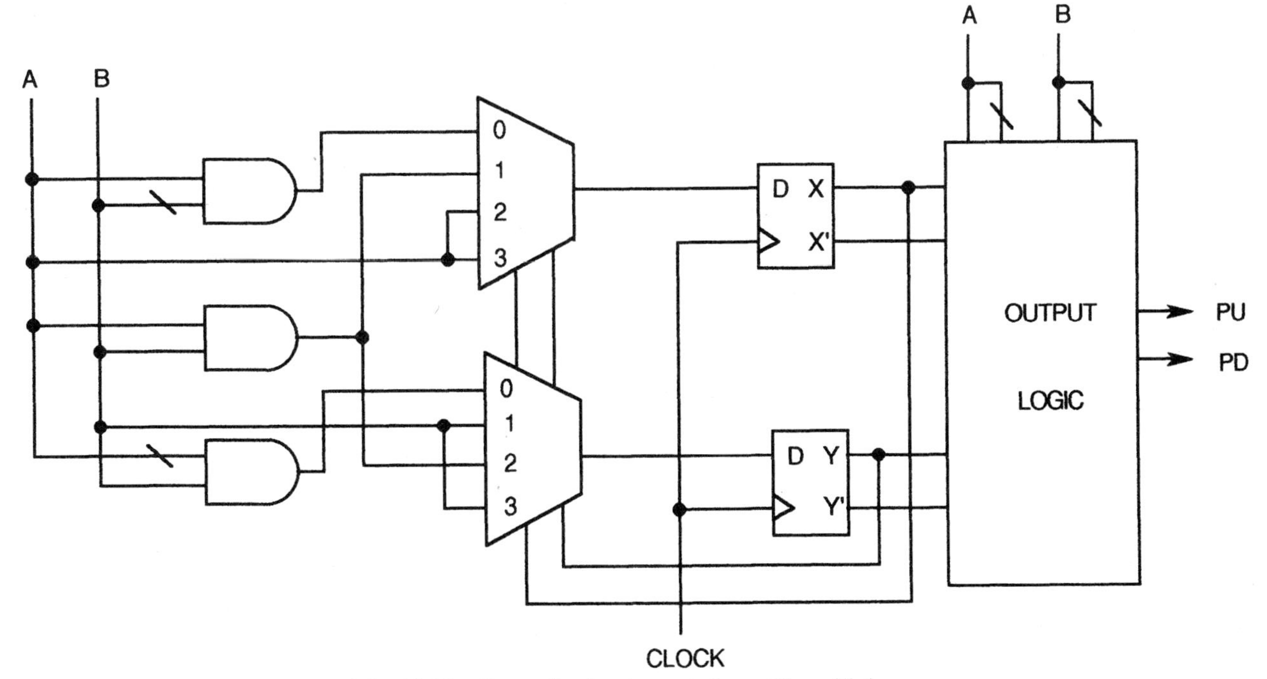

Fig. 10-23 Controller implementation with multiplexers.

If we are interested in fabricating the incremental encoder using SSI and MSI chips, it is useful to compare chip counts for the one-hot and multiplexer implementations (Figs. 10-11 and 10-23, respectively).

One-hot implementation:

Output logic	2 four-input gates	(2/chip)	1 chip
	8 two-input gates	(4/chip)	2 chips
Flip-flops	6 D flip-flops*	(4/chip)	2 chips
A,B decoder	4 two-input gates	(4/chip)	1 chip
	2 inverters	(6/chip)	1 chip
			7 chips

Multiplexer implementation:

Output logic	10 four-input gates	(2/chip)	5 chips
Controller	2 two-select MUX	(2/chip)	1 chip
Flip-flops	4 D flip-flops*	(4/chip)	1 chip
MUX gates	3 two-input gates	(4/chip)	1 chip
			8 chips

10.8 Given the ASM chart shown in Fig. 10-24, list the design equations for the state variables and for the outputs. Assume that the design uses a minimum number of D flip-flops.

There are eight states; therefore, three flip-flops are required which we will label R, S, and T. The states have been designated F, G, H, J, K, L, M, and N as shown in the figure.

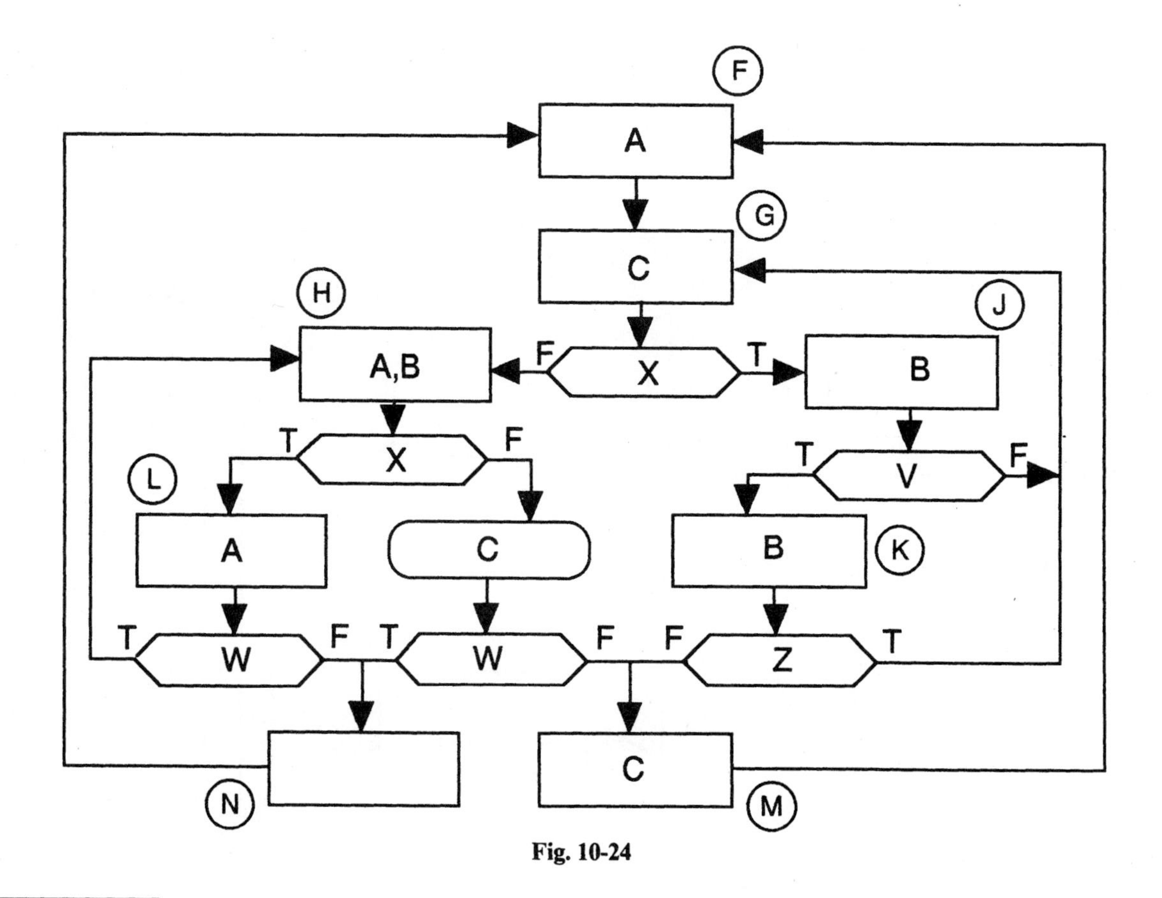

Fig. 10-24

* Includes synchronizing flip-flops for inputs A and B, as discussed in Example 10.5.

Construct a state-transition table (Table 10.15) as per step 2 in Sec. 10.6. The R, S, T state assignments are arbitrarily made in alphabetical order.

Table 10.15

PRESENT STATE	R S T	NEXT STATE	R S T	CONDITIONS
F	0 0 0	G	0 0 1	Unconditional
G	0 0 1	H	0 1 0	X'
		J	0 1 1	X
H	0 1 0	L	1 0 1	X
		M	1 1 0	X' W'
		N	1 1 1	X' W
J	0 1 1	K	1 0 0	V
		G	0 0 1	V'
K	1 0 0	G	0 0 1	Z
		M	1 1 0	Z'
L	1 0 1	N	1 1 1	W'
		H	0 1 0	W
M	1 1 0	F	0 0 0	Unconditional
N	1 1 1	F	0 0 0	Unconditional

Next, write the controller equations using the method described in step 3, Sec. 10.6.

$$
\begin{aligned}
d_R &= H + JV + KZ' + LW' \\
&= R'ST' + R'STV + RS'T'Z' + RS'TW' \\
d_S &= G + HX' + KZ' + L \\
&= R'S'T + R'ST'X' + RS'T'Z' + RS'T \\
d_T &= F + GX + H(X + W) + JV' + KZ + LW' \\
&= R'S'T' + R'S'TX + R'ST'(X + W) + R'STV' + RS'T'Z + RS'TW'
\end{aligned}
$$

Note that some Boolean simplification has been used. For example, in the third equation, the term $H(X + W)$ is a reduction of $H(X + X'W)$.

Finally, output equations may be written directly from the ASM chart. For example, we see that output A is produced when the system is in either states F, H, or L:

$$
\begin{aligned}
A &= F + H + L \\
&= R'S'T' + R'ST' + RS'T \\
B &= H + J + K \\
&= R'ST' + R'ST + RS'T' \\
C &= G + M + HX' \\
&= R'S'T + RST' + R'ST'X'
\end{aligned}
$$

10.9 Create an ASM chart corresponding to the synchronous state machine shown in Fig. 10-25.

The lower left-hand symbol in Fig. 10-25 is an equivalence gate which, in positive logic, produces a TRUE output when both its inputs are equal. In Boolean form, the equivalence of variables B and C is expressed as $F = B'C' + BC$.

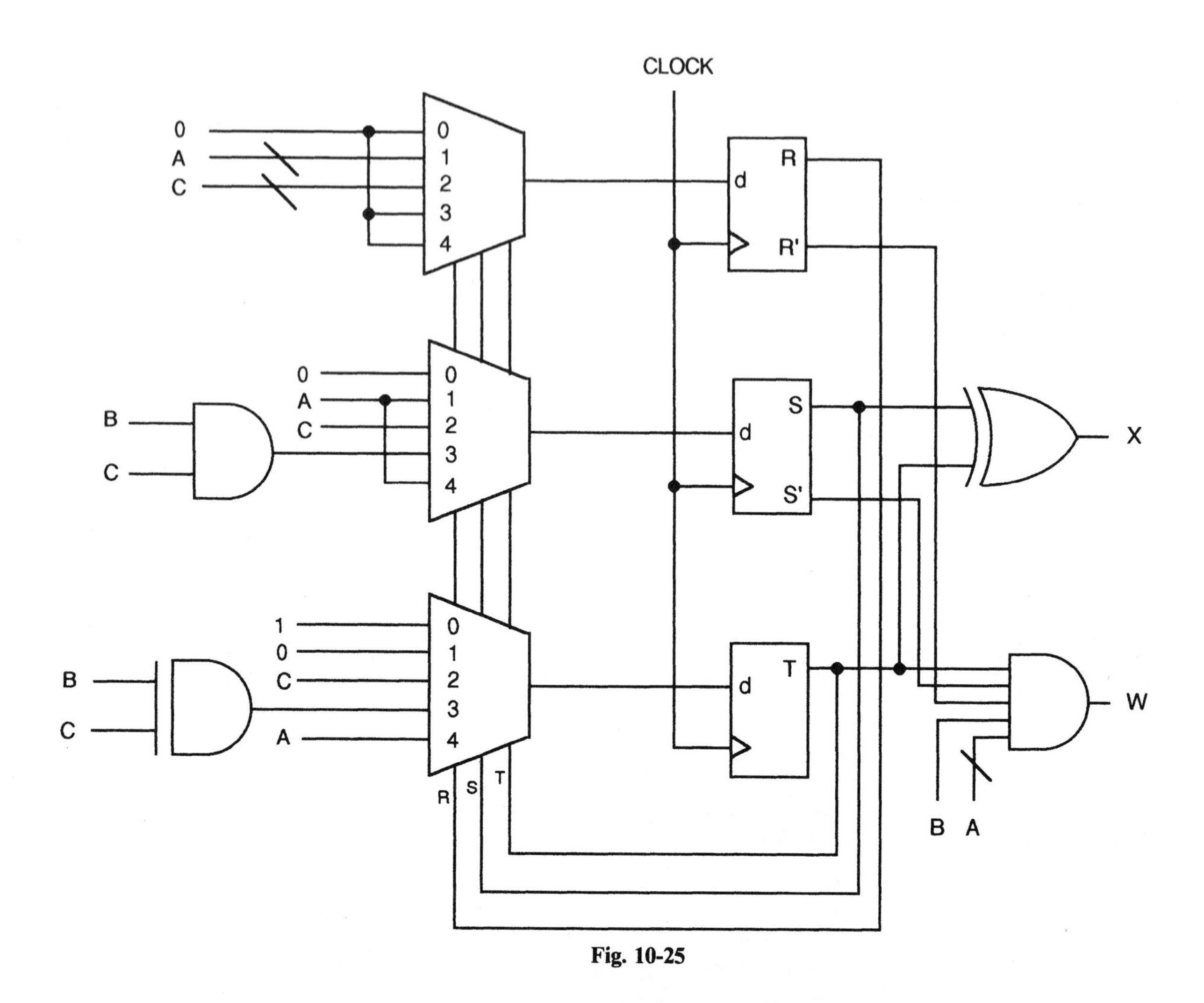

Fig. 10-25

Start at RST = 000 and determine each next state by observing inputs at the addressed multiplexer lines following each clock pulse:

When we are in state 000, it is clear that the next state will be 001.

When in state 001, the d inputs are A′A0 which means that the next state will be 010 (2) if A = 1 and 100 (4) if A = 0.

When in state 010, the d inputs are C′CC so that we will go to state 3 if C = 1 and state four if C = 0.

For state 3, the d inputs are $d_R = 0$, $d_S = BC$, and $d_T = (B'C' + BC)$, respectively. Analysis of the various possibilities is most easily achieved by means of a small truth table (Table 10.16).

Table 10.16

	B	C	R	S	T	STATE
	0	0	0	0	1	"1"
B′ C + BC′	0	1	0	0	0	"0"
	1	0	0	0	0	"0"
	1	1	0	1	1	"3"

State 4 produces d inputs of 0AA, and the next state will be 3 if A = 1 and 0 if A = 0.

From the above information, we may now construct a state-transition table starting at RST = 000 (Table 10.17).

Table 10.17

PRESENT STATE	RST	NEXT STATE	RST	CONDITIONS
"0"	000	"1"	001	Unconditional
"1"	001	"2"	010	A
		"4"	100	A'
"2"	010	"3"	011	C
		"4"	100	C'
"3"	011	"3"	011	B C
		"1"	001	B' C'
		"0"	000	B' C + B C'
"4"	100	"0"	000	A'
		"3"	011	A

Note that we pass from state 1 to state 4 if A is FALSE, regardless of the value of B.

From Fig. 10-25, we see that output X occurs when we are in states 001 or 010 (1 or 2); output W belongs to state 001, but only if A is FALSE and B is TRUE. The outputs may be expressed in Boolean form as

$$W = R'S'TBA' \text{ [state "1" and } BA' \text{ (conditional output)]}$$
$$X = ST' + S'T \text{ [state "1" or state "2"]}$$

The ASM chart may now be drawn as shown in Fig. 10-26.

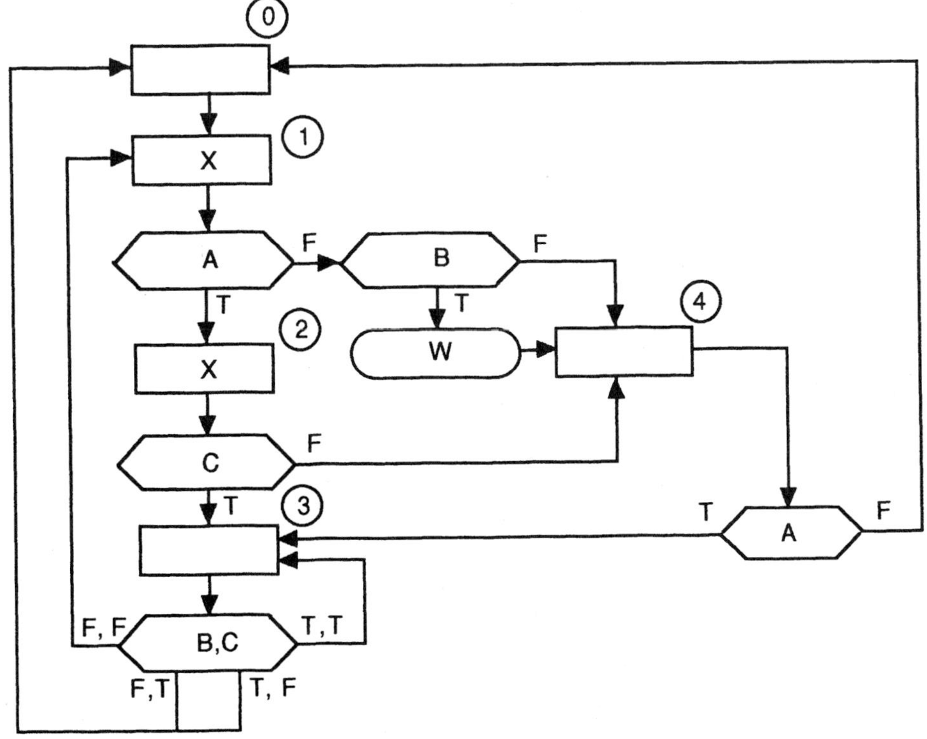

Fig. 10-26

10.10 Using the ASM chart shown in Fig. 10-27, implement a corresponding one-hot controller.

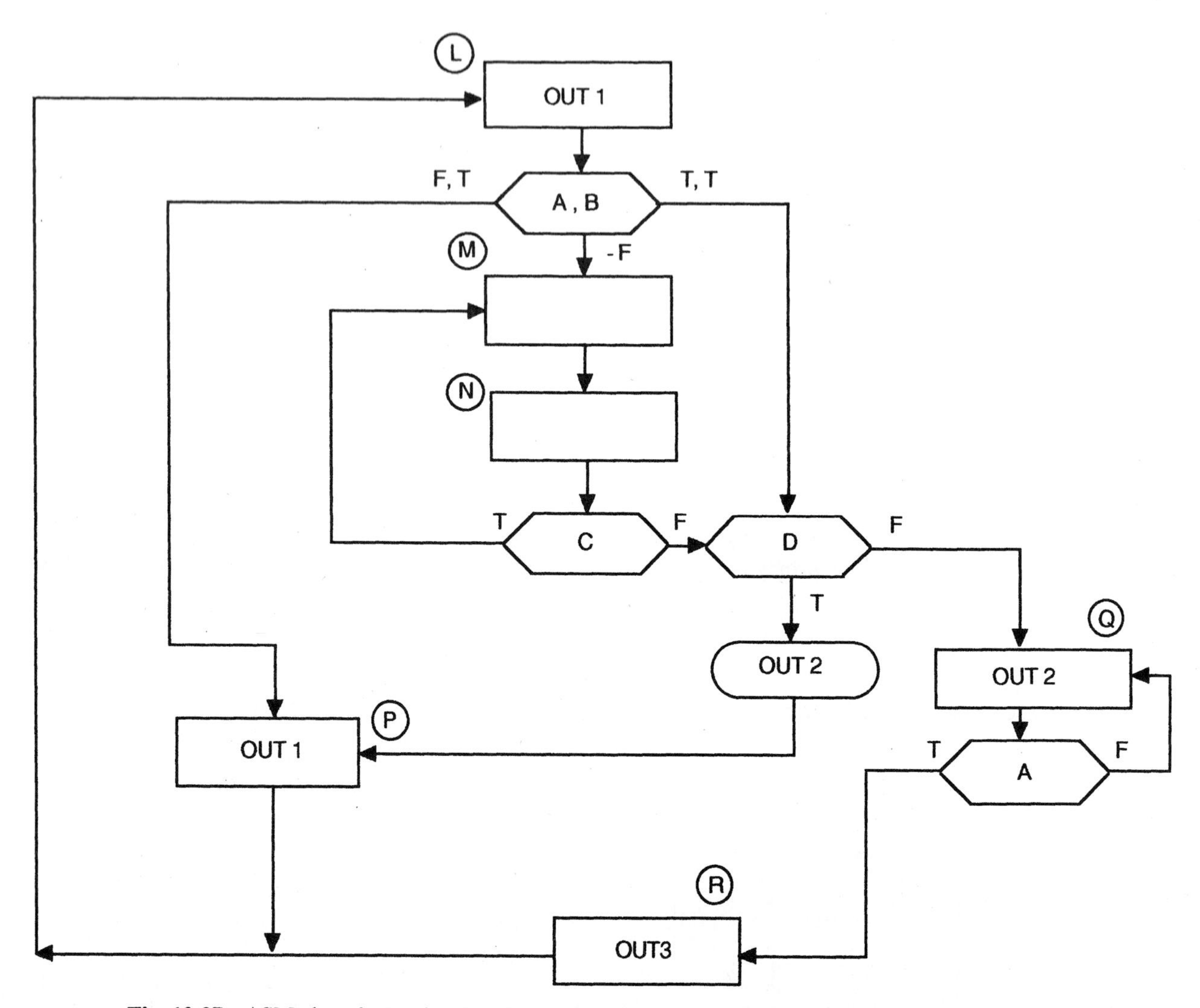

Fig. 10-27 ASM chart for Prob. 10.10. The "F" notation at the lower exit of decision block A, B indicates that when B is FALSE, A is a "don't care"; i.e., it can be either TRUE or FALSE.

For each of the six states, trace back through the chart to determine the previous state or states and the transition conditions associated with each of them. The results of this analysis are organized and presented in Table 10.18.

Controller equations may be written for each present state directly from the table:

$$L = R + P$$
$$M = LB' + NC$$
$$N = M$$
$$P = L(A'B + ABD) + NC'D$$
$$Q = LABD' + NC'D' + QA'$$
$$R = QA$$

The complete one-hot design is shown in Fig. 10-28. Note the intialization modification, involving flip-flop L, which is described in Sec. 10.10

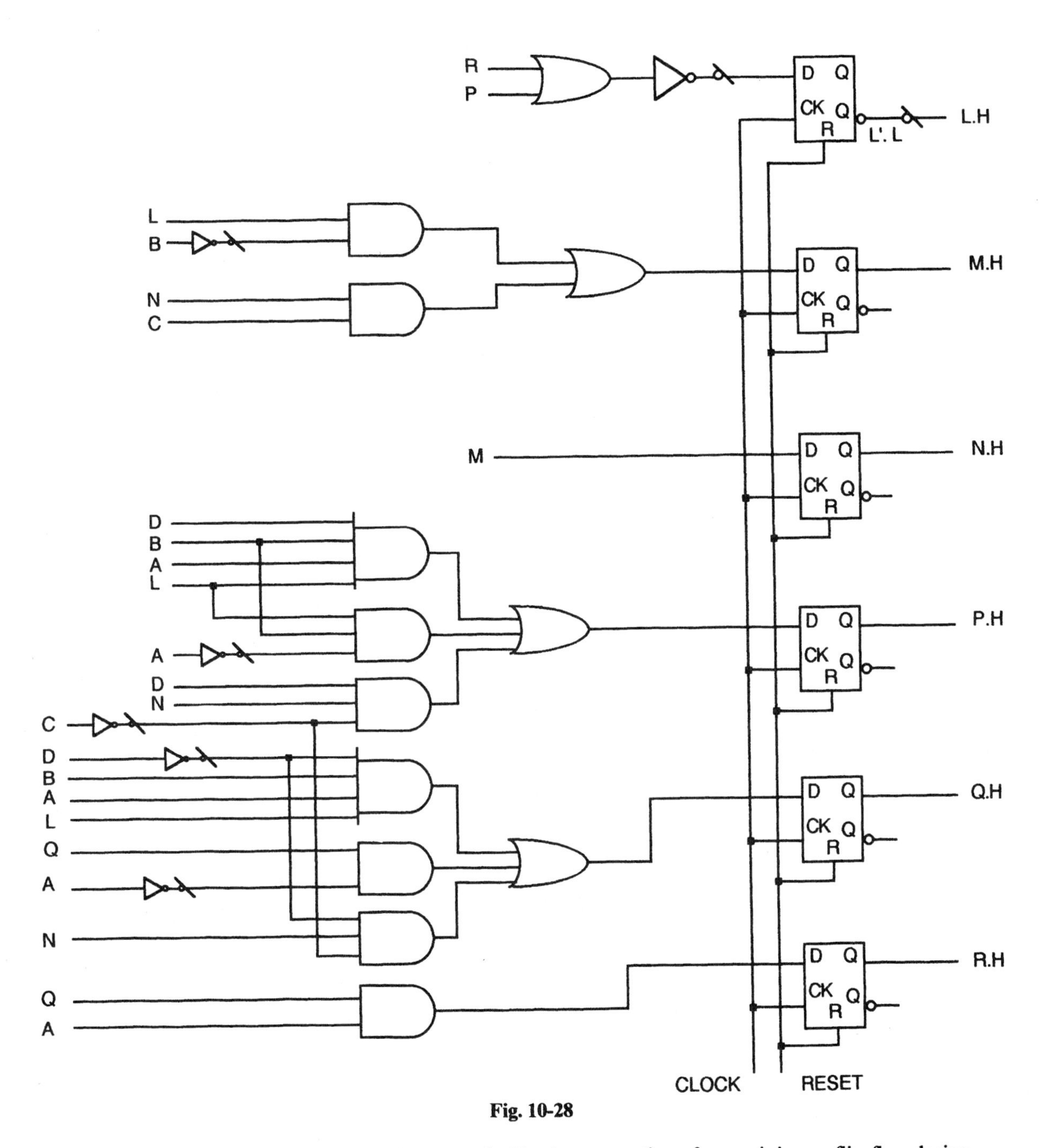

Fig. 10-28

10.11 With reference to Prob. 10.10, create the Boolean equations for a minimum flip-flop design.

Three flip-flops are required to encode the five states. Calling them U, V, and W, we create the state-transition table by tracing the sequential logic in the ASM chart of Fig. 10-27 or by converting previous state to present state and present state to next state in Table 10.18 and rearranging the entries. The result is shown in Table 10.19.

Controller design equations are obtained by writing down, for each of the D flip-flops, all the conditions for which the next state is a logic 1. The resulting equations that follow have been simplified by applying identity A.13 (Appendix A) in appropriate places:

$$D_U = LABD' + NC'D' + Q = U'V'W'ABD' + U'VW'C'D' + UV'W'$$

$$D_V = L(A'B + ABD) + M + NC'D = U'V'W'B(A' + D) + U'V'W + U'VW'C'D$$

$$D_W = L(B' + A'B + ABD) + N(C + C'D) + QA = L(B' + B(A' + AD)) + N(C + D) + QA$$
$$= U'V'W'(B' + A' + D) + U'VW'(C + D) + UV'W'A$$

Table 10.18

PREVIOUS STATE	PRESENT STATE	CONDITIONS
R	L	Unconditional
P		Unconditional
L	M	B'
N		C
M	N	Unconditional
L	P	A' B + A B D
N		C' D
L	Q	A B D'
N		C' D'
Q		A'
Q	R	A

Table 10.19

PRESENT STATE	UVW	NEXT STATE	UVW	CONDITIONS
L	000	M	001	B'
		Q	100	A B D'
		P	011	A' B + A B D
M	001	N	010	Unconditional
N	010	M	001	C
		P	011	C' D
		Q	100	C' D'
P	011	L	000	Unconditional
Q	100	Q	100	A'
		R	101	A
R	101	L	000	Unconditional

The output equations may be written directly from the ASM chart:

$$\text{OUT1} = L + P = U'V'W' + U'VW$$
$$\text{OUT2} = Q + LABD + NC'D = UV'W' + U'V'W'ABD + U'VW'C'D$$
$$\text{OUT3} = R = UV'W$$

10.12 Convert the solution of Prob. 10.10 to a listing for state table entry.

The essential part of the state table entry program consists of the state matrix and list of conditions for each state, as obtained from Table 10.19.

STATES	[Q_U	Q_V	Q_W]
L	[0	0	0]
M	[0	0	1]
N	[0	1	0]
P	[0	1	1]

STATES	[Q_U	Q_V	Q_W]
Q	[1	0	0]
R	[1	0	1]

```
L:
     IF B' THEN M
     IF A'*B + A*B*D THEN P
     IF A*B*D' THEN Q
     ASSERT:
         OUT1
         IF A*B*D THEN OUTPUT 2
M:
     N
N:
     IF C THEN M
     IF C'*D THEN P
     IF C'*D' THEN Q
     ASSERT:
         IF C'*D THEN OUT2
P:
     L
     ASSERT:
         OUT1
Q:
     IF A THEN R
     IF A' THEN Q
     ASSERT:
         OUT2
R:
     L
     ASSERT:
         OUT3
```

10.13 Using the ASM chart of Prob. 10.10 and the waveforms shown in Fig. 10-29, sketch the timing diagram for all states and outputs assuming a prior reset.

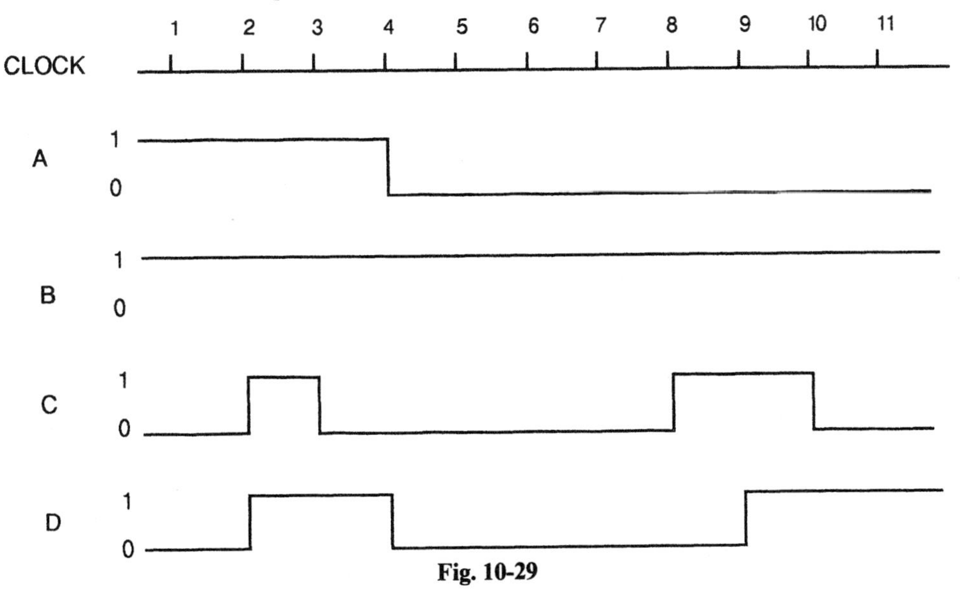

Fig. 10-29

Note that input transitions are coincident with clock pulses indicating that this is a synchronous machine. When reset, the machine is initially in state L, as specified.

Refer to the ASM chart of Fig. 10-27 to follow state sequencing:

1. Since, prior to the first clock pulse, A and B = 1 and D = 0, we see that the machine will go to state Q.
2. At pulse 2, A = 1 and we move to state R.
3. The next state is unconditionally L.
4. A = 1, B = 1, D = 1; state = P.
5. Unconditional transfer to L.
6. A = 0, B = 1; state = P.
7. Unconditional return to L.
8. Since A and B alone determine the state which follows L and they remain unchanged following clock pulse 4, the state machine will oscillate between states P and L indefinitely. From the solution to Prob. 10.11,

$$\text{OUT1} = \text{L} + \text{P} \text{ (unconditional: OUT1} = 1 \text{ if the machine is in states L or P)}$$
$$\text{OUT2} = \text{Q} + \text{LABD} + \text{NC'D}$$
$$\text{OUT3} = \text{R (unconditional)}$$

Output waveforms are shown in Fig. 10-30.

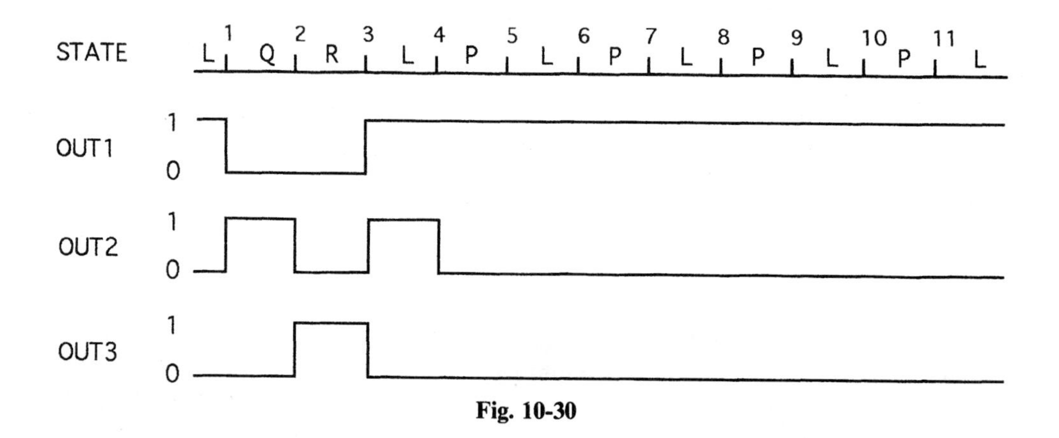

Fig. 10-30

Output 2 remains 0 after pulse 4 even though state L recurs because it is conditional upon input A which is seen to remain at 0 over this range.

10.14 In the system of Example 7.5 (Fig. 7-18), the two flip-flops are intended to be clocked at the same time. Because of differences in signal path length, however, clock pulses arrive at each flip-flop at slightly different times; they are said to be skewed. In addition to the data given in Example 7.5, *minimum* timing parameters are provided as follows:

For Flip-Flops	For Gates
$T_{pd,\,HL} = 7$ ns $T_{pd,\,LH} = 5$ ns $T_{su} = 0$ ns $T_h = 0$ ns	$T_{pd,\,HL} = 4$ ns $T_{pd,\,LH} = 4$ ns

Determine the maximum allowable clock skew.

Consider the timing sequence following the arrival of the earliest clock pulse. It triggers an associated flip-flop which causes a change in the logic inputs which, in turn, causes a change in the logic outputs. The process is shown graphically in Fig. 10-31. After $5 + 4 + 4 = 13$ ns, the D inputs could contain erroneous data which would be latched by the late clock pulse. The maximum allowable clock skew is thus 13 ns.

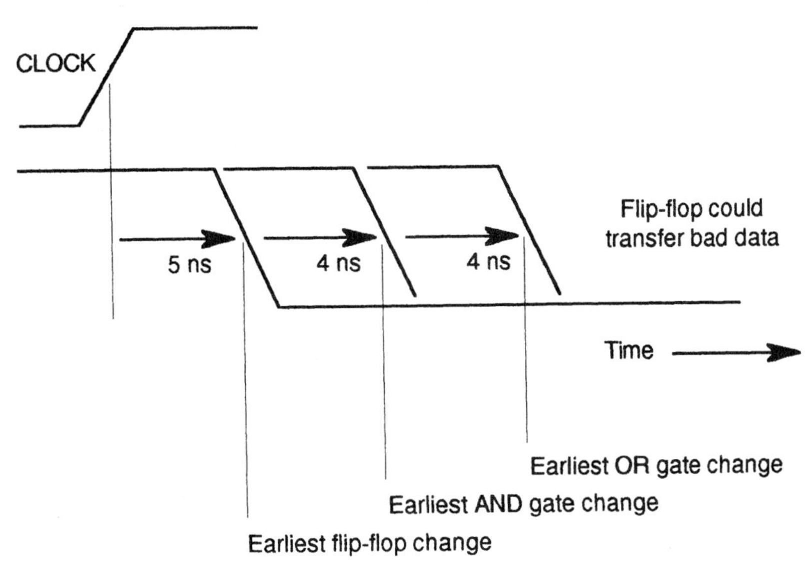

Fig. 10-31

Supplementary Problems

10.15 An abbreviated schematic diagram of a state machine counter with ROM combinational logic is shown in Fig. 10-32. It is desired to insert, in each line marked with an X, a box containing logic which will permit loading data in parallel into all the D flip-flops when an input $L = 1$. When $L = 0$, the count is to proceed. Design the logic for a typical box (Fig. 10-33), using as few gates as possible.

10.16 Use JK flip-flops and such gates as are needed to implement the state machine counter defined by the sequence in Table 10.20. Attempt to minimize the number of gates required and define your design with logic equations (refer to Fig. 10-19).

Table 10.20

A	B	C
0	0	1
1	0	0
0	1	1
1	1	0
1	1	1

repeats

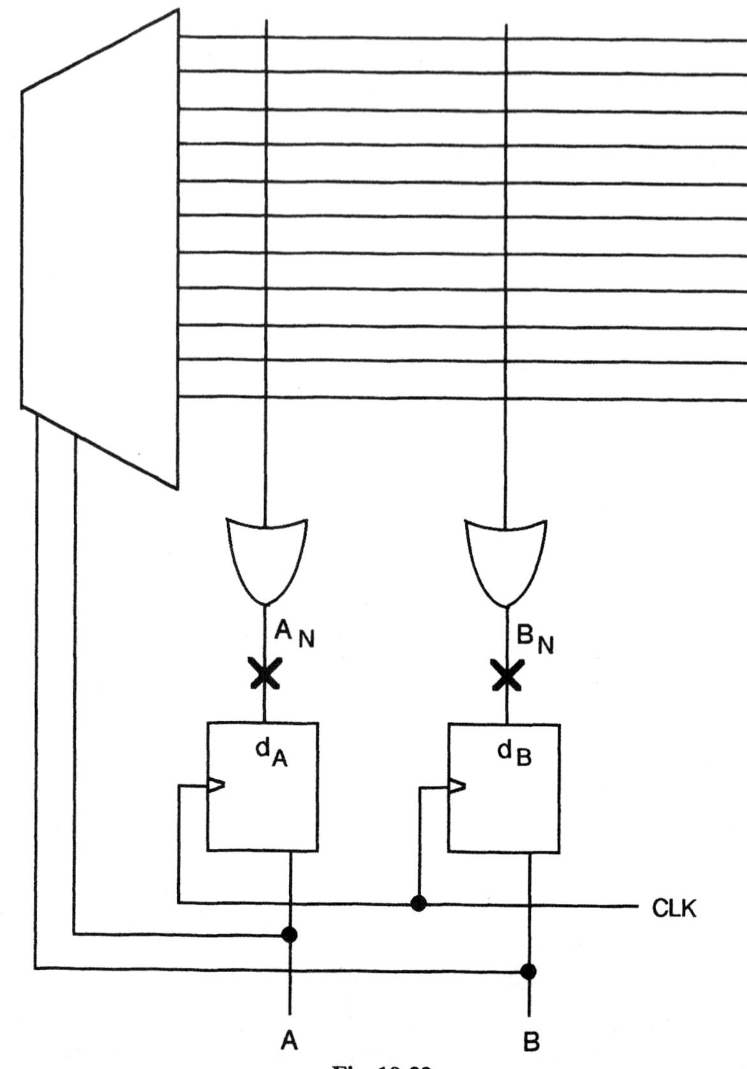

Fig. 10-32

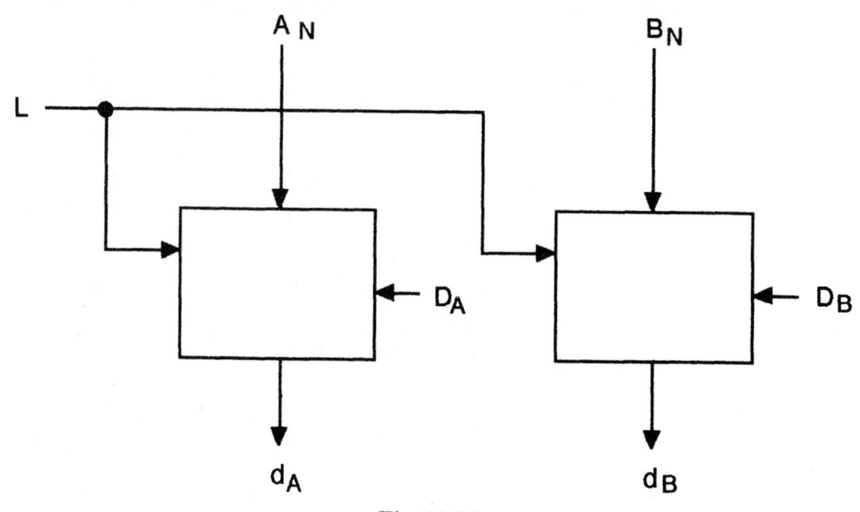

Fig. 10-33

10.17 Repeat Prob. 10.16 for the sequence given in Table 10.21.

Table 10.21

D	C	B	A
0	0	0	0
0	0	0	1
0	0	1	0
0	0	1	1
0	1	0	0
0	1	0	1
0	1	1	0
0	1	1	1
1	0	0	0
1	0	0	1

repeats

10.18 For comparison purposes, implement Prob. 9.17 using SSI gates and JK flip-flops.

10.19 Determine the Boolean design equations for a state machine which counts in the sequence of Table 10.22. Assume that D flip-flops are used to register the states.

Table 10.22

C	B	A
0	0	0
0	1	0
1	1	0
1	0	1
0	0	1
1	1	1
0	1	1
1	0	0

10.20 Shown below is a portion of a state table entry program. Create the corresponding chart. Note: Only those parts of the program which are pertinent to the required chart are listed.

```
A:
    IF X + Y THEN B
    A
B:
    C
    ASSERT:
    OUT2
C:
    CASE
        X : A
    ENDCASE
    B
    ASSERT:
        IF X' THEN OUT1
        IF X THEN OUT2
END$
```

10.21 Using the ASM chart shown in Fig. 10-34, draw the state transition table for a one-hot controller and write the design equations.

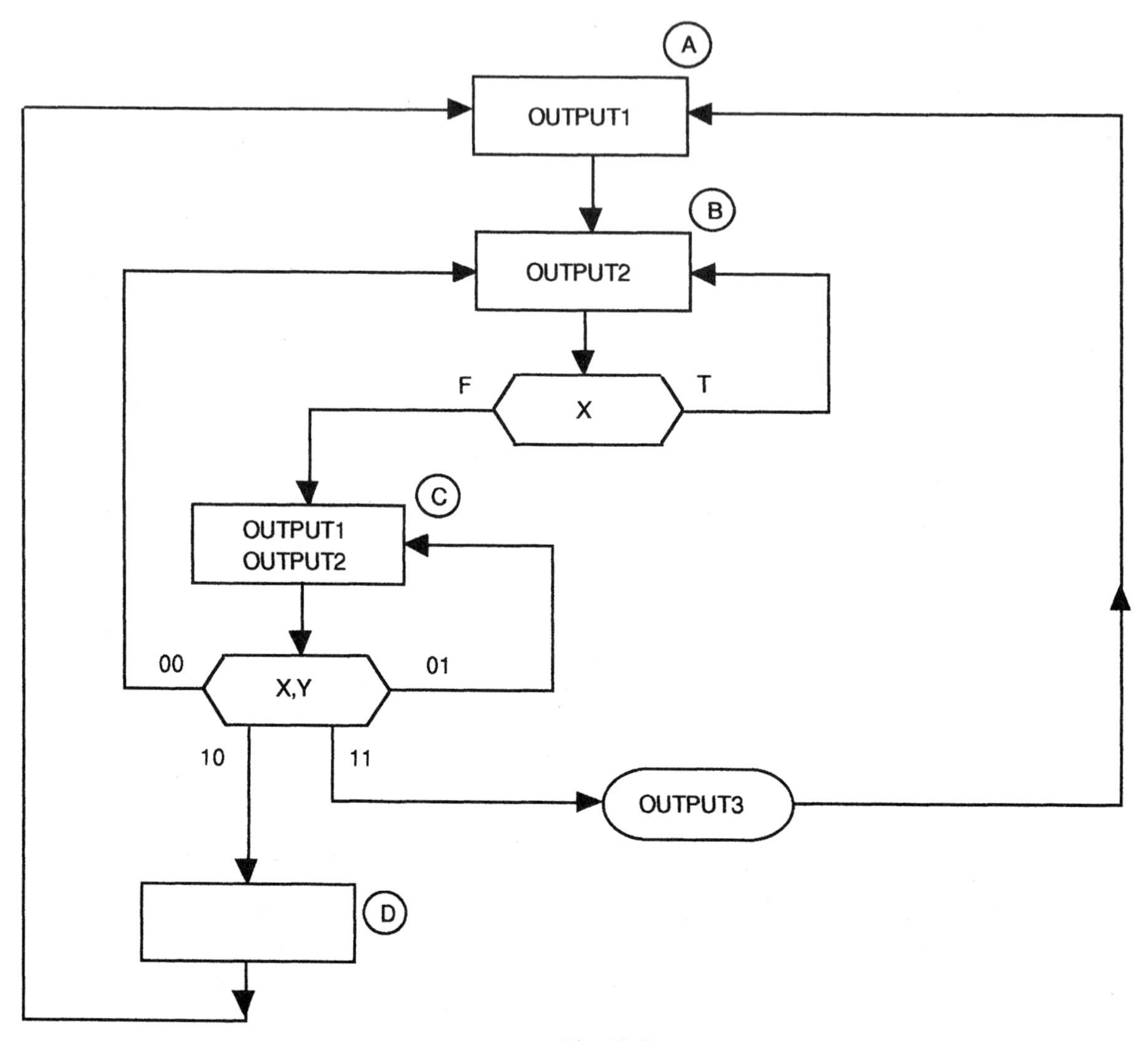

Fig. 10-34

10.22 Use the ASM chart of the previous problem to create the design equations for a minimal flip-flop state machine.

10.23 Repeat Prob. 10.22, this time writing a program for state table entry using iPLSII software.

10.24 Draw the state-transition table corresponding to the one-hot controller shown in Fig. 10-35.

10.25 Rearrange the state table of Prob. 10.24 into a form suitable for obtaining a minimal flip-flop realization, and write design equations for the states.

10.26 A small state machine is shown in Fig. 10-36. Create the corresponding ASM chart, designating A and B as the state flip-flops. It will be helpful to start at some arbitrary state and examine the performance as clocking proceeds.

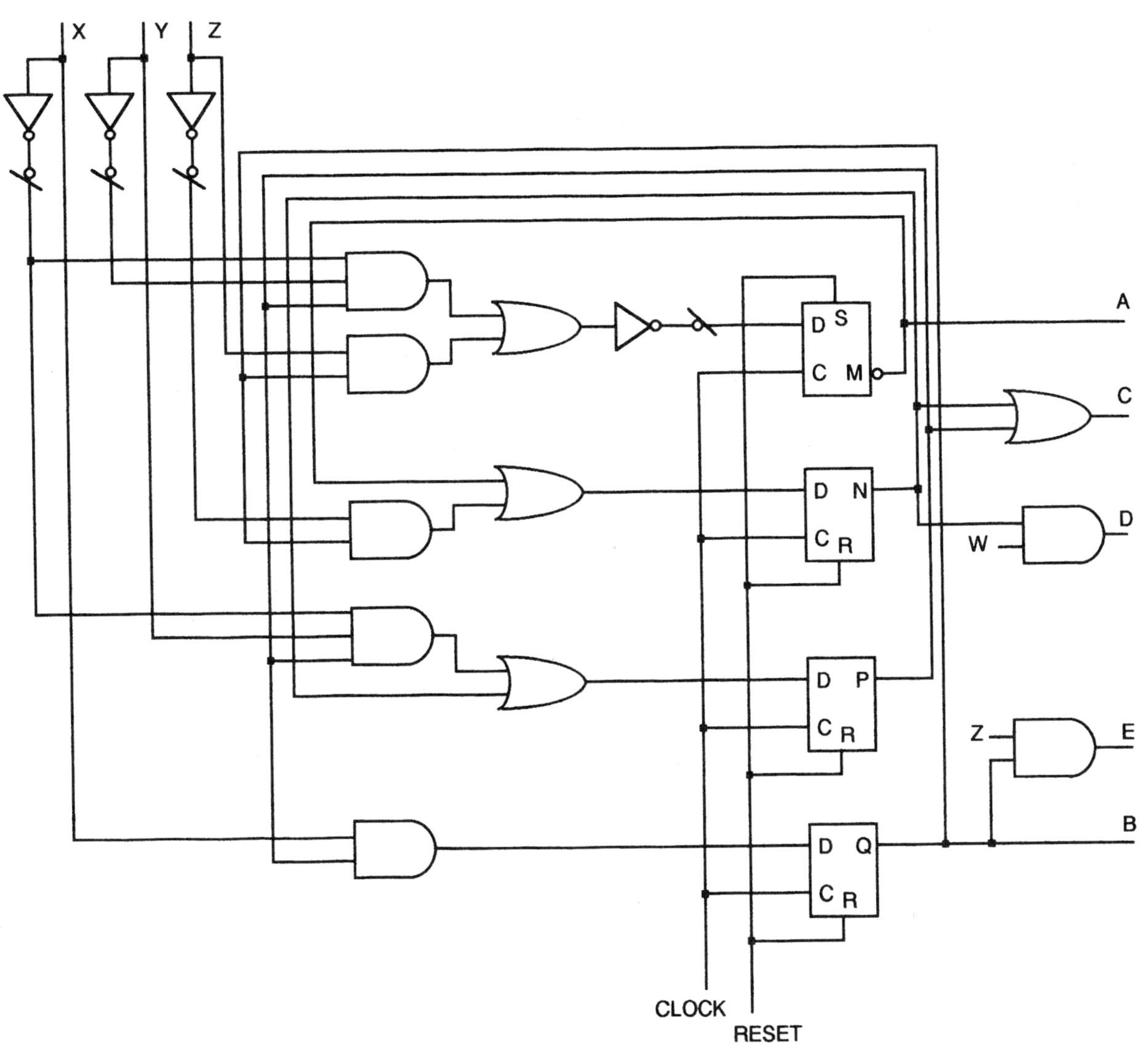

Fig. 10-35

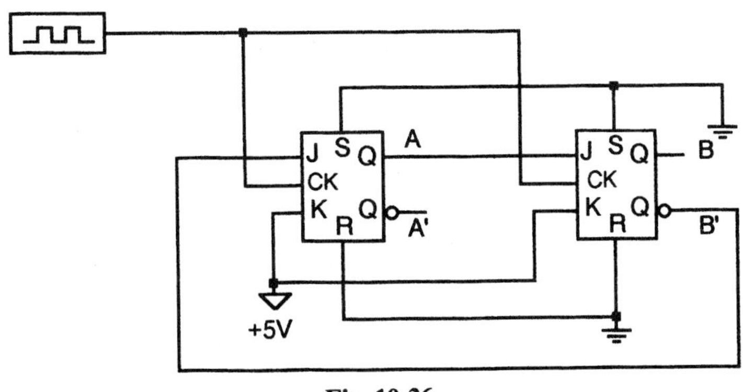

Fig. 10-36

10.27 Given the ASM chart shown in Fig. 10-37, draw a state transition table and produce a set of design equations for a minimal flip-flop realization.

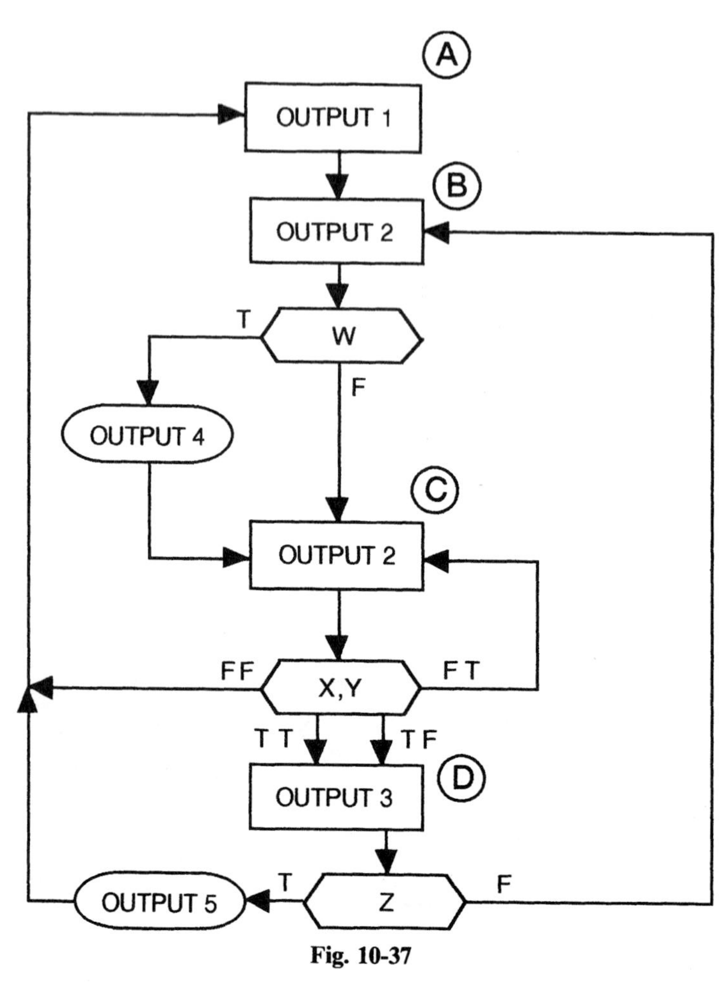

Fig. 10-37

10.28 Using the ASM chart of Prob. 10.27, complete the timing diagram, shown in Fig. 10-38, assuming that the system is initially in state A.

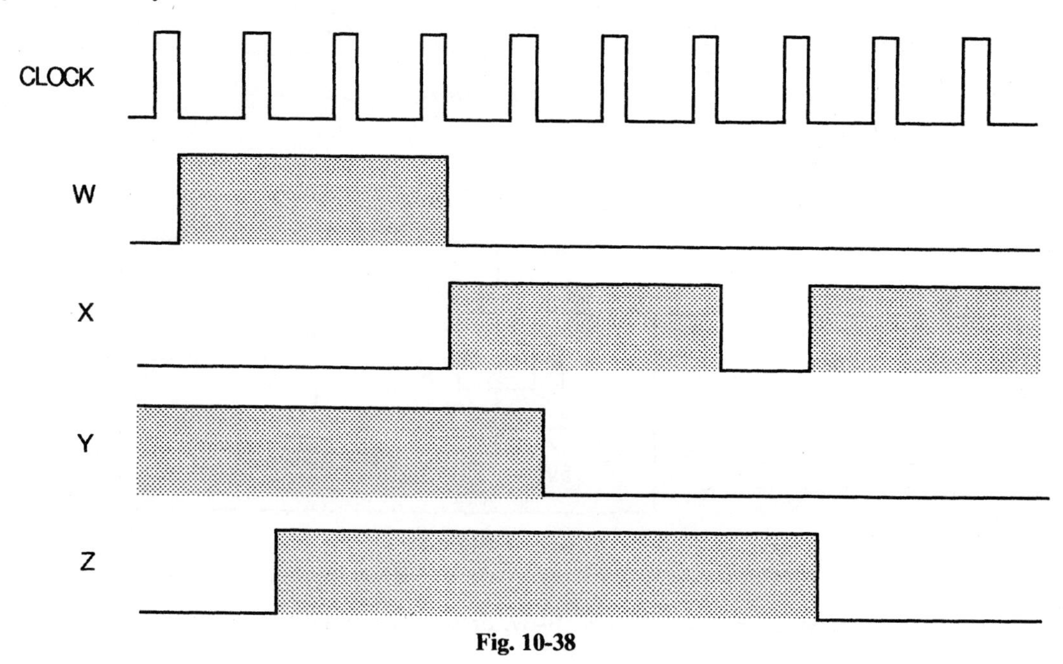

Fig. 10-38

10.29 Repeat Prob. 10.8 for a one-hot design. Use F as the initial state.

10.30 Repeat Prob. 10.29 using JK flip-flops instead of D flip-flops.

Answers to Supplementary Problems

10.15 See Fig. 10-39. Note that if L′ is not available, the hardware inverter required can produce a glitch condition which must be accounted for by the designer.

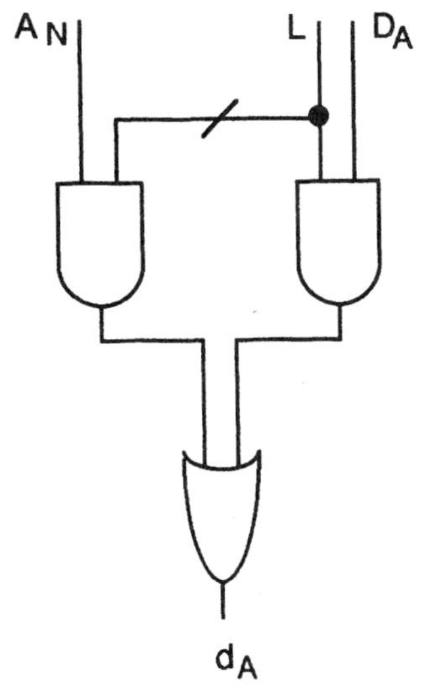

Fig. 10-39

10.16 Three JK flip-flops are required, having the following inputs:

$$J_A = 1 \qquad K_A = C + B'$$
$$J_B = A \qquad K_B = AC$$
$$J_C = 1 \qquad K_C = A'$$

Only two gates are needed if both outputs of the flip-flops are available.

10.17 The JK inputs are

$$J_D = ABC \qquad K_D = A$$
$$J_C = AB \qquad K_C = AB$$
$$J_B = AD' \qquad K_B = A$$
$$J_A = 1 \qquad K_A = 1$$

10.18 A reasonable approach is to create a K and a J map for each of the count variables A, B, C, and D (F = 0) and to repeat the process for F = 1. From the 16 maps, we obtain:

$$J_{AN} = F'B + FB'C'$$
$$K_{AN} = FAB'$$

$$J_{BN} = F'A + F'D + FC' + FA$$
$$K_{BN} = F'C + FA' + F \oplus D$$

$$J_{CN} = F'AB + F'A'D' + FA'B'$$
$$K_{CN} = F'AB + FA'B' + FAD$$

$$J_{DN} = F'A' + F \oplus B$$
$$K_{DN} = FA + F \oplus B$$

10.19 The unsimplified equations are

$$d_A = AB'C' + AB'C + A'BC + ABC$$
$$d_B = A'B'C' + AB'C' + A'BC' + ABC$$
$$d_C = AB'C' + A'BC' + ABC' + A'BC$$

10.20 See Fig. 10-40.

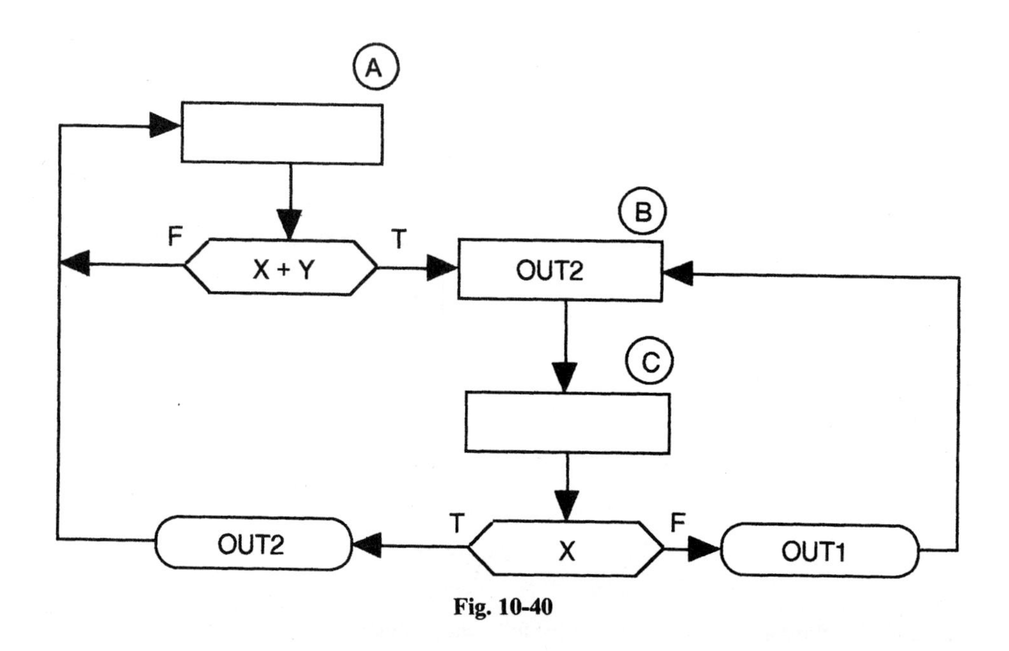

Fig. 10-40

10.21 See Table 10.23.

Table 10.23

PREVIOUS STATE	PRESENT STATE	CONDITIONS
C	A	X Y
D		UNCONDITIONAL
A	B	UNCONDITIONAL
B		X
C		X' Y'
B	C	X'
C		X' Y
C	D	X Y'

Design equations:

dA = C X Y + D	OUTPUT1 = A + C
dB = A + B X + C X' Y'	OUTPUT2 = B + C
dC = B X' + C X' Y	OUTPUT3 = C X Y
dD = C X Y'	

10.22 There are four states, so three flip-flops will be minimal. See Table 10.24.

Table 10.24

PRESENT STATE M N		NEXT STATE M N		CONDITIONS
A	0 0	B	0 1	UNCONDITIONAL
B	0 1	B	0 1	X
		C	1 0	X'
C	1 0	A	0 0	X Y
		B	0 1	X' Y'
		C	1 0	X' Y
		D	1 1	X Y'
D	1 1	A	0 0	UNCONDITIONAL

Design equations:

$$D_M = B\,X' + C\,X'\,Y + C\,X\,Y' = M'N\,X' + M\,N'\,(X'\,Y + X\,Y')$$

$$D_N = A + B\,X + C\,(X'\,Y' + X\,Y') = M'N' + M'\,N\,X + M\,N'\,Y'$$

OUTPUT1 = A + C = M' N' + M N' = N'

OUTPUT2 = B + C = M' N + M N'

OUTPUT3 = C X Y = M N' X Y

10.23 The following answer includes only the pertinent parts.

```
STATES:   [QM  QN]
  A       [ 0   0 ]
  B       [ 0   1 ]
  C       [ 1   0 ]
  D       [ 1   1 ]
A:
    B
    ASSERT:
            OUTPUT1
B:
    IF X THEN B
    IF X' THEN C
    ASSERT:
            OUTPUT2
```

```
C:
   IF X'*Y THEN C
   IF X*Y THEN A
   IF X'*Y' THEN B
   IF X*Y' THEN D
   ASSERT:
            OUTPUT1
            OUTPUT2
           IF X*Y THEN OUTPUT3
D:
   A
```

10.24 See Table 10.25.

Table 10.25

PREVIOUS STATE	PRESENT STATE	CONDITIONS
P Q	M	X' Y' Z
M Q	N	TRUE Z'
N P	P	TRUE X' Y
P	Q	X

10.25 See Table 10.26.

Table 10.26

PRESENT STATE	NEXT STATE [U V]	CONDITIONS
M	N 0 1	TRUE
N	P 1 0	TRUE
P	P 1 0 Q 1 1 M 0 0	X' Y X X' Y'
Q	N 0 1 M 0 0	Z' Z

Design equations:

$$dU = N + P(X + X'Y) = U'V + UV'(X + Y)$$

$$dV = M + PX + QZ' = U'V' + UV'X + UVZ'$$

10.26 See Table 10.27 and Fig. 10-41.

Table 10.27

PRESENT STATE A	PRESENT STATE B	J_A	K_A	J_B	K_B	NEXT STATE A	NEXT STATE B
0	0	1	1	0	1	1	0
1	0	1	1	1	1	0	1
0	1	0	1	0	1	0	0
1	1	0	1	1	1	0	0

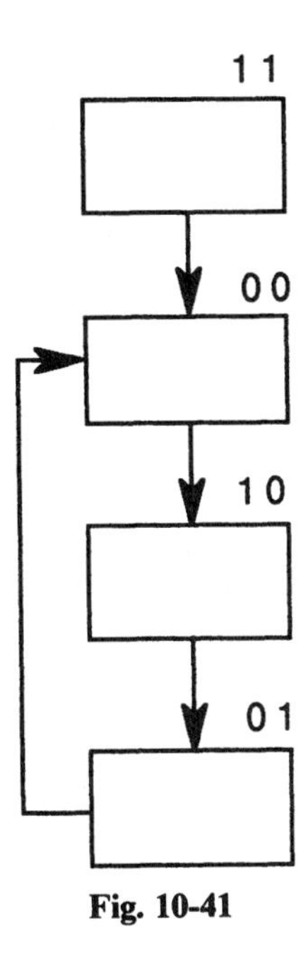

Fig. 10-41

Note that the state machine cannot assume state 11 by itself; it can only get to this state if forced by an external input.

10.27 Two flip-flops, designated M and N, are required (see Table 10.28).

Table 10.28

PRESENT STATE [M N]		NEXT STATE [M N]		CONDITION
A	0 0	B	0 1	TRUE
B	0 1	C	1 0	TRUE
C	1 0	A	0 0	X' Y'
		C	1 0	X' Y
		D	1 1	X
D	1 1	A	0 0	Z
		B	0 1	Z'

$$M = B + CX'Y + CX = M'N + MN'X'Y + MN'X$$
$$N = A + CX + DZ' = M'N' + MN'X + MNZ'$$

10.28 See Fig. 10-42.

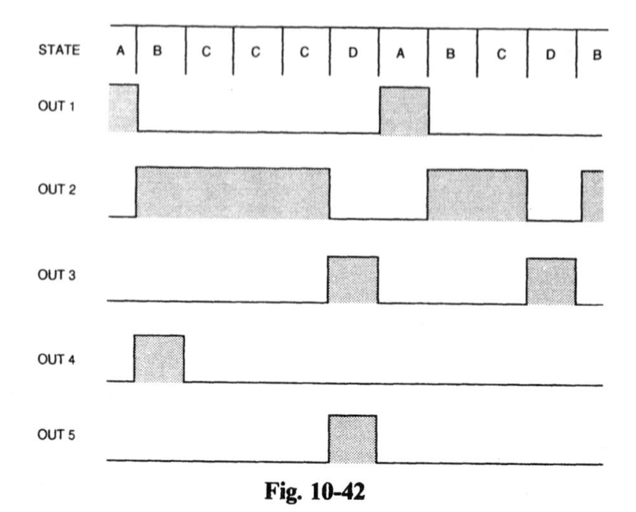

Fig. 10-42

10.29 Controller equations:

$$D_F = N + M'$$
$$D_G = F$$
$$D_H = GX' + LW$$
$$D_J = GX$$
$$D_K = JV$$
$$D_L = HX$$
$$D_M = HX'W' + KZ'$$
$$D_N = LW' + HX'W$$

Output equations:

$$A = F + H + L$$
$$B = H + J + K$$
$$C = G + M + HX'$$

10.30 The transitions in a one-hot circuit are a 0-to-1 for the selected state and 1-to-0 when it is deselected. Otherwise, the flip-flops default to 0. Make K always equal to 1, and then make J = 1 when a 0-to-1 transition is desired. For the initial state, the roles of J and K are reversed.
Controller equations:

$$K_F = N + M \qquad J_F = 1$$
$$J_G = F \qquad K_G, \ldots, K_N = 1$$
$$J_H = GX' + LW$$
$$J_J = GX$$
$$J_K = JV$$
$$J_L = HX$$
$$J_M = HX'W' + KZ'$$
$$J_N = LW' + HX'W$$

Output equations:

$$A = F + H + L$$
$$B = H + J + K$$
$$C = G + M + HX'$$

Chapter 11

Electronically Programmable Functions

11.1 INTRODUCTION

In computer systems, use is often made of circuitry which is capable of having its function altered by means of one or more *control signals*. In contrast with the programmable logic devices discussed in Chap. 9, electrical inputs alter circuit function without causing any physical change to occur in the hardware.

11.2 BASIC COMPONENTS

The Exclusive OR

The XOR gate is often used as a *controlled inverter* in which data is applied to one of its two inputs and a control signal is applied to the other as shown in Fig. 11-1. The corresponding logic equation and related truth table are also indicated. We see that when the control signal is 0, data out and data in are identical, and, when the control is 1, the data out become the logical inverse of the data in. The control input can thus be viewed as determining whether the input is passed through the device in direct or in logically inverted form.

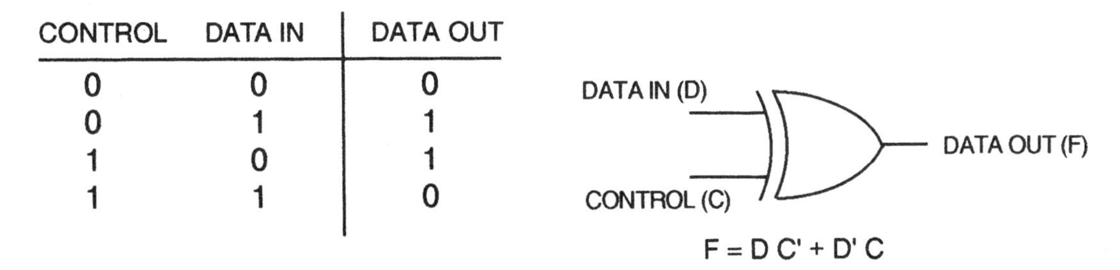

CONTROL	DATA IN	DATA OUT
0	0	0
0	1	1
1	0	1
1	1	0

Fig. 11-1 Exclusive OR as a controlled inverter.

Note that though operation of this gate can be described in conventional Boolean form, when the design goal is to implement a programmable device, *it is more useful to shift one's perspective and view it functionally* as, in the present case, a programmable logic inverter.

The Two-Input Logic Gate

This basic *AND gate* can be viewed as a *controlled switch.* As before, one of the inputs is designated for control and the other for data. Figure 11-2 shows the configuration, the logic equation, and the truth table. If the control input is 1, then data passes through the gate in direct form; if the control signal is 0, the output is 0 and no data is transmitted. In positive logic, two-input NAND hardware can be viewed as a controlled *inverting* switch.

A two-input *OR gate* can be used as a controlled switch in a similar manner. In this case, $C = 0$ yields a transmitted data signal and $C = 1$ produces a 1 at the output regardless of the data input.

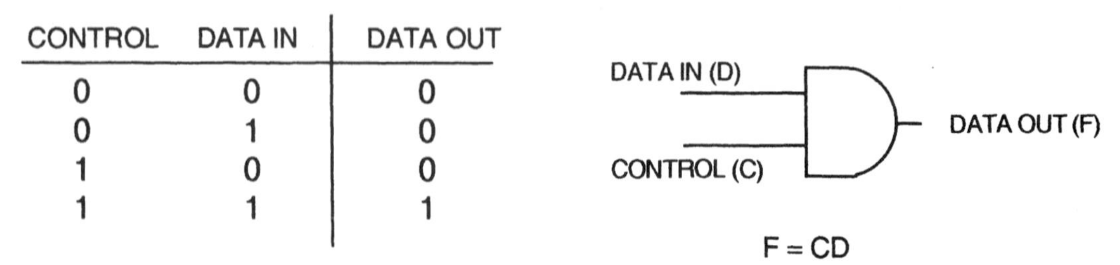

CONTROL	DATA IN	DATA OUT
0	0	0
0	1	0
1	0	0
1	1	1

Fig. 11-2 AND gate viewed as a controlled switch.

The Multiplexer

The use of the multiplexer as a controlled switch has been fully described in Chap. 5. This device may be used to choose from among multiple inputs when appropriate control signals are applied to its select lines.

The TRI-STATE® Device

The TRI-STATE® device (previously described in Chap. 9) is also used as a controlled switch. It can be configurated to function as either an inverter or a buffer (no inversion) when enabled and will always appear as an open circuit at its output when not enabled. Additional flexibility can be obtained if the status of the enable input is determined by some logic as in Fig. 11-3. In this case, an LT enable TRI-STATE® device is controlled by two signals (A and B) such that data will appear on the output line only when A and B are not simultaneously TRUE.

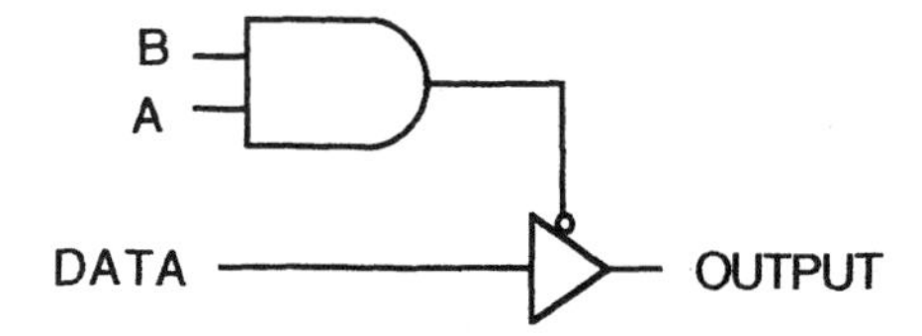

Fig. 11-3 Logically controlled TRI-STATE® output device.

RAM Lookup Tables

In Sec. 5.4, the use of read only memory (ROM) for doing logic was described. We learned that in such an application, the ROM is basically used as a lookup table with truth table inputs serving as addresses and the corresponding memory contents providing appropriate outputs. From a user's standpoint, the device may be viewed as doing combinational logic. Recall also that the contents of RAM memory, as described in Chap. 8, can be modified at electronic speeds.

We may combine these two functions by setting aside a block of RAM and treating its address bits as logic inputs and storing the outputs corresponding to a desired logic function. If the logic is to be changed, the RAM can simply be reloaded electronically with data corresponding to the modified logic. In essence, the RAM can be considered as an electronically programmed EPROM doing logic.

11.3 PROGRAMMABLE GATE ARRAYS

As briefly described in Chap. 9, there is a product which is somewhat intermediate between PAL devices, PLAs, and the "full" gate array. It is called the *field programmable gate array* (FPGA), sometimes referred to as a logic cell array (LCA™).* The FPGA differs from many configurable devices in that it may be programmed without the requirement for any physical changes in circuit interconnections.

* LCA is a trademark of Xilinx, Inc.

These devices contain a very long shift register (SRAM) which, during programming, is loaded with a long serial bit stream produced by the programming software. This embedded bit stream circulates in the background and is used to control connections among the various components composing the FPGA. The device can be reconfigured easily by entering a different bit stream.

As an example of such an FPGA, consider the LCA offered by Xilinx, Inc. It consists of three types of structured components:

1. Configurable logic blocks (CLB)
2. Input/output blocks (I/OB)
3. Programmable interconnect points (PIP)

Configurable Logic Block

Shown in Fig. 11-4 is a basic functional element which contains a combinatorial logic section, two D flip-flops with a common clock, and multiplexers used as selector switches. There is also an internal control section (not shown) made up of RAM elements which are serviced by the embedded bit stream. The combinatorial logic unit, depicted as a large rectangle, can produce various combinations of the five external logic variables and fed-back signals from the two flip-flops. The CLB can be programmed to produce either one or two output variables comprising several groupings of its inputs.

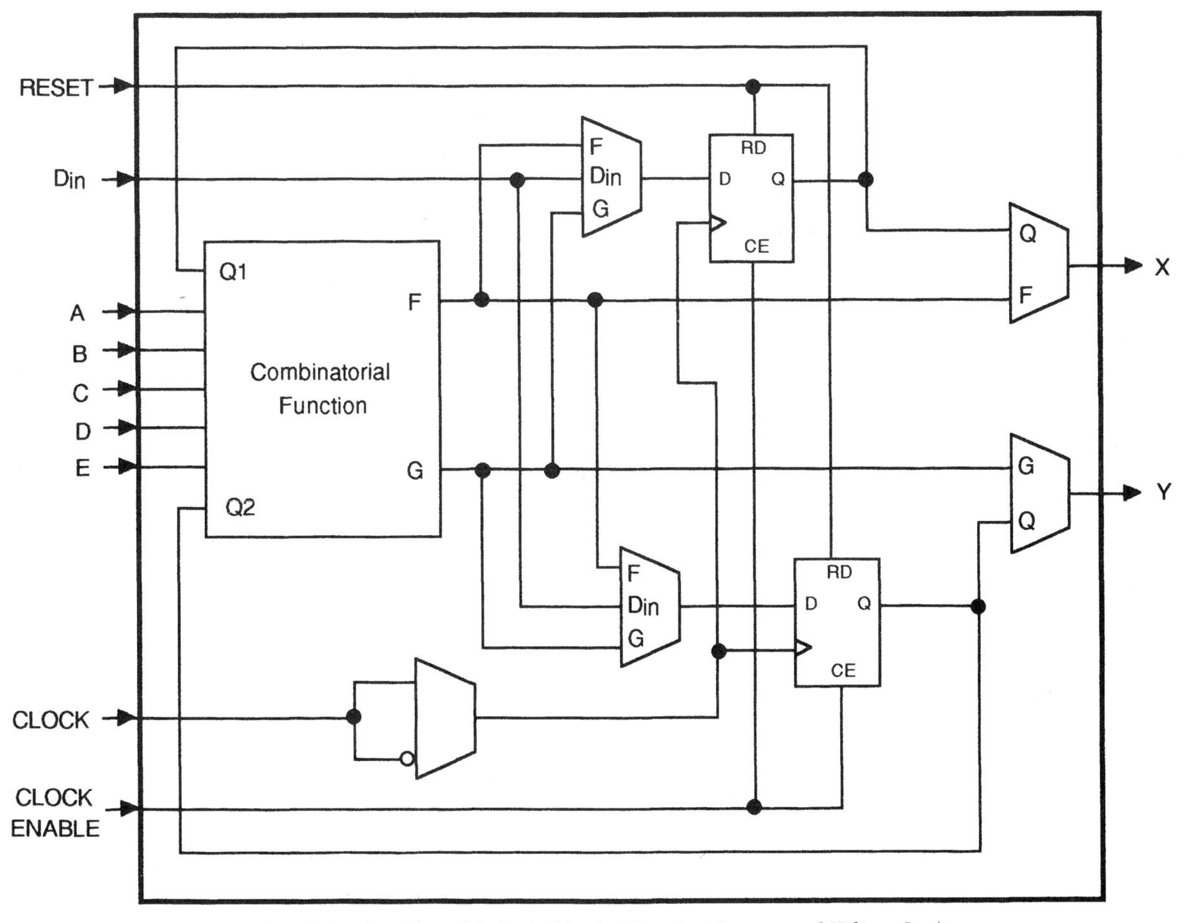

Fig. 11-4 Configurable logic block. (*Reprint Courtesy of Xilinx, Inc.*)

One option permits the creation of any single function of five of the variables; another permits the creation of any two functions of up to four variables (with some restriction as to how the variables are combined). In actual practice, the combinatorial logic is accomplished by utilizing a portion of the embedded bit stream as a RAM lookup table derived from the truth table of the logic to be implemented. This bit stream is loaded during device programming, and it is only read (like a ROM) during execution. The CLB also contains a direct data input for the flip-flops which bypasses the combinatorial logic.

The D input of each flip-flop is buffered by a multiplexer so that it can be connected to direct data input or to either of the possible combinatorial logic outputs. Multiplexers on the output side of the flip-flops permit selection of either direct or registered CLB outputs from the combinatorial-logic outputs. The clock signal can be applied directly, or it may be inverted. External reset signals, both controllable and global, can also be routed to the flip-flops. The multiplexers described above are either 1- or 2-bit address types whose select inputs are all under control of the embedded bit stream data. The CLB output signal can be routed to other CLBs for use as inputs, or they can be sent to I/O blocks and leave the chip. Each CLB is capable of direct connection to any adjacent CLB without entering a switching matrix.

Input/Output Block

This is a bidirectional port attached to an I/O pin. Under control of the input bit stream, it can be configured either as a standard output port, a three state output port, or as an input port. Through the use of multiplexers (as controlled switches) and XORs (as controlled inverters), an output signal can be presented to the port in either direct or inverted form, registered or not. When the I/O block is configured as an input port, the input signal appears in both direct and registered form, and either or both of these signals can be routed to appropriate CLBs.

Figure 11-5 shows the internal logical structure of an input/output block. The five blocks at the top of the figure represent states of background RAM elements which control the programming of this specific element.

The OUT INV (output inversion) control flip-flop is connected to an exclusive-OR gate (G_1) whose other input is connected to an FPGA output. Thus, the OUT INV flip-flop is used as a control element for the output data; when it is 1, D flip-flop FF1 is loaded with the inverted output variable and, when it is 0, FF1 is loaded with the direct version of the output variable. In a similar manner, the TS INV (output enable invert) flip-flop when set to 1, allows a designer to use the inverted form of the TS (three state) control signal to enable the output buffer, thereby making the pad serve as an output pin. We see that use is made of basic elements to electronically control both a connection and its nature.

The Output Source RAM variable determines whether the output buffer, when enabled, will pass the direct or registered version of the output variable through the I/O block.

Slew rate refers to the speed with which circuitry will respond to a logic-level transition and is meaured in volts per microsecond (or nanosecond). In the present case, this response time is switchable between two predetermined values.

Passive pull-up, when set to 1, causes the output pad to be connected, via a resistance, to the supply voltage source and thereby "pulled up" to a positive logic TRUE value when the output buffer is an open circuit (not enabled).

The lower section of the I/O block in Fig. 11-5 involves input signals which are to be routed to various places within the FPGA. Since output and input signal paths share a common connection at the pad, outputs can be used for internal feedback to CLBs or other I/O blocks. Input data is passed on in both direct and registered forms; the latter via FF2. Any selection between them is made outside the I/O block.

Programmable Interconnect

The programmable interconnect (shown schematically in Fig. 11-6) consists of a multiple-layer grid of metal segments. Transistor switches, under control of the embedded bit stream, form programmable

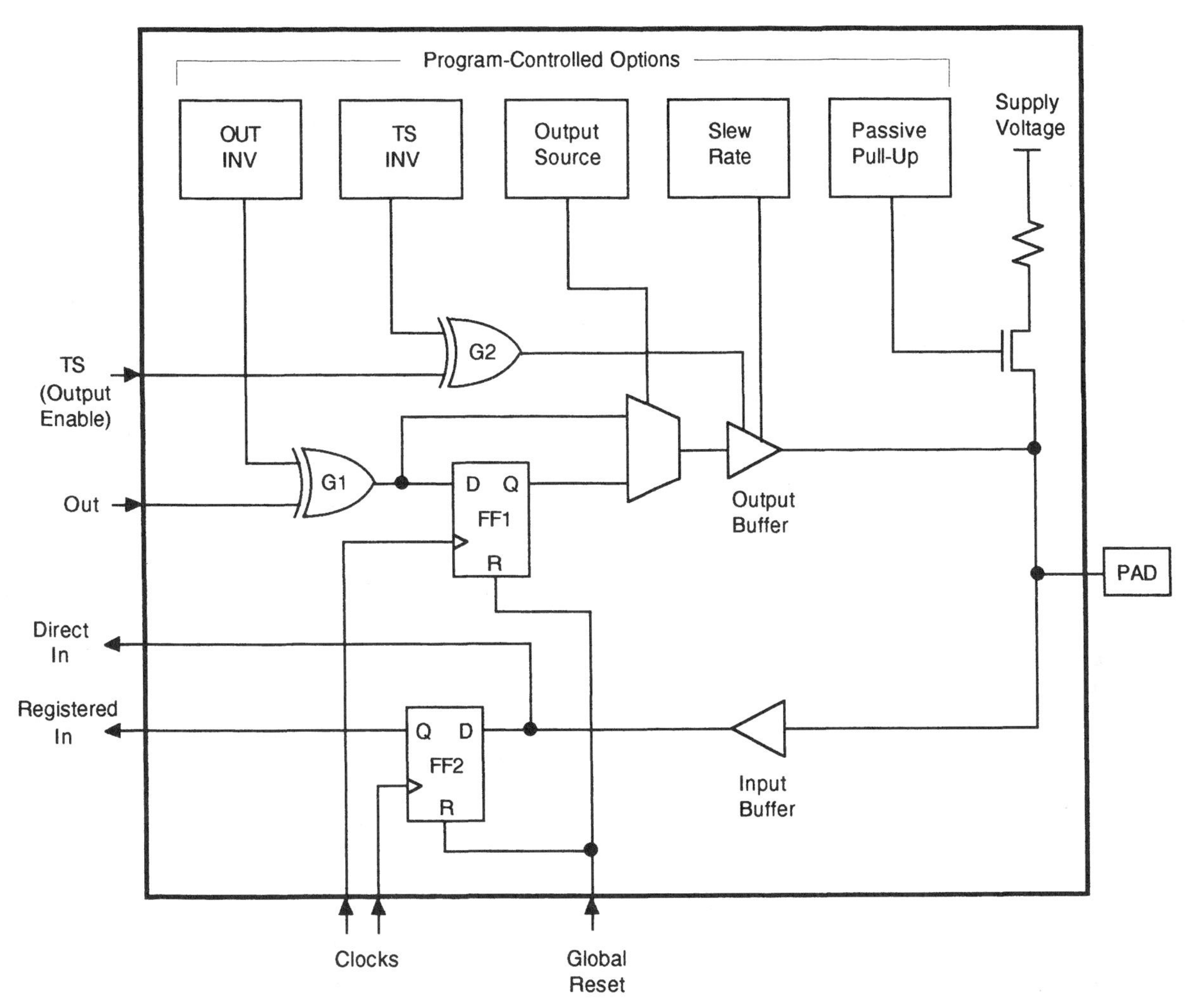

Fig. 11-5 Configurable I/O block. (*Courtesy of Xilinx, Inc.*)

interconnect points (PIPs) between selected metal segments and the block pins. Three types of metal resources are available:

1. General-purpose interconnect
2. Direct interconnect
3. Long lines

The *general-purpose interconnect* is a switching matrix which consists of five vertical and five horizontal metal segments which can be connected by PIPs. Each metal segment can be connected to perpendicular segments or to adjacent segments in its own direction.

Direct interconnections provide pathways between adjacent CLBs and I/OBs.

Long lines bypass the switching matrices and run both vertically and horizontally the length of the interconnect area. They minimize switching delays for signals which must travel a long distance, and they serve to equalize delays between signals which must arrive at various destinations with minimum skew.

Programming

The FPGA is programmed under computer control. Logic to be realized can be input to the computer as a captured logic schematic or in the form of logic equations for logical groupings (equivalent

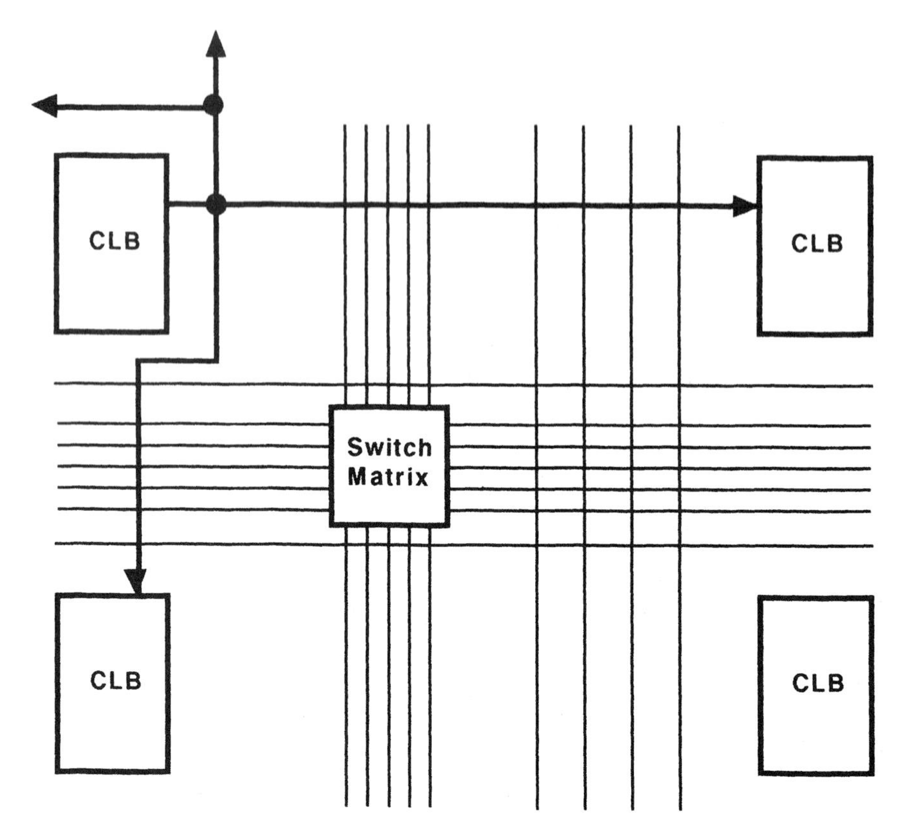

Fig. 11-6 Programmable interconnect. (*Courtesy of Xilinx, Inc.*)

PAL devices). The FPGA program requires additional steps called *placement and routing*. In the placement step, the logic and registers to be implemented are assigned to various CLBs and I/OBs using a software placement algorithm. The computer then attempts to route the interconnects, i.e., develop a set of allowable connection paths. On an FPGA which uses a large percentage of its resources, this process is often iterative and sometimes requires interactive manual intervention by a skilled human. When fewer resources are needed, the automatic place and route is usually satisfactory. When programming is completed, the computer has generated the necessary bit stream which can be fed directly into the hardware device or, more commonly, loaded into a PROM or EPROM which is located close to the FPGA.

An FPGA can lose its bit stream when power is removed. To accommodate this situation, the host system can be designed to generate a signal at power-up which causes the FPGA to load itself from a PROM. This mode of operation offers the possibility of "field changing" the FPGA configuration by changing the PROM. It also makes possible the use of several PROMs with one FPGA to permit reconfiguring during operation.

11.4 ARITHMETIC LOGIC UNITS

The *arithmetic logic unit* (*ALU*) commonly used in microprocessors is another example of an electronically programmed device. It is one of a family of building blocks which are used to perform a variety of controlled functions on a small number of bits which, in the present example, include addition, 2s complement subtraction, incrementing, decrementing, and logical inversion. Such building blocks are often called *bit slices* indicating that they may be cascaded to process larger numbers of bits. The ALU, as currently available, is quite complex internally. Its operation, however, can be easily understood by considering the simplified version shown in Fig. 11-7.

Figure 11-7*a* shows the logic of a single cell of the ALU, and Fig. 11-7*b* shows the internal interconnection of identical cells into a four-cell ALU bit slice. Note that the carry-out of cell i serves as the carry-in for cell i + 1. The carry-in for cell 0 is connected to an external pin as is the carry-out of cell 3.

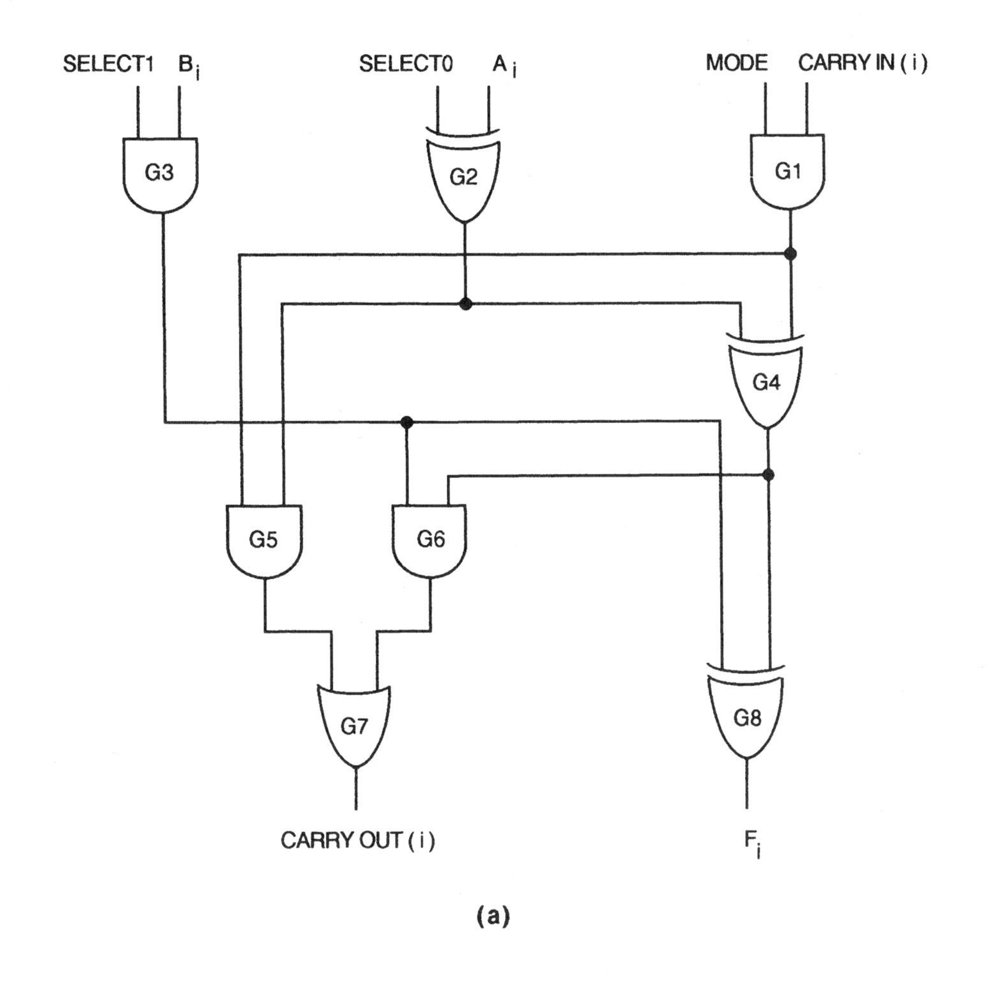

(a)

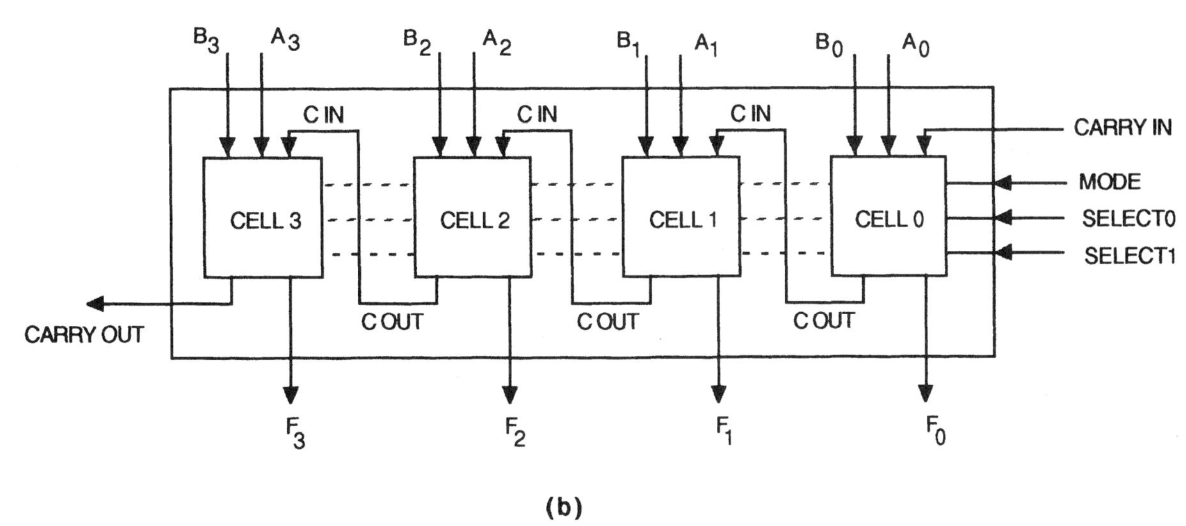

(b)

Fig. 11-7 (*Adapted from Herbert Taub, "Digital Circuits and Microprocessors," McGraw-Hill, Inc., New York, NY, 1982.*)

Control inputs (MODE, SELECT0, and SELECT1) are connected internally to each of the cells which have two inputs, A_i and B_i and produce an output F_i. Inputs and outputs are connected to external pins on the device. If two such devices are available, the two 4-bit slices can be converted to an 8-bit slice by connecting the control inputs together and routing the carry-out of the lower-order device to the carry-in of the higher-order device. The relationship between inputs and outputs depends upon the applied control signals. The *mode signal* controls the internal connection between cells; if it is 0, the cells are isolated, and if it is 1, the cells are linked with the carry-out of cell i being connected to the carry-in of cell i + 1. When MODE = 1, the circuit functions in the arithmetic mode and may be configured as an adder (refer to Example 11.1).

SELECT0 is connected to a controlled inverter (XOR) which determines whether A_i or its logical inverse is used in the cell. SELECT1 is connected to an AND gate which determines if B_i is to enter the cell (SELECT1 = 1) or if B_i is to be omitted (SELECT1 = 0). The carry-in to cell 0 can also be considered as a special control input. If the system is configured as an adder (MODE = 1), the carry-in to cell 0 will appear as an initial carry and will increment the result by 1.

EXAMPLE 11.1 **Configuring the ALU to arithmetically ADD.** Consider the following external connections:

$$\text{MODE} = 1$$
$$\text{SELECT0} = 0$$
$$\text{SELECT1} = 1$$
$$\text{Carry-in (0)} = 0$$

The following step-by-step analysis is presented with reference to Fig. 11-8:

1. MODE = 1 ensures that the output of G_1 is C_i.
2. SELECT0 = 0 ensures that G_2 produces A_i at its output.
3. SELECT1 = 1 ensures that the output of G_3 is B_i.
4. The inputs to G_4 are A_i and C_i, so its output is $A_i \oplus C_i$.
5. The inputs to G_5 are also A_i and C_i, so its output is $A_i C_i$.
6. The inputs to G_6 are B_i and $A_i \oplus C_i$, so the output is $B_i(A_i \oplus C_i)$.
7. G_7 ORs the outputs of G_5 and G_6, so the output (carry-out) is given by

   $\text{Carry-out (i)} = A_i C_i + B_i(A_i \oplus C_i)$
8. The output of G_8 is the exclusive OR of the outputs of G_3 and G_4, so

$$F_i = A_i \oplus B_i \oplus C_i.$$

Create the truth table for the expressions for carry-out and F.

This truth table (Table 11.1) is identical to that for an adder as developed in Chap. 2. Thus, the system functions as an adder.

Table 11.1

Ai	Bi	Ci	Fi	Carry Out(i)
0	0	0	0	0
0	0	1	1	0
0	1	0	1	0
0	1	1	0	1
1	0	0	1	0
1	0	1	0	1
1	1	0	0	1
1	1	1	1	1

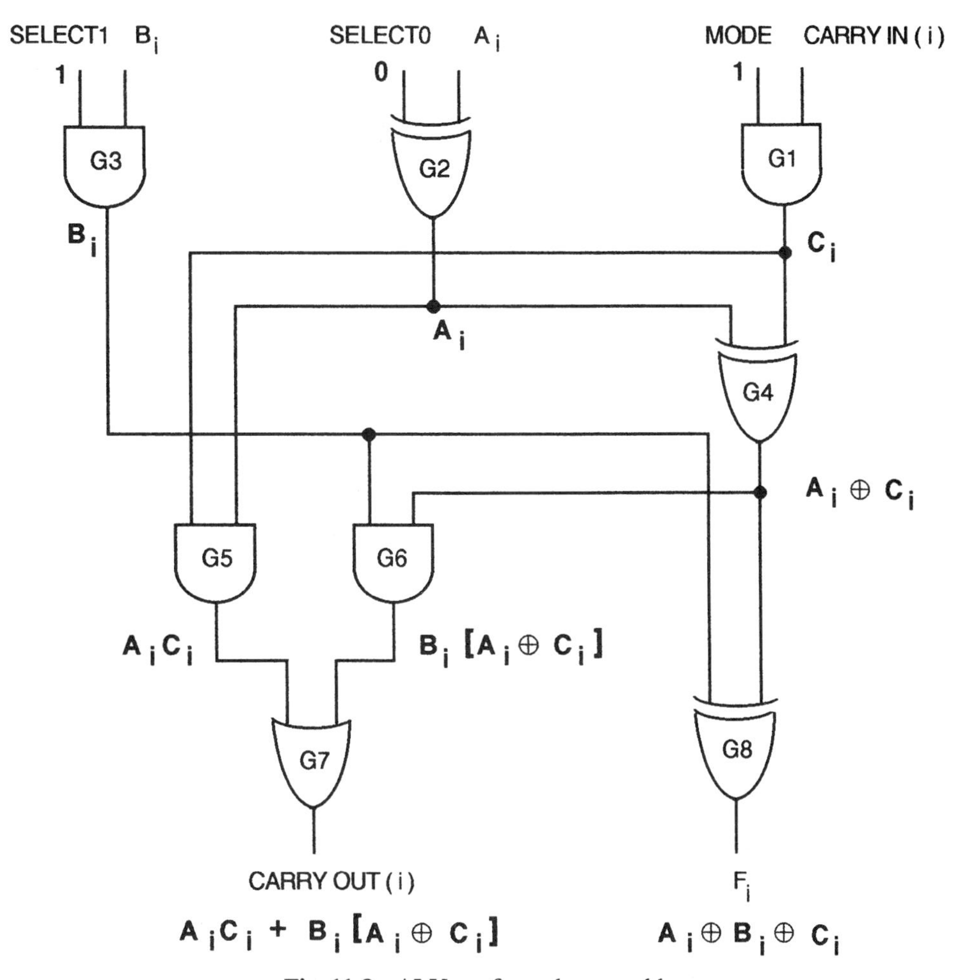

Fig. 11-8 ALU configured as an adder.

EXAMPLE 11.2 Configuring the ALU to form the 1s complement of a binary number A.

The 1s complement is formed by inverting each bit of the number. We will see that this is accomplished when the ALU control inputs are set as follows:

$$\text{MODE} = 0$$
$$\text{SELECT1} = 0$$
$$\text{SELECT0} = 1$$

Carry-in is irrelevant

Figure 11-9 can be used for a step-by-step analysis of the circuit's behavior:

1. MODE = 0 ensures that the output of G_1 is 0 (isolates the bits).
2. SELECT1 = 0 ensures that the output of G_3 is 0 (eliminates B_i from the circuit).
3. SELECT0 = 1 ensures that G_2 produces A_i' at its output.
4. The inputs to G_4 are A_i' and 0, so its output is A_i'.
5. The inputs to G_5 are A_i' and 0, so its output is 0.
6. The inputs to G_6 are A_i' and 0, so its output is 0.
7. G_7 ORs the outputs of G_5 and G_6, so its output (carry-out) is 0.
8. The output of G_8 is given by the XOR of the outputs of G_3 and G_4, so

   $F_i = A_i' \oplus 0 = A_i'$

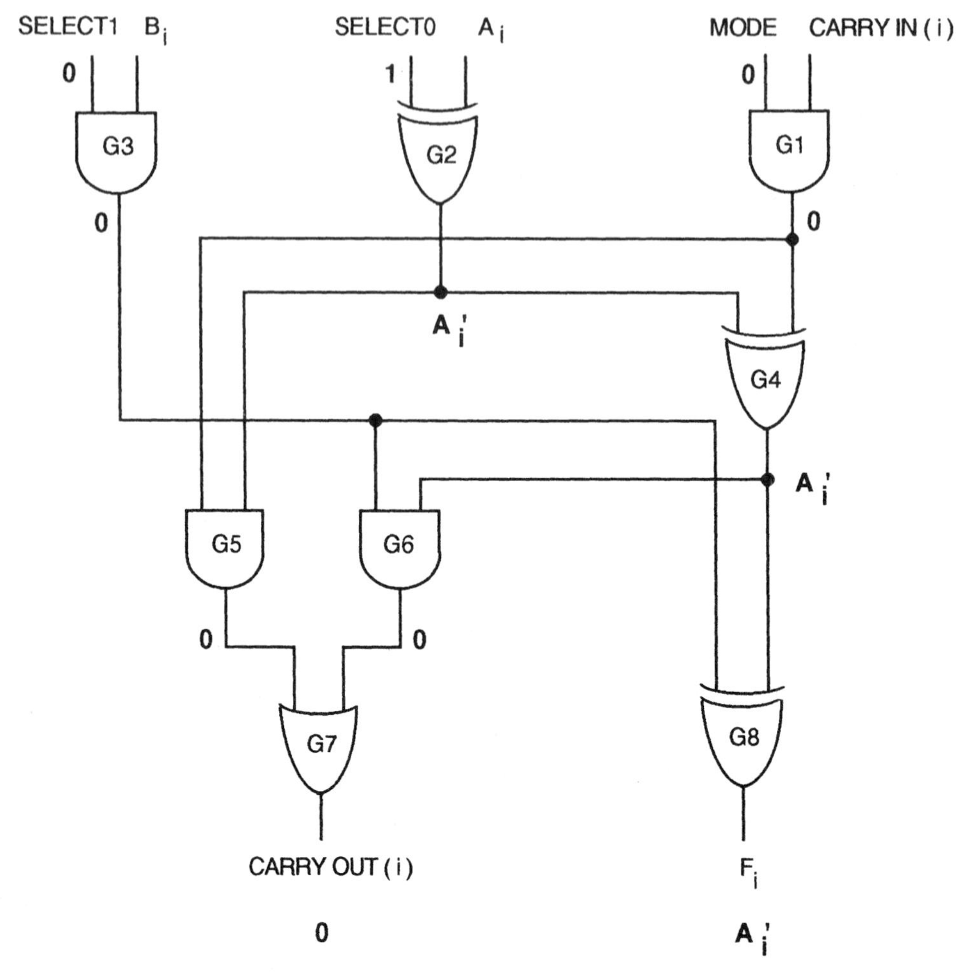

Fig. 11-9 ALU configured as 1s complementer.

11.5 PROGRAMMABLE REGISTERS

Computers often make use of registers which will perform different operations in response to various commands. Such a register, in which programming is done with AND gates, is shown in Fig. 11-10. Control and signal lines are shown bold. The function of this register is to take information from a bus, store it, possibly operate on it, and place it back on the bus under appropriate command. The two bus lines shown in the figure will, quite often, be the same physical conductor.

Analysis of the register's performance is facilitated by noting that only one operation is attempted at a time; that is, only one command is TRUE at any time. For example, when the transfer in signal is TRUE, gates G_1 and G_2 of each state are primed. Because the increment signal is 0, G_3 and G_4 are both producing 0s at their outputs. Thus, the bus content is applied to the J input of FF(i), while its complement is applied to the K input. In this situation, the next clock pulse produces $Q = J$. Similarly, if the reset line alone is TRUE, all the outputs of the AND gates are 0, as is the complement signal. The J and K inputs are 0 and 1, respectively, and the next clock pulse sets Q to 0. Notice that the transfer out signal is the enable for the TRI-STATE® buffers which connect flip-flop outputs to the bus.

It is worth describing the increment operation in some detail. When 1 is added to a binary number, the least significant bit will always toggle and, in progressing toward higher-order bits, each will successively toggle until a 0 is encountered. At this point, after the 0 toggles, the process ceases. The incrementing process occurs for all bits at the same time, coincident with the arrival of a clock pulse, and we see that the toggling of a given bit is controlled by the value of its next lowest neighbor prior to

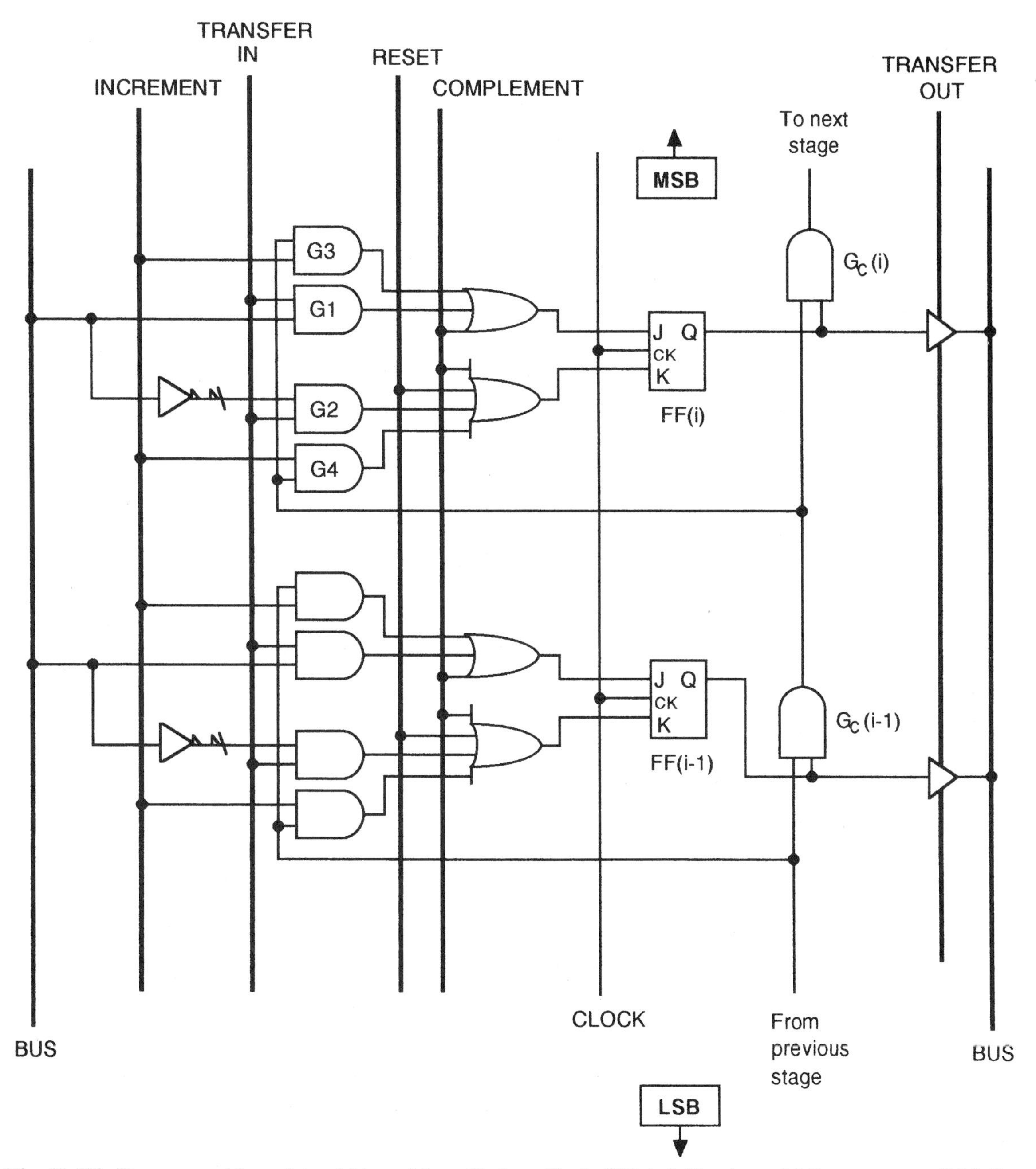

Fig. 11-10 Programmable register. (*Adapted from Herbert Taub, "Digital Circuits and Microprocessors." McGraw Hill, Inc., New York, NY, 1982.*)

clocking. In the circuit of Fig. 11-10, this process is accomplished by *Increment control gates* [G_c(i)] which permit each successive high-order stage (bit) to toggle only if the preceding bit is a 1. It should be obvious that the previous stage input of the lowest-order control gate must be tied to a 1.

EXAMPLE 11.3 Increment 1001111.

$$\begin{array}{r} 1001111 \\ +\quad\quad\;\; 1 \\ \hline 1010000 \end{array}$$

We see that the result is equivalent to examining the number from its lowest-order bit toward the highest-order bit and toggling all bits up to and including the first 0 encountered.

Solved Problems

11.1 In the ALU of Fig. 11-7*a*, with SELECT0, SELECT1, MODE, and carry-in (0) all equal to 1, show that the 4-bit slice accomplishes 2s complement subtraction.

This arrangement is almost identical to that of Example 11.1 where it is demonstrated that the circuit functions as a binary adder. In the present case, the differences lie in the fact that SELECT0 and carry-in 0 are now 1s.

The use of SELECT0 = 1 as an input to the controlled inverter means that the variable A_i' rather than A_i will be added to B_i. Furthermore, setting carry-in (0) = 1 causes the result to be incremented so that we have

$$\begin{aligned} F_i &= B_i(+)A_i'(+)1 \\ &= B_i(+)(\text{2s complement of } A_i) \\ &= B - A \end{aligned}$$

The symbol "(+)" is used to distinguish arithmetic addition from the logical OR function. See Prob. 11.5 for the implementation of A − B.

11.2 Demonstrate the complement action of the register shown in Fig. 11-10.

There is only one control signal TRUE at any time, so

$$\begin{aligned} &\text{Complement (C)} = 1 \\ &\text{Increment (I)} = 0 \\ &\text{Transfer in (TI)} = 0 \\ &\text{Reset (R)} = 0 \\ &\text{Transfer out (TO)} = 0 \end{aligned}$$

The Boolean equation for the J input is given by

$$\begin{aligned} J &= C + (TI)(BUS) + G_c(i-1)(I) = 1 + 0 \cdot BUS + G_c \cdot 0 = 1 \\ K &= C + R + (TI)(BUS') + G_c(i-1)(I) = 1 + 0 + 0 \cdot BUS' + G_c \cdot 0 = 1 \end{aligned}$$

With both J and K TRUE, the flip-flop will toggle, thus complementing the value stored in the register.

11.3 Modify the register in Fig. 11-10 so that the circuit will decrement on command.

Decrement example: Subtract 1 from a number.

$$\begin{array}{r} 00111001000 \\ -1 \\ \hline 00111000111 \end{array}$$

Note that the register contents are toggled from right to left up to and including the first 1 encountered. The circuit can be modified to accumulate 0s in the same way that it currently accumulates 1s. We add two additional gates to the AND gate trees associated with the JK inputs of each stage and obtain inputs for them from additional *decrement control gates* [Gd(i)] as shown in Fig. 11-11. The presence of 0s in a stored binary number is determined by obtaining signals from the Q′ side of the flip-flops. The addi-

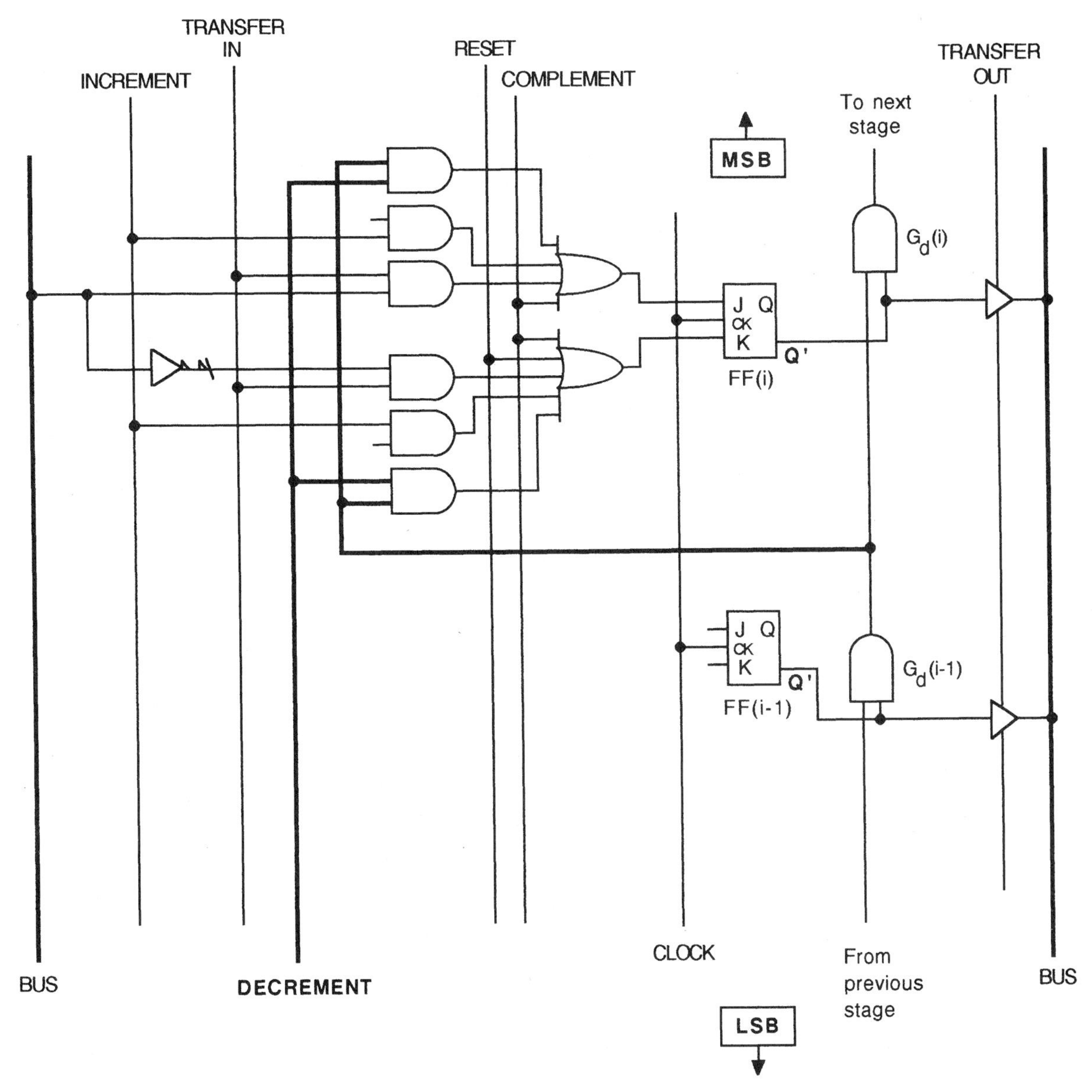

Fig. 11-11

tional circuitry (indicated by bold lines) does not interfere with other operations because only one control signal is present at any given time. In Fig. 11-11, the *increment control gates* have been deleted to enhance clarity.

11.4 Using the register shown in Fig. 11-10, determine the timing of control signal inputs which will cause the following sequence of events:

1. Load in data from the bus.
2. Take the 2s complement of the data.
3. Transfer the processed data out onto the bus.

Noting that the 2s complement of a binary number is accomplished by complementing each bit and adding 1 (incrementing), we develop the detailed sequence as follows:

1.0	Transfer in goes TRUE.
1.1	After at least two gate delays plus the flip-flop setup time, the clock pulse goes TRUE. Data loaded.
2.0	Transfer in goes FALSE, clock pulse goes FALSE, and complement goes TRUE.
2.1	After at least one gate delay plus the flip-flop setup time, the clock pulse goes TRUE. Data complemented.
3.0	Complement goes FALSE, clock pulse goes FALSE, and increment goes TRUE.
3.1	After at least two gate delays plus the flip-flop setup time, the clock pulse goes TRUE. Data incremented.
4.0	Increment goes FALSE, clock pulse goes FALSE, and transfer out goes TRUE. Data on bus.

11.5 A modified version of the ALU cell is shown in Fig. 11-12 in which,

SELECT0 = 0
SELECT1 = 1
SELECT2 = 1
MODE = 1
Carry-in (0) = 1

Determine the operation performed and the maximum delay from input to output.

The only difference between the given circuit and that of Prob. 11.1 is the addition of an exclusive-OR gate (controlled by SELECT2) through which input B_i is passed.

If SELECT2 = 1, B_i is complemented and the circuit produces $A_i(+)B_i'(+)1$ which is equal to A − B.

The maximum delay occurs in the carry-out portion where the signal passes through four gates. Note that since carry-out of one stage becomes carry-in for the next, a ripple effect occurs and the total delay in an n-bit slice circuit is equal to 4n gate delays.

11.6 Consider the ALU with the control signal values set as follows:

SELECT0 = 1
SELECT1 = 1
Carry-in (0) = 1
MODE = 0

Show that there is an output of F = 1 only if A_i and B_i are identical.

MODE = 0 means that there is no carry propagation. Thus, the operation proceeds bitwise. SELECT0 = 1 means that A_i appears in complemented form A_i'. SELECT1 = 1 means that B_i is used in direct form. Since carry-in (0) is irrelevant, carry-out is irrelevant also.

Reference to Fig. 11-13 shows that $F_i = A_i' \oplus B_i$. Construction of an appropriate truth table shows that F can only be TRUE if AB = 00 or 11.

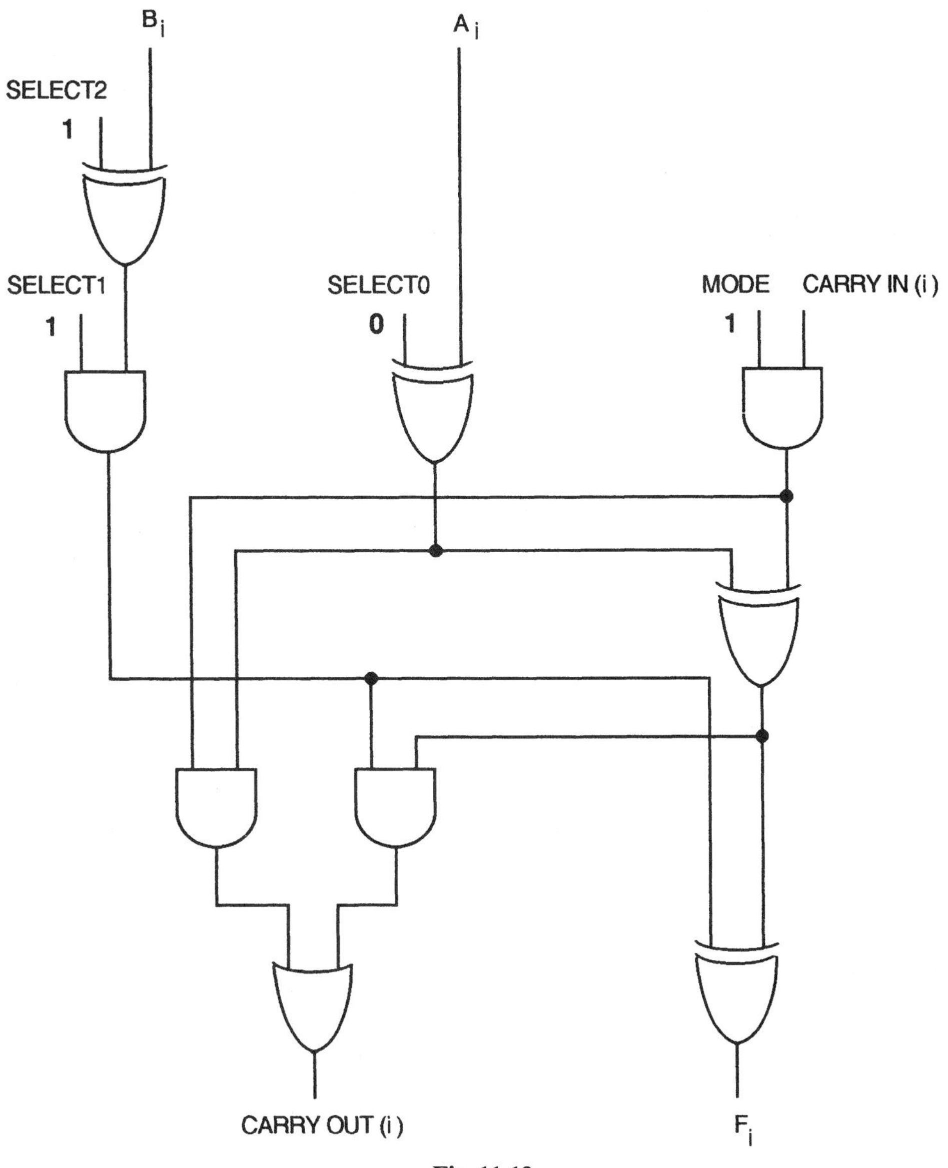

Fig. 11-12

11.7 Using a multiplexer, devise a circuit which can be used to select whether a signal or its complement appears at the output. Compare your result with the exclusive-OR approach discussed in Sec. 11.2.

When an inverter is used as shown in Fig. 11-14, the two inputs of a single-select multiplexer are derived from the same variable, one being direct (A) and the other inverted (A′).

The two-input multiplexer is composed of two dual-input AND gates and an OR gate (see Sec. 5.2). The logic is shown in Fig. 11-15.

The XOR Boolean equation is given by $F = AS' + A'S$, where S = select. We see that this yields exactly the same logic as the multiplexer solution shown in Fig. 11-15. The distinction between the multiplexer and XOR approaches is seen to exist in the designer's mind only and not in silicon fabricaton.

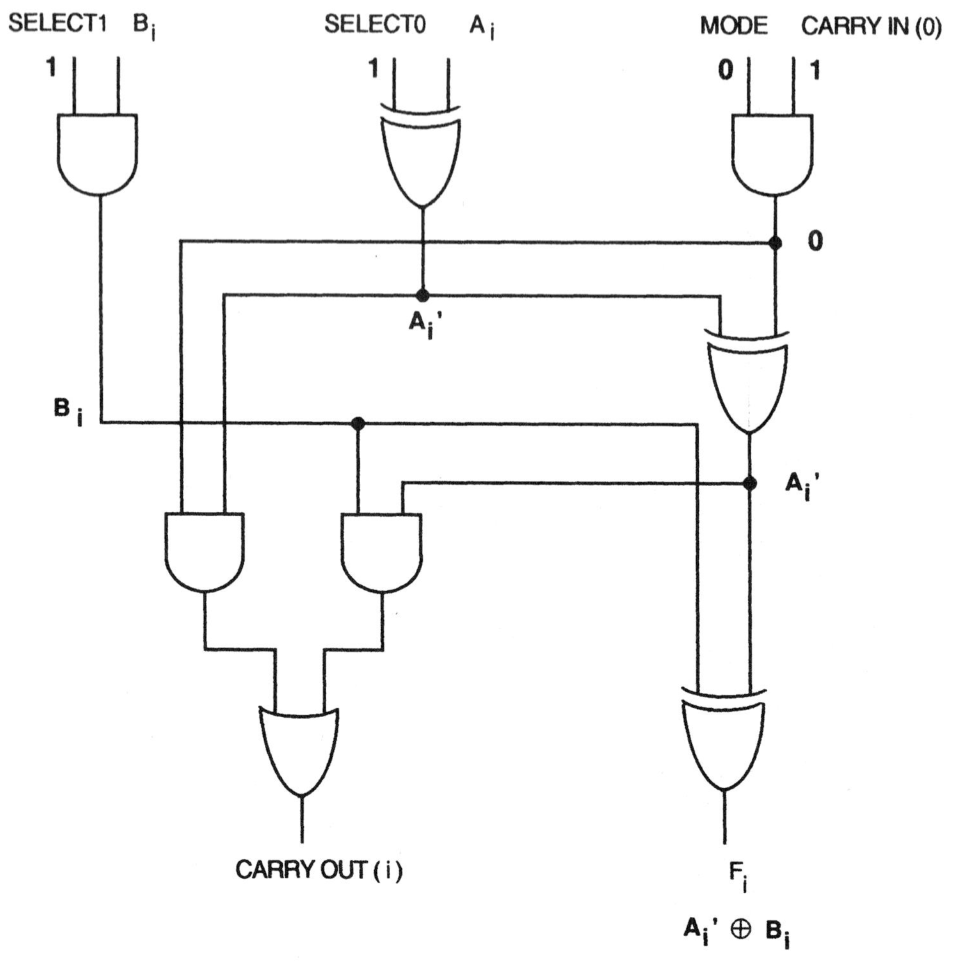

Fig. 11-13

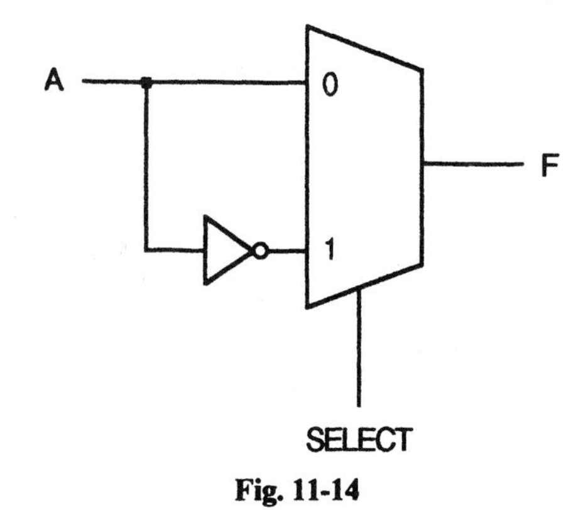

Fig. 11-14

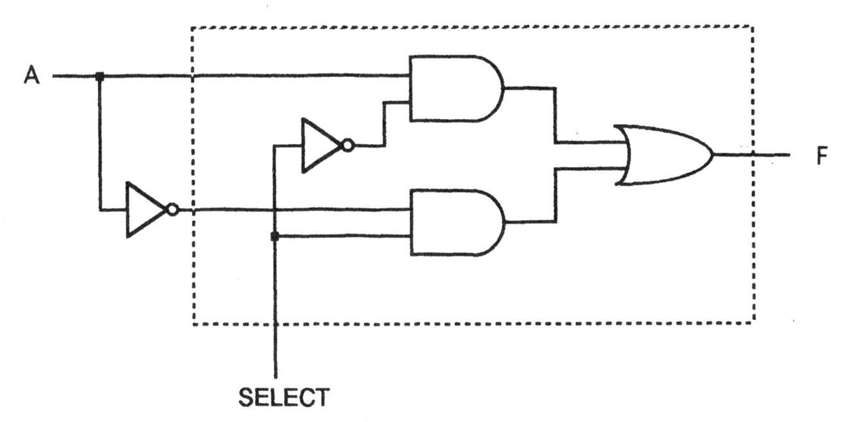

Fig. 11-15

11.8 A student has designed the electronically programmed circuit shown in Fig. 11-16. Determine the operations it is capable of performing and comment on its efficiency.

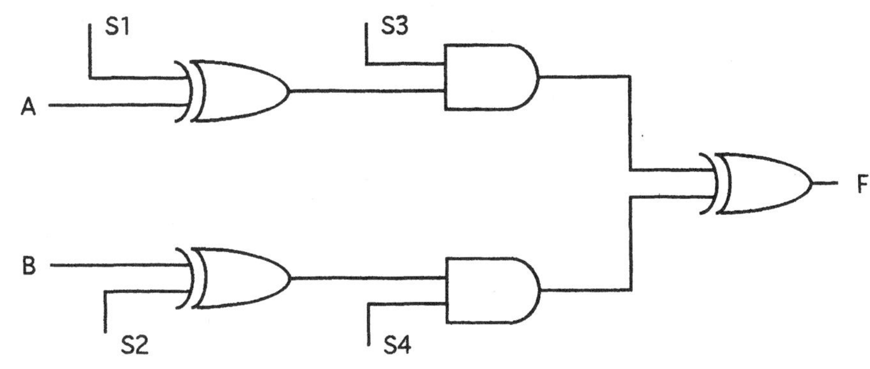

Fig. 11-16

Table 11.2

S1	S2	S3	S4	F
0	0	0	0	0
0	0	0	1	B
0	0	1	0	A
0	0	1	1	$A \oplus B$
0	1	0	0	0
0	1	0	1	B'
0	1	1	0	A
0	1	1	1	$A \oplus B'$
1	0	0	0	0
1	0	0	1	B
1	0	1	0	A'
1	0	1	1	$A' \oplus B$
1	1	0	0	0
1	1	0	1	B'
1	1	1	0	A'
1	1	1	1	$A' \oplus B'$

S_1 selects between A and A′. S_2 selects between B and B′, S_3 determines whether A or A′ is used at all, S_4 determines whether B or B′ is used at all.

Tabulating the outputs for all control configurations (Table 11.2), we see that there are nine unique outputs.

If it is necessary to use all of them, then all four selects must be used because 4 bits are required to distinguish between nine things. If one output (say, $F = 0$) could be dispensed with, then it should be possible to realize the function with three state variables instead of four.

11.9 Design a circuit which will accomplish the operations defined in Prob. 11.8, except for the 0 output.

Strategy: Since eight outputs are required and three select variables are to be used, each select configuration must be equated with a desired output.

Note, from Table 11.3, that S_1 divides the table into two groups. Use this variable to distinguish between the XORed outputs and the single-variable outputs. If S_2 and S_3 can be used to create unique combinations of A and B and their complements, they can be sorted to create the single variables and combined to create XORs.

Table 11.3

S1	S2	S3
0	0	0
0	0	1
0	1	0
0	1	1
1	0	0
1	0	1
1	1	0
1	1	1

If S_2 is used as the control for an XOR with A as the other input, then A will appear when S_2 is 0 and A′ when S_2 is 1. Similarly, S_3 can be used to select between B and B′. Another combination of S_2 and S_3 is needed to distinguish between the variables. A little trial and error leads us to use S_2 XORed with S_3 (mid-output Y). When $Y = 1$, W is routed to G_7 (variable A selected and B suppressed), and when $Y = 0$, X is routed to G_7 (variable B selected and A suppressed). See Table 11.4.

Table 11.4

S1	S2	S3	W = S2 ⊕ A	X = S3 ⊕ B	Y = S2 ⊕ S3
0	0	0	A	B	0
0	0	1	A	B'	1
0	1	0	A'	B	1
0	1	1	A'	B'	0
1	0	0			
1	0	1			
1	1	0			
1	1	1			

The "fenced" portion of the circuitry shown in Fig. 11-17 implements the logic described above. Gates G_1, G_2, and G_3 are the XORs which produce mid-outputs W, X, and Y. The inversion of $Y = S_2 \oplus S_3$ required for controlling the selection between variables A and B is provided by G_4. AND gate G_5 passes variable B and G_6 passes A; they are never primed simultaneously, so their outputs may be combined on a single line via G_7 which provides the selected variable to the final output logic.

Note that elements G_4, G_5, G_6, and G_7 make up a circuit which selects between two input lines (a single-select multiplexer) as shown in Fig. 11-15. This structure is repeated to select between single-variable

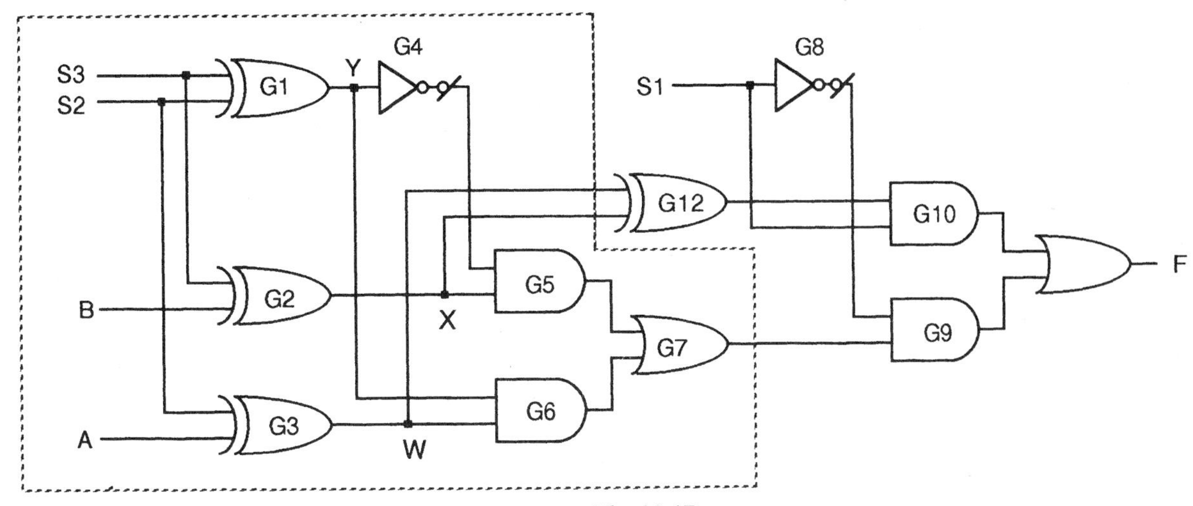

Fig. 11-17

or XORed combinations of A and B. In this case, when control variable $S_1 = 1$, the XOR output of G_{12} is passed to the output F, and when $S_1 = 0$, G_9 is primed to pass a single variable to the output.

Note that though this design requires only three control variables, it uses a much larger amount of logic than the four-control-variable version of Prob. 11.8. This is often the case. A reduction in the number of control or select variables usually increases the complexity of the logic required.

Attention should also be drawn to the design technique used in the solution of this problem. Insight and the use of functional blocks has greatly simplified what would otherwise have been a relatively tedious logic design by route using a five-variable K-map approach. In general, synthesis of a circuit from a given specification yields many valid designs, and there is often no single best solution. The experience and skill of the designer is obviously a key factor in the utility and efficiency of any design approach.

11.10 Figure 11-18 shows a generalized register structure. Assuming that only one of the inputs is TRUE at any time, show analytically, by determining expressions for J and K, how the register functions for each of the following operations:

1. Transfer in
2. Reset
3. Complement
4. Increment

In general:

$$J_i = C + WD + I[G_c(i-1)]$$
$$K_i = C + Z + D'W + I[G_c(i-1)]$$

Case 1. *Transfer in:* $W = 1, C = I = Z = 0.$

$$J_i = D \qquad K_i = D' \qquad \text{After clock pulse, } Q_i = D$$

Case 2. *Reset* (*clear*): $Z = 1, W = C = I = 0.$

$$J_i = 0 \qquad K_i = 1 \qquad \text{After clock pulse, } Q_i = 0$$

Case 3. *Complement:* $C = 1, Z = W = I = 0.$

$$J_i = 1 \qquad K_i = 1 \qquad \text{After clock pulse, } Q_i = Q_i' \text{ (toggle)}$$

Case 4. *Increment:* $I = 1, C = W = Z = 0.$

$$J_i = G_c(i-1) \qquad K_i = G_c(i-1)$$

Clock will cause Q_i to toggle if $G_c(i-1) = 1$, which will be the case if all lower-order bits in the register are 1.

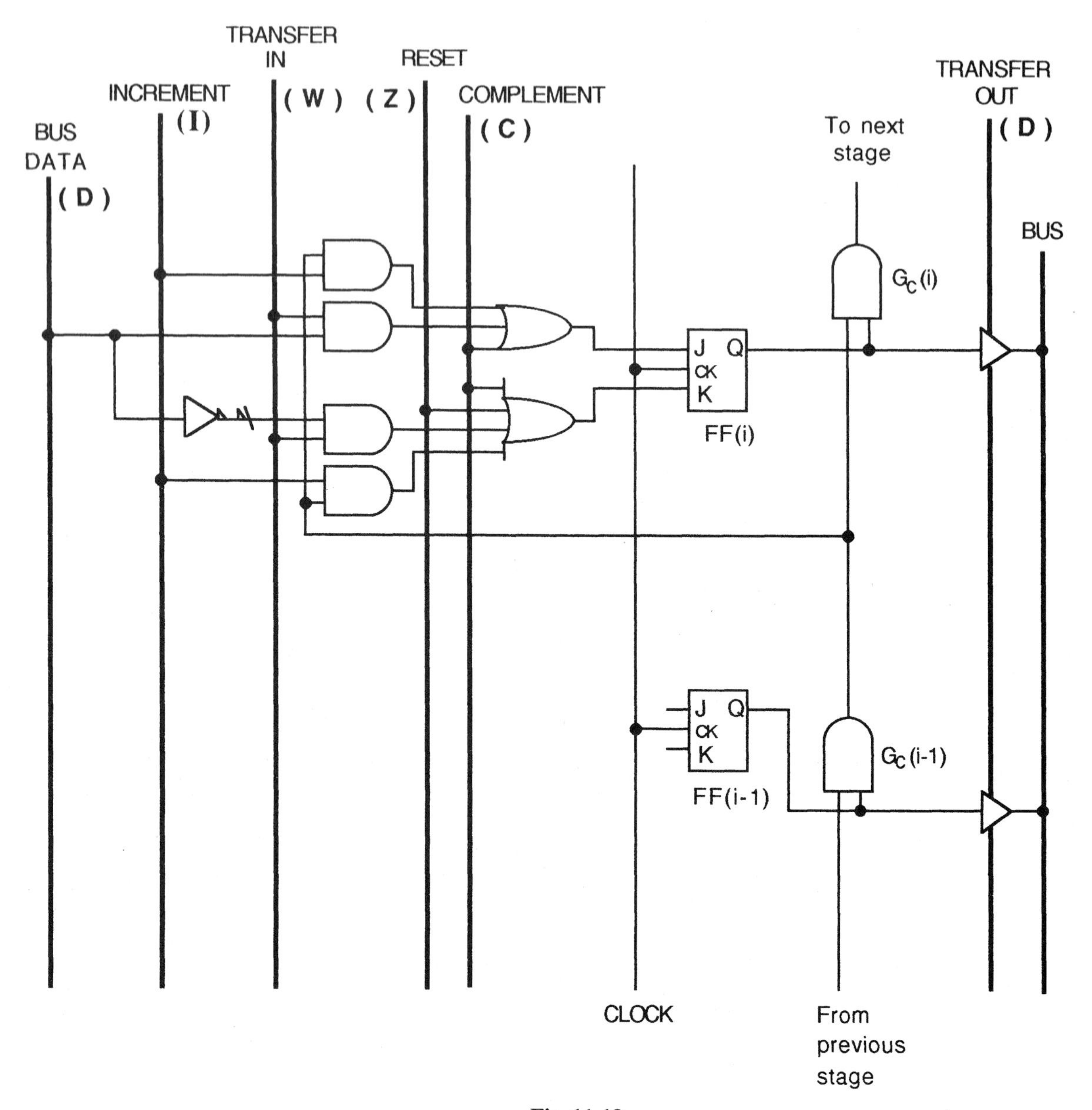

Fig. 11-18

11.11 Shown in Fig. 11-19 is logic circuitry which is controlled by information stored in programmable RAM cells. Cell 1 controls S_1, cell 2 controls S_2, etc. Assuming that the data input is stable just prior to and immediately following a system clock pulse, determine the output and feedback signals for the following configuration: cells 1, 2, 3, 4 = 0010.

Since cell 1 and cell 2 both contain 0s, $S_1 = S_2 = 0$ and the output follows the data input, unaffected by clocking.

Since cell 3 = 1 and cell 4 = 0, multiplexer line 2 is selected and the feedback is equal to both direct and complemented values of the data which has most recently been clocked into and stored by the D flip-flop.

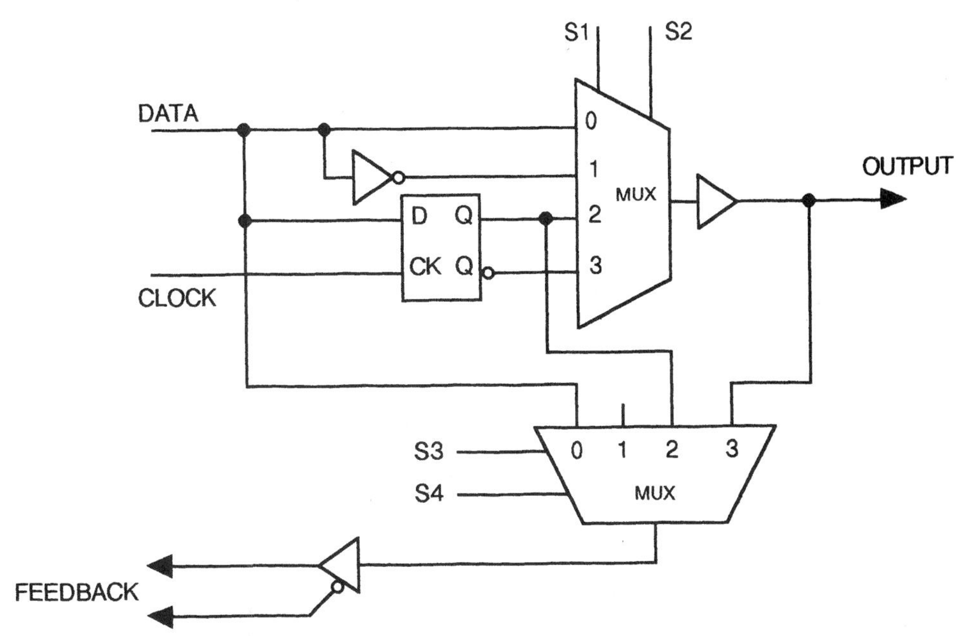

Fig. 11-19 An amplifier of the type shown in Fig. 9-5 is used for the feedback output.

11.12 Consider the logic circuitry of Prob. 11.1. The dual amplifier at the feedback output is replaced by an XOR gate, one of whose inputs is the feedback multiplexer output and the other is connected to an additional RAM cell (5). Determine the cell contents if the output is to be the logical inverse of the *registered* input data and the feedback signal is to be the logical inverse of the *unregistered* data. Assume that the data input is stable immediately before and after the system clock.

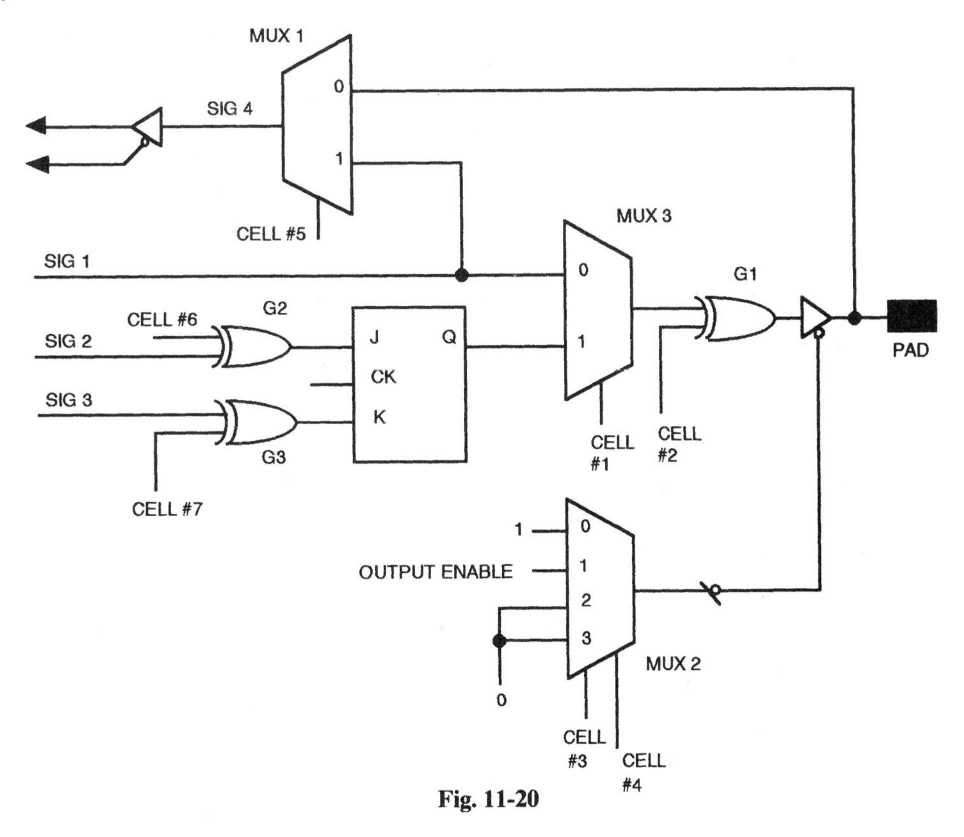

Fig. 11-20

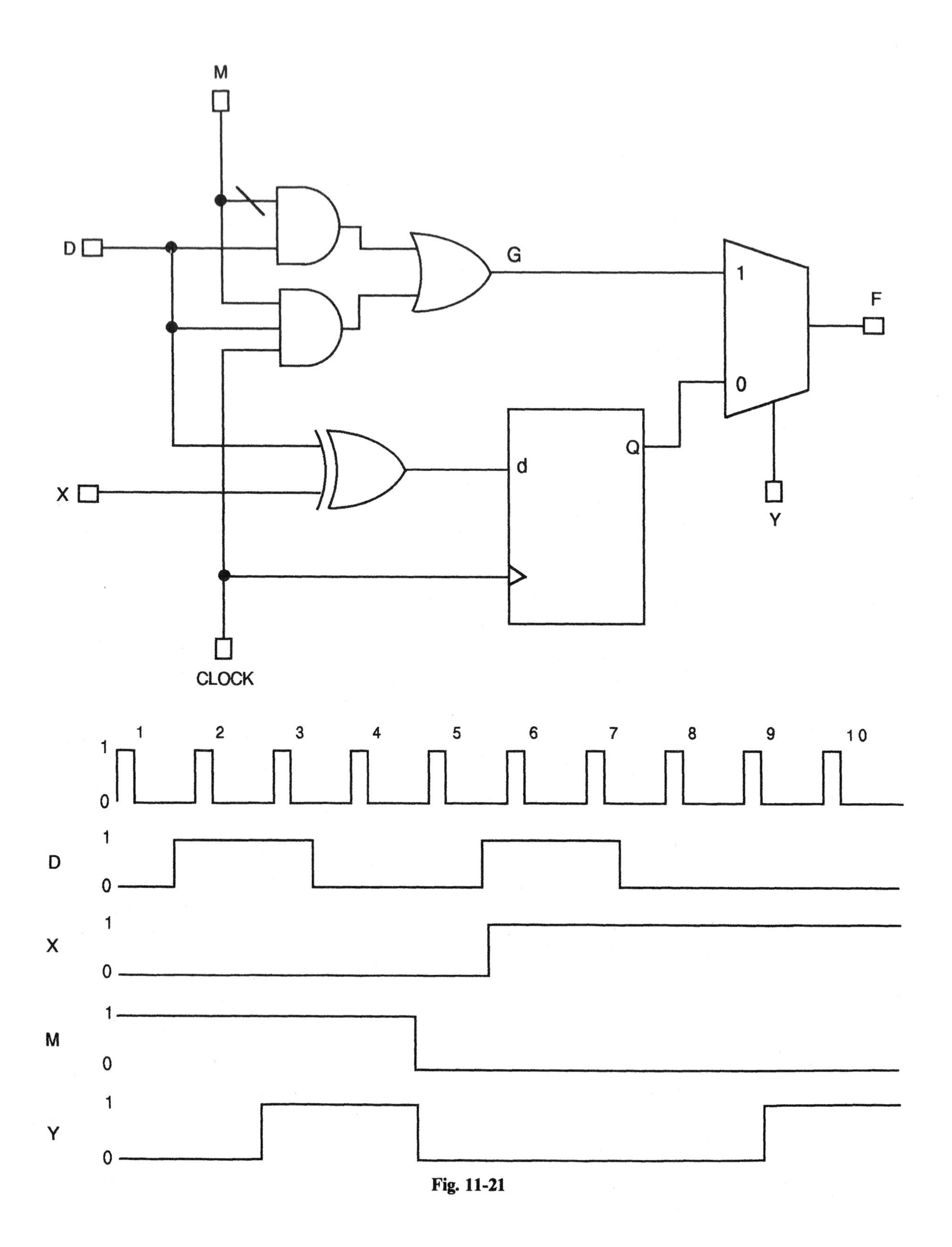

Fig. 11-21

Line 3 of the output multiplexer receives the inverse of the registered input data. Thus, S_1 and S_2 (or cells 1 and 2) must both be 1. Unregistered data appears at input 0 of the feedback multiplexer and therefore, S_3 and S_4 (cells 3 and 4) must be programmed to 0. The required inversion can be obtained if the XOR control input (cell 5) is set to 1.

11.13 Figure 11-20 shows an input/output macrocell which is controlled by a hidden RAM. Determine what the seven-digit RAM cell contents must be in order for signal 4 to equal an input signal from the pad.

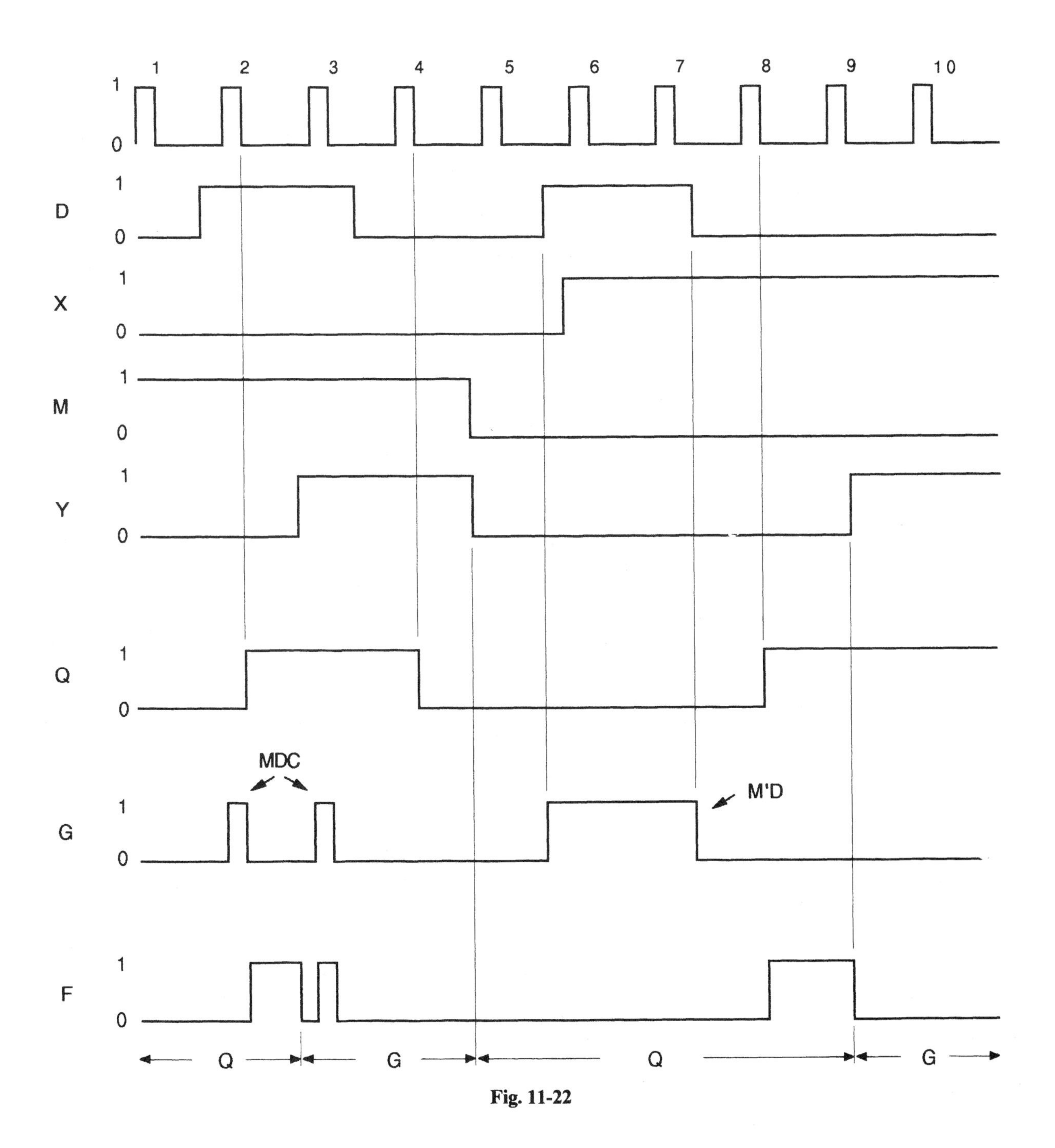

Fig. 11-22

If cell 3 is 1, MUX 2 will disable the TRI-STATE® buffer driving the pad and cause it to look like an open circuit. If cell 5 is 0, the pad will be connected to signal 4. Thus, the cell contents (cell 1, ..., 7) should be XX1X0XX, where the Xs are "don't cares".

11.14 In Prob. 11.13, determine the function if cell 1, ..., 7 = 1100100.

Cell 1 = 1	The flip-flop's Q output is coupled to an input of XOR gate G_1.
Cell 2 = 1	Q is inverted to form Q′.
Cells 3, 4 = 00	The TRI-STATE® buffer is enabled to couple Q′ to the pad.
Cell 5 = 1	Signal 1 is fed back as signal 4.
Cells 6, 7 = 00	The status of Q is determined by the direct values of signal 2 and signal 3.

11.15 Figure 11-21 shows a programmable circuit along with the input waveshapes. Describe its operation and sketch the waveshape at output F.

Q is the registered value of input D when X = 0 and is the registered value of D′ when X = 1. G, the OR gate output, is equal to D when M = 0 and to D ANDed with the clock when M = 1. F = G when Y = 1 and F = Q when Y = 0.

In order to draw waveform F, it will be helpful to also construct intermediate waveforms G and Q as shown in Fig. 11-22.

Supplementary Problems

11.16 The following values are stored in the internal RAM of the FPGA I/O block shown in Fig. 11-5.

OUT INV = 1	Output source = 1
TS INV = 0	Passive pull-up = 1

If the TS signal input is 1 and we assume that the multiplexer input closest to the select is line 0,

(*a*) What TS logic value enables the output?

(*b*) If the output is not enabled, what positive logic value appears at the pad?

(c) Is the output direct or registered?

(*d*) Is the output normal or inverted?

11.17 For the circuit shown in Fig. 11-23, find an expression for the output if the control signals are set to $S_1 = 1$, $S_2 = 0$, and $S_3 = 1$.

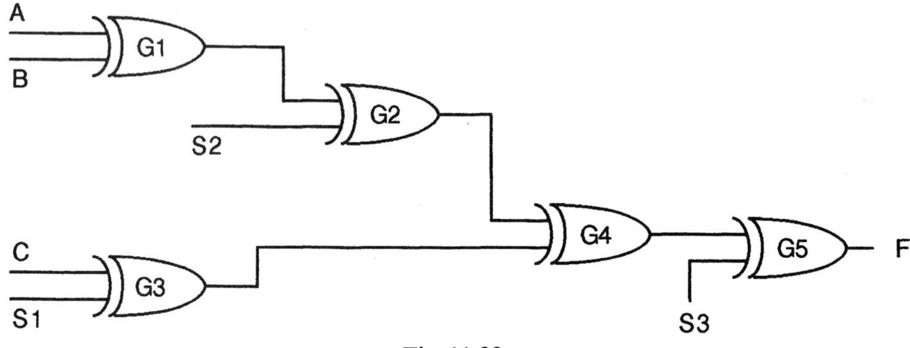

Fig. 11-23

11.18 In the circuit shown in Fig. 11-24, the control word A is $A_1, A_2, A_3, A_4, A_5, A_6, A_7$. What is the effect of the word A = 0111110?

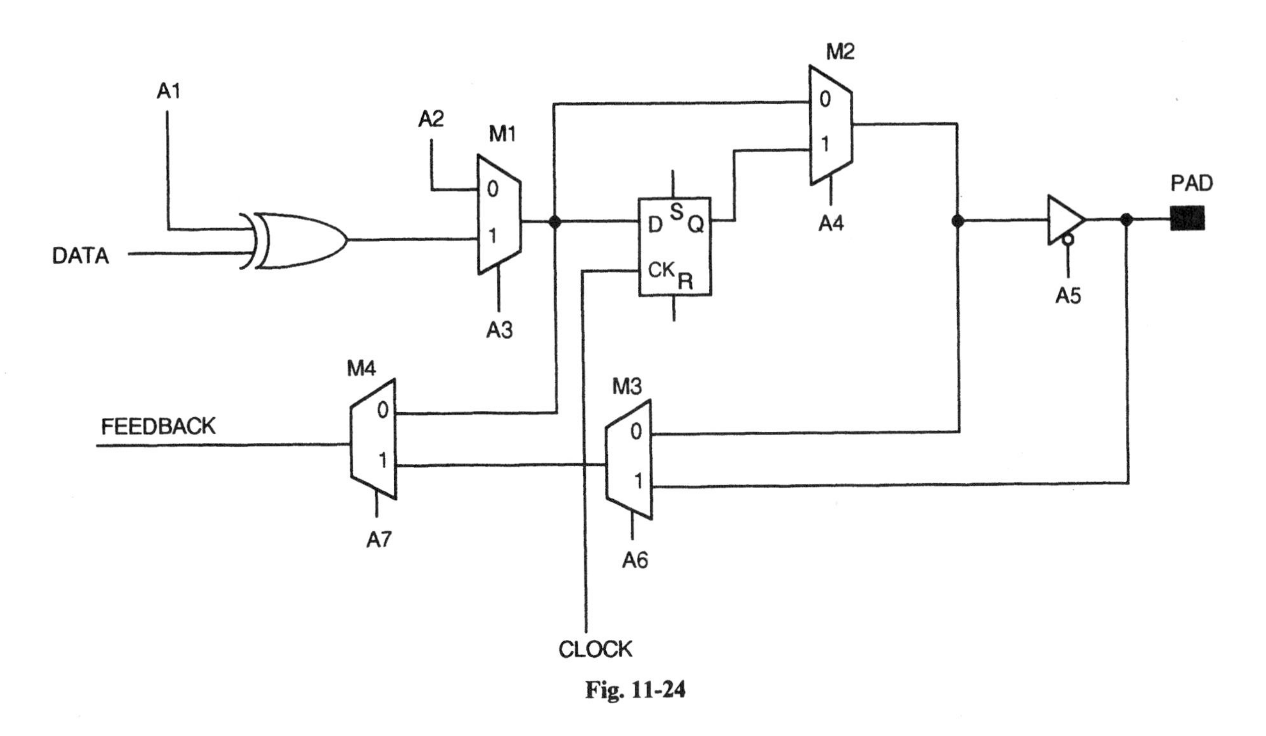

Fig. 11-24

11.19 Using the logic of Prob. 11.18, determine a control word which will provide registered, inverted data at the pad as an output and also feed back the unregistered data.

11.20 For testing purposes, it is often desirable to have a known value (as in Fig. 11-24) placed on the output pad after a clock pulse and to feed back the same value. Determine a suitable control word using the logic of Prob. 11.18.

11.21 How could the logic of Prob. 11.18 be modified so that it would be possible to feed back the inverted pad value?

11.22 Control signal values for the ALU of Fig. 11-25 are set as follows: $S_1 = 1$, $S_0 = 0$, $M = 1$, and carry-in $(0) = 1$. What function does the circuit perform?

11.23 A student, in attempting to create an ALU, has miswired a few connections as shown in Fig. 11-26. Given that $M = S_0 = 1$ and carry-in $= S_1 = 0$, determine an expression for F and C_0.

11.24 Design a circuit where, dependent on a control word, a signal present at an input pad can be passed on as registered or unregistered and inverted or noninverted.

11.25 With reference to the given solution to Prob. 11.24, create a truth table for all possible values of the control word variables.

11.26 Use TRI-STATE® devices to modify the solution to Prob. 11.24 so that the pad can be used for either input or output.

11.27 Use an AND gate to modify the circuit of Fig. 11-27 so that the flip-flop in the feedback path can be synchronously cleared.

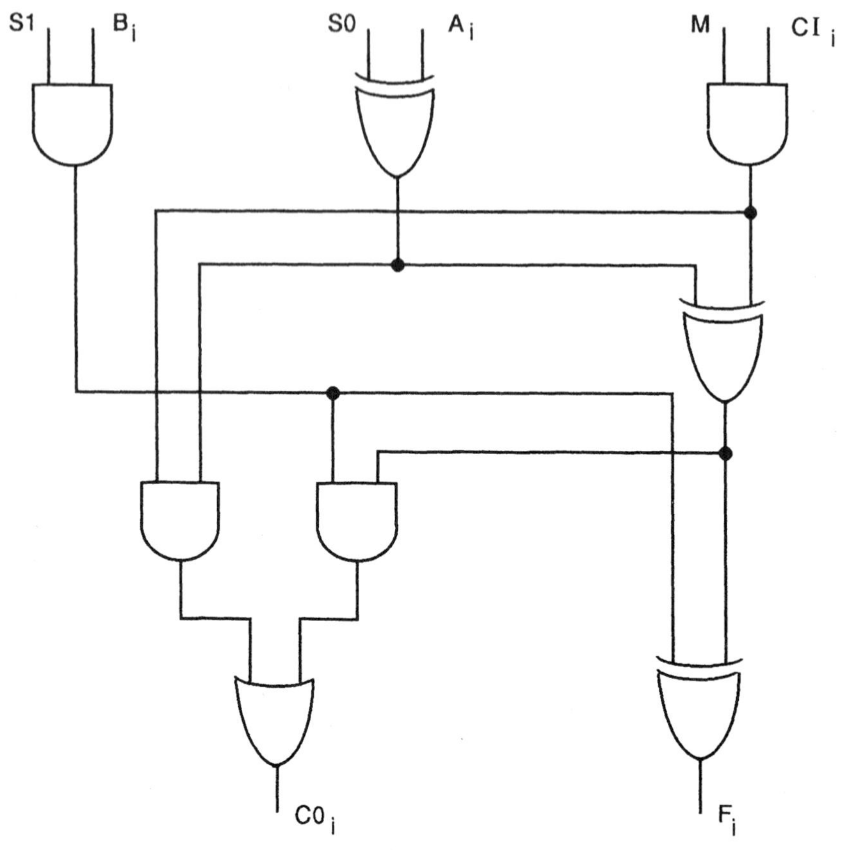

Fig. 11-25

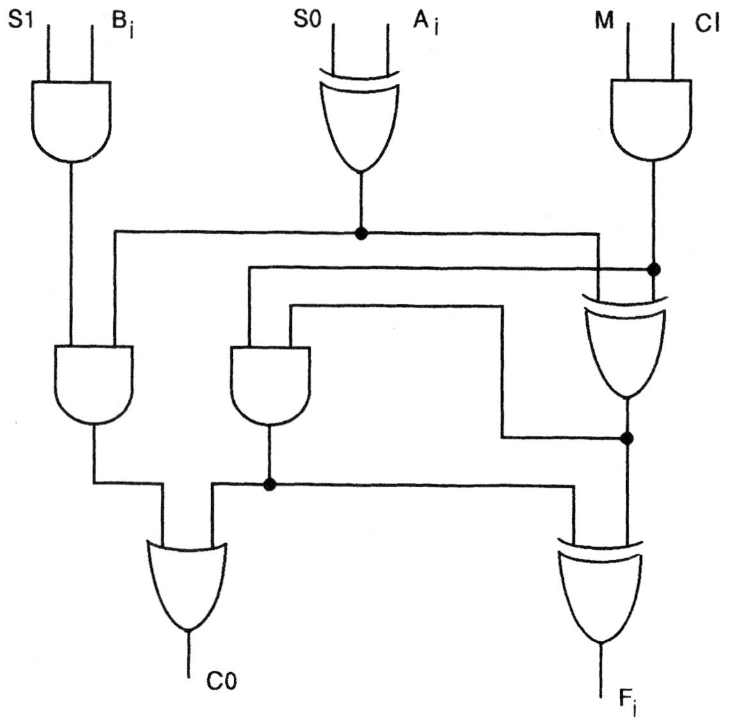

Fig. 11-26

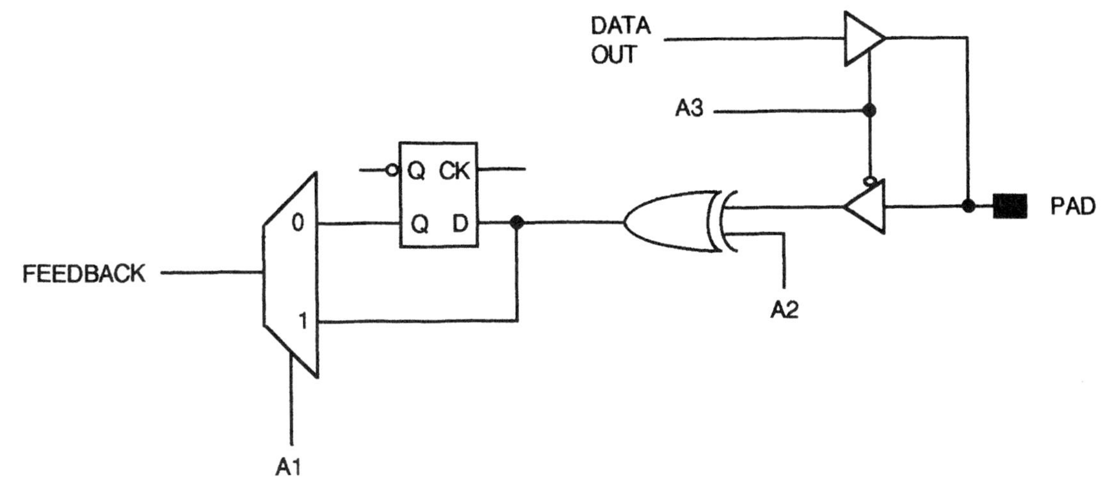

Fig. 11-27

Answers to Supplementary Problems

11.16 (*a*) The enable is HT, and TS = 1 will enable the output.

(*b*) If the output is not enabled, it looks like an open circuit. The passive pull-up is 1 causing a high voltage to appear, representing TRUE.

(*c*) Multiplexer line 1 is selected and the output is registered.

(*d*) Inverted.

The RAM commands create an output which is inverted, registered, and enabled.

11.17 $F = [(A \oplus B) \oplus C']'$

11.18 $A_1 = 0$ Data is not inverted.
$A_2 = 1$ Irrelevant since $A_3 = 1$.
$A_4 = 1$ Registered data is selected by the second multiplexer.
$A_5 = 1$ Disables the pad driver; the pad is configured as an input.
$A_6 = 1$ Selects data input from the pad.
$A_7 = 0$ Selects the first multiplexer output for feedback and ignores pad input.

Summary: The pad is disconnected and uninverted input data is fed back directly (unregistered).

11.19 A = 1X110X0 where X represents irrelevant or "don't care" bits.

11.20 $A = X(A_2)01011$.

11.21 A controlled inverter (XOR) could be added with a control variable A_8. It could be placed at the output of either multiplexer M_3 or M_4.

11.22 $A_i (+) B_i (+) 1$. The symbol (+) indicates arithmetic addition.

11.23 $C_0 = 0$ and $F = A_i'$.

11.24 One of many possible solutions is shown in Fig. 11-28.

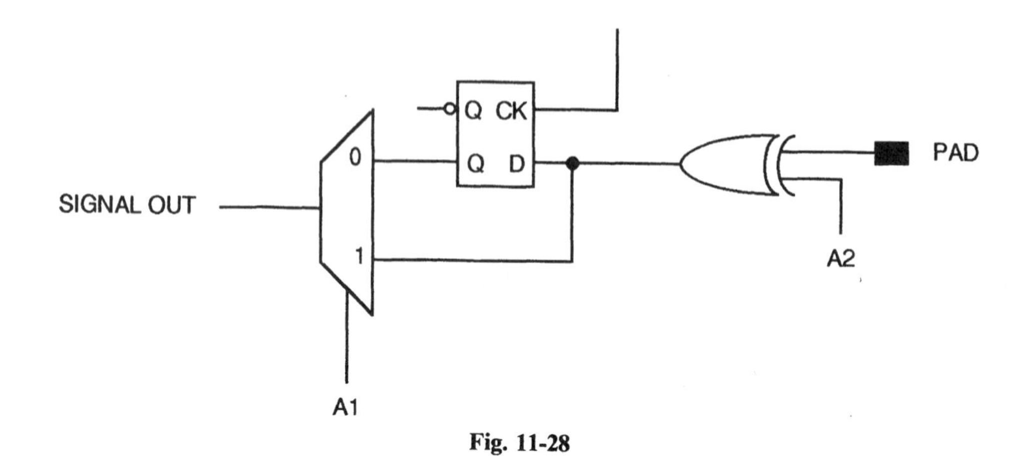

Fig. 11-28

11.25 See Table 11.5.

Table 11.5

A1	A2	INVERTED DATA	REGISTERED DATA
0	0	NO	YES
0	1	YES	YES
1	0	NO	NO
1	1	YES	NO

11.26 Refer to Fig. 11-27. In this solution, one TRI-STATE® device is enabled HIGH and the other LOW so that a common signal A_3 will always enable one or the other.

11.27 See Fig. 11-29. When A_4 is 0, the flip-flop will be cleared on the next clock pulse.

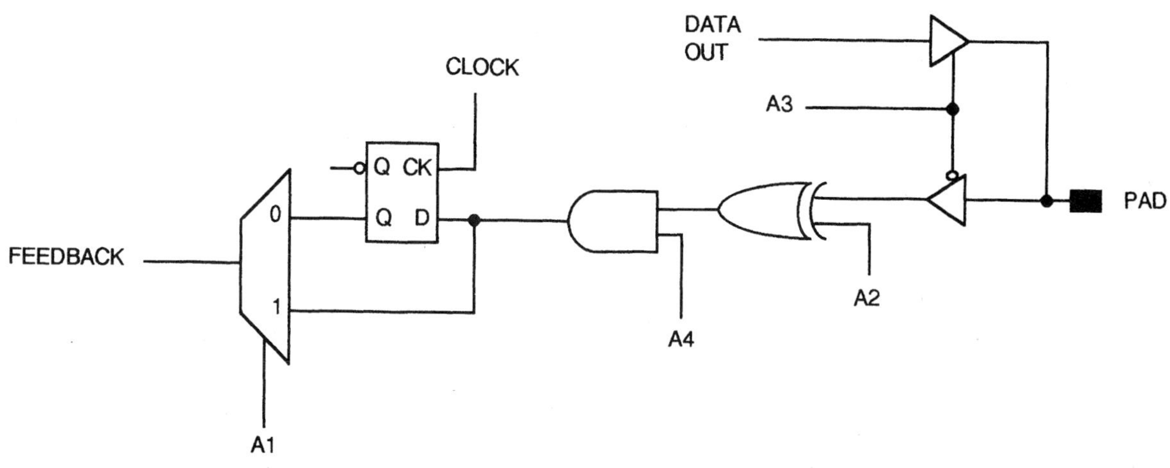

Fig. 11-28

Appendix A

Basic Boolean Theorems and Identities

$$A \cdot 1 = A \tag{A.1}$$

$$A \cdot 0 = 0 \tag{A.2}$$

$$A \cdot A = A \tag{A.3}$$

$$A \cdot A' = 0 \tag{A.4}$$

$$(AB)' = A' + B' \qquad \text{(De Morgan's theorem)} \tag{A.5}$$

In Boolean algebra, *duality applies*. This means that if, in the above set of equations, we change each AND to OR, each OR to AND, and interchange the 1s and 0s, we will obtain a second set of valid identities:

$$A + 0 = A \tag{A.6}$$

$$A + 1 = 1 \tag{A.7}$$

$$A + A = A \tag{A.8}$$

$$A + A' = 1 \tag{A.9}$$

$$(A + B)' = A'B' \qquad \text{(De Morgan's theorem)} \tag{A.10}$$

Other Useful Boolean Identities

$$(A')' = A \tag{A.11}$$

$$A + AB = A \tag{A.12}$$

$$A + A'B = A + B \tag{A.13}$$

$$AB + A'B = B \tag{A.14}$$

The dual identities corresponding to Eqs. (A.11) through (A.14) are

$$A(A + B) = A \tag{A.15}$$

$$A(A' + B) = AB \tag{A.16}$$

$$(A + B)(A' + B) = B \tag{A.17}$$

Useful Three-Variable Theorems

$$A + BC = (A + B)(A + C) \tag{A.18}$$

$$AB + A'C = (A + C)(A' + B) \tag{A.19}$$

$$(A + B + C)' = A'B'C' \qquad \text{(De Morgan's theorem)} \tag{A.20}$$

Duals of Eqs. (A.18) through (A.20):

$$A(B + C) = AB + AC \qquad \text{(Distributive law)} \tag{A.21}$$

$$(A + B)(A' + C) = AC + A'B \tag{A.22}$$

$$(ABC)' = A' + B' + C' \qquad \text{(De Morgan's theorem)} \tag{A.23}$$

Appendix B

Standard Logic Symbols

B.1 INTRODUCTION

At this writing (late 1991), the logic symbols used by designers and vendors have not been completely standardized, and, consequently, there continues to be some confusion. This appendix represents an attempt to illustrate typical symbols of various sorts that occur in the literature and to point out some general properties. Publications of many digital logic hardware vendors contain positive logic symbols as well as the latest revised IEEE/IEC symbolism. For example, the *Advanced CMOS Logic Data Book*, 1990 edition, published by Texas Instruments, Inc., shows alternative logic symbols for a broad array of digital integrated circuits.

B.2 GROUPS OF INDIVIDUAL HARDWARE GATES

NAND Logic (Example, 74 × 00 quadruple two-input positive NAND gate)

A conventional pin-out symbol is shown in Fig. B-1. Gate symbols shown within the package outline designate hardware type and are usually of the "generic" positive logic variety. Numbers in the small boxes pertain to package pin numbers as viewed from the top. V_{cc} represents the electric supply voltage, usually +5 volts, while GND (ground) indicates the common supply return path. Mixed-logic symbols for NAND hardware are described in Chap. 4.

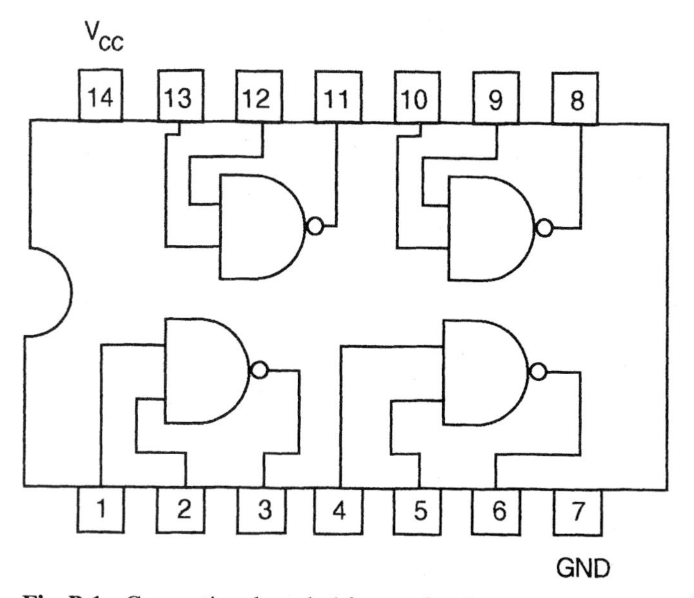

Fig. B-1 Conventional symbol for quadruple two-input NAND.

The symbol for the same integrated circuit using the *IEEE/IEC convention* is shown in Fig. B-2. This type of symbol arose from the deliberations of an international committee, and it is proposed that it become the industry standard. In the present illustration, the "&" symbol designates AND logic and the numbers in parentheses represent the package pins. The four rectangular sections (having lead groups labeled 1, 2, 3, and 4) signify that there are four identical circuits in the package, each having two inputs labeled A and B and an output labeled Y. The half arrows signify LOW-TRUE outputs. Power pins are not indicated on the diagram.

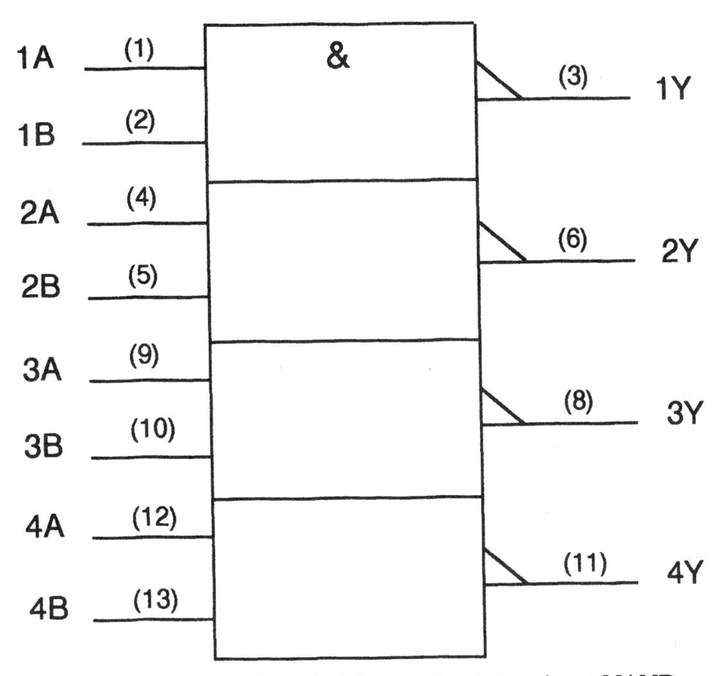

Fig. B-2 IEEE/IEC symbol for quadruple two-input NAND.

NOR Logic (Example, 74 × 02 quadruple two-input positive NOR gate)

A conventional pin-out diagram is shown in Fig. B-3. The internal symbols designate NOR gates, and, as with the previous example, the box numbers represent package pin numbers as viewed from the top.

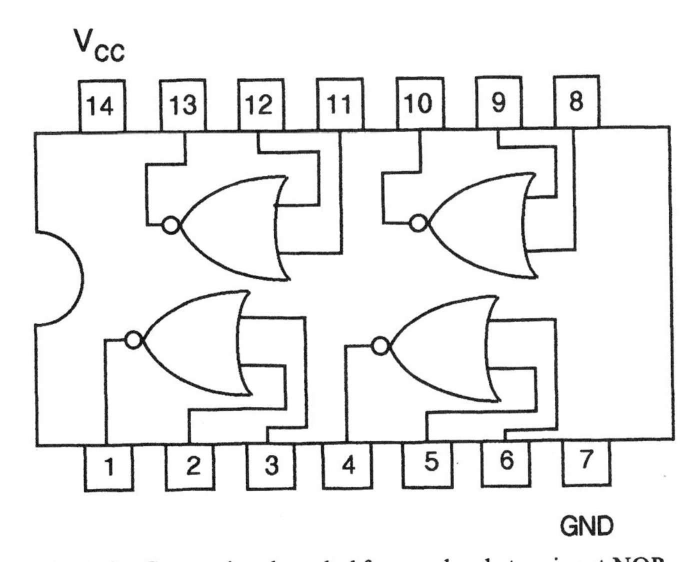

Fig. B-3 Conventional symbol for quadruple two-input NOR.

The IEEE/IEC logic symbol for a quadruple two-input NOR package is shown in Fig. B-4. The "≥ 1" designates the logic as (inclusive) OR since a 1 is produced every time the sum of the inputs to any one of the four identical sections equals or exceeds 1. Again, the half arrows signify LOW-TRUE oututs and power pins are not indicated on the diagram.

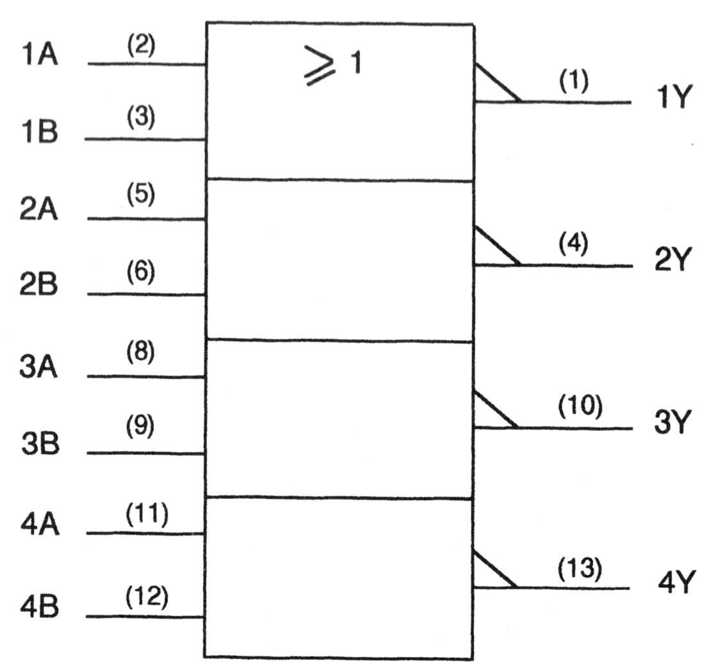

Fig. B-4 IEEE/IEC symbol for quadruple two-input NOR.

B.3 MSI LOGIC

Multiplexers (Example, 74 × 153 dual four-line to one-line data multiplexer)

This component has two identical 4-to-1 multiplexers which have common select inputs (A and B) and are individually strobed by active-LOW signals (1G′ and 2G′). The outputs are generically named Y, while the data inputs are designated by the letter C. The generic component can be indicated in a logic diagram with conventional notation, as shown in Fig. B-5*a*, although some authors prefer the triangular version shown in Fig. B-5*b*. In many cases, a circle is used in place of the half arrow.

The IEEE/IEC notation for the same circuit is depicted in Fig. B-6 where the common select signals are shown entering at a uniquely shaped portion of the symbol. The strobe inputs are labeled "EN" which stands for "ENable."

Flip-flops (Example, 7478 dual JK flip-flop)

This component contains two JK flip-flops which are individually presetable but have internally wired common clear and clock lines. A widely used generic flip-flop symbol is shown in Fig. B-7. In the IEEE/IEC symbol of Fig. B-8, the characteristically shaped "block" for common signal inputs appears at the top, as it did in the multiplexer. Note also, the "right-angle" symbols at the flip-flop outputs which indicate that these outputs are registered.

B.4 EXTENSION OF IEC NOTATION TO MORE COMPLEX CIRCUITS

Shift Registers (Example, 74 × 163 parallel-load 8-bit shift register)

The IEC notation is extended to more complex circuits in a straightforward manner as shown in Fig. B-9. As previously, signals common to all stages enter the uniquely shaped block at the top.

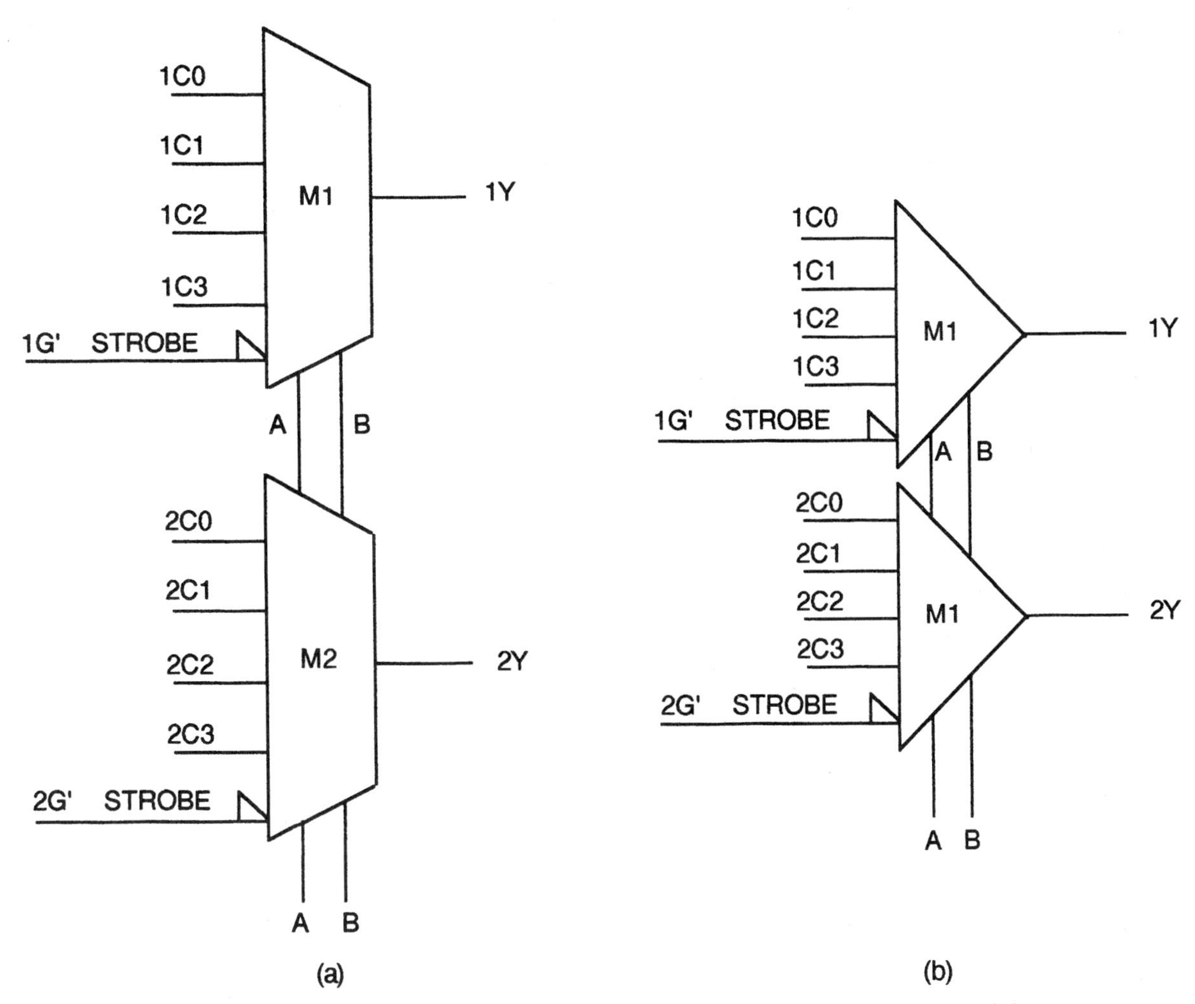

Fig. B-5 Two alternative conventional notations for a dual, strobed, four-line multiplexer.

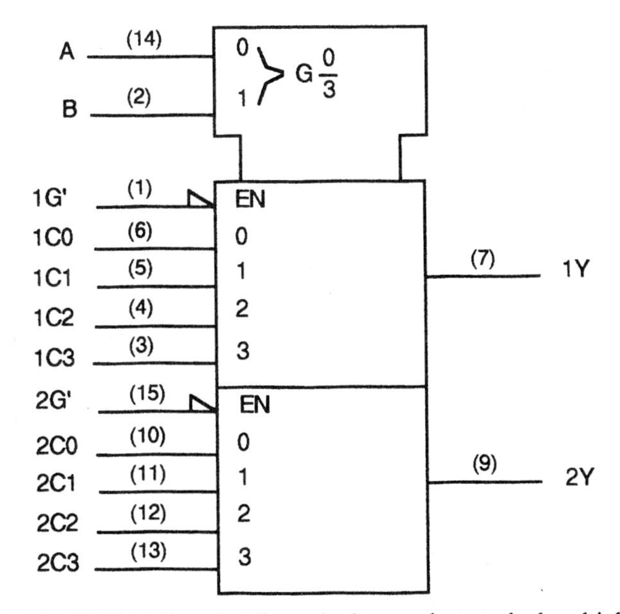

Fig. B-6 IEEE/IEC symbol for a dual two-select strobed multiplexer.

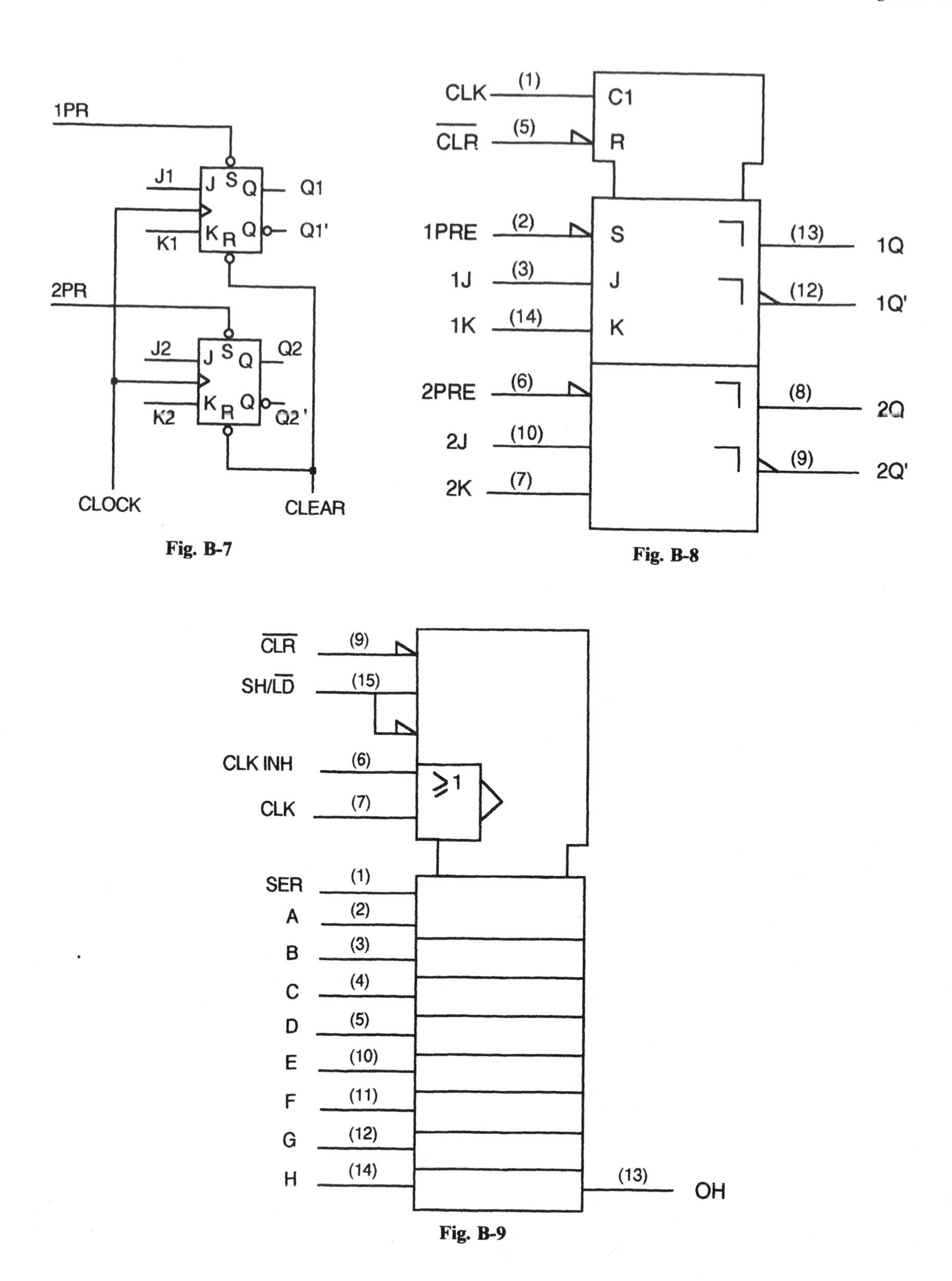

Fig. B-7

Fig. B-8

Fig. B-9

The common clear signal is active-LOW (half arrow) and is attached to pin 9. The shift/load signal changes the operating mode by causing a clocked shift to occur when it is HIGH and loading parallel data when it is LOW. Two clock inputs are ORed such that if either or both of them remain TRUE (HIGH), clocking is inhibited; thus one clock line can be used for synchronizing pulses, while the other serves as an enable.

There are eight shift register stages, each indicated by a rectangular section. The parallel data inputs are denoted by the letters A through H, and the shift register stages are linked in alphabetical order. Section A has a second input (SER) which is used if serial data input to the register is required. Only the last stage (section H) has an output pin connection from which serial data is extracted.

Counters (Example, 74 × 160 synchronous 4-bit decade counter)

Refer to Fig. B-10 for the IEEE/IEC symbol. This device has several inputs which are common to its four stages. Along with a common clock, there is a common clear ($\overline{CLR}$) which is active-LOW, a common mode-control signal ($\overline{LOAD}$) which, when LOW, selects mode 1 (M_1) for parallel data loading, and, when HIGH, selects the counting mode (M_2). There are also two common HIGH-TRUE enables (ENT and ENP), both of which must be HIGH to count. ENT alone enables the ripple carry-out (RCO). Each of the four stages has an input for loading data and an output pin (labeled Q).

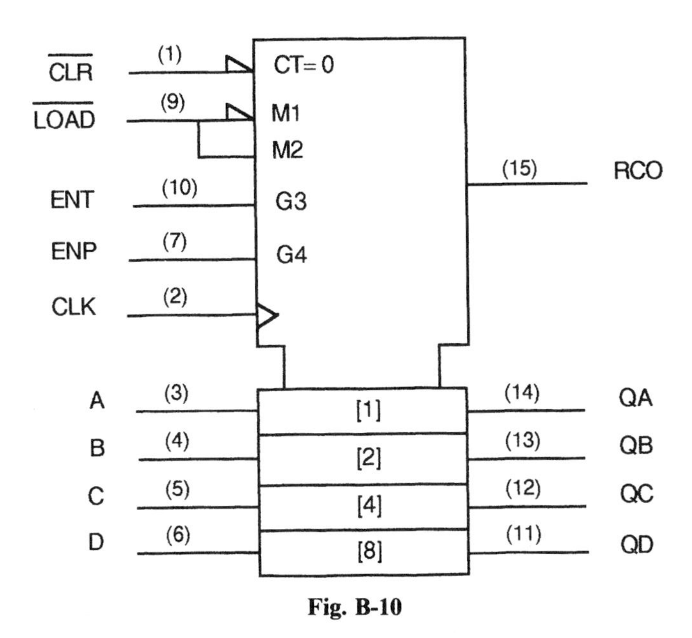

Fig. B-10

Appendix C

Some Comments on Digital Logic Simulation

C.1 PROBLEMS IN THE SIMULATION OF CIRCUITS WITH FEEDBACK

The use of computer simulation to test logic circuits is a powerful tool, but there are pitfalls which, if unrecognized, can cause serious errors. For example, consider the gated latch design of Fig. 7-8 whose LogicWorks™ screen image is reproduced in Fig. C-1.

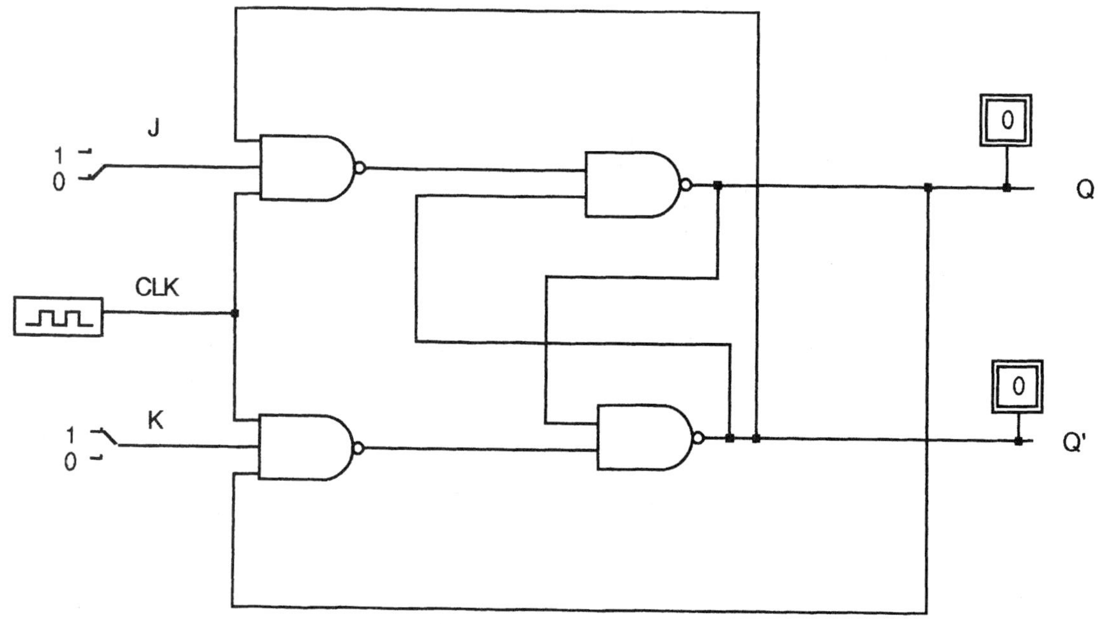

Fig. C-1

As discussed in Sec. 7.5, we expect an oscillatory condition to occur, but what are we to make of the result shown in Fig. C-2? Note that the oscillations will not terminate regardless of what we do with the input switches, and they persist event if we reset the timing to zero in an attempt to start over. Worse yet, if we save the circuit and restart the program, the problem remains—almost as if a part had been damaged!

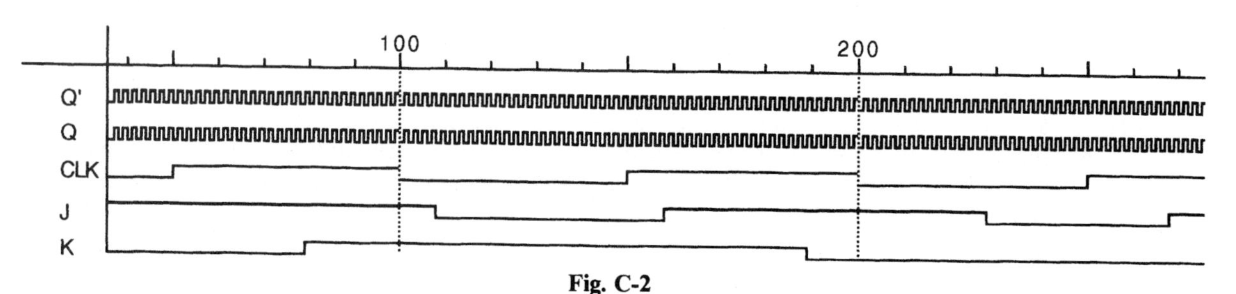

Fig. C-2

The difficulty arises from the fact that *simulators like LogicWorks, in order to function properly, need to start from a condition where the state of all gates is unambiguously determinable.* Cross-coupled latch circuits (and many others with feedback connections) can, for certain input combinations, have two output states which are equally likely to occur. Thus, confusion can arise and the simulation can go bananas, so to speak. A strong indication that things are amiss is the occurrence of impossible behavior such as two 0s at latch outputs (Fig. C-1) or an inverter having identical input and output logic levels. As with the testing of real circuits, the engineer must not be too gullible; oscilloscopes and, yes, even computers can lie.

The Solution

A reset circuit should be added, most likely within the critical feedback path, to force a unique state on the offending circuit. In the present example, a single AND gate and switch (as shown in Fig. C-3) will suffice nicely. The reset gate should have its time delay set to zero so that when its switched input is at logic 1, it will be totally transparent to circuit operation.

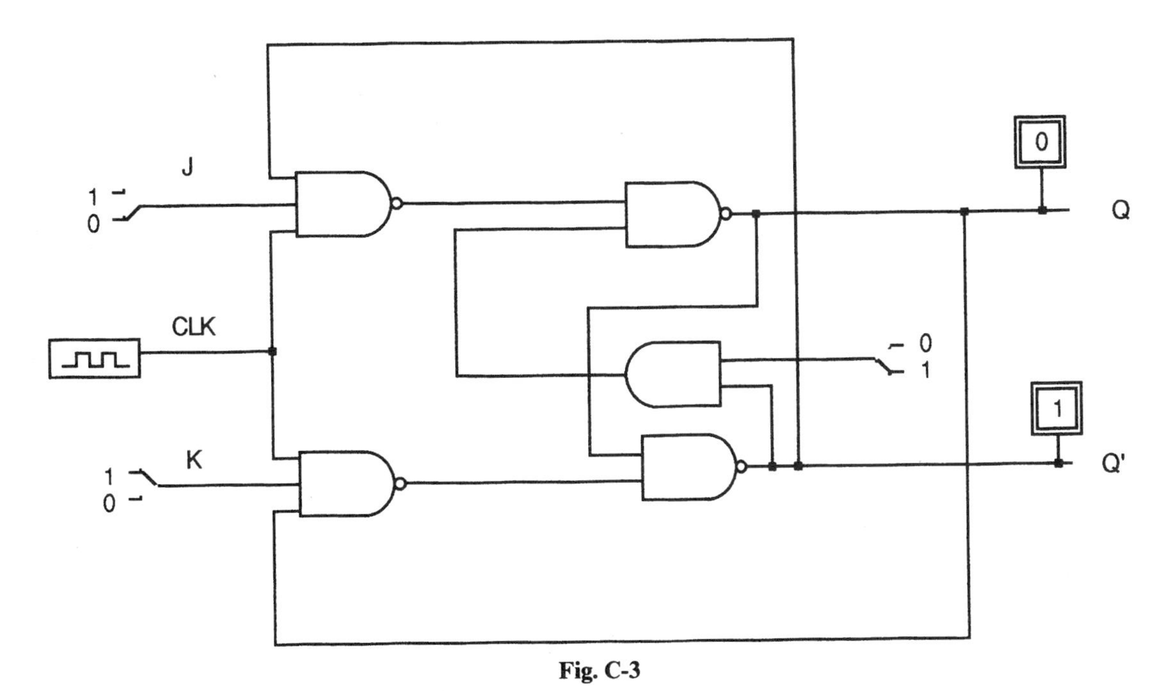

Fig. C-3

C.2 GENERATION OF CLOCK-SYNCHRONIZED INPUT WAVEFORMS

The INPUT signal in Prob. 7.8 starting from an initial logic 1, is required to change level at the trailing edge of every fourth clock pulse. This may be easily achieved in simulation by using two clock sources (one for each waveform) and adjusting the on and off times of each, as required. In the present instance, the clock has been arbitrarily set for 10 time units high and 40 time units low, followed by defining the INPUT source to have equal high and low times of 200 time units.

The arrangement is repeated in Prob. 8.21 with some interesting modification. As pointed out in the solution to Prob. 7.8, simultaneous signal transitions create a zero set-up and hold time situation, making it impossible for flip-flop A to respond to an input change until the next succeeding clock pulse.* If this condition is to be avoided, the INPUT signal can be synchronized to the clock's leading

* In LogicWorks™, D and JK flip-flops have a setup time of 1 time unit and zero hold time. Consequently, if the data and clock signals change simultaneously, the immediately prior data value will be used.

edge instead of its trailing edge. This is achieved in the simulator by adding 10 time units of delay to the clock signal which effectively shifts it to the right, repositioning its *leading* edge in alignment with an INPUT signal transition. As shown in Fig. 8-39, two inverters are used for the purpose, so that there is no net inversion.

Problem 7.8 calls for an initial value of 1 for the input waveform. In Prob. 8.21, this could have easily been achieved in the LogicWorks™ simulator by manually drawing a high initial level in the timing diagram window when the simulation is paused. The authors have chosen, instead, to start with INPUT = 0 because it permits important features of the timing diagram to more easily fit into the space available in Fig. 8-40.

Index

Note: Numbers in parenthesis refer to *problem numbers* not page numbers.

Absorption Theorem, 52
  application to K-map simplification, 55
Active low (defined), 88
Adder (*See also*, half-adder), 360
  binary, 5, 25, 30, (1.9, 2.13, 4.4)
  mod-2, (8.5)
Addition:
  binary, 5, 30, 360
  hexadecimal, (1.8)
Address, 131
Address decoder, 202, 242
Adjacency, 56, 59
Algebra, Boolean, 26
Algorithmic state machine chart (*See also*, ASM chart), 313
Alphanumeric code, 9
ALU, 358
  propagation delay in, (11.5)
Amplifier pair, 275
AND function, defined, 27
AND gate:
  as a controlled switch, 353
  defined, 30
Application specific devices, introduction to, 369
Arithmetic logic unit (*See* ALU)
Arithmetic, binary, 5
ASCII code, 9, 10, (1.22)
ASICs, introduction to, 269
ASM chart, 313
  construction, 314
  design for minimum number of flip-flops, 316
  design of incremental encoder, 314, 319
  design of one-hot controller, 318
  design using state table entry, 319
  from logic diagram, (10.9)
  symbols, 313
  output design equations, 318
  timing diagram from, (10.13)
Asynchronism in rate multipliers, (8.12)
Asynchronous:
  mode of register operation, 236
  operation of JK flip-flop, 199

Balancing half arrows, 91, (4.14)
Base, numerical, 1, 4–6
Bases, conversion between, 4–5
BCD code, 7
BCD-to-seven segment display, 58, (2.2)
Bidirectional port, 356
Binary:
  adder, PLA design, (9.7)
  addition, 5, 30, (2.13)
  arithmetic, 5
  coded decimal (BCD), 7
  conversion to Gray code, 9, 135, 271
  counter, 238
  numbers, 1
  sequence generator, (8.14, 8.15)
  subtraction, 6
  system, 1, 2
Bit, 1
  slice, 358
  stream, embedded, 355
Boolean algebra, 26
  fundamental identities and theorems, 50, Appendix A
  hierarchy of operations, 28
Boolean AND function, defined, 27
Boolean entry for programmable ICs
Boolean equations:
  defined, 26
  from logic diagram, (2.14, 4.11)
  from positive logic diagram (4.11)
  reading of, 28, 94
  relation to truth tables, 29
  stepping motor controller (9.3)
  structure of, 26
Boolean expressions:
  recovered from multiplexer diagram, (5.14)
  recovered from ROM, (5.15)
  recovered form mixed logic diagram, 94, (5.14)
  simplification by manipulation, 53, 54
Boolean OR function, defined, 27
Branching, conditional, 309
Bubble, use of, 87, 88, 90
Buffer, three state, 243, 244, 272
Bus (*See* data bus)

Carry-in, 5, 25, 359
Carry-out, 5, 25, 359
Chatter-free switch, 195
Checkerboard pattern (*See* Karnaugh maps)
Clear, 200
Clock, 196, 200
  frequency, 207
  period, 206
Clock (*continued*)
  pulse leading edge, 203, Appendix C.2
  pulse timing in synchronous machines, 314, 321
  signal delay in simulation, Appendix C.2
  skew (*See also*, timing), 321
  state machine operation with, 306, 322
  synchronized input waveforms, simulation of, Appendix C.2
Clocked RS flip-flop, 196
Code:
  alphanumeric, 9
  ASCII, 9
  BCD, 7
  error correcting, 11
  Gray, 7
  Hamming, 11, 12
  positionally weighted, 7
  types, 7
Combinational logic:
  design of, 24, 50
  with multiplexers, 130, (5.11)
Commutative Law, 52
Complementary addition, 361
Computer aided reduction of K-maps, 61
Computer circuit simulation, 172, 174
Conditional:
  branching, 313
  output, 314
Configurable logic block, 355
Control gate (in programmable register), 363
Control input, 360
  mode, 360
  timing, (11.4)
Control signals for electronically programmable functions, 353, 356
Control variables (*See also*, programmable logic and ALU), (11.8–11.14)
Controlled:
  inverter, 353, 354
  switch, 353, 354
Controller, 318
  design equations, 318
Conversion:
  between binary and Gray code, 9, 135, (1.15, 5.8)
  between number systems, 1, 3–5, (1.1–1.7, 1.16)
  parallel-serial, 236

Count sequence, (8.4, 8.9, 8.11, 8.14)
Counter:
 binary, 238
 determination of macrotiming diagram for, (8.1–8.3)
 excess-three, (10.2)
 Gray code, 307, 312
 initialization of, (9.4)
 Johnson, 245
 maximum pulse rate, (8.8)
 Mobius, 245
 ripple, 237, 238, (8.8)
 sequence form PAL IC connection diagram, (9.6)
 state machine design of, 307–309
 symbols for, Appendix B.5
 synchronous, 307
 up/down, 309
Custom IC, 269

D flip-flop, 203, 233, 234, (7.8, 7.12)
 JK implementation of, 234
 state table, 204
Data:
 bus, 198, 243
 parallel, 236
 serial, 236
 synchronous input, 236, Appendix C.2
 unsyncronized, 207
Decimal number system, properties of, 1–3
Decision block, 313
Decoder, 133
 address, 133
 full, 133, 270
 seven segment, 58, (2.2, 2.11)
Decrement, (11.3)
Delay flip-flop, 203
DeMorgan's Theorem, defined, 52
Demultiplexer, 133
Destination state, 316
Device library, 269, 279, 280
Digital watch, 239
Distributive Law, 32
 proof by exhaustion, 53
Divider, frequency, 239
Don't cares, 57, 58
Duality (*See also*, Boolean algebra), Appendix A

Edge triggering (*See also*, flip-flop), 200, 203, (7.6, 7.7)
EEPROM, 270
Eights complement addition, (1.14)
Electronically programmable functions (*See also*, ALU), 353
Embedded bit stream, 355
Enable, 128, 197, 243, 244
Encoder, incremental, 314
Encoding wheel, 8
EPROM, 270
Equivalence, logical, 27, (2.5)
Eraseable memory devices, 321
Error detection and correction, 11, (1.19–1.21, 124–1.27)
Errors in sequential circuits (*See also*, metastability), 207
Excess-three counter, (10.2, 10.3)
Exclusive OR function (*See* XOR function)
Exhaustion, proof by, 53

Fan in, 134
Feedback in combinational logic circuits, 174, 272
Field programmable gate array programming, 357
Field programmable gate arrays (FPGA), 281, 354
Flip-flop (*See also*, D, JK, latch, RS and T):
 arrays, microtiming diagram for, (8.1–8.3)
 circuits, analysis of sequential operation (*See also*, state machines), (7.9, 7.10, 8.2, 8.4, 8.9, 8.11, 9.4)
 combinations of, 233
 hold time, 205
 introduction to, 194
 master-slave, 199
 propagation delay time, 201, 205, (7.1, 7.3)
 set-up time, 205
 symbols, Appendix B.4
 triggering and timing, 202, 203, 205, (7.6, 7.7)
 types, 203
Forbidden state, 198, 201
Fraction (in number systems), 1, 4, 5
Frequency divider, 239
Full:
 adder implementation, (4.5–4.9)
 custom design, 281
 decoder, 133
 gate arrays, 279
Fusible links, 269

Gate (*See also*, NAND and NOR), 86
 arrays, 279, 354
 arrays, programmable, 281, 354
 defined, 30
 delay, 171
 hardware implications, 54, 86
 pin-out, Appendix B.2
 primed, 242
 recognition, 239, (8.5)
 symbols, Appendix B.2
Gated latch:
 flawed design, 199
 problem in simulation, Appendix C.1
Glitch, 167, 169, 171, 173, 239
Global reset, 356
Gray code, 7
 application to K-maps, 54
 conversion to binary, 24, (1.15, 1.16, 5.8)
 counter, 307, 312
 unit distance property, 7
 use in position encoding wheel, 8, 314

Half arrow:
 symbol (defined), 88, 91–94
 use of, 88
Half-adder, (2.15)
Hamming code, 11, (1.19–1.21)
 truth table for, (2.1)
Hang-up in sequential circuits (*See also*, lock-out), (8.9, 8.11)
Hardware (*See also* NAND and NOR)
 gate, 86
 mixed logic convention, 86
Hazard, 169, (6.2)
 covering in a K-map, 170, 174, (6.2)
Hex inverter, 94
Hexadecimal number system, 3
Hierarchy of operations in Boolean algebra, 28
High-TRUE, 87–89
Hold time, 205

I/O, 172, 243
 block, 356
 register, 243, 244
Incremental encoder interface, design with ASM chart, 314, 319
Incrementing, 363
Initial state in simulation, Appendix C.1
Initialization:
 of a counter, (9.4)
 of a one-hot controller, 322
 of state machines, 322
Input:
 control, 353, 357
 synchronized, 236, 316, Appendix C.2
Input/Output (*See* I/O)
Instability, (6.7)
Inverse Exclusive Or (*See* logical equivalence)
Inversion, logical, 27, 29
Inverter, 90–92
 controlled, 353, 354

JK flip-flop:
 as a D flip-flop, 236
 asynchronous and synchronous

JK flip-flop (*continued*)
operation of, 198–200, 310
clearing of, (6.9)
edge triggered, (7.6, 7.7, 7.9)
initial conditions, (6.9, 8.11)
microtiming diagram, 201, (6.9)
operation of, 199
preset and clear, 200
response to clock transition, 201
state table, 197, 309
Johnson counter, 245

K-map (*See* Karnaugh map)
Karnaugh map:
adjacency, 56, 61
checkerboard pattern, 61, (4.17)
depiction of present state/next state, 310
don't cares, handling of, 57, (4.17)
five and six variable, 59
four variable, 56
Gray code with, 54, 55
hazard display with, 169, 174, (6.2)
introduction to, 54
mirror symmetry in, 59, 60
multiplexer relation to, (5.5)
ones covering, 57
overlapping coverings, 57
reduction of (computer aided), 61
sequential circuit analysis with, (8.4, 9.6, 10.1)
simplification rules, 56, 57
six variable map slices, 60
XOR representation with, 61
zeros covering, 57

Latch (*See also*, RS flip-flop), 194, (6.4, 6.6)
gated, 196, 199
implementation of, 197
set dominant, (7.5)
strobed, 197
Latching, 233
register, 233, (8.13)
Leading edge (of clock pulse), 203
Leading zeroes, 6
Library, component, 269
Lock-out (*See also*, hang), 322
Logic cell array, 277, 356
Logic:
conventions, 94
diagram, construction from logic equations, 31, 32, 91
diagram, creation of ASM chart from, (10.9)
diagram, defined, 30
diagram, recovery of mixed logic form, 94
functions, 26
gate, defined, 30, 86
inverter, 90, 91
levels, 86, 87
positive (defined), 87
symbols, 31, Appendix B.1
Logical:
equivalence, 27, (2.5)
inversion, 27, 96
operators, 28
variables, 27
LogicWorks™ (digital simulation software), 172, 176, Appendix C.1
Look-up table, 135, 354
Low-TRUE, 87, 93

Macros (in gate array programming), 280
Macrotiming diagram, 171, (6.8, 7.9–7.11)
for counter circuits, (8.1–8.3)
Master-slave flip-fop (*See also*, JK flip-flop), 199
Maximal length PRBS, (8.14, 8.15)
Mean time between failure (MTBF), 207, 208, (8.13)
Memory (*See also*, RAM and ROM), 195
Metastability, 207, (8.13)
Metastable behavior, 207
Microtiming diagram, 250
preparation of, 167, (6.1, 6.3, 6.5)
to display hazards, 169, (6.1)
voltage inverters in, 168
Mixed logic (*See also*, positive logic), 86
as a design tool, 87
Boolean expression recovery from, 94
diagram, construction of, 91, (4.16)
diagram, conversion to wiring diagram, 92, 94
logical inversion in, 90, 91
symbols, 88
trouble-shooting, applications in, 96
Mobius counter, 245
Mod-2 adder, (8.15)
Mode control signal, 360
MSI and LSI elements, 128
Multiplexer, 128
address, 131
combinational logic device, use as, 130, (5.2–5.4, 5.13)
controlled switch, use as, 130, 354
logic, 128, (5.1)
logic design, 130, 132
sequential circuit applications of, (7.14)
select between a signal and its complement, used to, (11.7)
select logic, 32
selects driven from a counter, (8.3)
state machine design, use in, (10.7)
symbols, Appendix B.3
tree, 130
Multiplier, (5.7)

NAND gate:
hardware, 86
controlled inverting switch, 353
voltage inverter, 90
Negative edge triggering, 203
Net list, 172
Next state, 307
Nines complement addition, (1.10)
Noise:
immunity, 1, 2
margin, 2, (1.23)
NOR gate:
controlled inverting switch, 353
hardware, (4.1)
voltage inverter, 90
Number systems, 1, 4
bases, 1
conversion from one base to another, 1, 4
conversion of fractions, 4
weighting in, 1, 3, 5

Octal, 3
conversion to binary, 3
conversion to Gray code, (1.17)
number system, 3
subtraction, (1.14)
One-hot controller, 318
design from ASM chart, (10.10)
initialization, 322
Ones:
complement, 361
covering, 57, 59
Operators, logical, 26
OR function, defined, 27
OR gate
controlled switch, use as, 353
defined, 30
Oscillation, 174, (6.6, 6.7)
in simulations, Appendix C.1
OUT INV control signal, 356
Output:
conditional, 314
design equations (ASM chart), 318
enable (read), 244
pin truth values (PAL 16R6), 275
registered, 275, 356
source control signal, 356
unconditional, 313
unregistered, 272

PAL device, 272
architecture, 272
connection diagram, (9.5, 9.6)
design from a truth table, (9.8)
programming, 272
PAL 16R6, 272
reversible shift register, design with, (9.2)

state machine excess-three counter, design with, (10.3)
stepping motor controller, design with, (9.3)
up/down counter, design with, (9.4)
PALASM® device programming language, 401
Parallel data input, 236
Parallel-serial conversion, 236
Parentheses, use of in Boolean equations, 28
Parity, 11, (1.18)
checker, (5.12)
generator, (2.1, 2.12)
group, 11, 12
Passive pull-up control signal, 356
Path length, 169–171
Pin-out, Appendix B.2
PLA, 277
architecture, 277
binary adder, design with, (9.7)
design from a truth table, (9.7)
Placement and routing, 358
PLDs, 269
Port:
bidirectional, 356
input, 356
Position encoding wheel, 8
Positional notation, 1
Positive logic, 87, 95
conversion to mixed logic form, (4.10)
Present state, 307
Present state/next state truth table, 307
Preset, 201
Primed gate, 242
Programmable:
array logic (PAL®), 272
gate arrays (PGA), 279–281
interconnect, 356, 357
logic, 269
logic array (PLA), 277
logic device, state table entry, 319
logic in state machine design, 309
logic, programming techniques, 319, 357
Programmable registers, 362
Programming:
gate arrays, 279
software, 217, 280, 320, 358
the FPGA, 357
the PAL device, 272
ROM, 134
PROM, 271
burner, 271
Proof by exhaustion, 53
Propagation delay time, 167, 201, 205, 321
accumulation of, 321
in ALU, (11.5)
Propagation delays, flip-flop, 201, 205, 321
Pseudo random binary sequence generator, (8.14, 8.15)
Pulse rate, maximum, (8.8)

RAM, 241
address, 241–245
look-up table, 354
Ramdom Access Memory (*See* RAM)
Rate multiplier, 240
asynchronism in, (8.12)
Reading:
Boolean equations, 28
data from RAM, 243–245
Recognition gate, 239, (8.5)
Reflection axis, 7
Register (*See also*, flip-flop arrays and shift register), 194, 206, 233, 243
I/O, 243, 244
latching, 233
programmable, 362
shift, 234
Registered output, 356
Reset, 194, 201
in simulation of circuits with feedback, Appendix C.1
Reversible shift register, 234, (9.2)
Ripple counter, 237, 238, (8.3, 8.5, 8.8)
glitch problems in, 238, 239, (8.5, 8.6)
transient states in, 239
ROM, 134, 135, 241
address, 134, 135, (5.16)
memory, use as, 135
programming, 135
relation to logic diagrams, (5.9, 5.10)
relation to truth tables, 135
use in sequential circuits, (8.7)
use in state machines, (10.2)
Routing and placement, 358
RS flip-flop:
clocked, 196, (7.1)
microtiming diagram for, (7.1)
NAND implementation of, (7.1)
NOR implementation of, (7.4)
RS latch:
forbidden state, 197
state table, 197

Schematic capture, 172, 280
Select (*See also*, address), 128, 130, 132
Sequential digital circuits (*See also*, state machines), (7.9, 7.10, 8.1–8.15, 9.4–9.6)
Sequential state change table (for JK flip-flop), 311
Serial input, 235
Set, 194, 201, 205
dominant latch, (7.5)
Setup time, 205
Seven segment display decoder, (2.2, 2.11)
Shaft encoder, 8, 314
Shift register, 234–236, 245
reversible, 236
symbols for, Appendix B.4
Sign bit, 6
Signal propagation within a flip-flop, 201, 206
Simplification:
of Boolean expressions by manipulation, 53
of Boolean expressions using K-maps, 55
rules for K-maps, 56
Simulation:
clock-synchronized waveforms in, Appendix C.2
of circuits with feedback, problems, Appendix C.1
of JK flip-flop response, 201
Slash mark, use of, 89, 90, 92
Slew rate control signal, 356
Software for state table entry, 319
SRAM, 355
Stage, 233
Standard cell design, 281
State, 313
destination, 316
State change table (for JK flip-flop), 311
State machine:
system specification, 306
clock skew in, 321
State machine design:
excess-three counter, (10.2)
excess-three counter using PAL 16R6, (10.3)
Gray code counter, (10.1)
external input, 307
initialization, 322
introduction to, 306
lock-out in, 322
conversion from D to JK implementation, 310
with D flip-flops, 307, 310
with JK flip-flops, 309, 310
with multiplexers, (10.7)
with ROMs, (10.2)
State table, 197, 307
State table entry, 319
to programmable logic device, 319
software for, 317
State transition table
use in state machine design, 316
State variable:
design equations from ASM chart, (10.7)
Stepping motor controller, (9.3)
Strobe, 197
Strobed data latch, 197
Subtraction
binary, 6
octal, (1.14)
two's complement, 6, 361

using complementary addition, 6, 361
Switch:
chatter-free, 195
electronically controlled, 353, 354
Switches, used to visualize basic Boolean relations, 51
Symbols
ASM chart
logic, 31, Appendix B.1
mixed logic, 88
PALASM®, 273
Synchronized input waveforms, simulation of, Appendix C.2
Synchronous:
counter, state machine design of
data input, 236
operation of digital systems, 196
operation of JK flip-flop, 199
state machine (*See* state machine)

T flip-flop, 204, (7.8, 7.13)
state table, 204
Theorems, of Boolean algebra, 50, 52
Time delay accumulation, 239
Time division multiplexing, 133
Timing diagram, 167, (6.1, 6.3–6.5)
from ASM chart, (10.13)
Timing:
control inputs, (11.4)
errors (*See also*, clock skew), 207, (6.7)
simulation, 172, (6.1)
flip-flop (*See also*, triggering), 205
state machine, 307, 321
Toggle, 199, 200, (8.5)
flip-flop, 204
Transfer out, 362
Transient states, 239
Transparent operation (strobed latch), 198, 203
Tree, multiplexer, 130
TRI-STATE® device
as a buffer amplifier, 243, 272
as a controlled switch, 354
Triggering, 200, 205
edge, (7.6, 7.7)
Trouble-shooting, 96
Truth table:
construction of, 24, 25
introduction to, 24
present state/next state, 307
relation to Boolean equations, 29, (2.3, 2.6, 2.7)
relation to ROM, 134
Truth values, assignment of, 87, 275
Two's complement, 6, (1.11)

Unbalanced half arrows, 92
Unconditional outputs, 314
Unit distance property, 7
Universal logic module, 128
Unregistered output, 272
Up/down counter, (9.4)

Variables:
assignment of truth values to, 87
control (*See also*, programmable logic, and ALU), (11.8–11.14)
logical, 26, 88
state, 317
Voltage inverter, 90
in microtiming diagrams, 167
hardware forms, 88
Voltage waveforms, 96, (4.12)

Waveforms:
design from, (6.10)
determination from PAL IC connection diagram, (9.5, 9.6)
generation of, (8.7)
simulation of clock-synchronized input, Appendix C.2
voltage, 96, (4.12)
Weighting:
in codes, 9
in number systems, 1, 5, 7
Wheel, position encoding, 8
Writing to RAM, 242, 244, 245

XOR function:
Boolean equivalent, (2.4)
defined, 28
mapping of, 61
truth table representation, (2.4, 2.10)
XOR gate:
as a controlled inverter, 353
for logical inversion, use of, 30, 353
defined, 30
hardware implementation, 91

Zeroes:
covering, 57, (4.14)
leading, 6